SCHAUM'S
OUTLINE OF

ELEMENTARY

ALGEBRA

SCHAUM'S
OUTLINE OF

Theory and Problems of
ELEMENTARY ALGEBRA

Third Edition

BARNETT RICH, Ph.D.

PHILIP A. SCHMIDT, Ph.D.

Program Coordinator, Mathematics and Science Education
The Teachers College, Western Governors University
Salt Lake City, Utah

Schaum's Outline Series

McGRAW-HILL
New York   Chicago   San Francisco   Lisbon
London   Madrid   Mexico City   Milan   New Delhi
San Juan   Seoul   Singapore   Sydney   Toronto

DR. BARNETT RICH held a doctor of philosophy degree (Ph.D.) from Columbia University and a doctor of jurisprudence (J.D.) from New York University. He began his professional career at Townsend Harris Hall High School of New York City and was one of the prominent organizers of the High School of Music and Art, where he served as the Administrative Assistant. Later he taught at CUNY and Columbia University and held the post of Chairman of Mathematics at Brooklyn Technical High School for 14 years. Among his many achievements are the six degrees that he earned and the 23 books that he wrote, among them Schaum's Outlines of *Review of Elementary Mathematics* and *Geometry*.

PHILIP A. SCHMIDT, Ph.D., has a B.S. from Brooklyn College (with a major in mathematics), an M.A. in mathematics, and a Ph.D. in mathematics education from Syracuse University. He is currently Program Coordinator in Mathematics and Science Education at The Teachers College of Western Governors University in Salt Lake City, Utah. He is also the coauthor of the *Schaum's Outlines of College Mathematics* as well as the reviser of Schaum's Outlines of *Geometry* and *Review of Elementary Mathematics*. Among his many achievements are numerous grants and scholarly publications in mathematics education.

Schaum's Outline of Theory and Problems of
ELEMENTARY ALGEBRA

3 4 5 6 7 8 9 10 11 12 13 14 15 16 17 18 19 20 VFM VFM 0 9 8 7 6 5 4

ISBN 0-07-141083-X

PREFACE

In the third edition of *Elementary Algebra*, I have maintained the pacing, philosophy, and point of view of the prior two editions. Students who are interested in learning algebra for the first time, or who are engaged in the study of algebra and are seeking a study guide or supplementary problems and solutions, will find virtually all major topics in elementary algebra in this text. In addition, there is an ample review of arithmetic, as well as a serious introduction to geometry, trigonometry, problem solving, and mathematical modeling.

In crafting this edition, I have removed material that is no longer a part of the standard algebra curriculum, added topics as part of that effort as well, modernized terminology and notation when necessary, and incorporated the use of calculators throughout the text. Mathematical modeling is introduced as an important component to problem solving in mathematics. Pedagogy has been revamped to match current teaching methods.

My thanks must be expressed to Barbara Gilson and Andrew Littell of McGraw-Hill. They have been supportive of this project from its earliest stages. I must also thank Dr. Marti Garlett, Dean of the Teachers College of Western Governors University, for her professional support as I worked to meet important publishing deadlines. I thank my wife, Dr. Jan Zlotnik Schmidt, for her love and support during this project. And I dedicate this edition to the late Jean (Mrs. Barnett) Rich. Her drive to make certain that this volume was written was a substantial part of my own drive to do this work.

PHILIP A. SCHMIDT
New Paltz, NY

CONTENTS

SCHAUM'S
OUTLINE OF

ELEMENTARY
ALGEBRA

Elementary Algebra
(Third Edition)
By Schmidt and Rich

Table of Contents

CHAPTER 1

From Arithmetic to Algebra

1. REPRESENTING NUMBERS BY LETTERS

In this chapter, we are going to lead you from arithmetic to algebra. Readers wanting an arithmetic review should turn to Appendix A. Underlying algebra as well as arithmetic are the four fundamental operations:

1. Addition (sum)
2. Subtraction (difference)
3. Multiplication (product)
4. Division (quotient)

The answer for each operation is the word enclosed in parentheses.

In algebra, **letters may be used to represent numbers**. By using letters and mathematical symbols, short algebraic statements replace lengthy verbal statements. Note how this is done in the following examples:

Verbal Statements	**Algebraic Statements**
1. Seven times a number reduced by the same number equals 6 times the number.	1. $7n - n = 6n$
2. The sum of twice a number and 3 times the same number equals 5 times that number.	2. $2n + 3n = 5n$
3. The perimeter of a square equals 4 times the length of one of its sides.	3. $p = 4s$

In the first example, $7n$ is used instead of "seven times a number." When you multiply a number by a letter, the multiplication sign may be omitted. Multiplication may also be indicated by using a multiplication sign, a raised dot, or parentheses.

Thus, 7 times a number may be shown by $7 \times n$, $7 \cdot n$, $7(n)$, or $7n$.

Omitting the multiplication sign, as in $7n$, is the preferred method of indicating multiplication. However, the multiplication sign may not be omitted when two numbers are multiplied. "Seven times four" may be written as 7×4, $7 \cdot 4$, or $7(4)$, but *never* as 74.

1

1.1. STATING PRODUCTS WITHOUT MULTIPLICATION SIGNS

$a)$ $7 \times y$	*Ans.*	$7y$	$e)$ $b \times c \times d$	*Ans.*	bcd
$b)$ $3 \times 5 \times a$		$15a$	$f)$ $7 \times 11 \times h \times k$		$77hk$
$c)$ $l \times w$		lw	$g)$ $\frac{1}{2} \times 8 \times n$		$4n$
$d)$ $10 \times r \times s$		$10rs$	$h)$ $0.07 \times p \times q \times t$		$0.07pqt$

1.2. CHANGING VERBAL STATEMENTS TO ALGEBRAIC EQUATIONS

Using letters and symbols, replace each verbal statement by an algebraic equation:

$a)$ If 6 times a number is reduced by the same number, the result must be 5 times the number.

Ans. $6n - n = 5n$

$b)$ The sum of twice a number, 3 times the same number, and 4 times the same number is equivalent to 9 times the number.

Ans. $2n + 3n + 4n = 9n$

$c)$ Increasing a number by itself and 20 is the same as doubling the number and adding 20.

Ans. $n + n + 20 = 2n + 20$

$d)$ The area of a rectangle is equal to the product of its length and width.

Ans. $A = l\,w$

2. INTERCHANGING NUMBERS IN ADDITION

Addends are numbers being added. Their sum is the answer obtained.

Thus, in $5 + 3 = 8$, the addends are 5 and 3. Their sum is 8.

Numerical addends are numbers used as addends.

Thus, in $3 + 4 + 6 = 13$, 3, 4, and 6 are numerical addends.

Literal addends are letters used to represent numbers being added.

Thus, in $a + b = 8$, a and b are literal addends.

RULE: Interchanging addends does not change their sum (the commutative law for addition).

Thus, $2 + 3 = 3 + 2$ and $3 + 4 + 6 = 4 + 6 + 3$.

In general, $a + b = b + a$ and $b + c + a = a + b + c$.

Interchanging addends may be used

(1) **To simplify addition**

Thus, $25 + 82 + 75$ by interchanging becomes $25 + 75 + 82$.

The sum is $100 + 82 = 182$.

(2) **To check addition**

Thus, numbers may be added downward and checked upward:

Add down	Check up
148	148
357	357
762	762
1267	1267

(3) **To rearrange addends in a preferred order**

Thus, $b + c + a$ becomes $a + b + c$ if the literal addends are to be arranged alphabetically. Also, $3 + x$ becomes $x + 3$ if the literal addend is to precede the numerical addend.

2.1. INTERCHANGING ADDENDS TO SIMPLIFY ADDITION

Simplify each addition by interchanging addends:

a) $20 + 73 + 280$ c) $\frac{3}{4} + 2\frac{1}{2} + 1\frac{1}{4}$ e) $1.95 + 2.65 + 0.05 + 0.35$

b) $42 + 113 + 58$ d) $1\frac{1}{2} + 2\frac{2}{7} + \frac{1}{2} + \frac{1}{7}$ f) $9.4 + 18.7 + 1.3 + 0.6$

Ans. a) $20 + 280 + 73$ c) $\frac{3}{4} + 1\frac{1}{4} + 2\frac{1}{2}$ e) $1.95 + 0.05 + 2.65 + 0.35$

$300 + 73 = 373$ $2 + 2\frac{1}{2} = 4\frac{1}{2}$ $2 + 3 = 5$

b) $42 + 58 + 113$ d) $1\frac{1}{2} + \frac{1}{2} + 2\frac{2}{7} + \frac{1}{7}$ f) $9.4 + 0.6 + 18.7 + 1.3$

$100 + 113 = 213$ $2 + 2\frac{3}{7} = 4\frac{3}{7}$ $10 + 20 = 30$

2.2. REARRANGING ADDENDS

Rearrange the addends so that literal addends are arranged alphabetically and precede numerical addends:

a) $3 + b$ c) $d + 10 + e$ e) $15 + x + 10$ g) $w + y + x$

b) $c + a$ d) $c + 12 + b$ f) $20 + s + r$ h) $b + 8 + c + a$

Ans. a) $b + 3$ c) $d + e + 10$ e) $x + 25$ g) $w + x + y$

b) $a + c$ d) $b + c + 12$ f) $r + s + 20$ h) $a + b + c + 8$

3. INTERCHANGING NUMBERS IN MULTIPLICATION

Factors are numbers being multiplied. Their product is the answer obtained.

Thus, in $5 \times 3 = 15$, the factors are 5 and 3. Their product is 15

Numerical factors are numbers used as factors.

Thus, in $2 \times 3 \times 5 = 30$,   2, 3, and 5 are numerical factors.

Literal factors are letters used to represent numbers being multiplied.

Thus, in $ab = 20$, a and b are literal factors.

RULE: Interchanging factors does not change their product (the commutative law for multiplication).

Thus, $2 \times 5 = 5 \times 2$ and $2 \times 4 \times 5 = 2 \times 5 \times 4$.

In general, $ab = ba$ and $cba = abc$.

Interchanging factors may be used

(1) **To simplify multiplication**

Thus, $4 \times 13 \times 25$ by interchanging becomes $4 \times 25 \times 13$.

The product is $100 \times 13 = 1300$.

(2) **To check multiplication**

Thus, since $24 \times 75 = 75 \times 24$,

	24	Check:	75
	$\times 75$		$\times 24$
	120		300
	168		150
	1800		1800

(3) **To rearrange factors in a preferred order**

Thus, bca becomes abc if the literal factors are arranged alphabetically. Also $x3$ becomes $3x$ if the numerical factor is to precede the literal factor.

3.1. SIMPLIFYING MULTIPLICATION

Simplify each multiplication by interchanging factors:

a) $2 \times 17 \times 5$ c) $7\frac{1}{2} \times 7 \times 4$ e) $1.25 \times 4.4 \times 4 \times 5$

b) $25 \times 19 \times 4 \times 2$ d) $33\frac{1}{3} \times 23 \times 3$ f) $0.33 \times 225 \times 3\frac{1}{3} \times 4$

Ans. a) $2 \times 5 \times 17$ c) $7\frac{1}{2} \times 4 \times 7$ e) $1.25 \times 4 \times 4.4 \times 5$
$\qquad 10 \times 17 = 170 \qquad\qquad 30 \times 7 = 210 \qquad\qquad 5 \times \qquad 22 = 110$

b) $25 \times 4 \times 19 \times 2$ d) $33\frac{1}{3} \times 3 \times 23$ f) $0.33 \times 3\frac{1}{3} \times 225 \times 4$
$\qquad 100 \times 38 = 3800 \qquad\qquad 100 \times 23 = 2300 \qquad\qquad 1.1 \times 900 = 990$

3.2. REARRANGING FACTORS

Rearrange the factors so that literal factors are arranged alphabetically and follow numerical factors:

a) $b3$ b) ca c) $d10e$ d) $c12b$ e) $15x10$ f) $20sr$ g) wyx h) $b35ca$

Ans. a) $3b$ b) ac c) $10de$ d) $12bc$ e) $150x$ f) $20rs$ g) wxy h) $35abc$

4. SYMBOLIZING THE OPERATIONS IN ALGEBRA

The symbols for the fundamental operations are as follows:

1. Addition: $+$
2. Subtraction: $-$
3. Multiplication: $\times$, $(\ )$, $\cdot$, no sign
4. Division: $\div$, $:$, fraction bar

Thus, $n + 4$ means "add n and 4." $4 \times n$, $4(n)$, $4 \cdot n$, and $4n$ mean "multiply n and 4." $n - 4$ means "subtract 4 from n." $n \div 4$, $n: 4$, and $\dfrac{n}{4}$ mean "n divided by 4."

RULE: Division by zero is an impossible operation. Hence, $\dfrac{3}{0}$ is meaningless.

Thus, $4 \div 0$ or $x \div 0$ is impossible.

Hence, $\dfrac{4}{0}$ or $\dfrac{x}{0}$ is meaningless.

Also, $\dfrac{4}{n}$ is meaningless if $n = 0$. See Sec. 6.2 for further explanation.

4.1. SYMBOLS IN MULTIPLICATION

Symbolize each, using multiplication signs:

a) 8 times 11   c) b times c e) d divided by the product of 7 and e

b) 8 times x d) 5 multiplied by a and the result divided by b

Ans. a) 8×11, $8 \cdot 11$, $8(11)$, or $(8)(11)$ c) $b \cdot c$ or bc (avoid $b \times c$)   e) $\dfrac{d}{7e}$

b) $8 \cdot x$ or $8x$ (avoid $8 \times x$)   d) $\dfrac{5a}{b}$

4.2. DIVISION BY ZERO

When is each division impossible?

a) $\dfrac{10}{b}$ b) $\dfrac{a}{3c}$ c) $\dfrac{8}{x - 5}$ d) $\dfrac{7}{xy}$

Ans. a) if $b = 0$ b) if $c = 0$ c) if $x = 5$ d) if $x = 0$ or $y = 0$

5. EXPRESSING ADDITION AND SUBTRACTION ALGEBRAICALLY

In algebra, changing verbal statements into algebraic expressions is of major importance. The operations of addition and subtraction are denoted by words such as the following:

Words Denoting Addition		Words Denoting Subtraction	
Sum	More than	Difference	Less than
Plus	Greater than	Minus	Smaller than
Gain	Larger than	Lose	Fewer than
Increase	Enlarge	Decrease	Shorten
Rise	Grow	Drop	Shrink
Expand	Augment	Lower	Diminish

In adding two numbers, the numbers (addends) may be interchanged.

Thus, "the sum of n and 20" may be represented by $n + 20$ or $20 + n$.

But in subtracting one number from another, the numbers may not be interchanged.

Thus, "a number less 20" may be represented by $n - 20$ but **not** by $20 - n$.

Also "20 minus a number" may be represented by $20 - n$ but **not** by $n - 20$.

5.1. EXPRESSING ADDITION ALGEBRAICALLY

If n represents a number, express algebraically:

a) the sum of the number and 7   *c*) the number increased by 9   *e*) 20 enlarged by the number

b) the number plus 8   *d*) 15 plus the number   *f*) 25 augmented by the number

Ans.   *a*) $n + 7$ or $7 + n$ *c*) $n + 9$ or $9 + n$ *e*) $20 + n$ or $n + 20$

   b) $n + 8$ or $8 + n$ *d*) $15 + n$ or $n + 15$ *f*) $25 + n$ or $n + 25$

5.2. EXPRESSING SUBTRACTION ALGEBRAICALLY

If n represents a number, express algebraically:

		Ans.
a)	the difference if the number is subtracted from 15	$15 - n$
b)	the number diminished by 20	$n - 20$
c)	25 less than the number	$n - 25$
d)	25 less the number	$25 - n$
e)	the difference if 15 is subtracted from the number	$n - 15$
f)	50 subtracted from the number	$n - 50$
g)	the number subtracted from 50	$50 - n$
h)	the number reduced by 75	$n - 75$

5.3. CHANGING VERBAL STATEMENTS INTO ALGEBRAIC EXPRESSIONS

Express algebraically:

		Ans.
a)	the number of kilograms (kg) of a weight that is 10 kg heavier than w kg	*a*) $w + 10$
b)	the number of kilometers (km) in a distance that is 40 km farther than d km	*b*) $d + 40$
c)	the number of degrees in a temperature $50°$ hotter than $t°$	*c*) $t + 50$
d)	the number of dollars in a price $60 cheaper than p dollars	*d*) $p - 60$

e) the number of miles per hour in a speed 30 miles per hour (mi/h) faster than *r* mi/h   *e*) $r + 30$
f) the number of meters (m) in a length of *l*m expanded 6 m   *f*) $l + 6$
g) the number of grams (g) in a weight that is 10 g lighter than *w* g   *g*) $w - 10$
h) the number of meters in a distance that is 120 cm shorter than *d* m   *h*) $d - 1.2$

6. EXPRESSING MULTIPLICATION AND DIVISION ALGEBRAICALLY

Words Denoting Multiplication		Words Denoting Division	
Multiplied by	Double	Divided by	Ratio
Times	Triple or treble	Quotient	Half
Product	Quadruple		
Twice	Quintuple		

In multiplying two numbers, the numbers (factors) may be interchanged.
   Thus, "the product of *n* and 10" may be represented by *n*10 or 10*n*. The latter is preferred.
In dividing one number by another, the numbers may not be interchanged, unless they are the same number.

   Thus, "a number divided by 20" may be represented by $\dfrac{n}{20}$ but **not** by $\dfrac{20}{n}$.

   Also, "20 divided by a number" may be represented by $\dfrac{20}{n}$ but **not** by $\dfrac{n}{20}$.

6.1. REPRESENTING MULTIPLICATION OR DIVISION

What statement may be represented by each?

   a) 5*x*   *b*) $\dfrac{y}{5}$ *c*) $\dfrac{5w}{7}$

Ans.   *a*) 1. 5 multiplied by *x*   *b*) 1. *y* divided by 5   *c*) 1. five-sevenths of *w*
           2. 5 times *x*   2. quotient of *y* and 5   2. 5*w* divided by 7
           3. product of 5 and *x*   3. ratio of *y* to 5   3. quotient of 5*w* and 7
                                           4. one-fifth of *y*   4. ratio of 5*w* to 7

6.2. DIVISION AND ITS RELATIONSHIP TO MULTIPLICATION

The relationship between the operations of multiplication and division is important to note:

   If $xy = n$, then $x = \dfrac{n}{y}$ and $y = \dfrac{n}{x}$ (with the understanding that *x* and *y* are not zero). Notice that if

$xy = n$ and either *x* or *y* is zero (or both), then *n* must be zero.

   Notice also that if $a = \dfrac{b}{c}$ and *c* is zero, then $a \cdot 0 = b$, which means $b = 0$. Thus, if $a = \dfrac{5}{0}$, then 5

must be 0, which is absurd. This is a fairly easy way to understand the impossibility of division by

zero.

7. EXPRESSING TWO OR MORE OPERATIONS ALGEBRAICALLY

Parentheses ( ) are used to treat an expression as a single number.
Thus, to double the sum of 4 and *x*, write $2(4 + x)$.

7.1. EXPRESSING TWO OPERATIONS ALGEBRAICALLY

Express algebraically: *Ans.*

a) *a* increased by twice *b*   a) $a + 2b$
b) twice the sum of *a* and *b*   b) $2(a + b)$
c) 30 decreased by 3 times *c*   c) $30 - 3c$
d) 3 times the difference of 30 and *c*   d) $3(30 - c)$
e) 50 minus the product of 10 and *p*   e) $50 - 10p$
f) the product of 50 and the sum of *p* and 10   f) $50(p + 10)$

g) 100 increased by the quotient of *x* and *y*   g) $100 + \dfrac{x}{y}$

h) the quotient of *x* and the sum of *y* and 100   h) $\dfrac{x}{y + 100}$

i) the average of *s* and 20   i) $\dfrac{s + 20}{2}$

7.2. MORE DIFFICULT EXPRESSIONS

Express algebraically:

                                                                                          Ans.

a) half of *a* increased by the product of 25 and *b*   a) $\dfrac{a}{2} + 25b$

b) 4 times *c* decreased by one-fifth of *d*   b) $4c - \dfrac{d}{5}$

c) half the sum of *m* and twice *n*   c) $\dfrac{m + 2n}{2}$

d) the average of *m*, *r*, and 80   d) $\dfrac{m + r + 80}{3}$

e) 60 diminished by one-third the product of 7 and *x*   e) $60 - \dfrac{7x}{3}$

f) twice the sum of *e* and 30 diminished by 40   f) $2(e + 30) - 40$

g) two-thirds the sum of *n* and three-sevenths of *p*   g) $\dfrac{2}{3}\left(n + \dfrac{3p}{7}\right)$

h) the product of *a* and *b* decreased by twice the difference of *c* and *d*   h) $ab - 2(c - d)$

7.3. CHANGING VERBAL STATEMENTS INTO ALGEBRAIC EXPRESSIONS

Express algebraically:

                                                                                          Ans.

a) a speed in feet per second (ft/s) that is 30 ft/s faster than twice another of *r* ft/s   a) $2r + 30$

b) a weight in kilograms that is 20 kg less than 3 times another of *w* kg   b) $3w - 20$

c) a temperature in degrees that is 15° colder than two-thirds another of *t*°   c) $\dfrac{2t}{3} - 15$

d) a price in cents that is 25¢ cheaper than another of *D* dollars   d) $100D - 25$

e) a length in centimeters (cm) that is 8 cm longer than another of *f* m   e) $100f + 8$

8. ORDER IN WHICH FUNDAMENTAL OPERATIONS ARE PERFORMED

In evaluating or finding the value of an expression containing numbers, the operations involved must be performed in a certain order. Note in the following how **multiplication and division precede addition and subtraction**.

To Evaluate a Numerical Expression Not Containing Parentheses

Evaluate: *a*) $3 + 4 \times 2$ *b*) $5 \times 4 - 18 \div 6$

Procedure:

1. Do **multiplications and divisions (M & D)** in order from left to right:
2. Do remaining **additions and subtractions (A & S)** in order from left to right:

Solutions:

1. $3 + 4 \times 2$ 1. $5 \times 4 - 18 \div 6$
   $3 + 8$ $20 - 3$

2. 11 *Ans.*   2. 17 *Ans.*

To Evaluate an Algebraic Expression Not Containing Parentheses

Evaluate $x + 2y - \dfrac{z}{5}$ when $x = 5, y = 3, z = 20$.

Procedure:

1. **Substitute** the value given for each letter:

2. Do **multiplications and divisions (M & D)** in order from left to right:

3. Do remaining **additions and subtractions (A & S)** in order from left to right:

Solutions:

1. $x + 2y - \dfrac{z}{5}$

   $5 + 2(3) - \dfrac{20}{5}$

2. $5 + 6 - 4$

3. 7 *Ans.*

8.1. EVALUATING NUMERICAL EXPRESSIONS

Evaluate:

Procedure: *a*) $24 \div 4 + 8$ *b*) $24 + 8 \div 4$ *c*) $8 \times 6 - 10 \div 5 + 12$

1. Do **M & D**: $6 \quad + 8$ $24 + \quad 2$ $48 \quad - 2 \quad + 12$
2. Do **A & S**: 14 *Ans.* 26 *Ans.* 58 *Ans.*

8.2. EVALUATING ALGEBRAIC EXPRESSIONS

Evaluate if $a = 8, b = 10, x = 3$:

Procedure: *a*) $4b - \dfrac{a}{4}$ *b*) $12x + ab$ *c*) $\dfrac{3a}{4} + \dfrac{4b}{5} - \dfrac{2x}{3}$

1. **Substitute:** $4 \times 10 - \dfrac{8}{4}$ $12 \times 3 + 8 \times 10$ $\dfrac{3}{4} \times 8 + \dfrac{4}{5} \times 10 - \dfrac{2}{3} \times 3$

2. Do **M & D**: $40 \quad - 2$ $36 \quad + \quad 80$ $6 \quad + 8 \quad - 2$
2. Do **A & S**: 38 *Ans.* 116 *Ans.* 12 *Ans.*

8.3. EVALUATING WHEN A LETTER REPRESENTS 0

Evaluate if $w = 4, x = 2$, and $y = 0$:

a) $wx + y$ *b*) $w + xy$ *c*) $\dfrac{w + y}{x}$ *d*) $\dfrac{xy}{w}$ *e*) $\dfrac{x}{w + y}$

Solutions:

$a)\ 4 \times 2 + 0$	$b)\ 4 + 2 \times 0$	$c)\ \dfrac{4 + 0}{2}$	$d)\ \dfrac{2 \times 0}{4}$	$e)\ \dfrac{2}{4 + 0}$
$8\ \ \ + 0$	$4 + 0$	$\dfrac{4}{2}$	$\dfrac{0}{4}$	$\dfrac{2}{4}$
$8\ $ *Ans.*	$4\ $ *Ans.*	$2\ $ *Ans.*	$0\ $ *Ans.*	$\frac{1}{2}\ $ *Ans.*

9. THE USES OF PARENTHESES: CHANGING THE ORDER OF OPERATIONS

Parentheses may be used

(1) **To treat an expression as a single number**

Thus, $2(x + y)$ represents twice the sum of x and y.

(2) **To replace the multiplication sign**

Thus, $4(5)$ represents the product of 4 and 5.

(3) **To change the order of operations in evaluating**

Thus, to evaluate $2(4 + 3)$, **add** 4 and 3 in the parentheses **before multiplying**;

that is, $2(4 + 3) = 2 \cdot 7 = 14$. Compare this

with $2 \cdot 4 + 3 = 8 + 3 = 11$.

To Evaluate an Algebraic Expression Containing Parentheses

Evaluate: $2(a + b) + 3a - \dfrac{b}{2}$ if $a = 7$ and $b = 2$.

Procedure:	**Solution:**
1. **Substitute** the value given for each letter:	1. $2(7 + 2) + 3 \cdot 7 - \dfrac{2}{2}$
2. **Evaluate inside parentheses:**	2. $2 \cdot 9 + 3 \cdot 7 - \dfrac{2}{2}$
3. Do **multiplications and divisions (M & D)** in order from left to right:	3. $18 + 21 - 1$
4. Do remaining **additions and subtractions (A & S)** in order from left to right:	4. $38\ \ $ *Ans.*

9.1. EVALUATING NUMERICAL EXPRESSIONS CONTAINING PARENTHESES

Evaluate:

Procedure:	$a)\ 3(4 - 2) + 12$	$b)\ 7 - \frac{1}{2}(14 - 6)$	$c)\ 8 + \frac{1}{3}(4 + 2)$	$d)\ 20 - 5(4-1)$
1. Do ( ):	$3 \cdot 2\ \ \ \ + 12$	$7 - \frac{1}{2} \cdot 8$	$8 + \frac{1}{3} \cdot 6$	$20 - 5 \cdot 3$
2. Do **M & D:**	$6\ \ \ \ \ \ \ \ \ \ + 12$	$7 - 4$	$8 + 2$	$20 - 15$
3. Do **A & S:**	$18\ \ $ *Ans.*	$3\ \ $ *Ans.*	$10\ \ $ *Ans.*	$5\ \ $ *Ans.*

9.2. EVALUATING ALGEBRAIC EXPRESSIONS CONTAINING PARENTHESES

Evaluate if $a = 10$, $b = 2$, and $x = 12$:

Procedure:	$a)\ 3(x + 2b) - 30$	$b)\ 8 + 2\left(\dfrac{a}{b} + x\right)$	$c)\ 3x - \frac{1}{2}(a + b)$
1. **Substitute:**	$3(12 + 2 \cdot 2) - 30$	$8 + 2\left(\dfrac{10}{2} + 12\right)$	$3 \cdot 12 - \frac{1}{2}(10 + 2)$

2. Do ( ):	$3 \cdot 16 - 30$	$8 + 2 \cdot 17$	$36 - \frac{1}{2} \cdot 12$
3. Do **M & D:**	$48 - 30$	$8 + 34$	$36 - 6$
4. Do **A & S:**	18 *Ans.*	42 *Ans.*	30 *Ans.*

9.3. **EVALUATING WHEN A LETTER REPRESENTS ZERO**

Evaluate if $w = 1$, $y = 4$, and $x = 0$:

a) $x(2w + 3y)$ *b*) $y(wx + 5)$ *c*) $\frac{1}{2}\left(y + \frac{x}{w}\right) + 5$ *d*) $\frac{y}{w + x} - w(y + x)$

    $0(2 \cdot 1 + 3 \cdot 4)$ $4(1 \cdot 0 + 5)$ $\frac{1}{2}\left(4 + \frac{0}{1}\right) + 5$ $\frac{4}{1 + 0} - 1(4 + 0)$

    $0 \cdot 14$ $4 \cdot 5$ $\frac{1}{2} \cdot 4 + 5$ $4 - 4$

    0 *Ans.* 20 *Ans.* 7 *Ans.* 0 *Ans.*

10. MULTIPLYING FACTORS IN TERMS: NUMERICAL AND LITERAL COEFFICIENTS

A term is a number or letter representing a number or the product of numbers or letters representing numbers.

    Thus, 5, $8y$, cd, $3wx$, and $\frac{2}{3}rst$ are terms.

    Also the expression $8y + 5$ consists of two terms, $8y$ and 5.

A factor of a term is each of the numbers multiplied to form the term.

    Thus, 8 and y are factors of the term $8y$; 3, w, and x are factors of the term $3wx$.

    Also, 5 and $a + b$ are the factors of the term $5(a + b)$.

Any factor or group of factors of a term is a **coefficient** of the product of the remaining factors.

    Thus, in $3abc$, 3 is the **numerical coefficient** of abc, while abc is the **literal coefficient** of 3. Other examples are the following:

Term	Numerical Coefficient	Literal Coefficient
xy	1	xy
$\dfrac{3y}{7}$	$\dfrac{3}{7}$	y
$\dfrac{3ab}{5c}$	$\dfrac{3}{5}$	$\dfrac{ab}{c}$

    An **expression** contains one or more terms connected by plus or minus signs. Thus, $2x$, $3ab + 2$, and $5a - 2b - 7$ are expressions.

10.1. **EXPRESSIONS CONTAINING TERMS**

    State the number of terms and the terms in each expression.

a) $8abc$ *Ans.* 1 term: $8abc$
b) $8 + a + bc$ 3 terms: 8, a, and bc
c) $8a + bc$ 2 terms: $8a$ and bc
d) $3b + c + d$ 3 terms: $3b$, c, and d
e) $3 + bcd$ 2 terms: 3 and bcd
f) $3(b + c) + d$ 2 terms: $3(b + c)$ and d

10.2. FACTORS OF TERMS

State the factors of the following, disregarding 1 and the product itself.

a) 21	*Ans.* 3 and 7		e) $\frac{1}{3}m$	*Ans.* $\frac{1}{3}$ and *m*	
b) 121	11 and 11		f) $\frac{n}{5}$	$\frac{1}{5}$ and *n*	
c) *rs*	*r* and *s*		g) $\frac{n+3}{5}$	$\frac{1}{5}$ and *n* + 3	
d) 5*cd*	5, *c*, and *d*		h) 3(*x* + 2)	3 and *x* + 2	

10.3. NUMERICAL AND LITERAL COEFFICIENTS

State each numerical and literal coefficient:

	a) *y*	b) $\frac{4x}{5}$	c) $\frac{w}{7}$	d) 0.7*abc*	e) 8(*a* + *b*)
Numerical coefficient:	1	$\frac{4}{5}$	$\frac{1}{7}$	0.7	8
Literal coefficient:	*y*	*x*	*w*	*abc*	*a* + *b*

11. REPEATED MULTIPLYING OF A FACTOR: BASE, EXPONENT, AND POWER

$$\boxed{\text{Base}^{\text{exponent}} = \text{Power}}$$

In $2 \cdot 2 \cdot 2 \cdot 2 \cdot 2$, the factor 2 is being multiplied repeatedly. This may be written in a shorter form as 2^5, where the repeated factor 2 is the **base** while the small 5 written above and to the right of 2 is the **exponent**. The answer 32 is called the fifth **power** of 2.

An **exponent** is a number which indicates how many times another number, the **base**, is being used as a repeated factor. The **power** is the answer thus obtained. Thus, since $3 \cdot 3 \cdot 3 \cdot 3$ or $3^4 = 81$, 3 is the base, 4 is the exponent, and 81 is the fourth power of 3.

Table of Powers

The table of powers in Fig. 1–1 contains the first five powers of the most frequently used numerical bases, 1, 2, 3, 4, 5, and 10. It will be very useful to learn these, each of which may be verified using a calculator.

		Exponent				
		1	2	3	4	5
	1	1	1	1	1	1
	2	2	4	8	16	32
	3	3	9	27	81	243
Base	4	4	16	64	256	1,024
	5	5	25	125	625	3,125
	10	10	100	1,000	10,000	100,000

Fig. 1–1

Literal Bases: Squares and Cubes

The area of a square with a side of length s is found by multiplying s by s. This may be written as $A = s^2$ and read "area equals s squared." Here A is the second power of s. See Fig 1–2.

The volume of a cube with a side of length s is found by multiplying s three times; that is, $s \cdot s \cdot s$. This may be written as $V = s^3$ and read "volume equals s cubed." Here V is the third power of s. See Fig. 1–3.

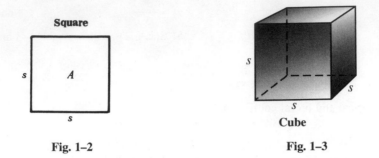

Square

Cube

Fig. 1–2   Fig. 1–3

Reading Powers

b^2 is read as "b squared," "b to the second power," "b second," or "b to the second."
x^3 is read as "x cubed," "x to the third power," "x third," or "x to the third."

11.1.   WRITING AS BASES AND EXPONENTS

Write each, using bases and exponents:

a) $5 \cdot 5 \cdot 5$ c) $2 \cdot 8 \cdot 8 \cdot 8 \cdot 8$ e) $bbccc$ g) $(3y)(3y)(3y)$ i) $7rr(s-8)$ k) $\dfrac{7w}{xx}$

b) $3 \cdot 3 \cdot 7 \cdot 7$ d) $bbbbb$ f) $12bccd$ h) $2(a+b)(a+b)$ j) $\dfrac{yy}{x}$ l) $\dfrac{2ttt}{5vvvv}$

Ans. a) 5^3 c) $2 \cdot 8^4$ e) b^2c^3 g) $(3y)^3$ i) $7r^2(s-8)$ k) $\dfrac{7w}{x^2}$

b) $3^2 7^2$ d) b^5 f) $12bc^2d$ h) $2(a+b)^2$ j) $\dfrac{y^2}{x}$ l) $\dfrac{2t^3}{5v^4}$

11.2.   WRITING WITHOUT EXPONENTS

Write each without exponents:

a) 2^6 d) x^5 g) $(2x)^3$ j) $\dfrac{a^5}{b^2}$

b) $3 \cdot 4^2$ e) $10y^4z^2$ h) $6(5y)^2$ k) $\dfrac{2(a+b)^2}{c^5}$

c) $5 \cdot 7^3 \cdot 8$ f) $8rs^2t^3$ i) $4(a-b)^2$ l) $\dfrac{a^2+b^2}{c^3-d^3}$

Ans. a) $2 \cdot 2 \cdot 2 \cdot 2 \cdot 2 \cdot 2$ d) $xxxxx$ g) $(2x)(2x)(2x)$ j) $\dfrac{aaaaa}{bb}$

b) $3 \cdot 4 \cdot 4$ e) $10yyyyzz$ h) $6(5y)(5y)$ k) $\dfrac{2(a+b)(a+b)}{ccccc}$

c) $5 \cdot 7 \cdot 7 \cdot 7 \cdot 8$ f) $8rssttt$ i) $4(a-b)(a-b)$ l) $\dfrac{aa+bb}{ccc-ddd}$

11.3. EVALUATING POWERS

Evaluate (the table of powers or a calculator may be used to check values):

a) 3^5 *d*) $2^2 + 3^2$ *g*) $1^2 \cdot 1^3 \cdot 1^4$ *j*) $10 + 3 \cdot 2^2$ *m*) $(3 + 4^2)(3^3 - 5^2)$

b) 5^4 *e*) $2^3 + 3^3$ *h*) $2^3 5^2$ *k*) $8 \cdot 10^2 - 3^3$ *n*) $\dfrac{4^4}{2^5}$

c) 10^3 *f*) $10^4 - 4^4$ *i*) $\frac{1}{2} \cdot 2^4 \cdot 3^2$ *l*) $\frac{1}{2} \cdot 4^2 - \frac{1}{3} \cdot 3^2$ *o*) $\dfrac{3^3 + 2^5}{10^2}$

Ans.   *a*) 243   *d*) $4 + 9 = 13$ *g*) $1 \cdot 1 \cdot 1 = 1$ *j*) $10 + 3 \cdot 4 = 22$ *m*) $19 \cdot 2 = 38$

b) 625   *e*) $8 + 27 = 35$ *h*) $8 \cdot 25 = 200$ *k*) $8 \cdot 100 - 27 = 773$ *n*) $\dfrac{256}{32} = 8$

c) 1000   *f*) $10,000 - 256$ *i*) $\frac{1}{2} \cdot 16 \cdot 9 = 72$ *l*) $\frac{1}{2} \cdot 16 - \frac{1}{3} \cdot 9 = 5$ *o*) $\dfrac{27 + 32}{100} = \dfrac{59}{100}$
$= 9744$

11.4. EVALUATING POWERS OF FRACTIONS AND DECIMALS

Evaluate:

a) 0.2^4 *d*) $1000(0.2^3)$ *g*) $(\frac{1}{2})^3$ *j*) $100(\frac{1}{5})^2$

b) 0.5^2 *e*) $\frac{1}{3}(0.3^2)$ *h*) $(\frac{2}{3})^2$ *k*) $32(\frac{3}{2})^3$

c) 0.01^3 *f*) $200(0.4^4)$ *i*) $(\frac{5}{3})^4$ *l*) $80(\frac{5}{2})^3$

Ans.   *a*) 0.0016   *d*) $1000(0.008) - 8$ *g*) $\frac{1}{2} \cdot \frac{1}{2} \cdot \frac{1}{2} - \frac{1}{8}$ *j*) $100(\frac{1}{25}) = 4$

b) 0.25   *e*) $\frac{1}{3}(0.09) = 0.03$ *h*) $\frac{2}{3} \cdot \frac{2}{3} = \frac{4}{9}$ *k*) $32(\frac{27}{8}) = 108$

c) 0.000001   *f*) $200(0.0256) = 5.12$ *i*) $\frac{5}{3} \cdot \frac{5}{3} \cdot \frac{5}{3} \cdot \frac{5}{3} = \frac{625}{81}$ *l*) $80(\frac{125}{8}) = 1250$

11.5. EVALUATING POWERS OF LITERAL BASES

Evaluate if $a = 5$, $b = 1$, and $c = 10$:

a) a^3 *d*) $2a^2$ *g*) $(\frac{1}{2}c)^2$ *j*) $a^2 + c^2$ *m*) $c^2(a + b)$

b) b^4 *e*) $(2a)^2$ *h*) $\left(\dfrac{b}{3}\right)^2$ *k*) $(c + 3b)^2$ *n*) $c(a - b^2)$

c) c^2 *f*) $(a + 2)^2$ *i*) $\dfrac{4a^2}{c}$ *l*) $5(a^2 - b^2)$ *o*) $3b(a^3 - c^2)$

Ans.   *a*) $5 \cdot 5 \cdot 5 = 125$ *d*) $2 \cdot 25 = 50$ *g*) $5^2 = 25$ *j*) $25 + 100 = 125$ *m*) $100 \cdot 6 = 600$

b) $1 \cdot 1 \cdot 1 \cdot 1 = 1$ *e*) $10^2 = 100$ *h*) $(\frac{1}{3})^2 = \frac{1}{9}$ *k*) $13^2 = 169$ *n*) $10 \cdot 4 = 40$

c) $10 \cdot 10 = 100$ *f*) $7^2 = 49$ *i*) $\frac{100}{10} = 10$ *l*) $5 \cdot 24 = 120$ *o*) $3 \cdot 25 = 75$

12. COMBINING LIKE AND UNLIKE TERMS

Like terms or **similar terms** are terms having the same literal factors, each with the same base and same exponent.

Thus:

Like Terms	Unlike Terms
$7x$ and $5x$	$7x$ and $5y$
$8a^2$ and a^2	$8a^2$ and a^3
$5rs^2$ and $2rs^2$	$5rs^2$ and $2r^2s$

NOTE: Like terms must have a common literal coefficient.

To Combine Like Terms Being Added or Subtracted

Combine: a) $7x + 5x - 3x$ b) $8a^2 - a^2$

Procedure:

1. Add or subtract numerical coefficients:
2. Keep common literal coefficient:

Solutions:

1. $7 + 5 - 3 = 9$ 1. $8 - 1 = 7$
2. $9x$ *Ans.*   2. $7a^2$ *Ans.*

12.1.   COMBINE LIKE TERMS

Combine:

a) $8 + 5 - 3$ *Ans.* 10   e) $7x - 4x - x$ *Ans.* $2x$ i) $12c^2 + c^2 - 7c^2$ *Ans.* $6c^2$

b) $7 - 4 - 1$ 2   f) $20r + 30r - 40r$ $10r$ j) $8ab + 7ab$ $15ab$

c) $20 + 30 - 45$ 5   g) $2a^2 + a^2$ $3a^2$ k) $6r^2s - 2r^2s$ $4r^2s$

d) $8b + 5b - 3b$ $10b$ h) $13y^2 - 2y^2$ $11y^2$ l) $30x^3y^2 - 25x^3y^2$ $5x^3y^2$

12.2.   SIMPLIFYING EXPRESSIONS BY COMBINING LIKE TERMS

Simplify each expression by combining like terms:

a) $18a + 12a - 10$ *Ans.* $30a - 10$ f) $2x^2 + 3y^2 - y^2$ *Ans.* $2x^2 + 2y^2$

b) $18a + 12 - 10$ $18a + 2$ g) $6b + 20b + 2c - c$ $26b + c$

c) $18a + 12 - 10a$ $8a + 12$ h) $6b + 20c + 2b - c$ $8b + 19c$

d) $2x^2 + 3x^2 - y^2$ $5x^2 - y^2$ i) $6c + 20b + 2b - c$ $22b + 5c$

e) $2x^2 + 3y^2 - x^2$ $x^2 + 3y^2$

12.3.   COMBINING LIKE TERMS WITH FRACTIONAL AND DECIMAL COEFFICIENTS

Combine like terms:

a) $8x + 3\frac{1}{4}x$ e) $\frac{2}{3}b^2 + \frac{1}{2}b^2$ i) $4.8w + 6.5w - 6.3w$

b) $7x - 2\frac{1}{3}x$ f) $1\frac{2}{3}c^2 - 1\frac{1}{3}c^2$ j) $1\frac{1}{2}y^2 + 3\frac{3}{4}y^2 - 2\frac{3}{4}y^2$

c) $\frac{5}{3}b - b$ g) $2.3ab + 11.6ab$ k) $3xy + 1.2xy - 0.6xy$

d) $3c - \frac{2}{3}c$ h) $0.58cd^2 - 0.39cd^2$ l) $1.7x^2y + 0.9x^2y - 2.05x^2y$

Ans. a) $11\frac{1}{4}x$ e) $1\frac{1}{6}b^2$ i) $5w$

b) $4\frac{2}{3}x$ f) $\frac{1}{3}c^2$ j) $2\frac{1}{2}y^2$

c) $\frac{2}{3}b$ g) $13.9ab$ k) $3.6xy$

d) $2\frac{1}{3}c$ h) $0.19cd^2$ l) $0.55x^2y$

12.4. COMBINING LIKE TERMS REPRESENTING LINE SEGMENTS

Represent the length of each entire line:

a) $\underline{a \mid a}$

b) $\underline{a \mid b \mid a}$

c) $\underline{a \mid a \mid a \mid b \mid b}$

d) $\underline{x \mid 1.5x \mid 2x}$

e) $\underline{y \mid y \mid 0.7y \mid 1.3y}$

f) $\underline{1\frac{1}{2}c \mid \frac{3}{4}c \mid c}$

g) $\underline{a \mid a \mid a \mid a \mid a \mid b \mid b \mid c}$

h) $\underline{a \mid a \mid b \mid b \mid 2b \mid 4a}$

i) $\underline{b \mid a \mid b \mid 3c \mid 8c \mid c}$

Ans. a) $2a$

 b) $2a + b$

 c) $3a + 2b$

 d) $4.5x$

 e) $4y$

 f) $3\frac{1}{4}c$

 g) $5a + 2b + c$

 h) $6a + 4b$

 i) $a + 2b + 12c$

Supplementary Problems

The numbers in parentheses at the right indicate where to find the same type of example in this chapter. Refer to these examples for any help you may require.

1.1.   State each product without multiplication signs: **(1.1)**

 a) $8 \times w$ c) $b \times c \times d$ e) $0.5 \times y \times z$ g) $2\frac{1}{2} \times g \times h \times n$

 b) $2 \times 8 \times a$ d) $10 \times l \times m$ f) $\frac{2}{3} \times 12 \times t$ h) $0.15 \times 100 \times q \times r$

 Ans. a) $8w$ b) $16a$ c) bcd d) $10lm$ e) $0.5yz$ f) $8t$ g) $2\frac{1}{2}ghn$ h) $15qr$

1.2.   Using letters and symbols, replace each verbal statement by an algebraic equation: **(1.2)**

 a) Three times a number added to 8 times the same number is equivalent to 11 times the number.
    Ans. $3n + 8n = 11n$

 b) The difference between 10 times a number and one-half of the same number is exactly the same as $9\frac{1}{2}$ times the number.   *Ans.* $10n - \frac{1}{2}n = 9\frac{1}{2}n$

 c) The perimeter of an equilateral triangle is equal to 3 times the length of one of the sides.
    Ans. $p = 3s$

 d) The area of a square is found by multiplying the length of a side by itself.
    Ans. $A = ss$

1.3.   Simplify each addition by interchanging addends: **(2.1)**

 a) $64 + 138 + 36$ c) $1\frac{1}{3} + \frac{3}{5} + 6\frac{2}{3}$ e) $12\frac{1}{2}\% + 46\% + 87\frac{1}{2}\%$

 b) $15 + 78 + 15 + 170$ d) $2\frac{1}{8} + \frac{13}{16} + \frac{3}{16} + \frac{3}{8}$ f) $5.991 + 1.79 + 0.21 + 0.009$

 Ans. a) $64 + 36 + 138$ c) $1\frac{1}{3} + 6\frac{2}{3} + \frac{3}{5}$ e) $12\frac{1}{2}\% + 87\frac{1}{2}\% + 46\%$

          $100 + 138 = 238$ $8 + \frac{3}{5} = 8\frac{3}{5}$ $100\% + 46\% = 146\%$

       b) $15 + 15 + 170 + 78$ d) $2\frac{1}{8} + \frac{3}{8} + \frac{13}{16} + \frac{3}{16}$ f) $5.991 + 0.009 + 1.79 + 0.21$

          $200 + 78 = 278$ $2\frac{1}{2} + 1 = 3\frac{1}{2}$ $6 + 2 = 8$

1.4.   Rearrange the addends so that literal addends are arranged alphabetically and precede numerical addends: **(2.2)**

 a) $10 + d$ c) $15 + g + f$ e) $d + b + e + a$ g) $5 + m + 4 + j + 11$

 b) $y + x$ d) $17 + r + q + 13$ f) $s + 12 + p + 48$ h) $v + w + 16 + t + 50$

 Ans. a) $d + 10$ c) $f + g + 15$ e) $a + b + d + e$ g) $j + m + 20$

       b) $x + y$ d) $q + r + 30$ f) $p + s + 60$ h) $t + v + w + 66$

1.5. Simplify each multiplication by interchanging factors: **(3.1)**

a) $5 \times 26 \times 40$ c) $10\frac{1}{2} \times 7 \times 2$ e) $3.75 \times 0.15 \times 20 \times 4$

b) $17 \times 12 \times 6$ d) $303 \times 8 \times 1\frac{2}{3}$ f) $66\frac{2}{3}\% \times 50 \times 27$

Ans.   a) $5 \times 40 \times 26$ c) $10\frac{1}{2} \times 2 \times 7$ e) $3.75 \times 4 \times 0.15 \times 20$

         $200 \times 26 = 5200$ $21 \times 7 = 147$ $15 \times 3 = 45$

         b) $17 \times 6 \times 12$ d) $303 \times 1\frac{2}{3} \times 8$ f) $66\frac{2}{3}\% \times 27 \times 50$

         $102 \times 12 = 1224$ $505 \times 8 = 4040$ $\frac{2}{3} \times 27 \times 50 = 18 \times 50 = 900$

1.6. Rearrange the factors so that literal factors are arranged alphabetically and follow numerical factors: **(3.2)**

a) $r8$ b) cab c) $qp7$ d) $4v3t$ e) $x5y7w$ f) $def\,13c$ g) $2h5k10$ h) $r11sm4$ i) $cd13ab$

Ans.   a) $8r$ b) abc c) $7pq$ d) $12tv$ e) $35wxy$ f) $13cdef$ g) $100hk$ h) $44mrs$ i) $13abcd$

1.7. Express each, using symbols of operation: **(4.1)**

a) 10 added to r e) the product of 8, m, and n i) f divided by 7

b) 10 subtracted from r f) 5 times p times q j) 25 divided by x

c) r subtracted from 20   g) two-thirds of c k) the product of u

d) the sum of 7, x, and y h) one-half of b multiplied by h and v divided by 9

Ans.   a) $r + 10$ c) $20 - r$ e) $8mn$ g) $\frac{2}{3}c$ i) $\dfrac{f}{7}$ k) $\dfrac{uv}{9}$

         b) $r - 10$ d) $x + y + 7$ f) $5pq$ h) $\frac{1}{2}bh$ j) $\dfrac{25}{x}$

1.8. When is each division impossible?   **(4.2)**

a) $\dfrac{7}{d}$ c) $\dfrac{3}{4x}$ e) $\dfrac{b}{7 - c}$ g) $\dfrac{50}{w - y}$ i) $\dfrac{45}{x - 2y}$

b) $\dfrac{r}{t}$ d) $\dfrac{5}{a - 8}$ f) $\dfrac{10}{2x - 4}$ h) $\dfrac{100}{pq}$

Ans.   a) if $d = 0$ c) if $x = 0$ e) if $c = 7$ g) if $w = y$ i) if $x = 2y$

         b) if $t = 0$ d) if $a = 8$ f) if $x = 2$ h) if $p = 0$ or $q = 0$

1.9. If n represents a number, express algebraically: **(5.1, 5.2)**

a) 25 more than the number   g) 30 less than the number

b) 30 greater than the number   h) 35 fewer than the number

c) the sum of the number and 35   i) 40 less the number

d) the number increased by 40   j) 45 decreased by the number

e) 45 plus the number   k) 50 minus the number

f) 50 added to the number   l) 55 subtracted from the number

Ans.   a) $n + 25$ or $25 + n$ d) $n + 40$ or $40 + n$ g) $n - 30$ j) $45 - n$

         b) $n + 30$ or $30 + n$ e) $n + 45$ or $45 + n$ h) $n - 35$ k) $50 - n$

         c) $n + 35$ or $35 + n$ f) $n + 50$ or $50 + n$ i) $40 - n$ l) $n - 55$

1.10. Express algebraically: **(5.3)**

a) the number of kilograms of a weight that is 15 kg lighter than w kg   *Ans.* $w - 15$

b) the number of centimeters in a length that is 50 cm shorter than l cm $l - 50$

c) the number of seconds (s) in a time interval that is 1 min less than t seconds (s) $t - 60$

d) the number of cents in a price that is \$1 more than p cents $p + 100$

e) the number of feet per second (ft/s) in a speed that is 20 ft/s slower than r ft/s $r - 20$

f) the number of meters in a distance that is 10 cm farther than d m $d + 0.1$

    g) the number of square feet in an area that is 30 ft² greater than *A* ft²   *Ans.* *A* + 30

    h) the number of degrees in a temperature that is 40° colder than *t°*   *t* − 40

    i) the number of floors in a building that is eight floors higher than *f* floors   *f* + 8

    j) the number of years in an age 5 yr younger than *a* yr   *a* − 5

1.11. Express algebraically: **(6.1)**

    a) *x* times 3   *c)* product of 12 and *y*   *e)* 10 divided by *y*

    b) one-eighth of *b*   *d)* three-eighths of *r*   *f)* quotient of *y* and 10

    Ans. *a)* 3*x*   *b)* $\dfrac{b}{8}$ or $\dfrac{1}{8}b$ *c)* 12*y*   *d)* $\dfrac{3}{8}r$ or $\dfrac{3r}{8}$ *e)* $\dfrac{10}{y}$ *f)* $\dfrac{y}{10}$

1.12. Express algebraically: **(7.1, 7.2)**

    a) *b* decreased by one-half *c*   *f)* 8 more than the product of 5 and *x*

    b) one-third of *g* decreased by 5   *g)* 4 times the sum of *r* and 9

    c) 4 times *r* divided by 9   *h)* the average of 60, *m*, *p*, and *q*

    d) the average of *m* and 60   *i)* the ratio of *b* to 3 times *c*

    e) twice *d* less 25

    Ans. *a)* $b - \dfrac{c}{2}$ *c)* $\dfrac{4r}{9}$ *e)* $2d - 25$ *g)* $4(r+9)$ *i)* $\dfrac{b}{3c}$

    b) $\dfrac{g}{3} - 5$ *d)* $\dfrac{m+60}{2}$ *f)* $5x + 8$ *h)* $\dfrac{m+p+q+60}{4}$

1.13. Express algebraically: **(7.3)**

    a) a distance in meters that is 25 m shorter than three times another of *d* m   *Ans.* $3d - 25$

    b) a weight in ounces that is 5 oz more than twice another of *w* oz $2w + 5$

    c) a temperature in degrees that is 8° warmer than 5 times another of *T°* $5T + 8$

    d) a price in dollars that is \$50 less than one-half another of *p* dollars $\dfrac{p}{2} + 50$

    e) a price in cents that is 50¢ cheaper than one-third another of *p* cents $\dfrac{p}{3} - 50$

    f) a length in feet that is 2 ft longer than *y* yd $3y + 2$

1.14. Evaluate: **(8.1)**

    a) $40 - 2 \times 5$ *c)* $40 \div 2 + 5$ *e)* $16 \div 2 - \tfrac{1}{2} \cdot 10$ *g)* $40 \times 2 - 40 \div 2$

    b) $3 \times 8 - 2 \times 5$ *d)* $3 + 8 - 2 \times 5$ *f)* $3 + 8 \times 2 \times 5$ *h)* $3 + 8 \times 2 - 5 \div 10$

    Ans. *a)* 30   *b)* 14   *c)* 25   *d)* 1   *e)* 3   *f)* 83   *g)* 60   *h)* $18\tfrac{1}{2}$

1.15. Evaluate if *a* = 5, *b* = 6, and *c* = 10:   **(8.2)**

    a) $a + b - c$ *Ans.* 1   *f)* $3 + \dfrac{c}{a}$ *Ans.* 5   *k)* $5a + 4b - 2c$ *Ans.* 29

    b) $a + 2b$ 17   *g)* $\dfrac{4}{5}c$ or $\dfrac{4c}{5}$ 8   *l)* $6c - 2ab$ 0

    c) $a + \dfrac{b}{2}$ 8   *h)* $\dfrac{2}{3}b + \dfrac{3}{2}c$ 19   *m)* $a + \dfrac{c-b}{2}$ 7

    d) $a + \dfrac{b}{2}$ $5\tfrac{1}{2}$ *i)* $\dfrac{a+b}{c-9}$ 11   *n)* $a + c - \dfrac{b}{2}$ 12

    e) $\dfrac{3c}{a}$ 6   *j)* $\dfrac{ab}{\varepsilon}$ 3   *o)* $\dfrac{a+c-b}{3}$ 3

1.16. Evaluate if $x = 3$, $y = 2$, and $z = 0$:   **(8.3)**

a) $x + y + z$	Ans. 5	f) $\dfrac{z}{x}$	Ans. 0	k) $xz + yz$	Ans. 0
b) $x - y - z$	1	g) $\dfrac{x}{z}$	meaningless	l) $\dfrac{z}{x + y}$	0
c) $x(y + z)$	6	h) xyz	0	m) $\dfrac{x}{y + z}$	$1\frac{1}{2}$
d) $z(x + y)$	0	i) $xy + z$	6	n) $x + \dfrac{z}{y}$	3
e) $y(x + z)$	6	j) $x + zy$	3	o) $\dfrac{y + z}{x}$	$\frac{2}{3}$

1.17. Evaluate: **(9.1)**

a) $5(8 + 2)$	Ans. 50	e) $8 \cdot 2(5 - 3)$	Ans. 32	i) $4(4 \cdot 4 - 4)$	Ans. 48
b) $5(8 - 2)$	30	f) $3(6 + 2 \cdot 5)$	48	j) $(4 + 4)4 - 4$	28
c) $8 + 2(5 - 3)$	12	g) $(3 \cdot 6 + 2)5$	100	k) $(4 + 4)(4 - 4)$	0
d) $8(2 \cdot 5 - 3)$	56	h) $3(6 + 2)5$	120	l) $4 + 4(4 - 4)$	4

1.18. Evaluate if $a = 4$, $b = 3$, and $c = 5$:   **(9.2)**

a) $a(b + c)$	Ans. 32	e) $\frac{1}{2}(a + b) + c$	Ans. $8\frac{1}{2}$	i) $3a + 2(c - b)$	Ans. 16
b) $b(c - a)$	3	f) $3(b + 2c)$	39	j) $3(a + 2c) - b$	39
c) $c(a - b)$	5	g) $3(b + 2)c$	75	k) $3(a + 2c - b)$	33
d) $\frac{1}{2}(a + b + c)$	6	h) $(3b + 2)c$	55	l) $3(a + 2)(c - b)$	36

1.19. Evaluate if $x = 6$, $y = 4$, $w = 2$, and $z = 0$:   **(9.3)**

a) $x(y + w)$	Ans. 36	d) $wx(y + z)$	Ans. 48	g) $z \div (x + w)$	Ans. 0
b) $z(w + x)$	0	e) $x + y(w + z)$	14	h) $wy \div (z + x)$	$\frac{4}{3}$
c) $w(x - y)$	4	f) $x + z(y - w)$	6	i) $(w + x)(y - z)$	32

1.20. State the number of terms and the terms in each expression: **(10.1)**

a) $5xyz$	Ans. 1 term: $5xyz$	d) $3a + bc$	Ans. 2 terms: $3a$ and bc
b) $5 + xyz$	2 terms: 5 and xyz	e) $3ab + c$	2 terms: $3ab$ and c
c) $5 + x + y + z$	4 terms: 5, x, y, and z	f) $3a(b + c)$	1 term: $3a(b + c)$

1.21. State the factors of the following, disregarding 1 and the product itself: **(10.2)**

     a) 77   b) 25   c) pq d) $\dfrac{3}{4}x$ e) $\dfrac{w}{10}$ f) $8(x - 5)$ g) $\dfrac{y - 2}{4}$

Ans.   a) 7 and 11   b) 5 and 5   c) p and q d) $\frac{3}{4}$ and x e) $\frac{1}{10}$ and w f) 8 and $(x - 5)$ g) $\frac{1}{4}$, $y - 2$

1.22. State each numerical and literal coefficient: **(10.3)**

     a) w b) $\frac{1}{8}x$ c) $\dfrac{n}{10}$ d) $0.03ab$ e) $\dfrac{3y}{10}$ f) $\dfrac{2a}{3b}$ g) $\frac{3}{5}(a - b)$

	(a)	(b)	(c)	(d)	(e)	(f)	(g)
Ans.							
Numerical coefficient:	1	$\frac{1}{8}$	$\frac{1}{10}$	0.03	$\frac{3}{10}$	$\frac{2}{3}$	$\frac{3}{5}$
Literal coefficient:	w	x	n	ab	y	$\dfrac{a}{b}$	$a - b$

1.23. Write each, using bases and exponents:	**(11.1)**

      a) $7 \cdot 3 \cdot 3$ *b)* $7xyyy$ *c)* $\dfrac{7x}{yyy}$ *d)* $(7x)(7x)$ *e)* $(a+5)(a+5)$ *f)* $\dfrac{2rrw}{5stvv}$

    Ans. *a)* $7 \cdot 3^2$ *b)* $7xy^3$ *c)* $\dfrac{7x}{y^3}$ *d)* $(7x)^2$ *e)* $(a+5)^2$ *f)* $\dfrac{2r^2w}{5stv^2}$

1.24. Write each without exponents:	**(11.2)**

      a) $4 \cdot 7^2$ *b)* $\frac{1}{2}y^4$ *c)* $\dfrac{5a}{b^4}$ *d)* $(ab)^3$ *e)* $(x+2)^2$ *f)* $\dfrac{a^2-b^3}{c+d^2}$

    Ans. *a)* $4 \cdot 7 \cdot 7$ *b)* $\frac{1}{2}yyyy$ *c)* $\dfrac{5a}{bbbb}$ *d)* $(ab)(ab)(ab)$ *e)* $(x+2)(x+2)$ *f)* $\dfrac{aa-bbb}{c+dd}$

1.25. Evaluate (the table of powers may be used to check values):	**(11.3)**

    a) $3^3 - 2^3$ *c)* $10^3 + 5^4$ *e)* $1^2 + 2^2 + 3^2$ *g)* $1^5 + 1^4 + 1^3 + 1^2$ *i)* $5 \cdot 1^3 - 3 \cdot 1^5$

    b) $5^2 \cdot 2^5$ *d)* $10^2 \div 2$ *f)* $5^3 \div 5$ *h)* $2^5 - 4 \cdot 2^2$ *j)* $\frac{1}{2} \cdot 2^2 + \frac{1}{3} \cdot 3^3$

    Ans. *a)* 19   *b)* 800   *c)* 1625   *d)* 50   *e)* 14   *f)* 25   *g)* 4   *h)* 16   *i)* 2   *j)* 11

1.26. Evaluate:	**(11.4)**

      a) $0.1^2 \cdot 9^2$ *b)* $0.3 \cdot 4^2$ *c)* $3^2 4^2$ *d)* $40(\frac{1}{2})^3$ *e)* $(\frac{2}{5})^3$ *f)* $\dfrac{2^3}{5^2}$ *g)* $\dfrac{10}{0.1^2}$

    Ans. *a)* 0.81   *b)* 4.8   *c)* 144   *d)* 5   *e)* $\frac{8}{125}$ *f)* $\frac{8}{25}$ *g)* 1000

1.27. Evaluate if $a = 3$ and $b = 2$:	**(11.5)**

    a) a^2b *Ans.* 18   *e)* $(a+b)^2$ *Ans.* 25   *i)* $a^3 - b^3$ *Ans.* 19

    b) ab^2 12   *f)* $a^2 + b^2$ 13   *j)* $(a-b)^3$ 1

    c) $(ab)^2$ 36   *g)* a^3b 54   *k)* a^2b^3 72

    d) $a + b^2$ 7   *h)* $(ab)^3$ 216   *l)* a^3b^2 108

1.28. Evaluate if $w = 1$, $x = 3$, and $y = 4$:	**(11.5)**

    a) $2w^2$ *Ans.* 2   *d)* $y^2 + x^2$ *Ans.* 25   *g)* $(y-x)^2$ *Ans.* 1   *j)* y^3x *Ans.* 192

    b) $(2w)^2$ 4   *e)* $(y+x)^2$ 49   *h)* $(w+x+y)^2$ 64   *k)* yx^3 108

    c) $(x+2)^2$ 25   *f)* $y^2 - x^2$ 7   *i)* $w^2 + x^2 + y^2$ 26   *l)* $(yx)^3$ 1728

1.29. Combine:	**(12.1)**

      a) $20 + 10 - 18$ *d)* $6x^2 + 5x^2 + x^2$ *g)* $3pq + 11pq - pq$ *j)* $8(a+b) - 2(a+b)$

      b) $10 - 6 - 1$ *e)* $13y^2 - y^2 + 10y^2$ *h)* $2abc + abc - 3abc$ *k)* $11(x^2 + y^2) + 4(x^2 + y^2)$

      c) $6x + 5x + x$ *f)* $27w^5 - 22w^5$ *i)* $36a^2b^2c - 23a^2b^2c$ *l)* $5(x+y)^2 - (x+y)^2$

    Ans. *a)* 12   *d)* $12x^2$ *g)* $13pq$ *j)* $6(a+b)$

      b) 3   *e)* $22y^2$ *h)* 0   *k)* $15(x^2 + y^2)$

      c) $12x$ *f)* $5w^5$ *i)* $13a^2b^2c$ *l)* $4(x+y)^2$

1.30. Simplify each expression by combining like terms: (12.2)

a) $13b + 7b - 6$ Ans. $20b - 6$ f) $5y^2 + 3y^2 + 10y - 2$ Ans. $8y^2 + 10y - 2$

b) $13b + 7 - 6$ $13b + 1$ g) $5y^2 + 3y + 10y - 2y$ $5y^2 + 11y$

c) $13b + 7b - 6b$ $14b$ h) $5y^2 + 3y + 10y^2 - 2y$ $15y^2 + y$

d) $13b^2 + 7b^2 - 6b$ $20b^2 - 6b$ i) $5y + 3y^2 + 10y^2 - 2y^2$ $11y^2 + 5y$

e) $13b^2 + 7b - 6b^2$ $7b^2 + 7b$ j) $5 + 3y + 10y^2 - 2$ $10y^2 + 3y + 3$

1.31. Combine like terms: (12.3)

a) $3y + 7\frac{2}{3}y$ d) $5d - \dfrac{d}{5}$ g) $7.1ab + 3.9ab - 2.7ab$

b) $20\frac{1}{2}a - 11a$ e) $\dfrac{x^2}{2} - \dfrac{x^2}{3}$ h) $2.12c^2 - 1.09c^2 - 0.55c^2$

c) $\frac{7}{3}c - \frac{1}{3}c$ f) $\dfrac{y^2}{2} + \dfrac{y^2}{6}$ i) $3\frac{5}{12}xy^2 + 4\frac{1}{6}xy^2 - 2xy^2$

Ans.   a) $10\frac{2}{3}y$ d) $4\frac{4}{5}d$ g) $8.3ab$

b) $9\frac{1}{2}a$ e) $\dfrac{x^2}{6}$ h) $0.48c^2$

c) $2c$ f) $\dfrac{2}{3}y^2$ i) $5\frac{7}{12}xy^2$

1.32. Represent the length of each entire line: (12.4)

a) d | d | d d) u | $\frac{3}{4}u$ g) $2a$ | $6a$ | $4a$ | $3b$

b) p | q | q | q e) x | $0.5x$ | $1.5x$ h) $3b$ | $5b$ | $7c$ | $4b$

c) r | t | r | t f) y | $2y$ | $3y$ i) $a + b$ | $a + c$ | $b + c$

Ans.   a) $3d$ d) $1\frac{3}{4}u$ g) $12a + 3b$

b) $p + 3q$ e) $3x$ h) $12b + 7c$

c) $2r + 2t$ f) $6y$ i) $2a + 2b + 2c$

CHAPTER 2

Simple Equations and Their Solutions

1. KINDS OF EQUALITIES: EQUATIONS AND IDENTITIES

An **equality** is a mathematical statement that two expressions are equal, or have the same value.

Thus, $2n = 6$, $2n + 3n = 5n$, and $16 = 16$ are equalities.

In an equality, the expression to the left of the equals sign is called the **left member** or **left-hand side** of the equality; the expression to the right is the **right member** or **right-hand side** of the equality.

Thus, in $6n = 3n - 9$, $6n$ is the left member or left-hand side.

While $3n - 9$ is the right member or right-hand side.

An **equation** is an equality in which the unknown or unknowns may have only a particular value or values. An equation is a conditional equality.

Thus, $2n = 12$ is an equation since n may have only one value, 6.

An **identity** is an equality in which a letter or letters may have any value. An identity is an unconditional equality.

Thus, $2n + 3n = 5n$ and $x + y = y + x$ are identities since there is no restriction on the values that n, x, and y may have.

A solution to **an equation** is any number which when substituted for the unknown, will make both sides of the equation equal. A solution is said to **satisfy the equation.**

Thus, 6 is a solution of $2n = 12$, while 5 or any other number is not.

Checking an equation is the process of substituting a particular value for an unknown to see if the value will make both sides equal.

Thus, check in $2x + 3 = 11$ for $x = 4$ and $x = 5$ as follows:

$2x + 3 = 11$	$2x + 3 = 11$	**NOTE.**
$2(4) + 3 \stackrel{?}{=} 11$	$2(5) + 3 \stackrel{?}{=} 11$	(*1*) The symbol $\stackrel{?}{=}$ is read "should equal."
$8 + 3 \stackrel{?}{=} 11$	$10 + 3 \stackrel{?}{=} 11$	(*2*) The symbol $\neq$ is read "does not equal."
$11 = 11$	$13 \neq 11$	

Hence, 4 is a solution to $2x + 3 = 11$ since it satisfies the equation.

NOTE. A solution to an equation is also called a *root* of the equation.

21

1.1. CHECKING AN EQUATION

By checking, determine which value is a solution to each equation:

a) Check $2n + 3n = 25$ for $n = 5$ and $n = 6$ *b)*   Check $8x - 14 = 6x$ for $x = 6$ and $x = 7$

Check: $n = 5$ $n = 6$ **Check:** $x = 6$ $x = 7$

a) $2n + 3n = 25$ $2n + 3n = 25$ *b)* $8x - 14 = 6x$ $8x - 14 = 6x$

$2(5) + 3(5) \overset{?}{=} 25$ $2(6) + 3(6) \overset{?}{=} 25$ $8(6) - 14 \overset{?}{=} 6(6)$ $8(7) - 14 \overset{?}{=} 6(7)$

$10 + 15 \overset{?}{=} 25$ $12 + 18 \overset{?}{=} 25$ $48 - 14 \overset{?}{=} 36$ $56 - 14 \overset{?}{=} 42$

$25 = 25$ $30 \neq 25$ $34 \neq 36$ $42 = 42$

Ans.   5 is a solution to $2n + 3n = 25$.   *Ans.*   7 is a solution to $8x - 14 = 6x$.

1.2. CHECKING AN IDENTITY

By checking the identity $4(x + 2) = 4x + 8$, show that x may have any of the following values:

a) $x = 10$ *b)* $x = 6$ *c)* $x = 4\frac{1}{2}$ *d)* $x = 3.2$

Check:

a) $4(x + 2) = 4x + 8$ *b)* $4(x + 2) = 4x + 8$ *c)* $4(x + 2) = 4x + 8$ *d)* $4(x + 2) = 4x + 8$

$4(10 + 2) \overset{?}{=} 4(10) + 8$ $4(6 + 2) \overset{?}{=} 4(6) + 8$ $4(4\frac{1}{2} + 2) \overset{?}{=} 4(4\frac{1}{2}) + 8$ $4(3.2 + 2) \overset{?}{=} 4(3.2) + 8$

$4(12) \overset{?}{=} 40 + 8$ $4(8) \overset{?}{=} 24 + 8$ $4(6\frac{1}{2}) \overset{?}{=} 18 + 8$ $4(5.2) \overset{?}{=} 12.8 + 8$

$48 = 48$ $32 = 32$ $26 = 26$ $20.8 = 20.8$

2. TRANSLATING VERBAL STATEMENTS INTO EQUATIONS

In algebra, a verbal problem is solved when the value of its unknown (or unknowns) is found. In the process it is necessary to "translate" verbal statements into equations. The first step is to choose a letter to represent the unknown.

Thus, by letting n represent the unknown number, "twice what number equals 12" becomes $2n = 12$.

2.1. TRANSLATING STATEMENTS INTO EQUATIONS

Translate into an equation, letting n represent the number. (You need not find the value of the unknown.)

a) Four less than what number equals 8?   *Ans.* *a)* $n - 4 = 8$

b) One-half of what number equals 10?   *b)* $\dfrac{n}{2} = 10$

c) Ten times what number equals 20?   *c)* $10n = 20$
d) What number increased by 12 equals 17?   *d)* $n + 12 = 17$
e) Twice what number added to 8 is 16?   *e)* $2n + 8 = 16$
f) Fifteen less than 3 times what number is 27?   *f)* $3n - 15 = 27$
g) The sum of what number and twice the same number is 18?   *g)* $n + 2n = 18$
h) What number plus 4 more equals 5 times the number?   *h)* $n + 4 = 5n$
i) Twice the sum of a certain number and 5 is 24. What is the number?   *i)* $2(n + 5) = 24$

2.2. MATCHING STATEMENTS AND EQUATIONS

Match the statements in column 1 with the equations in column 2:

Column 1	**Column 2**
1. The product of 8 and a number is 40.	*a)* $n - 8 = 40$
2. A number increased by 8 is 40.	*b)* $8(n + 8) = 40$
3. Eight less than a number equals 40.	*c)* $8n = 40$
4. Eight times a number less 8 is 40.	*d)* $\dfrac{n}{8} = 40$
5. Eight times the sum of a number and 8 is 40.	*e)* $8n - 8 = 40$
6. One-eighth of a number is 40.	*f)* $n + 8 = 40$

Ans.   *1* and *c,*   *2* and *f,*   *3* and *a,*   *4* and *e,*   *5* and *b,*   *6* and *d.*

2.3. REPRESENTING UNKNOWNS

Represent the unknown by a letter and obtain an equation for each problem. (You need not solve each equation.)

a) A woman worked for 5 hours (h) and earned $130. What was her hourly wage?

b) How old is Henry now if 10 years ago he was 23 years old?

c) After gaining 12 kg, Mary weighed 80 kg. What was her previous weight?

d) A baseball team won 4 times as many games as it lost. How many games did it lose if it played a total of 100 games?

Ans. *a)* Let w = hourly wage in dollars. Then, $5w = 130$.

b) Let H = Henry's age now. Then, $H - 10 = 23$.

c) Let M = Mary's previous weight in kilograms. Then $M + 12 = 80$.

d) Let n = number of games lost and $4n$ = number of games won. Then $n + 4n = 100$.

3. SOLVING SIMPLE EQUATIONS BY USING INVERSE OPERATIONS

For the present, we will study simple equations containing a single unknown having one value. **To solve a simple equation** is to find this value of the unknown. This value is the solution to the equation.

Thus, the equation $2n = 12$ is solved when n is found to equal 6.

To solve an equation, think of it as asking a question such as in each of the following equations:

Equation	Question Asked by Equation	Finding Root of Equation
1. $n + 4 = 12$	What number plus 4 equals 12?	$n = 12 - 4 = 8$
2. $n - 4 = 12$	What number minus 4 equals 12?	$n = 12 + 4 = 16$
3. $4n = 12$	What number multiplied by 4 equals 12?	$n = 12 \div 4 = 3$
4. $\dfrac{n}{4} = 12$	What number divided by 4 equals 12?	$n = 12 \cdot 4 = 48$

Note the two operations involved in each of the above cases:

1. The equation $n + 4 = 12$ involving **addition** is solved by **subtracting** 4 from 12.
2. The equation $n - 4 = 12$ involving **subtraction** is solved by **adding** 4 to 12.
3. The equation $4n = 12$ involving **multiplication** is solved by **dividing** 4 into 12.

4.   The equation $\dfrac{n}{4} = 12$ involving **division** is solved by **multiplying** 4 by 12.

Inverse operations are two operations such that if one is involved with the unknown in the equation, then the other is used to solve the equation.

Rule 1. **Addition and subtraction are inverse operations.**

Thus, in $n + 6 = 10$, n and 6 are **added**. To find n, **subtract** 6 from 10.

In $n - 3 = 9$, 3 is **subtracted** from n. To find n, **add** 3 to 9.

Rule 2. **Multiplication and division are inverse operations.**

Thus, in $7n = 35$, n and 7 are **multiplied**. To find n, **divide** 7 into 35.

In $\dfrac{n}{11} = 4$, n is **divided** by 11. To find n, **multiply** 11 by 4.

NOTE. In 3.1, rule 1 is applied to such equations as $x - 10 = 2$ and $w - 20 = 12$. Later in the chapter, the solution of such equations as $10 - x = 2$ and $20 - w = 12$ will be considered. In 3.2, rule 2 is applied to such equations as $\dfrac{x}{3} = 12$ and $\dfrac{y}{5} = 10$. Later in the chapter, the solution of such equations as $\dfrac{3}{x} = 12$ and $\dfrac{5}{y} = 10$ will be considered.

3.1.   RULE 1: ADDITION AND SUBTRACTION ARE INVERSE OPERATIONS

Solve each equation:

Equations Involving Addition of Unknown	Solutions Requiring Subtraction *Ans.*	Equations Involving Subtraction from Unknown	Solutions Requiring Addition *Ans.*
a) $x + 3 = 8$	a) $x = 8 - 3$ or 5	e) $x - 10 = 2$	e) $x = 2 + 10$ or 12
b) $5 + y = 13$	b) $y = 13 - 5$ or 8	f) $w - 20 = 12$	f) $w = 12 + 20$ or 32
c) $15 = a + 10$	c) $a = 15 - 10$ or 5	g) $18 = a - 13$	g) $a = 18 + 13$ or 31
d) $28 = 20 + b$	d) $b = 28 - 20$ or 8	h) $21 = b - 2$	h) $b = 21 + 2$ or 23

3.2.   RULE 2: MULTIPLICATION AND DIVISION ARE INVERSE OPERATIONS

Solve each equation:

Equations Involving Multiplication of Unknown	Solutions Requiring Division *Ans.*	Equations Involving Division of Unknown	Solutions Requiring Multiplication *Ans.*
a) $3x = 12$	a) $x = \frac{12}{3}$ or 4	e) $\dfrac{x}{3} = 12$	e) $x = 12 \cdot 3$ or 36
b) $12y = 3$	b) $y = \frac{3}{12}$ or $\frac{1}{4}$	f) $\dfrac{y}{12} = 3$	f) $y = 3 \cdot 12$ or 36
c) $35 = 7a$	c) $a = \frac{35}{7}$ or 5	g) $4 = \dfrac{a}{7}$	g) $a = 4 \cdot 7$ or 28
d) $7 = 35b$	d) $b = \frac{7}{35}$ or $\frac{1}{5}$	h) $7 = \dfrac{b}{4}$	h) $b = 7 \cdot 4$ or 28

3.3.   SOLVING BY USING INVERSE OPERATIONS

Solve each equation, showing operation used to solve:

a) $x + 5 = 20$	*Ans.* $x = 20 - 5$ or 15	d) $\dfrac{x}{5} = 20$	*Ans.* $x = 20(5)$ or 100
b) $x - 5 = 20$	$x = 20 + 5$ or 25	e) $10 + y = 30$	$y = 30 - 10$ or 20
c) $5x = 20$	$x = \frac{20}{5}$ or 4	f) $10y = 30$	$y = \frac{30}{10}$ or 3

g) $\dfrac{y}{10} = 30$ *Ans.* $y = 30(10)$ or 300 l) $b - 8 = 2$ *Ans.* $b = 2 + 8$ or 10

h) $14 = a + 7$ $a = 14 - 7$ or 7 m) $8b = 2$ $b = \frac{2}{8}$ or $\frac{1}{4}$

i) $14 = a - 7$ $a = 14 + 7$ or 21 n) $\dfrac{b}{8} = 2$ $b = 2(8)$ or 16

j) $14 = 7a$ $a = \frac{14}{7}$ or 2 o) $24 = 6 + c$ $c = 24 - 6$ or 18

k) $14 = \dfrac{a}{7}$ $a = 14(7)$ or 98 p) $6 = 24c$ $c = \frac{6}{24}$ or $\frac{1}{4}$

4. RULES OF EQUALITY FOR SOLVING EQUATIONS

1. **The addition rule of equality:**
   To maintain an equality, equal numbers may be **added to** both sides of an equation.
2. **The subtraction rule of equality:**
   To maintain an equality, equal numbers may be **subtracted from** both sides of an equation.
3. **The multiplication rule of equality:**
   To maintain an equality, both sides of an equation may be **multiplied by** equal numbers.
4. **The division rule of equality:**
   To maintain an equality, both sides of an equation may be **divided by** equal numbers, except by zero.

These four rules may be summarized in one rule:

The Rule of Equality for All Operations

To maintain an equality, the same operation, using equal numbers, may be performed on both sides of an equation, except division by zero.

To understand these **rules of equality**, think of an **equality** as a **scale in balance.**

If only one side of a balanced scale is changed, the scale becomes unbalanced. To balance the scale, exactly the same change must be made on the other side. Similarly, if only one side of an equality is changed, the two sides are no longer equal. To maintain an equality, exactly the same change must be made on both sides.

Balanced Scales

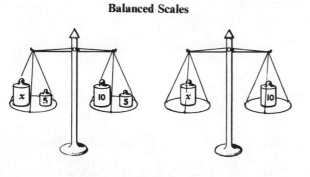

Thus, if 5 is subtracted from both sides of the balanced scale in Fig. 2–1, the scale is still in balance.

Fig. 2–1

Equalities

$$
\begin{array}{rcl}
x + 5 &=& 15 \\
-5 &=& -5 \\
\hline
x &=& 10
\end{array}
$$

If 5 is subtracted from both sides of an equality, an equality remains.

4.1. USING RULES OF EQUALITY

State the equality rule used to solve each equation:

a) $x + 15 = 21$

$$\underline{-15 = -15}$$
$$x = 6$$

b) $40 = r - 8$

$$\underline{+8 = +8}$$
$$48 = r$$

c) $25 = 5m$

$$\frac{25}{5} = \frac{5m}{5}$$
$$5 = m$$

d) $\dfrac{n}{8} = 3$

$$8 \cdot \frac{n}{8} = 8 \cdot 3$$
$$n = 24$$

e) $24x = 8$

$$\frac{24x}{24} = \frac{8}{24}$$
$$x = \frac{1}{3}$$

Ans. *a)* subtraction rule *b)* addition rule *c)* division rule *d)* multiplication rule *e)* division rule.

5. USING DIVISION TO SOLVE AN EQUATION

Division Rule of Equality

To maintain an equality, both sides in an equation may be divided by equal numbers, except by zero.

To Solve an Equation by Using the Division Rule of Equality

Solve: *a)* $2n = 16$ *b)* $16n = 2$

Procedure:

1. Divide both sides of the equation by the coefficient or multiplier of the unknown:

2. Check the original equation:

Solutions:

$\mathbf{D_2}$ $\dfrac{2n}{2} = \dfrac{16}{2}$ $\mathbf{D_{16}}$ $\dfrac{16n}{16} = \dfrac{2}{16}$

Ans. $n = 8$ *Ans.* $n = \dfrac{1}{8}$

Check: $2n = 16$ $16n = 2$
$2(8) \overset{?}{=} 16$ $16(\frac{1}{8}) \overset{?}{=} 2$
$16 = 16$ $2 = 2$

Note 1. **D** is a convenient symbol for "divide both sides."
$\mathbf{D_2}$ means "divide both sides by 2."

Note 2. A common factor may be eliminated in $\dfrac{\overset{1}{\cancel{2}}n}{\cancel{2}}$ and $\dfrac{\overset{1}{\cancel{16}}n}{\cancel{16}}$

5.1 SOLVING EQUATIONS WITH INTEGRAL COEFFICIENTS

Solve each equation:

a) $7x = 35$

$\mathbf{D_7}$ $\dfrac{7x}{7} = \dfrac{35}{7}$

Ans. $x = 5$

Check:

$7x = 35$
$7(5) \overset{?}{=} 35$
$35 = 35$

b) $35y = 7$

$\mathbf{D_{35}}$ $\dfrac{35y}{35} = \dfrac{7}{35}$

Ans. $y = \frac{1}{5}$

Check:

$35y = 7$
$35(\frac{1}{5}) \overset{?}{=} 7$
$7 = 7$

c) $33 = 11z$

$\mathbf{D_{11}}$ $\dfrac{33}{11} = \dfrac{11z}{11}$

Ans. $3 = z$

Check:

$33 = 11z$
$33 \overset{?}{=} 11(3)$
$33 = 33$

d) $11 = 33w$

$\mathbf{D_{33}}$ $\dfrac{11}{33} = \dfrac{33w}{33}$

Ans. $\frac{1}{3} = w$

Check:

$11 = 33w$
$11 \overset{?}{=} 33(\frac{1}{3})$
$11 = 11$

5.2. Division in Equations with Decimal Coefficients

Solve each equation:

a) $0.3a = 9$

$D_{0.3}$ $\dfrac{0.3a}{0.3} = \dfrac{9}{0.3}$

Ans. $a = 30$

Check:

$\qquad 0.3a = 9$

$\qquad 0.3(30) \overset{?}{=} 9$

$\qquad\qquad 9 = 9$

b) $1.2b = 48$

$D_{1.2}$ $\dfrac{1.2b}{1.2} = \dfrac{48}{1.2}$

Ans. $b = 40$

Check:

$\qquad 1.2b = 48$

$\qquad 1.2(40) \overset{?}{=} 48$

$\qquad\quad 48 = 48$

c) $15 = 0.05c$

$D_{0.05}$ $\dfrac{15}{0.05} = \dfrac{0.05c}{0.05}$

Ans. $300 = c$

Check:

$\qquad 15 = 0.05c$

$\qquad 15 \overset{?}{=} 0.05(300)$

$\qquad 15 = 15$

5.3. Solving Equations with Percents as Coefficients

Solve each equation. (*Hint: First replace each percent by a decimal.*)

a) $22\%s = 88$
Since $22\% = 0.22$,

$D_{0.22}$ $\dfrac{0.22s}{0.22} = \dfrac{88}{0.22}$

Ans. $s = 400$
Check:

$\qquad 22\%s - 88$

$\qquad (0.22)(400) \overset{?}{=} 88$

$\qquad\qquad 88 = 88$

b) $75\%t = 18$
Since $75\% = 0.75$,

$D_{0.75}$ $\dfrac{0.75t}{0.75} = \dfrac{18}{0.75}$

Ans. $t = 24$
Check:

$\qquad 75\%t = 18$

$\qquad (0.75)(24) \overset{?}{=} 18$

$\qquad\qquad 18 = 18$

c) $72 = 2\%n$
Since $2\% = 0.02$,

$D_{0.02}$ $\dfrac{72}{0.02} = \dfrac{0.02n}{0.02}$

Ans. $3600 = n$
Check:

$\qquad 72 = 2\%n$

$\qquad 72 \overset{?}{=} (0.02)(3600)$

$\qquad 72 = 72$

5.4. Solving Equations with Like Terms on One Side

Solve each equation. (*Hint: First, collect like terms.*)

a) $60 = 7x - x$
$\qquad 60 = 6x$

D_6 $\dfrac{60}{6} = \dfrac{6x}{6}$

Ans. $10 = x$

Check:

$\qquad 60 = 7x - x$

$\qquad 60 \overset{?}{=} 70 - 10$

$\qquad 60 = 60$

b) $3x + 5x = 48$
$\qquad\quad 8x = 48$

D_8 $\dfrac{8x}{8} = \dfrac{48}{8}$

Ans. $x = 6$

Check:

$\qquad 3x + 5x = 48$

$\qquad 18 + 30 \overset{?}{=} 48$

$\qquad\quad 48 = 48$

c) $7x - 2x = 55$
$\qquad\quad 5x = 55$

D_5 $\dfrac{5x}{5} = \dfrac{55}{5}$

Ans. $x = 11$

Check:

$\qquad 7x - 2x = 55$

$\qquad 77 - 22 \overset{?}{=} 55$

$\qquad\quad 55 = 55$

5.5. Division Rule in a Wage Problem

John worked 7 h and earned $210.21. What was his hourly wage?

Solution: Let h = hourly wage in dollars.

$\qquad$ Then $7h = 210.21$

D_7 $\dfrac{7h}{7} = \dfrac{210.21}{7}$

$\qquad\qquad h = 30.03$

Ans.   John's hourly wage was $30.03.

Check (the problem):

In 7 h, John should earn $210.21

Hence, $7(\$30.03) \overset{?}{=} \$210.21$

$\$210.21 = \$210.21$

5.6. **DIVISION RULE IN A COMMISSION PROBLEM**

Ms. Wang's commission rate was 6 percent. If she earned $66 in commission, how much did she sell?

Solution: Let s = Wang's sales in dollars.   **Check** (the problem):

Then, $6\%s$ or $0.06s = 66$ At 6 percent, Wang's commission
                                                         should be $66. Hence,

$\mathbf{D}_{0.06}$ $\dfrac{0.06s}{0.06} = \dfrac{66}{0.06}$ 6% of $\$1100 \overset{?}{=} \$66$

$(0.06)(\$1100) \overset{?}{=} \$66$

$s = 1100$ $\$66 = \$66$

Ans.   Wang's sales were $1100.

6.   USING MULTIPLICATION TO SOLVE AN EQUATION

Multiplication Rule of Equality

To maintain an equality, both sides in an equation may be multiplied by equal numbers.

To Solve an Equation by Using the Multiplication Rule of Equality

Solve: *a)* $\dfrac{w}{3} = 5$ *b)* $10 = \dfrac{x}{7}$

Procedure: **Solutions:**

1. **Multiply both sides of the equation** $\mathbf{M}_3$ $3 \cdot \dfrac{w}{3} = 5 \cdot 3$ $\mathbf{M}_7$ $7 \cdot 10 = \dfrac{x}{7} \cdot 7$
   by the divisor of the unknown:

   Ans. $w = 15$ *Ans.* $70 = x$

2. **Check** the original equation: Check: $\dfrac{w}{3} = 5$ Check: $10 = \dfrac{x}{7}$

   $\dfrac{15}{3} \overset{?}{=} 5$ $10 \overset{?}{=} \dfrac{70}{7}$

   $5 = 5$ $10 = 10$

Note 1.   **M** is a convenient symbol for "multiply both sides."
              $\mathbf{M}_3$ means "multiply both sides by 3."

Note 2.   A common factor may be eliminated in $\overset{1}{\cancel{3}} \cdot \dfrac{w}{\cancel{3}}$ and $\dfrac{x}{\cancel{7}} \cdot \overset{1}{\cancel{7}}$

Dividing by a Fraction

To **divide by a fraction**, invert the fraction and multiply. Thus, $8 \div \frac{2}{3} = 8 \times \frac{3}{2} = 12$. Hence, multiplying by $\frac{3}{2}$ is equivalent to dividing by $\frac{2}{3}$.

To Solve an Equation Whose Unknown Has a Fractional Coefficient

Solve: *a)* $\frac{2}{3}x = 8$ *b)* $\frac{5}{3}y = 25$

Procedure: **Solutions:**

1. **Multiply both sides of the equation** $\mathbf{M}_{3/2}$ $\frac{3}{2} \cdot \frac{2}{3}x = 8 \cdot \frac{3}{2}$ $\mathbf{M}_{3/5}$ $\frac{3}{5} \cdot \frac{5}{3}y = 25 \cdot \frac{3}{5}$
   by the fractional coefficient inverted   *Ans.* $x = 12$ *Ans.* $y = 15$
   (instead of dividing by the fractional coefficient):

2. **Check** the original equation: Check: $\frac{2}{3}x = 8$ Check: $\frac{5}{3}y = 25$

   $\frac{2}{3} \cdot 12 \overset{?}{=} 8$ $\frac{5}{3} \cdot 15 \overset{?}{=} 25$

   $8 = 8$ $25 = 25$

6.1. SOLVING EQUATIONS WITH INTEGRAL DIVISORS

Solve each equation:

a) $\dfrac{x}{8} = 4$ *b)* $\dfrac{1}{3}y = 12$ *c)* $20 = \dfrac{z}{10}$ *d)* $0.2 = \dfrac{w}{40}$

$\mathbf{M_8}$ $8 \cdot \dfrac{x}{8} = 4 \cdot 8$ $\mathbf{M_3}$ $3 \cdot \dfrac{1}{3}y = 12 \cdot 3$ $\mathbf{M_{10}}$ $10 \cdot 20 = \dfrac{z}{10} \cdot 10$ $\mathbf{M_{40}}$ $40(0.2) = \dfrac{w}{40} \cdot 40$

Ans. $x = 32$ *Ans.* $y = 36$ *Ans.* $200 = z$ *Ans.* $8 = w$

Check: Check: Check: Check:

$\dfrac{x}{8} = 4$ $\dfrac{1}{3}y = 12$ $20 = \dfrac{z}{10}$ $0.2 = \dfrac{w}{40}$

$\dfrac{32}{8} \overset{?}{=} 4$ $\dfrac{1}{3}(36) \overset{?}{=} 12$ $20 \overset{?}{=} \dfrac{200}{10}$ $0.2 \overset{?}{=} \dfrac{8}{40}$

$4 = 4$ $12 = 12$ $20 = 20$ $0.2 = 0.2$

6.2. SOLVING EQUATIONS WITH DECIMAL DIVISORS

Solve each equation:

a) $\dfrac{a}{0.5} = 4$ *b)* $\dfrac{b}{0.08} = 400$ *c)* $1.5 = \dfrac{c}{1.2}$

$\mathbf{M_{0.5}}$ $0.5\left(\dfrac{a}{0.5}\right) = 4(0.5)$ $\mathbf{M_{0.08}}$ $0.08\left(\dfrac{b}{0.08}\right) = 400(0.08)$ $\mathbf{M_{1.2}}$ $1.2(1.5) = \left(\dfrac{c}{1.2}\right)1.2$

Ans. $a = 2$ *Ans.* $b = 32$ *Ans.* $1.8 = c$

(Check your answers.)

6.3. SOLVING EQUATIONS WITH FRACTIONAL COEFFICIENTS

Solve each equation. (*Hint:* Multiply by the fractional coefficient inverted.)

a) $\frac{2}{5}x = 10$ *b)* $1\frac{1}{3}w = 30$ *c)* $c - \frac{1}{4}c = 24$

Solutions: $\frac{4}{3}w = 30$ $\frac{3}{4}c = 24$

$\mathbf{M_{5/2}}$ $\frac{5}{2} \cdot \frac{2}{5}x = 10(\frac{5}{2})$ $\mathbf{M_{3/4}}$ $\frac{3}{4} \cdot \frac{4}{3}w = 30(\frac{3}{4})$ $\mathbf{M_{4/3}}$ $\frac{4}{3} \cdot \frac{3}{4}c = 24(\frac{4}{3})$

Ans. $x = 25$ *Ans.* $w = 22\frac{1}{2}$ *Ans.* $c = 32$

(Check your answers.)

6.4. SOLVING EQUATIONS WITH PERCENTS AS COEFFICIENTS

Solve each equation. (*Hint:* Replace a percent by a fraction if the percent equals an easy fraction.)

a) $66\frac{2}{3}\%s = 22$ *b)* $87\frac{1}{2}\%t = 35$ *c)* $120\%w = 72$

Solutions:

$\frac{2}{3}s = 22$ $\frac{7}{8}t = 35$ $\frac{6}{5}w = 72$

$\mathbf{M_{3/2}}$ $\frac{3}{2} \cdot \frac{2}{3}s = 22(\frac{3}{2})$ $\mathbf{M_{8/7}}$ $\frac{8}{7} \cdot \frac{7}{8}t = 35(\frac{8}{7})$ $\mathbf{M_{5/6}}$ $\frac{5}{6} \cdot \frac{6}{5}w = 72(\frac{5}{6})$

Ans. $s = 33$ *Ans.* $t = 40$ *Ans.* $w = 60$

(Check your answers.)

6.5. MULTIPLICATION RULE IN DISTANCE PROBLEM

After traveling 84 miles (mi), Henry found that he had gone three-fourths of the entire distance to home. What is the total distance to his home?

Solution:

Let d = the total distance in miles. Then $\frac{3}{4}d = 84$

$\mathbf{M}_{4/3}$ $\frac{4}{3} \cdot \frac{3}{4}d = 84(\frac{4}{3})$

               $d = 112$

Ans.   The total distance is 112 mi.

Check (the problem)

The 84 mi traveled should be $\frac{3}{4}$ of the entire distance.
Hence,

84 mi $\overset{?}{=} \frac{3}{4}(112$ mi$)$

84 mi $=$ 84 mi

6.6. MULTIPLICATION RULE IN INVESTMENT PROBLEM

Ms. Fontanez receives 5 percent on a stock investment. If her interest at the end of 1 year was $140, how large was her investment?

Solution:

Let s = the sum invested in dollars.

Then, $5\%s$ or $\dfrac{s}{20} = 140$

$\mathbf{M}_{20}$ $20 \cdot \dfrac{s}{20} = 140(20)$

                  $s = 2800$

Ans.   The investment was $2800.

Check (the problem):

5 percent of the investment should be $140. Hence,

$5\%(\$2800) \overset{?}{=} \$140$

$\$140 = \$140$

7. USING SUBTRACTION TO SOLVE AN EQUATION

Subtraction Rule of Equality

To maintain an equality, equal numbers may be subtracted from both sides in an equation.

To Solve an Equation by Using the Subtraction Rule of Equality

	Solve:	*a)* $w + 12 = 19$	*b)*	$28 = 11 + x$
Procedure:	**Solutions:**	$w + 12 = 19$		$28 = 11 + x$
1. Subtract from both sides the number added to the unknown:	$\mathbf{S}_{12}$ *Ans.*	$\dfrac{-12 = -12}{w \quad = \quad 7}$	$\mathbf{S}_{11}$ *Ans.*	$\dfrac{-11 = -11}{17 = \quad x}$
2. Check the original equation:	Check:	$w + 12 = 19$ $7 + 12 \overset{?}{=} 19$ $19 = 19$		$28 = 11 + x$ $28 \overset{?}{=} 11 + 17$ $28 = 28$

NOTE. $\mathbf{S}$ is a convenient symbol for "subtract from both sides." $\mathbf{S}_{11}$ means "subtract 11 from both sides."

7.1. SUBTRACTION RULE IN EQUATIONS CONTAINING INTEGERS

Solve each equation:

a) $r + 8 = 13$
$\mathbf{S}_8$ $\dfrac{-8 = -8}{r = 5}$
Ans.

Check:

$r + 8 = 13$

$5 + 8 \overset{?}{=} 13$

$13 = 13$

b) $15 + t = 60$
$\mathbf{S}_{15}$ $\dfrac{-15 \quad = -15}{t = 45}$
Ans.

Check:

$15 + t = 60$

$15 + 45 \overset{?}{=} 60$

$60 = 60$

c) $110 = s + 20$
$\mathbf{S}_{20}$ $\dfrac{-20 = \quad -20}{90 = s}$
Ans.

Check:

$110 = s + 20$

$110 \overset{?}{=} 90 + 20$

$110 = 110$

7.2. Subtraction Rule in Equations Containing Fractions or Decimals

Solve each equation:

$a)$ $\quad b + \frac{1}{3} = 3\frac{2}{3}$

$\mathbf{S}_{\frac{1}{3}}$ $\quad \dfrac{-\frac{1}{3} = -\frac{1}{3}}{}$

Ans. $\quad b \quad = 3\frac{1}{3}$

Check:

$$b + \frac{1}{3} = 3\frac{2}{3}$$
$$3\frac{1}{3} + \frac{1}{3} \overset{?}{=} 3\frac{2}{3}$$
$$3\frac{2}{3} = 3\frac{2}{3}$$

$b)$ $\quad 2\frac{3}{4} + c = 8\frac{1}{2}$

$\mathbf{S}_{2\frac{3}{4}}$ $\quad \dfrac{-2\frac{3}{4} \quad = -2\frac{3}{4}}{}$

Ans. $\quad c = 5\frac{3}{4}$

Check:

$$2\frac{3}{4} + c = 8\frac{1}{2}$$
$$2\frac{3}{4} + 5\frac{3}{4} \overset{?}{=} 8\frac{1}{2}$$
$$8\frac{1}{2} = 8\frac{1}{2}$$

$c)$ $\quad 20.8 = d + 6.9$

$\mathbf{S}_{8.9}$ $\quad \dfrac{-6.9 = \quad -6.9}{13.9 = d}$

Ans.

Check:

$$20.8 = d + 6.9$$
$$20.8 \overset{?}{=} 13.9 + 6.9$$
$$20.8 = 20.8$$

7.3. Subtraction Rule in Problem Solving

After an increase of 22¢, the price of grade A eggs rose to $1.12. What was the original price?

Solution:

Let p = original price in cents.

Then $\quad p + 22 = 112$

$\mathbf{S}_{22}$ $\quad \dfrac{-22 = -22}{p \quad = \quad 90}$

Ans. The original price was 90¢.

Check (the problem):

After increasing 22¢, the new price should be 112¢.

Hence,

$$90¢ + 22¢ \overset{?}{=} 112¢$$
$$112¢ = 112¢$$

7.4. Subtraction Rule in Problem Solving

Pam's height is 5 ft 3 in. If she is 9 in taller than John, how tall is John?

Solution:

Let J = John's height in feet.

Then $\quad J + \frac{3}{4} = 5\frac{1}{4}$ $\qquad (9 \text{ in} = \frac{3}{4} \text{ ft})$

$\mathbf{S}_{3/4}$ $\quad \dfrac{-\frac{3}{4} = -\frac{3}{4}}{J \quad = \quad 4\frac{1}{2}}$

Ans. John is $4\frac{1}{2}$ ft, or 4 ft 6 in, tall.

Check (the problem):

9 in more than John's height should equal 5 ft 3 in.

$$4\frac{1}{2} \text{ ft} + \frac{3}{4} \text{ ft} \overset{?}{=} 5\frac{1}{4} \text{ ft}$$
$$5\frac{1}{4} \text{ ft} = 5\frac{1}{4} \text{ ft}$$

8. USING ADDITION TO SOLVE AN EQUATION

Addition Rule of Equality

To maintain an equality, equal numbers may be added to both sides.

To Solve an Equation by Using the Addition Rule of Equality

Solve: $a)$ $n - 19 = 21$ $\qquad$ $b)$ $\quad 17 = m - 8$

Procedure:

1. Add to both sides the number subtracted from the unknown:

2. Check the original equation:

Solutions:

$\mathbf{A}_{19}$ $\quad \dfrac{\begin{aligned} n - 19 &= \quad 21 \\ + 19 &= +19 \end{aligned}}{}$

Ans. $\quad n \quad = \quad 40$

Check: $\quad n - 19 = 21$
$$40 - 19 \overset{?}{=} 21$$
$$21 = 21$$

$\mathbf{A}_{8}$ $\quad \dfrac{\begin{aligned} 17 &= m - 8 \\ + 8 &= \quad + 8 \end{aligned}}{}$

Ans. $\quad 25 = m$

Check: $\quad 17 = m - 8$
$$17 \overset{?}{=} 25 - 8$$
$$17 = 17$$

NOTE: **A** is a convenient symbol for "add to both sides." $\mathbf{A}_{19}$ means "add 19 to both sides."

8.1. ADDITION RULE IN EQUATIONS CONTAINING INTEGERS

Solve each equation:

a)
$$w - 10 = 19$$
$\mathbf{A}_{10}$ $\quad + 10 = +10$
Ans. $\quad w \quad = 29$

b)
$$x - 19 = 10$$
$\mathbf{A}_{19}$ $\quad + 19 = +19$
Ans. $\quad x \quad = 29$

c)
$$7 = y - 82$$
$\mathbf{A}_{82}$ $\quad +82 = \quad +82$
Ans. $\quad 89 = y$

d)
$$82 = z - 7$$
$\mathbf{A}_7$ $\quad + 7 = \quad +7$
Ans. $\quad 89 = z$

Check:

$$w - 10 = 19$$
$$29 - 10 \stackrel{?}{=} 19$$
$$19 = 19$$

Check:

$$x - 19 = 10$$
$$29 - 19 \stackrel{?}{=} 10$$
$$10 = 10$$

Check:

$$7 = y - 82$$
$$7 \stackrel{?}{=} 89 - 82$$
$$7 = 7$$

Check:

$$82 = z - 7$$
$$82 \stackrel{?}{=} 89 - 7$$
$$82 = 82$$

8.2. ADDITION RULE IN EQUATIONS CONTAINING FRACTIONS OR DECIMALS

Solve each equation:

a)
$$h - \tfrac{3}{8} = 5\tfrac{1}{4}$$
$\mathbf{A}_{3\backslash 8}$ $\quad +\tfrac{3}{8} = \tfrac{3}{8}$
Ans. $\quad h \quad = 5\tfrac{5}{8}$

b)
$$j - 20\tfrac{7}{12} = 1\tfrac{1}{12}$$
$\mathbf{A}_{20\frac{7}{12}}$ $\quad +20\tfrac{7}{12} = 20\tfrac{7}{12}$
Ans. $\quad j \quad = 21\tfrac{2}{3}$

c)
$$12.5 = m - 2.9$$
$\mathbf{A}_{2.9}$ $\quad +2.9 = \quad +2.9$
Ans. $\quad 15.4 = m$

(Check your answers.)

8.3. ADDITION RULE IN PROBLEM SOLVING

A drop of 8° brought the temperature to 64°. What was the original temperature?

<u>Solution:</u>

Let t = original temperature in degrees.

Then $\quad t - 8 = \quad 64$
$\mathbf{A}_8$ $\qquad\quad + 8 = + 8$
$\qquad\qquad t \quad = \quad 72$

Ans. The original temperature was 72°.

Check (the problem):

The original temperature, dropped 8°, should become 64°.

Hence,
$$72° - 8° \stackrel{?}{=} 64°$$
$$64° = 64°$$

8.4. ADDITION RULE IN PROBLEM SOLVING

After giving 15 marbles to Sam, Mario has 43 left. How many did Mario have originally?

<u>Solution:</u>

Let m = the original number of marbles.
Then $\quad m - 15 = \quad 43$
$\mathbf{A}_{15}$ $\qquad\quad + 15 = + 15$
$\qquad\qquad m \qquad - \quad 58$

Ans. Sam had 58 marbles at first.

Check (the problem):

The original number of marbles, less 15, should be 43.

Hence,
$$58 \text{ marbles} - 15 \text{ marbles} \stackrel{?}{=} 43 \text{ marbles}$$
$$43 \text{ marbles} = 43 \text{ marbles}$$

9. USING TWO OR MORE OPERATIONS TO SOLVE AN EQUATION

In equations where two operations are performed upon the unknown, two inverse operations are needed to solve the equation.

Thus, in $2x + 7 = 19$, the two operations upon the unknown are **multiplication and addition.** To solve, use **division and subtraction,** performing subtraction first.

Also, in $\frac{x}{3} - 5 = 2$, the two operations upon the unknown are **division and subtraction.** To solve, use **multiplication and addition,** performing addition first.

To Solve Equations by Using Two Inverse Operations

Solve: *a)* $2x + 7 = 19$ *b)* $\frac{x}{3} - 5 = 2$

Procedure:	**Solutions:**			
		$2x + 7 = 19$		$\frac{x}{3} - 5 = 2$
1. Perform **addition** to undo subtraction, or **subtraction** to undo addition:	S_7	$\underline{\quad -7 = -7\quad}$ $2x \quad = 12$	A_5	$\underline{\quad +5 = +5\quad}$ $\frac{x}{3} = 7$
2. Perform **multiplication** to undo division, or **division** to undo **multiplication:**	D_2 *Ans.*	$\frac{2x}{2} = \frac{12}{2}$ $x = 6$	M_3 *Ans.*	$3 \cdot \frac{x}{3} = 3 \cdot 7$ $x = 21$
3. **Check** in the original equation:	Check:	$2x + 7 = 19$ $2(6) + 7 \stackrel{?}{=} 19$ $19 = 19$	Check:	$\frac{x}{3} - 5 = 2$ $\frac{21}{3} - 5 \stackrel{?}{=} 2$ $2 = 2$

9.1. USING TWO INVERSE OPERATIONS TO SOLVE AN EQUATION

Solve each equation:

a) $2x + 7 = 11$

S_7 $\underline{\quad -7 = -7\quad}$

    $2x \quad = 4$

D_2 $\frac{2x}{2} = \frac{4}{2}$

Ans. $x = 2$

Check:

    $2x + 7 = 11$

    $2(2) + 7 \stackrel{?}{=} 11$

    $4 + 7 \stackrel{?}{=} 11$

    $11 = 11$

b) $3x - 5 = 7$

A_5 $\underline{\quad +5 = +5\quad}$

    $3x \quad = 12$

D_3 $\frac{3x}{3} = \frac{12}{3}$

Ans. $x = 4$

Check:

    $3x - 5 = 7$

    $3(4) - 5 \stackrel{?}{=} 7$

    $12 - 5 \stackrel{?}{=} 7$

    $7 = 7$

c) $\frac{x}{3} + 5 = 7$

S_5 $\underline{\quad -5 = -5\quad}$

    $\frac{x}{3} \qquad 2$

M_3 $3 \cdot \frac{x}{3} = 3 \cdot 2$

Ans. $x = 6$

Check:

    $\frac{x}{3} + 5 = 7$

    $\frac{6}{3} + 5 \stackrel{?}{=} 7$

    $2 + 5 \stackrel{?}{=} 7$

    $7 = 7$

d) $\frac{x}{5} - 3 = 7$

A_3 $\underline{\quad +3 = 3\quad}$

    $\frac{x}{5} \quad = 10$

M_5 $5 \cdot \frac{x}{5} = 5 \cdot 10$

Ans. $x = 50$

Check:

    $\frac{x}{5} - 3 = 7$

    $\frac{50}{5} - 3 \stackrel{?}{=} 7$

    $10 - 3 \stackrel{?}{=} 7$

    $7 = 7$

9.2. SOLVING EQUATIONS WITH LIKE TERMS ON THE SAME SIDE

Solve each equation. (*Hint: Combine like terms first.*)

a) $8n + 4n - 3 = 9$

    $12n - 3 = 9$

A_3 $\underline{\quad +3 = 3\quad}$

    $12n \quad = 12$

b) $13n + 4 + n = 39$

    $14n + 4 = 39$

S_4 $\underline{\quad -4 = -4\quad}$

    $14n \quad = 35$

c) $10 = 7 + n - \frac{n}{2}$

    $10 = 7 + \frac{n}{2}$

S_7 $\underline{-7 = -7\quad}$

    $3 = \frac{n}{2}$

$$\mathbf{D_2} \qquad \frac{12n}{12} = \frac{12}{12} \qquad\qquad \mathbf{D_4} \qquad \frac{14n}{14} = \frac{35}{14} \qquad\qquad \mathbf{M_2} \; 2 \cdot 3 = \left(\frac{n}{2}\right)2$$

Ans. $\qquad\quad n = 1$ $\qquad\qquad$ *Ans.* $\qquad\quad n = 2\frac{1}{2}$ $\qquad\qquad$ *Ans.* $\quad 6 = n$

(Check your answers.)

9.3. SOLVING EQUATIONS WITH LIKE TERMS ON BOTH SIDES

Solve each equation. (*Hint: First add or subtract to collect like terms on the same side.*)

a) $\qquad 5n = 40 - 3n$ $\qquad\qquad$ *b)* $\quad 4u + 5 = 5u - 30$ $\qquad\qquad$ *c)* $\quad 3r + 10 = 2r + 20$

$\mathbf{A_{3n}} \quad \underline{+3n =} \quad \underline{+3n} \qquad\qquad \mathbf{A_{30}} \quad \underline{+30 =} \quad \underline{+30} \qquad\qquad \mathbf{S_{10}} \quad \underline{-10 =} \quad \underline{-10}$

$\qquad\qquad 8n = 40 \qquad\qquad\qquad\qquad 4u + 35 = 5u \qquad\qquad\qquad\qquad 3r = 2r + 10$

$\mathbf{D_8} \qquad \dfrac{8n}{8} = \dfrac{40}{8} \qquad\qquad \mathbf{S_{4u}} \quad \dfrac{-4u}{} = \dfrac{-4u}{} \qquad\qquad \mathbf{S_{2r}} \quad \dfrac{-2r}{} = \dfrac{-2r}{}$

Ans. $\qquad n = 5$ $\qquad\qquad$ *Ans.* $\qquad\quad 35 = u$ $\qquad\qquad$ *Ans.* $\qquad\quad r = 10$

(Check your answers.)

9.4. SOLVING EQUATIONS IN WHICH THE UNKNOWN IS A DIVISOR

Solve each equation. (*Hint: First multiply both sides by the unknown.*)

a) $\quad \dfrac{8}{x} = 2$ $\qquad$ *b)* $\quad 12 = \dfrac{3}{y}$ $\qquad$ *c)* $\quad \dfrac{7}{x} = \dfrac{1}{5}$ $\qquad$ *d)* $\quad \dfrac{1}{7} = \dfrac{3}{x}$

$\mathbf{M_x} \quad x\left(\dfrac{8}{x}\right) = 2x \qquad \mathbf{M_y} \quad 12\,y = \left(\dfrac{3}{y}\right)y \qquad \mathbf{M_x} \quad x \cdot \dfrac{7}{x} = \dfrac{1}{5}x \qquad \mathbf{M_x} \quad \dfrac{1}{7}x = \dfrac{3}{x} \cdot x$

$\qquad\qquad 8 = 2x \qquad\qquad\qquad 12y = 3 \qquad\qquad\qquad 7 = \dfrac{x}{5} \qquad\qquad\qquad \dfrac{x}{7} = 3$

$\mathbf{D_2} \quad \dfrac{8}{2} = \dfrac{2x}{2} \qquad \mathbf{D_{12}} \quad \dfrac{12y}{12} = \dfrac{3}{12} \qquad \mathbf{M_5} \; 5 \cdot 7 = \left(\dfrac{x}{5}\right)5 \qquad \mathbf{M_7} \quad 7 \cdot \dfrac{x}{7} = 3(7)$

Ans. $\quad 4 = x$ $\qquad$ *Ans.* $\quad y = \frac{1}{4}$ $\qquad$ *Ans.* $\quad 35 = x$ $\qquad$ *Ans.* $\quad x = 21$

(Check your answers.)

9.5. SOLVING EQUATIONS IN WHICH THE UNKNOWN IS BEING SUBTRACTED

Solve each equation. (*Hint: First add the unknown to both sides.*)

a) $\qquad 10 - w = 3$ $\qquad\qquad\qquad\qquad$ *b)* $\qquad\qquad 10 = 50 - n$

$\mathbf{A_w} \quad \underline{+ w =} \; \underline{w} \qquad\qquad\qquad\qquad \mathbf{A_n} \quad \underline{n} = \underline{+ n}$

$\qquad\qquad 10 = w + 3 \qquad\qquad\qquad\qquad\qquad n + 10 = 50$

$\mathbf{S_3} \quad \underline{-3} = \underline{- 3} \qquad\qquad\qquad \mathbf{S_{10}} \quad \underline{-10} = \underline{-10}$

Ans. $\qquad 7 = w \qquad\qquad\qquad\qquad$ *Ans.* $\quad n = 40$

c) $\quad 25 - 3n = 13$ $\qquad\qquad\qquad\qquad$ *d)* $\qquad\qquad 8 = 50 - 7n$

$\mathbf{A_{3n}} \quad \underline{+ 3n =} \; \underline{+3n} \qquad\qquad\qquad \mathbf{A_{7n}} \quad \underline{+7n} = \underline{+ 7n}$

$\qquad\qquad 25 = 3n + 13 \qquad\qquad\qquad\qquad\quad 7n + 8 = 50$

$\mathbf{S_{13}} \quad \underline{-13} = \underline{- 13} \qquad\qquad\qquad \mathbf{S_8} \quad \underline{-8} = \underline{-8}$

$\qquad\qquad 12 = 3n \qquad\qquad\qquad\qquad\qquad 7n = 42$

D₃ $\dfrac{12}{3} = \dfrac{3n}{3}$ **D₇** $\dfrac{7n}{7} = \dfrac{42}{7}$

Ans. $4 = n$ *Ans.* $n = 6$

(Check your answers.)

9.6. SOLVING EQUATIONS WHOSE UNKNOWN HAS A FRACTIONAL COEFFICIENT

a) Solve: $\dfrac{3}{8}x = 9$ *b*) Solve: $25 = \dfrac{5}{4}x$

Solution: **Solution:**

Using One Operation	**Using Two Operations**	**Using One Operation**	**Using Two Operations**

a) $\dfrac{3}{8}x = 9$ $\dfrac{3}{8}x = 9$ *b*) $25 = \dfrac{5}{4}x$ $25 = \dfrac{5}{4}x$

M₈/₃ $\dfrac{8}{3}\cdot\dfrac{3}{8}x = \dfrac{8}{3}\cdot 9$ **M₈** $8\cdot\dfrac{3}{8}x = 8\cdot 9$ **M₄/₅** $\dfrac{4}{5}\cdot 25 = \dfrac{4}{5}\cdot\dfrac{5}{4}x$ **M₄** $4(25) = 4\cdot\dfrac{5}{4}x$

Ans. $x = 24$ $3x = 72$ *Ans.* $20 = x$ $100 = 5x$

                          D₃ $\dfrac{3x}{3} = \dfrac{72}{3}$ **D₅** $\dfrac{100}{5} = \dfrac{5x}{5}$

                          Ans. $x = 24$ *Ans.* $20 = x$

(Check your answers.)

9.7. SOLVING EQUATIONS WHOSE UNKNOWN HAS A FRACTIONAL COEFFICIENT

a) $\dfrac{3}{4}y - 5 = 7$ *b*) $8 + \dfrac{2}{7}b = 20$ *c*) $48 - \dfrac{5}{3}w = 23$

A₅ $\dfrac{+5 \qquad\ +5}{\dfrac{3}{4}y \qquad 12}$ **S₈** $\dfrac{-8 \qquad\ -8}{\dfrac{2}{7}b \qquad 12}$ **A₅w** $\dfrac{+\frac{5}{3}w \quad - \quad +\frac{5}{3}w}{48 \qquad = \dfrac{5}{3}w + 23}$

                                                                    S₂₃ $\dfrac{-23 \qquad = \qquad -23}{25 \qquad = \dfrac{5}{3}w}$

M₄/₃ $\dfrac{4}{3}\cdot\dfrac{3}{4}y = \dfrac{4}{3}\cdot 12$ **M₇/₂** $\dfrac{7}{2}\cdot\dfrac{2}{7}b = \dfrac{7}{2}\cdot 12$ **M₃/₅** $\dfrac{3}{5}\cdot 25 = \dfrac{3}{5}\cdot\dfrac{5}{3}w$

Ans. $y = 16$ *Ans.* $b = 42$ *Ans.* $15 = w$

(Cheack your answers.)

9.8. USING TWO OPERATIONS IN PROBLEM SOLVING

a) How many boys are there in a class of 36 pupils if the number of girls is 6 more?   *b*) How many boys are there in a class of 36 pupils if the number of girls is 3 times as many?

Solution:

a)   Let b = number of boys   *b*)   Let b = number of boys
    Then $b + 6$ = number of girls   Then $3b$ = number of girls
        $b + b + 6 = 36$ $b + 3b = 36$
S₆ $2b + 6 = 36$ **D₄** $4b = 36$
D₂ $2b = 30$ $b = 9$
              $b = 15$

Ans. There are 15 boys.   *Ans.* There are 9 boys.

(Check your answers.)

9.9.   USING TWO OPERATIONS IN PROBLEM SOLVING

Paul had $260.00 in his bank. By adding equal deposits each week for 20 weeks, he hopes to have $780.00. How much should each weekly deposit be?

Solution:

Let d = number of dollars in each deposit.   **Check** (the problem):

$$\text{Then}\quad 20d + 260 = 780$$

$$\mathbf{S}_{260}\quad\underline{\quad\ -260 = -260\ }$$

$$\mathbf{D}_{20}\quad\underline{\ 20d\quad\ = \quad 520\ }$$

$$d\qquad\ = 26$$

20 deposits and $260.00 should
equal the total of $780.00

Hence,

20 ($26) + $260 $\overset{?}{=}$ $780.00

$520 + $260 $\overset{?}{=}$ $780.00

$780 = $780.00

Ans. He must deposit $26.00 a week.

9.10.   USING TWO OPERATIONS IN PROBLEM SOLVING

Ms. Richards sold her house for $90,000. Her loss amounted to two-fifths of her cost. What did the house cost?

Solution: **Check** (the problem):

Let c = the cost in dollars.

$$\text{Then}\qquad 90{,}000 = c - \tfrac{2}{5}c$$

$$90{,}000 = \tfrac{3}{5}c$$

$$\mathbf{M}_{5/3}\quad \tfrac{5}{3}\cdot 90{,}000 = \tfrac{5}{3}\cdot \tfrac{3}{5}c$$

$$150{,}000 = c$$

If $150,000 is the cost, the loss is
$\tfrac{2}{5}\cdot$ $150,000 or $60,000. The selling
price of $90,000 should be the
cost minus the loss. Hence,

$90,000 $\overset{?}{=}$ $150,000 − $60,000

$90,000 = $90,000

Ans.   The cost was $150,000.

Supplementary Problems

2.1.   By checking, determine which value is a solution to the equation: **(1.1)**

    a) $3x + 4x = 42$ for $x = 4, 6,$ and 8 *Ans.* $x = 6$
    b) $3n + 14 = 47$ for $n = 9, 10,$ and 11 $n = 11$
    c) $6y - 48 = 2y$ for $y = 8, 10,$ and 12 $y = 12$

2.2.   By checking, show that x may have any of the following values in the identity $2(x - 3) = 2x - 6$:

    a) $x = 10$ *b*) $x = 6$ *c*) $x = 4\tfrac{1}{2}$ *d*) $x = 31$ **(1.2)**

2.3.   Translate into an equation, letting n represent the number. (*You need not find the value of the unknown.*)   **(2.1)**

    a) What number diminished by 8 equals 13?   *Ans.* $n - 8 = 13$
    b) Two-thirds of what number equals 10? $\tfrac{2}{3}n = 10$
    c) Three times the sum of a number and 6 is 33. What is the number? $3(n + 6) = 33$
    d) What number increased by 20 equals 3 times the same number? $n + 20 = 3n$
    e) What number increased by 5 equals twice the same number $n + 5 = 2n - 4$
        decreased by 4?

2.4. Match the statement in column 1 with the equations in column 2: **(2.2)**

Column 1

1. The sum of 8 and twice a number is 18.

2. Twice a number less 8 is 18.

3. Twice the sum of a number and 8 is 18.

4. Eight times the difference of a number and 2 is 18.

5. One-half the difference of 8 and a number is 18.

6. Two more than one-eighth of a number is 18.

7. Eight less than half a number is 18.

Column 2

a) $\dfrac{n}{8} + 2 = 18$

b) $8(n - 2) = 18$

c) $\frac{1}{2}(8 - n) = 18$

d) $2(n + 8) = 18$

e) $2n + 8 = 18$

f) $\dfrac{n}{2} - 8 = 18$

g) $2n - 8 = 18$

Ans. *1* and *e,* *2* and *g,* *3* and *d,* *4* and *b,* *5* and *c,* *6* and *a,* *7* and *f.*

2.5. Letting *n* represent the number of games lost, obtain an equation for each problem. (*You need not solve each equation.*) **(2.3)**

a) A team won 3 times as many games as it lost.
It played a total of 52 games.

b) A team won 20 games more than it lost.
It played a total of 84 games.

c) A team won 15 games less than twice the number lost.
It played a total of 78 games.

Ans. a) $n + 3n = 52$

b) $n + n + 20 = 84$

c) $n + 2n - 15 = 78$

2.6. Solve each equation: **(3.1)**

a)	$a + 5 = 9$	*Ans.* $a = 4$	g)	$h - 6 = 14$	*Ans.* $h = 20$
b)	$7 + b = 15$	$b = 8$	h)	$k - 14 = 6$	$k = 20$
c)	$20 = c + 12$	$c = 8$	i)	$45 = m - 13$	$m = 58$
d)	$75 = 55 + d$	$d = 20$	j)	$22 = n - 50$	$n = 72$
e)	$x + 11 = 21 + 8$	$x = 18$	k)	$x - 42 = 80 - 75$	$x = 47$
f)	$27 + 13 = 18 + y$	$y = 22$	l)	$100 - 31 = y - 84$	$y = 153$

2.7. Solve each equation: **(3.2)**

a) $4p = 48$ d) $4n = 2$ g) $\dfrac{t}{5} = 6$ j) $\dfrac{y}{12} = \dfrac{3}{2}$

b) $10r = 160$ e) $12w = 4$ h) $\dfrac{u}{65} = 1$ k) $\dfrac{a}{10} = \dfrac{2}{5}$

c) $25s = 35$ f) $24x = 21$ i) $\dfrac{x}{15} = 4$ l) $\frac{1}{3}b = \frac{5}{6}$

Ans. a) $p = 12$ d) $n = \frac{1}{7}$ g) $t = 30$ j) $y = 18$

b) $r = 16$ e) $w = \frac{1}{3}$ h) $u = 65$ k) $a = 4$

c) $s = \frac{7}{5}$ or $1\frac{2}{5}$ f) $x = \frac{7}{8}$ i) $x = 60$ l) $b = \frac{5}{2}$ or $2\frac{1}{2}$

2.8. Solve each equation: **(3.3)**

a) $n + 8 = 24$ e) $3 + y = 15$ i) $16 = y - 20$ m) $x + \frac{1}{3} = 9$

b) $n - 8 = 24$ f) $15 = y - 3$ j) $16 = \dfrac{y}{20}$ n) $x - \frac{1}{3} = 9$

c) $8n = 24$ g) $15 = 3y$ k) $\dfrac{y}{20} = 16$ o) $\frac{1}{3}x = 9$

d) $\dfrac{n}{8} = 24$ h) $15 = \dfrac{y}{3}$ l) $16 + y = 20$ p) $\dfrac{x}{9} = \dfrac{1}{3}$

Ans. a) $n = 16$ c) $n = 3$ e) $y = 12$ g) $y = 5$ i) $y = 36$ k) $y = 320$ m) $x = 8\frac{2}{3}$ o) $x = 27$

b) $n = 32$ d) $n = 192$ f) $y = 18$ h) $y = 45$ j) $y = 320$ l) $y = 4$ n) $x = 9\frac{1}{3}$ p) $x = 3$

2.9. Solve each equation: **(3.3)**

a) $x + 11 = 14$ f) $h - 3 = 7\frac{1}{2}$ k) $11r = 55$ p) $6\frac{1}{2} = \dfrac{l}{2}$

b) $11 + y = 24$ g) $35 = m - 20\frac{1}{3}$ l) $44s = 44$ q) $1.7 = \dfrac{n}{3}$

c) $22 = 13 + a$ h) $17\frac{3}{4} = n - 2\frac{1}{4}$ m) $10t = 5$ r) $100 = \dfrac{h}{0.7}$

d) $45 = b + 33$ i) $x + 1.2 = 5.7$ n) $8x = 3$ s) $24 = \dfrac{t}{0.5}$

e) $z - 9 = 3$ j) $10.8 = y - 3.2$ o) $3y = 0$ t) $0.009 = \dfrac{x}{1000}$

Ans. a) $x = 3$ d) $b = 12$ g) $m = 55\frac{1}{3}$ j) $y = 14$ m) $t = \frac{1}{2}$ p) $l = 13$ s) $t = 12$
  b) $y = 13$ e) $z = 12$ h) $n = 20$ k) $r = 5$ n) $x = \frac{3}{8}$ q) $n = 5.1$ t) $x = 9$
  c) $a = 9$ f) $h = 10\frac{1}{2}$ i) $x = 4.5$ l) $s = 1$ o) $y = 0$ r) $h = 70$

2.10. State the equality rule used in each: **(4.1)**

a) $6r = 30$ c) $30 = \dfrac{r}{6}$ e) $100x = 5$

$\dfrac{6r}{6} = \dfrac{30}{6}$ $6(30) = 6 \cdot \dfrac{r}{6}$ $\dfrac{100x}{100} = \dfrac{5}{100}$

$r = 5$ $180 = r$ $x = \dfrac{1}{20}$

b) $30 = r - 6$ d) $30 = 6 + r$ f) $100 = \dfrac{y}{5}$

$\underline{+\ 6 = \ +\ 6}$ $\underline{-\ 6 = \ -\ 6}$ $5(100) = 5\left(\dfrac{y}{5}\right)$

$36 = r$ $24 = r$ $500 = y$

Ans.   Addition rule in (b), subtraction rule in (d), multiplication rule in (c) and (f), division rule in (a) and (e).

2.11. Solve each equation: **(5.1)**

a) $12x = 60$ c) $24 = 2z$ e) $6r = 9$ g) $10 = 4t$
b) $60y = 12$ d) $2 = 24w$ f) $9s = 6$ h) $4 = 10u$

Ans. a) $x = 5$ b) $y = \frac{1}{5}$ c) $12 = z$ d) $\frac{1}{12} = w$ e) $r = \frac{3}{2}$ f) $s = \frac{2}{3}$ g) $\frac{5}{2} = t$ h) $\frac{2}{5} = u$

2.12. Solve each equation: **(5.2)**

a) $0.7a = 21$ c) $24 = 0.06c$ e) $0.1h = 100$ g) $25.2 = 0.12k$
b) $1.1b = 55$ d) $18 = 0.009d$ f) $0.6j = 0.96$ h) $7.5 = 0.015m$

Ans. a) $a = 30$ c) $400 = c$ e) $h = 1000$ g) $210 = k$
  b) $b = 50$ d) $2000 = d$ f) $j = 1.6$ h) $500 = m$

2.13. Solve each equation: **(5.3)**

a) $10\%s = 7$ c) $18 = 3\%n$ e) $5\%m = 13$ g) $0.23 = 1\%y$
b) $25\%t = 3$ d) $14 = 70\%w$ f) $17\%x = 6.8$ h) $3.69 = 90\%z$

Ans. a) $s = 70$ c) $600 = n$ e) $m = 260$ g) $23 = y$
  b) $t = 12$ d) $20 = w$ f) $x = 40$ h) $4.1 = z$

2.14. Solve each equation: **(5.4)**

a) $14 = 3x - x$ c) $8z - 3z = 45$ e) $24 = 4\frac{1}{4}x - \dfrac{x}{2}$ g) $7\frac{1}{2}z - 7z = 28$
b) $7y + 3y = 50$ d) $132 = 10w + 3w - w$ f) $4y + 15y = 57$ h) $15w - 3w - 2w = 85$

Ans.   *a)* $7 = x$ *c)* $z = 9$ *e)* $6 = x$ *g)* $z = 56$

         b) $y = 5$ *d)* $11 = w$ *f)* $y = 3$ *h)* $w = 8\frac{1}{2}$

2.15.   Harry earned \$19.26. What was his hourly wage if he worked   *a)* 3 h,   *b)* 2 h,   *c)* $\frac{1}{2}$ h?   **(5.5)**

   Ans.   *a)*   \$6.42   *b)*   \$9.63   *c)*   \$38.52

2.16.   Ms. Hartung's commission rate is 5 percent. How much did she sell if her commissions were *a)* \$85, *b)* \$750,   *c)* \$6.20?   **(5.6)**

   Ans.   *a)*   \$1700   *b)*   \$15,000   *c)*   \$124

2.17.   Solve each equation: **(6.1)**

   a) $\dfrac{x}{3} = 2$ *b)* $\frac{1}{7}y = 12$ *c)* $16 = \dfrac{z}{5}$ *d)* $3 = \frac{1}{50}w$ *e)* $\dfrac{a}{2} = 3$ *f)* $\frac{1}{30}b = 20$ *g)* $0.6 = \dfrac{c}{10}$

   Ans.   *a)* $x = 6$ *b)* $y = 84$ *c)* $80 = z$ *d)* $150 = w$ *e)* $a = 6$ *f)* $b = 600$ *g)* $6 = c$

2.18.   Solve each equation: **(6.2)**

   a) $\dfrac{a}{0.7} = 10$ *b)* $\dfrac{b}{0.02} = 600$ *c)* $30 = \dfrac{c}{2.4}$ *d)* $11 = \dfrac{d}{0.05}$ *e)* $\dfrac{m}{0.4} = 220$ *f)* $\dfrac{n}{0.01} = 3$

   Ans. *a)* $a = 7$ *b)* $b = 12$ *c)* $72 = c$ *d)* $0.55 = d$ *e)* $m = 88$ *f)* $n = 0.03$

2.19.   Solve each equation: **(6.3)**

   a) $\frac{3}{4}x = 21$ *b)* $\frac{4}{3}y = 32$ *c)* $\dfrac{3x}{2} = 9$ *d)* $45 = \frac{5}{9}y$ *e)* $2\frac{1}{5}z = 55$ *f)* $2c + \frac{1}{2}c = 10$

   Ans.   *a)* $x = 28$ *b)* $y = 24$ *c)* $x = 6$ *d)* $y = 81$ *e)* $z = 25$ *f)* $c = 4$

2.20.   Solve each equation: **(6.4)**

   a) $37\frac{1}{2}\%s = 15$ *b)* $60\%t = 60$ *c)* $16\frac{2}{3}\%n = 14$ *d)* $150\%r = 15$ *e)* $83\frac{1}{3}\%w = 35$

   Hint: $37\frac{1}{2}\% = \frac{3}{8}$ $60\% = \frac{3}{5}$ $16\frac{2}{3}\% = \frac{1}{6}$ $150\% = 1\frac{1}{2}$ or $\frac{3}{2}$ $83\frac{1}{3}\% = \frac{5}{6}$

   Ans.   *a)* $s = 40$ *b)* $t = 100$ *c)* $n = 84$ *d)* $r = 10$ *e)* $w = 42$

2.21.   On a trip, John covered a distance of 35 mi. What was the total distance of the trip if the distance traveled was   *a)* five-sixths of the total distance,   *b)* 70 percent of the total distance?   **(6.5)**

   Ans.   *a)*   42 mi   *b)*   50 mi

2.22.   Mrs. Reynolds receives 7 percent per year on a stock investment. How large is her investment if, at the end of 1 yr, her interest is   *a)* \$28,   *b)* \$350,   *c)* \$4.27?   **(6.6)**

   Ans.   *a)*   \$400   *b)*   \$5000   *c)*   \$61

2.23.   Solve each equation: **(7.1)**

   a) $r + 25 = 70$ *c)* $18 = s + 3$ *e)* $x + 130 = 754$ *g)* $259 = s + 237$

   b) $31 + t = 140$ *d)* $842 = 720 + u$ *f)* $116 + y = 807$ *h)* $901 = 857 + w$

   Ans.   *a)* $r = 45$ *c)* $15 = s$ *e)* $x = 624$ *g)* $22 = s$

         b) $t = 109$ *d)* $122 = u$ *f)* $y = 691$ *h)* $44 = w$

2.24.   Solve each equation: **(7.2)**

   a) $b + \frac{2}{3} = 7\frac{2}{3}$ *c)* $35.4 = d + 23.2$ *e)* $f + \frac{5}{8} = 3\frac{1}{2}$ *g)* $7.28 = m + 0.79$

   b) $1\frac{1}{2} + c = 8\frac{3}{4}$ *d)* $87.4 = 80.6 + e$ *f)* $8\frac{1}{6} + g = 10\frac{5}{6}$ *h)* $15.87 = 6.41 + n$

   Ans.   *a)* $b = 7$ *c)* $12.2 = d$ *e)* $f = 2\frac{7}{8}$ *g)* $6.49 = m$

         b) $c = 7\frac{1}{4}$ *d)* $6.8 = e$ *f)* $g = 2\frac{2}{3}$ *h)* $9.46 = n$

2.25.   The price of eggs rose 29¢. What was the original price if the new price is   *a)* \$1.10,   *b)* \$1.30?

   Ans.   *a)*   81¢   *b)*   \$1.01   **(7.3)**

2.26. Will is 8 in taller than George. How tall is George if Will's height is   *a*) 5 ft 2 in,   *b*) 4 ft 3 in?

    Ans.   a)   4 ft 6 in   *b*)   3 ft 7 in   **(7.4)**

2.27. Solve each equation: **(8.1)**

        a) $w - 8 = 22$ *c*) $40 = y - 3$ *e*) $m - 140 = 25$ *g*) $158 = p - 317$
        b) $x - 22 = 8$ *d*) $3 = z - 40$ *f*) $n - 200 = 41$ *h*) $256 = r - 781$

    Ans. a) $w = 30$ *c*) $43 = y$ *e*) $m = 165$ *g*) $475 = p$
          b) $x = 30$ *d*) $43 = z$ *f*) $n = 241$ *h*) $1037 = r$

2.28. Solve each equation: **(8.2)**

        a) $h - \frac{7}{8} = 8\frac{3}{4}$ *c*) $28.4 = m - 13.9$ *e*) $p - 1\frac{5}{12} = 1\frac{7}{12}$ *g*) $0.03 = s - 2.07$
        b) $j - 34\frac{1}{2} = 65$ *d*) $0.37 = n - 8.96$ *f*) $r - 14\frac{2}{3} = 5\frac{1}{3}$ *h*) $5.84 = t - 3.06$

    Ans. a) $h = 9\frac{5}{8}$ *c*) $42.3 = m$ *e*) $p = 3$ *g*) $2.10 = s$
          b) $j = 99\frac{1}{2}$ *d*) $9.33 = n$ *f*) $r = 20$ *h*) $8.90 = t$

2.29. What was the original temperature if a drop of 12° brought the temperature to   **(8.3)**

        a) 75°   *b*) $14\frac{1}{2}°$ *c*) $6\frac{1}{4}°$

    Ans. a) 87°   *b*) $26\frac{1}{2}°$ *c*) $18\frac{1}{4}°$

2.30. Solve each equation: **(9.1)**

        a) $2x + 5 = 9$ *e*) $2x - 5 = 9$ *i*) $\dfrac{x}{4} + 3 = 7$ *m*) $\dfrac{x}{4} - 3 = 7$

        b) $4x + 11 = 21$ *f*) $4x - 11 = 21$ *j*) $\dfrac{x}{5} + 2 = 10$ *n*) $\dfrac{x}{5} - 2 = 10$

        c) $20 = 3x + 8$ *g*) $60 = 10x - 20$ *k*) $17 = \dfrac{x}{2} + 15$ *o*) $3 = \dfrac{x}{12} - 7\frac{1}{4}$

        d) $13 = 6 + 7x$ *h*) $11 = 6x - 16$ *l*) $25 = \dfrac{x}{10} + 2$ *p*) $5\frac{1}{2} = \dfrac{x}{8} - 4$

    Ans. a) $x = 2$ *e*) $x = 7$ *i*) $x = 16$ *m*) $x = 40$
          b) $x = 2\frac{1}{2}$ *f*) $x = 8$ *j*) $x = 40$ *n*) $x = 60$
          c) $x = 4$ *g*) $x = 8$ *k*) $x = 4$ *o*) $x = 123$
          d) $x = 1$ *h*) $x = 4\frac{1}{2}$ *l*) $x = 230$ *p*) $x = 76$

2.31. Solve each equation: **(9.2)**

        a) $10n + 5n - 6 = 9$ *d*) $35 = 6p + 8 + 3p$ *g*) $40 = 25t + 22 - 13t$
        b) $7m + 10 - 2m = 45$ *e*) $19n - 10 + n = 80$ *h*) $145 = 10 + 7.6s - 3.1s$
        c) $25 = 19 + 20n - 18n$ *f*) $3\frac{1}{2}r + r + 2 = 20$

    Ans. a) $n = 1$ *d*) $p = 3$ *g*) $y = \frac{3}{2}$ or $1\frac{1}{2}$
          b) $m = 7$ *e*) $n = 4\frac{1}{2}$ *h*) $s = 30$
          c) $n = 3$ *f*) $r = 4$

2.32. Solve each equation: **(9.3)**

        a) $5r = 2r + 27$ *d*) $18 - 5a = a$ *g*) $9u = 16u - 105$
        b) $2r = 90 - 7r$ *e*) $13b = 15 + 3b$ *h*) $5x + 3 - 2x = x + 8$
        c) $10r - 11 = 8r$ *f*) $100 + 3\frac{1}{2}t = 23\frac{1}{2}t$

Ans. *a)* $r = 9$ *d)* $a = 3$ *g)* $u = 15$

     b) $r = 10$ *e)* $b = 1\frac{1}{2}$ *h)* $x = 2\frac{1}{2}$

     c) $r = 5\frac{1}{2}$ *f)* $t = 5$

2.33. Solve each equation: **(9.4)**

     a) $\dfrac{40}{x} = 5$ *c)* $14 = \dfrac{28}{y}$ *e)* $\dfrac{32}{n} = 8$ *g)* $4 = \dfrac{15}{w}$

     b) $\dfrac{5}{x} = 40$ *d)* $28 = \dfrac{14}{y}$ *f)* $\dfrac{3}{n} = 2$ *h)* $15 = \dfrac{90}{w}$

Ans. *a)* $x = 8$ *b)* $x = \frac{1}{8}$ *c)* $y = 2$ *d)* $y = \frac{1}{2}$ *e)* $n = 4$ *f)* $n = 1\frac{1}{2}$ *g)* $w = 3\frac{3}{4}$ *h)* $w = 6$

2.34. Solve each equation: **(9.5)**

     a) $12 - w = 2$ *c)* $84 - r = 70$ *e)* $8\frac{1}{2} - t = 5\frac{1}{2}$ *g)* $8.7 - u = 7.8$

     b) $24 = 27 - n$ *d)* $90 = 105 - s$ *f)* $\frac{1}{2} = \frac{3}{4} - s$ *h)* $3.25 = 5.37 - v$

Ans. *a)* $w = 10$ *b)* $n = 3$ *c)* $r = 14$ *d)* $s = 15$ *e)* $t = 3$ *f)* $s = \frac{1}{4}$ *g)* $u = 0.9$ *h)* $v = 2.12$

2.35. Solve each equation: **(9.6, 9.7)**

     a) $\frac{7}{8}x = 21$ *d)* $1\frac{1}{2}w = 15$ *g)* $\frac{4}{5}n + 6 = 22$ *j)* $10 = \frac{2}{9}r + 8$

     b) $\frac{3}{4}x = 39$ *e)* $2\frac{1}{3}b = 35$ *h)* $10 + \frac{6}{5}m = 52$ *k)* $6 = 16 - \dfrac{5t}{3}$

     c) $\frac{5}{4}y = 15$ *f)* $2c + 2\frac{1}{2}c = 54$ *i)* $30 - \frac{3}{2}p = 24$ *l)* $3s + \dfrac{s}{3} - 7 = 5$

Ans. *a)* $x = 24$ *d)* $w = 10$ *g)* $n = 20$ *j)* $r = 9$

     b) $x = 52$ *e)* $b = 15$ *h)* $m = 35$ *k)* $t = 6$

     c) $y = 12$ *f)* $c = 12$ *i)* $p = 4$ *l)* $s = 3.6$

2.36. Solve each equation: **(9.1 to 9.7)**

     a) $20 = 3x - 10$ *Ans.* *a)* $x = 10$ *k)* $\dfrac{x}{2} + 27 = 30$ *Ans.* *k)* $x = 6$

     b) $20 = \dfrac{x}{3} - 10$ *b)* $x = 90$ *l)* $8x + 3 = 43$ *l)* $x = 5$

     c) $15 = \frac{3}{4}y$ *c)* $y = 20$ *m)* $21 = \frac{7}{5}w$ *m)* $w = 15$

     d) $17 = 24 - z$ *d)* $z = 7$ *n)* $60 = 66 - 12w$ *n)* $w = \frac{1}{2}$

     e) $3 = \dfrac{39}{x}$ *e)* $x = 13$ *o)* $10b - 3b = 49$ *o)* $b = 7$

     f) $\dfrac{15}{x} = \dfrac{5}{4}$ *f)* $x = 12$ *p)* $12b - 5 = 0.28 + b$ *p)* $b = 0.48$

     g) $5c = 2c + 4.5$ *g)* $c = 1.5$ *q)* $6d - 0.8 = 2d$ *q)* $d = 0.2$

     h) $0.30 - g = 0.13$ *h)* $g = 0.17$ *r)* $40 - 0.5h = 5$ *r)* $h = 70$

     i) $\frac{3}{4}n + 11\frac{1}{2} = 20\frac{1}{2}$ *i)* $n = 12s)$ *s)* $40 - \frac{3}{5}m = 37$ *s)* $m = 5$

     j) $6w + 5w - 8 = 8w$ *j)* $w = 2\frac{2}{3}$ *t)* $12t - 2t + 10 = 9t + 12$ *t)* $t = 2$

2.37. How many girls are there in a class of 30 pupils if *a)* the number of boys is 10 less, *b)* the number of boys is 4 times as many?   **(9.5)**

     Ans. *a)* 20 girls   *b)* 6 girls

2.38. Charles has $370.00 in his bank and hopes to increase this to $1000 by making equal deposits each week. How much should he deposit if he deposits money for 5 weeks?   **(9.9)**

Ans. $126.00

2.39. Mrs. Barr sold her house for $120,000. How much did the house cost her if her loss was
a) one-third of the cost, *b*) 20 percent of the cost?   **(9.10)**

Ans.   *a*) $180,000   *b*) $150,000

CHAPTER 3

Signed Numbers

1. UNDERSTANDING SIGNED NUMBERS: POSITIVE AND NEGATIVE NUMBERS

The temperatures listed in Fig. 3-1 are the Fahrenheit temperatures for five cities on a winter's day. Note how these temperatures have been shown on a **number scale**. On this scale, plus and minus signs are used to distinguish between temperatures above zero and those below zero. Such numbers are called **signed numbers**.

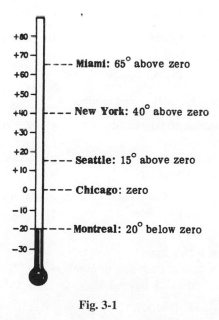

+80
+70 ---- Miami: 65° above zero
+60
+50
+40 ---- New York: 40° above zero
+30
+20
+10 ---- Seattle: 15° above zero
0 ---- Chicago: zero
−10
−20 ---- Montreal: 20° below zero
−30

Fig. 3-1

Signed numbers are positive or negative numbers used to represent quantities that are opposites of each other.

Thus, if +25 represents 25° above zero, −25 represents 25° below zero.

While a minus sign (−) is used to indicate a negative number, a positive number may be shown by a plus sign (+) or by no sign at all.

Thus, 35° above zero may be indicated by +35 or 35.

The table in Fig. 3-2 illustrates pairs of opposites which may be represented by +25 and −25.

+25	−25
$25 deposited	$25 withdrawn
25 mi/h faster	25 mi/h slower
25 lb gained	25 lb lost
25 mi to the north	25 mi to the south

Fig. 3-2

The **absolute value of a signed number** is the number which remains when the sign is removed. Thus, 25 is the absolute value of +25 or −25.

1.1. WORDS OPPOSITE IN MEANING

State the words that are opposite in meaning to the following:

a) Gain	e) North	i) Deposit	m) A.D.
b) Rise	f) East	j) Asset	n) Expand
c) Above	g) Right	k) Earnings	o) Accelerate
d) Up	h) Forward	l) Receipt	p) Clockwise

Ans.

a)	Loss	e) South	i) Withdrawal	m) B.C.
b)	Fall	f) West	j) Liability	n) Contract
c)	Below	g) Left	k) Spendings	o) Decelerate
d)	Down	h) Backward	l) Payment	p) Counterclockwise

1.2. EXPRESSING QUANTITIES AS SIGNED NUMBERS

State the quantity represented by each signed number:

a) by −10, if +10 means 10 yd gained   e) by −100, if +100 means 100 m east
b) by −5, if +5 means $5 earned   f) by −3, if +3 means 3 steps right
c) by +15, if −15 means 15 mi south   g) by 20, if −20 means 20 g underweight
d) by +8, if −8 means 8 h earlier   h) by 5, if −5 means 5 flights down

Ans.

a) 10 yd lost	c) 15 mi north	e) 100 m west	g) 20 g overweight
b) $5 spent	d) 8 h later	f) 3 steps left	h) 5 flights up

1.3. ABSOLUTE VALUE OF SIGNED NUMBERS

State (1) the absolute values of each pair of numbers and (2) the difference of these absolute values: a) +25 and −20   b) −3.5 and +2.4   c) $18\frac{3}{4}$, $-16\frac{3}{4}$

Ans.

a)	(1) 25, 20	b)	(1) 3.5, 2.4	c)	(1) $18\frac{3}{4}$, $16\frac{3}{4}$
	(2) 5		(2) 1.1		(2) 2

2. USING NUMBER SCALES FOR SIGNED NUMBERS

Examine both horizontal and vertical number scales in Figs. 3-3 and 3-4. Notice how positive and negative numbers are placed on opposite sides of zero. Since zero is the starting point, it is called the **origin**.

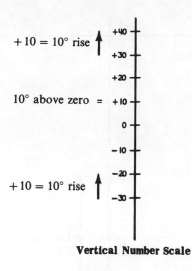

Vertical Number Scale

Fig. 3-3

Constructing Number Scales

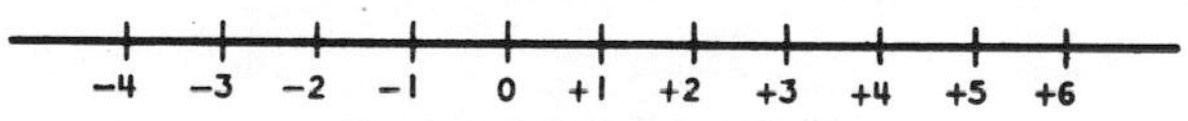

Horizontal Number Scale

Fig. 3-4

In marking off signed numbers, equal lengths on a scale must have equal values. On the horizontal scale shown, each interval has a value of 1, while the intervals on the vertical scale have a value of 10. In a problem, other values may be chosen for each interval, depending on the numbers in the problem.

Number scales may be used for the following purposes:

(1) To understand the meanings of signed numbers
(2) To show which of two signed numbers is greater
(3) To solve problems with signed numbers
(4) To understand operations on signed numbers
(5) To understand graphs in which horizontal and vertical number scales are combined (Such graphs are studied in Chap. 7.)

(1) **Using scales to understand the meanings of signed numbers:**
   A signed number may refer to (1) **a position** or (2) **a change in position**.
   Thus, +10 may refer to the 10° above zero position on a temperature scale.
   However, +10 may also show a 10° rise in temperature.
   Note on the vertical scale the use of +10 as a rise of 10°. The rise of 10° from any temperature is shown by arrows having the same size and direction. Note the two such arrows from +30 to +40 and from −30 to −20. These arrows, in mathematics, are called **vectors**. Since signed numbers involve direction, they are also known **as directed numbers**.

(2) **Using number scales to show which of two signed numbers is the greater** (see Figs. 3-5 and 3-6):
   On a horizontal scale, signed numbers increase to the right; on a vertical scale, the increase is upward. From this, we obtain the following **rules for comparing signed numbers:**

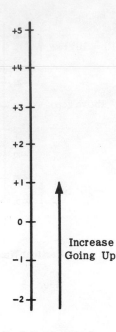

Vertical Number Scale

Fig. 3-5

Horizontal Number Scale

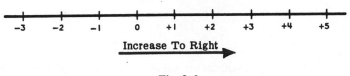

Fig. 3-6

Rule 1.   Any positive number is greater than zero.

Rule 2.   Any negative number is less than zero.

Rule 3.   Any positive number is greater than any negative number. Thus, $+1$ is greater than -1000.

Rule 4.   The greater of two positive numbers has the greater absolute value. Thus, $+20$ is greater than $+10$.

Rule 5.   The greater of two negative numbers has the smaller absolute value. Thus, since -10 is greater than -20, it
         has the smaller absolute value. It is better to owe $10 than to owe $20.

   (3) **Using number scales to solve problems:**
       Such solutions are shown in examples **2.3** to **2.6**.

   (4) **Using number scales to understand operations on signed numbers:**
       This is done throughout the remainder of this chapter.

   (5) **Using number scales to understand graphs:**
       In a later chapter, the horizontal number scale and vertical number scale are combined into the
       graph invented by Descartes in 1637. This graph is very important in mathematics and science.

2.1.   **TWO MEANINGS OF SIGNED NUMBERS**

   On a temperature scale, state two meanings of   *a*) $+25$,   *b*) -25,   *c*) 0.

      Ans.   *a*) $+25$ means (*1*) 25° above zero or (*2*) a rise of 25° from any temperature.

               b) -25 means (*1*) 25° below zero or (*2*) a drop of 25° from any temperature.

               c)   0 means (*1*) zero degrees or (*2*) no change in temperature.

NOTE.   As a number in arithmetic, zero means "nothing." Do not confuse this meaning of zero with the
         two in 2.1 *c*) above.

2.2. COMPARING SIGNED NUMBERS

Which is greater?

        a) $+\frac{1}{4}$ or 0   *b*) +50 or +30   *c*) −30 or 0   *d*) −30 or −10   *e*) +10 or −100

        (Refer to the rules for comparing signed numbers.)

Ans.   *a*) $+\frac{1}{4}$ *b*) +50   *c*) 0   *d*) −10   *e*) +10

            (Rule 1)   **(Rule 4)**   **(Rule 2)**   **(Rule 5)**   **(Rule 3)**

2.3 to 2.6. USE A NUMBER SCALE TO SOLVE PROBLEMS

2.3. Beginning with the main floor, an elevator went up 4 floors, then up 2 more, and then down 8. Find its location after making these changes.

        Solution: Each interval is one floor. The origin, 0, indicates the main floor. The three changes are shown by arrows.

                              (*1*) 4 floors up   (*2*) 2 floors up   (*3*) 8 floors down

      The final location is −2.

        Ans.   2 floors below the main floor. See Fig. 3-7.

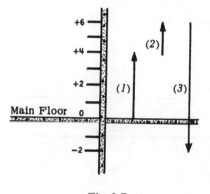

Fig. 3-7

2.4. Beginning with 3° below zero, the temperature changed by rising 9°, then dropped 12°, and finally rose 6°. Find the final temperature.

        Solution: Each interval is 3°. The temperature began at 3° below zero, −3 on the scale. Arrows show the temperature changes:

                              (*1*) 9° rise   (*2*) 12° drop   (*3*) 6° rise

      The final temperature reading is 0.

      Ans. Zero degrees. See Fig. 3-8.

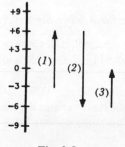

Fig. 3-8

2.5. A football player gained 50 yd and reached his opponent's 10-yd line. Where did he begin his gain?

> **Solution:** Each interval is 10 yd. The 50-yd gain is shown by an arrow. The origin, 0, indicates the center of the field. Since the arrow ends at +10 and has 50-yd length, it began at −40.
> **Ans.** He began his gain from the 40-yd line. See Fig. 3-9.

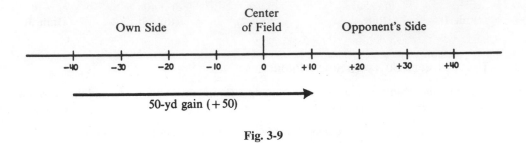

Fig. 3-9

2.6. A plane started from a point 150 km west of its base and flew directly east, reaching a point 150 km east of the base. How far did it travel?

> **Solution:** Each interval is 50 km. The origin indicates the base. The arrow begins at −150 and ends at +150. It has a length of 300 and points eastward.
> **Ans.** 300 km eastward. See Fig. 3-10.

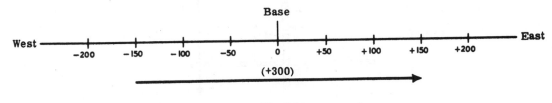

Fig. 3-10

3. ADDING SIGNED NUMBERS

Add **in algebra means "combine."**

In algebra, **adding signed numbers** means **combining them** to obtain a single number which represents the total or combined effect.

Thus, if Mary first gains 10 pounds (lb) and then loses 15 lb, the total effect of the two changes is a loss of 5 lb. This is shown by adding signed numbers: $+10 + (−15) = −5$.

Uses of the symbol $+$: The symbol $+$, when used in adding two signed numbers, has two meanings:

> (*1*) $+$ may mean "add" or
> (*2*) $+$ may mean "positive number."

Thus, $−8 + (+15)$ means **add positive** 15 to negative 8.

Rules for Adding Signed Numbers

Rule 1. To add two signed numbers with like signs, add their absolute values. To this result, prefix the common sign.
> Thus, to add $+7$ and $+3$ or to add $−7$ and $−3$, add the absolute values 7 and 3. To the result 10, prefix the common sign.
> Hence, $+7 + (+3) = +10$ and $−7 + (−3) = −10$.

Rule 2. To add two signed numbers with unlike signs, subtract the smaller absolute value from the other. To this result, prefix the sign of the number having the larger absolute value.

Thus, to add $+7$ and -3 or -7 and $+3$, subtract the absolute value 3 from the absolute value 7. To the result 4, prefix the sign of the number having the larger absolute value. Hence, $+7 + (-3) = +4$ and $-7 + (+3) = -4$.

Rule 3. Zero is the sum of two signed numbers with unlike signs and the same absolute value. Such signed numbers are **opposites** of each other.

Thus, $+27 + (-27) = 0$. The numbers $+27$ and -27 are opposites of each other.

3.1. COMBINING BY MEANS OF SIGNED NUMBERS
Using signed numbers, find each sum:
a) 10 m gained plus 5 m lost   *c*) $5 earned plus $5 spent
b) 8 steps right plus 10 steps left   *d*) 80 km south plus 100 km south

Solutions:

a) 10 m gained plus 5 m lost
$$+10 + (-5)$$

    Ans. $+5$ or 5 m gained

c) $5 earned plus $5 spent
$$+5 + (-5)$$

    Ans. 0 or no change

b) 8 steps right plus 10 steps left
$$+8 + (-10)$$

    Ans. -2 or 2 steps left

d) 80 km south plus 100 km south
$$-80 + (-100)$$

    Ans. -180 or 180 km south

3.2. RULE 1. ADDING SIGNED NUMBERS WITH LIKE SIGNS
Add: *a*) $+8, +2$ *b*) $-8, -2$ *c*) $-30, -14\frac{1}{2}$

Procedure: **Solutions:**

1. Add absolute values: *a*) $+8 + (+2)$ *b*) $-8 + (-2)$ *c*) $-30 + (-14\frac{1}{2})$
                            $8 + 2 = 10$ $8 + 2 = 10$ $30 + 14\frac{1}{2} = 44\frac{1}{2}$

2. Prefix common sign: **Ans.** $+10$ **Ans.** -10 **Ans.** $-44\frac{1}{2}$

3.3. RULE 2. ADDING SIGNED NUMBERS WITH UNLIKE SIGNS
Add: *a*) $+7, -5$ *b*) $-17, +10$

Procedure: **Solutions:**

1. Subtract absolute values: *a*) $+7 + (-5)$ *b*) $-17 + (+10)$
                                $7 - 5 = 2$ $17 - 10 = 7$

2. Prefix sign of number having   Sign of $+7$ is plus   Sign of -17 is minus
larger absolute value: **Ans.** $+2$ **Ans.** -7

3.4. RULE 3. ADDING SIGNED NUMBERS WHICH ARE OPPOSITES OF EACH OTHER
Add. *a*) $-18, +18$ *b*) $+30\frac{1}{2}, -30\frac{1}{2}$ *c*) $-1.75, +1.75$

Procedure: **Solutions:**

Sum is always zero: *a*) $-18 + (+18)$ *b*) $+30\frac{1}{2} + (-30\frac{1}{2})$ *c*) $-1.75 + (+1.75)$
                        Ans. 0   **Ans.** 0   **Ans.** 0

3.5.   RULES 1, 2, AND 3. ADDING SIGNED NUMBERS

Add +25 to   *a*) +30,   *b*) −30,   *c*) −25.   To −20 add   *d*) −30,   *e*) +10,   *f*) +20.

Ans.   *a*) +30 + (+25) = +55   **(Rule 1)**   *Ans.*   *d*) −20 + (−30) = −50   **(Rule 1)**
         b) −30 + (+25) = −5   **(Rule 2)**   *e*) −20 + (+10) = −10   **(Rule 2)**
         c) −25 + (+25) = 0   **(Rule 3)**   *f*) −20 + (+20) = 0   **(Rule 3)**

4.   SIMPLIFYING THE ADDITION OF SIGNED NUMBERS

To simplify the writing used in adding signed numbers:

(1) Parentheses may be omitted.
(2) The symbol + may be omitted when it means "add."
(3) If the first signed number is positive, its plus sign may be omitted.
     Thus, $8 + 9 - 10$ may be written instead of $(+8) + (+9) + (-10)$.

To simplify adding signed numbers   Add: $(+23), (-12), (-8), (+10)$

<u>Procedure:</u>	<u>Solutions:</u>
1. Add all the positive numbers: (Their sum is positive.)	**1.** Add pluses: $\begin{array}{r} 23 \\ 10 \\ \hline 33 \end{array}$
2. Add all the negative numbers: (Their sum is negative.)	**2.** Add minuses: $\begin{array}{r} -12 \\ -\ 8 \\ \hline -20 \end{array}$
3. Add the resulting sums:	**3.** Add sums: $33 - 20$

Ans.   13

4.1.   SIMPLIFYING THE ADDITION OF SIGNED NUMBERS

Express in simplified form, horizontally and vertically; then add:

a) $(+27) + (-15) + (+3)$ *b*) $-8 + (+13) + (-20)$ *c*) $+11.2 + (+13.5) + (-6.7)$
    $+ (-5)$ $+ (+9)$ $+ (+20.9)$

Simplified horizontal forms:

a) $27 - 15 + 3 - 5$ *b*) $-8 + 13 - 20 + 9$ *c*) $11.2 + 13.5 - 6.7 + 20.9$

Simplified vertical forms:

a)		*b*)		*c*)	
	27		− 8		11.2
	−15		13		13.5
	3		−20		−6.7
	$\underline{-\ 5}$		$\underline{9}$		$\underline{20.9}$
Ans.	10	*Ans.*	−6	*Ans.*	38.9

4.2.   ADDING POSITIVES AND NEGATIVES SEPARATELY

Add: *a*) +27, −15, +3, −5   *b*) −8, +13, −20, +9   *c*) +11.2, +13.5, −6.7, +20.9, −3.1

<u>Solutions:</u>

a) Simplify: *b*) Simplify: *c*) Simplify:
    $27 - 15 + 3 - 5$ $-8 + 13 - 20 + 9$ $+11.2 + 13.5 - 6.7 + 20.9 - 3.1$

Add pluses	Add minuses	Add pluses	Add minuses	Add pluses	Add minuses
27	−15	13	− 8	11.2	−6.7
3	− 5	9	−20	13.5	−3.1
30	−20	22	−28	20.9	−9.8
				45.6	

Add results: 30 − 20   Add results: 22 − 28   Add results: 45.6 − 9.8
Ans.   10   *Ans.*   −6   *Ans.*   35.8

4.3. USING SIGNED NUMBERS AND NUMBER SCALES TO SOLVE PROBLEMS

In May, Jovita added deposits of $30 and $20. Later, she withdrew $40 and $30. Find the change in her balance due to these changes, using   *a*) signed numbers and   *b*) number scale (see Fig. 3-11):

a) **Signed-Number Solution**

Add	Add pluses	Add minuses
+30	30	−40
+20	20	−30
−40	50	−70
−30		

$$50 - 70 = -20$$

Ans. $20 less

b) **Number Scale Solution**

The last of the four arrows ends at −20. This means $20 less than the original balance.

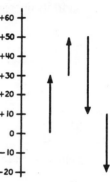

Fig. 3-11

5. SUBTRACTING SIGNED NUMBERS

The symbol −, used in subtracting signed numbers, has two meanings:
(1) − may mean "subtract" or
(2) − may mean "negative number."
Thus, $+8 - (-15)$ means **subtract negative** 15 from positive 8.

Using Subtraction to Find the Change from One Position to Another
Subtraction may be used to find the change from one position to another. (See **5.6** to **5.8**.)
Thus, to find a temperature change from 10° below zero to 20° above zero, subtract −10 from +20. The result is +30, meaning a rise of 30°.

Rule for Subtracting Signed Numbers

Rule 1. **To subtract** a positive number, **add** its opposite negative.
Thus, to subtract +10, add −10. For example, $+18 - (+\mathbf{10})$
$$= (+18) + (-\mathbf{10}) \qquad Ans. \quad +8$$
Rule 2. **To subtract** a negative number, **add** its opposite positive.
Thus, to subtract −10, add +10. For example, $+30 - (-\mathbf{10})$
$$= +30 + (+\mathbf{10}) \qquad Ans. \quad +40$$

5.1   RULE 1. SUBTRACTING A POSITIVE NUMBER

Subtract: *a*) $+8$ from $+29$ *b*) $+80$ from $+80$ *c*) $+18$ from -40

Procedure:	**Solutions:**		
To **subtract** a positive,	*a*) $+29 - (+8)$	*b*) $+80 - (+\mathbf{80})$	*c*) $-40 - (+\mathbf{18})$
add its opposite negative:	$+29 + (-8)$	$+80 + (-\mathbf{80})$	$-40 + (-\mathbf{18})$
	Ans. 21	**Ans.** 0	**Ans.** -58

5.2   RULE 2. SUBTRACTING A NEGATIVE NUMBER

Subtract   *a*) -7 from $+20$ *b*) -67 from -67 *c*) -27 from -87

Procedure:	**Solutions:**		
To **subtract** a negative,	*a*) $+20 - (-7)$	*b*) $-67 - (-\mathbf{67})$	*c*) $-87 - (-\mathbf{27})$
add its opposite positive:	$+20 + (+7)$	$-67 + (+\mathbf{67})$	$-87 + (+\mathbf{27})$
	Ans. 27	**Ans.** 0	**Ans.** -60

5.3.   SUBTRACTING VERTICALLY

Subtract the lower number from the upper:

a) $+40$ *b*) $+40$ *c*) -40
      $\underline{+10}$ $\underline{+55}$ $\underline{-75}$

Solutions: (To subtract a signed number, add its opposite.)

a) $+40$ $+40$ *b*) $+40$ $+40$ *c*) -40 -40
     $\underline{-(+10)\rightarrow}$ $\underline{+(-10)}$ $\underline{-(+55)\rightarrow}$ $\underline{+(-55)}$ $\underline{-(-75)\rightarrow}$ $\underline{+(+75)}$

Ans. $+30$ *Ans.* -15 *Ans.* $+35$

5.4.   RULES 1 AND 2. SUBTRACTING SIGNED NUMBERS

Subtract $+3$ from   *a*) $+11$,   *b*) -15.   From -8, subtract   *c*) -15,   *d*) -5.

Solutions:

a) $+11 - (+3)$	*b*) $-15 - (+3)$	*c*) $-8 - (-15)$	*d*) $-8 - (-5)$
$+11 + (-3)$	$-15 + (-3)$	$-8 + (+15)$	$-8 + (+5)$
$11 - \quad 3$	$-15 - \quad 3$	$-8 + \quad 15$	$-8 + \quad 5$
Ans. 8	*Ans.* -18	*Ans.* $+7$	*Ans.* -3

5.5.   COMBINING ADDITION AND SUBTRACTION OF SIGNED NUMBERS

Combine:

a) $+10 + (+6) - (-2)$ *b*) $+8 - (-12) - (+5) + (-3)$ *c*) $-11 - (+5) + (-7)$
      $+10 + (+6) + (+2)$ $+8 + (+12) + (-5) + (-3)$ $-11 + (-5) + (-7)$
      $10 + \quad 6 + \quad 2$ $8 + \quad 12 - \quad 5 - \quad 3$ $-11 - \quad 5 - \quad 7$

Ans. 18   *Ans.* 12   *Ans.* -23

5.6. **Finding the Change between Two Signed Numbers**

Find the change from +20 to −60, using   *a*) number scale and   *b*) signed numbers:

a) **Number Scale Solution**   *b*) **Signed Number Solution**

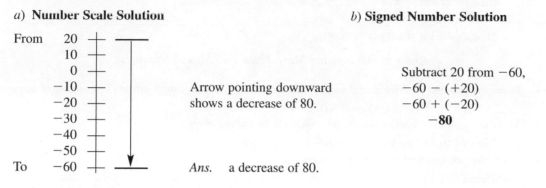

From   20

      10

       0

     −10   Arrow pointing downward   Subtract 20 from −60,
                                         shows a decrease of 80.   −60 − (+20)
     −20   −60 + (−20)
     −30   **−80**
     −40

     −50

To   −60   *Ans.*   a decrease of 80.

5.7. **Finding the Distance between Two Levels**

Using (*1*) a number scale and (*2*) signed numbers, find the distance from 300 ft below sea level to
a) 800 ft below sea level,   *b*) sea level,   *c*) 100 ft above sea level. See Fig. 3-12.

Signed-Number Solution

a) −800 − (−300)	*b*) 0 − (−300)	*c*) 100 − (−300)
−800 + (+300)	0 + (+300)	100 + (+300)
−500	+300	+400

Ans.   500 ft down   *Ans.*   300 ft up   *Ans.*   400 ft up

(Show how each arrow indicates an answer.)

Number Scale Solution

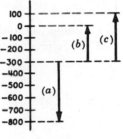

Fig. 3-12

5.8. **Finding a Temperature Change**

On Monday, the temperature changed from −10° to 20°. On Tuesday, the change was from 20° to
−20°. Find each change, using   *a*) a number scale and   *b*) signed numbers. (See Fig. 3-13.)

Signed Number Solution

Monday Change	Tuesday Change
20 − (−10)	−20 − (+20)
20 + (+10)	−20 + (−20)
+30	−40

Ans. Rise of 30°   *Ans.*   Drop of 40°

(Show how each arrow indicates an answer.)

Number Scale Solution

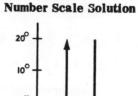
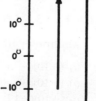

Fig. 3-13

6.   MULTIPLYING SIGNED NUMBERS

Rules for Multiplying Two Signed Numbers

Rule 1.   To multiply two signed numbers with like signs, multiply their absolute values and make the product positive.

Thus, $(+5)(+4) = +20$ and $(-5)(-4) = +20$.

Rule 2. To multiply two signed numbers with unlike signs, multiply their absolute values and make the product negative.

Thus, $(-7)(+2) = -14$ and $(+7)(-2) = -14$.

Rule 3. Zero times any signed number equals zero.

Thus, $0(-8\frac{1}{2}) = 0$ and $(44.7)(0) = 0$.

Rules for Multiplying More Than Two Signed Numbers

Rule 4. **Make the product positive** if all the signed numbers are positive or if there are an even number of negatives.

Thus, $(+10)(+4)(-3)(-5) = +600$.

Rule 5. **Make the product negative** if there are an odd number of negatives.

Thus, $(+10)(+4)(-3)(-5) = +600$.

Rule 6. **Make the product zero** if any number is zero.

Thus, $(-5)(+82)(0)(316\frac{1}{2}) = 0$.

6.1 RULE 1. MULTIPLYING SIGNED NUMBERS WITH LIKE SIGNS

Multiply: a) $(+5)(+9)$ b) $(-5)(-11)$ c) $(-3)(-2.4)$

Procedure: **Solutions:**

	a) $(+5)(+9)$	b) $(-5)(-11)$	c) $(-3)(-2.4)$
1. Multiply absolute values:	$5(9) = 45$	$5(11) = 55$	$3(2.4) = 7.2$
2. Make product positive:	**Ans.** $+45$	**Ans.** $+55$	**Ans.** $+7.2$

6.2. RULE 2. MULTIPLYING SIGNED NUMBERS WITH UNLIKE SIGNS

Multiply: a) $(+8)(-9)$ b) $(-20)(+5)$ c) $(+3\frac{1}{2})(-10)$

Procedure: **Solutions:**

	a) $(+8)(-9)$	b) $(-20)(+5)$	c) $(+3\frac{1}{2})(-10)$
1. Multiply absolute values:	$8(9) = 72$	$20(5) = 100$	$3\frac{1}{2}(10) = 35$
2. Make product negative:	**Ans.** -72	**Ans.** -100	**Ans.** -35

6.3. RULES 1 TO 3. MULTIPLYING SIGNED NUMBERS

Multiply $+10$ by a) $+18$, b) 0, c) -13 Multiply -20 by d) $+2.5$, e)0, f) $-2\frac{1}{4}$

Ans. a) $(+10)(+18) = +180$ **(Rule 1)**   d) $(-20)(+2.5) = -50$ **(Rule 2)**
b) $(+10)(0) = 0$ **(Rule 3)**   e) $(-20)\,0 = 0$ **(Rule 3)**
c) $(+10)(-13) = -130$ **(Rule 2)**   f) $(-20)(-2\frac{1}{4}) = +45$ **(Rule 1)**

6.4. RULES 4 TO 6. MULTIPLYING MORE THAN TWO SIGNED NUMBERS
Multiply:

a) $(+2)(+3)(+4)$ *Ans.* $+24$ **(Rule 4)**   f) $(-1)(-2)(-5)(-10)$ *Ans.* $+100$ **(Rule 4)**
b) $(+2)(+3)(-4)$ -24 **(Rule 5)**   g) $(-1)(-2)(+5)(+10)$ $+100$ **(Rule 4)**
c) $(+2)(-3)(-4)$ $+24$ **(Rule 4)**   h) $(-1)(+2)(+5)(+10)$ -100 **(Rule 5)**
d) $(-2)(-3)(-4)$ -24 **(Rule 5)**   i) $(-1)(-2)(-5)(+10)$ -100 **(Rule 5)**
e) $(+2)(+3)(-4)(0)$ 0 **(Rule 6)**   j) $(-1)(-2)(+5)(+10)(0)$ 0 **(Rule 6)**

6.5.   USING SIGNED NUMBERS TO SOLVE PROBLEMS

Complete each statement using signed numbers to obtain the answer:

a) If Tracy deposits $5 each week, then after 3 weeks her bank balance will be (   ).

> Let $+5$ = $5 weekly deposit   Then $(+5)(+3) = +15$
> and $+3$ = 3 weeks later   *Ans.* $15 more

b) If Henry has been spending $5 a day, then 4 days ago he had (   ).

> Let -5 = $5 daily spending   Then $(-5)(-4) = +20$
> and -4 = 4 days earlier   *Ans.* $20 more

c) If the temperature falls 6° each day, then after 3 days it will be (   ).

> Let -6 = 6° daily fall   Then, $(-6)(+3) = -18$
> and $+3$ = 3 days later   *Ans.*   18° lower

d) If a team has been gaining 10 yd on each play, then 3 plays ago it was (   ).

> Let $+10$ = 10-yd gain per play   Then, $(+10)(-3) = -30$
> and -3 = 3 plays earlier   *Ans.*   30 yd farther back

e) If a car has been traveling west at 30 mi/h, then 3 h ago it was (   ).

> Let -30 = 30 mi/h rate westward   Then $(-30)(-3) = +90$
> and -3 = 3 h earlier   *Ans.*   90 mi farther east

7.   FINDING POWERS OF SIGNED NUMBERS

Rules for Finding Powers of Signed Numbers

(Keep in mind that the power is the answer obtained.)

Rule 1.   For positive base, power is always positive.

> Thus, $(+2)^3$ or $2^3 = +8$ since $(+2)^3 = (+2)(+2)(+2)$.

Rule 2.   For negative base having even exponent, power is positive.

> Thus, $(-2)^4 = +16$ since $(-2)^4 = (-2)(-2)(-2)(-2)$.
>
> NOTE. -2^4 means "the negative of 2^4." Hence, $-2^4 = -16$.

Rule 3.   For negative base having odd exponent, power is negative.

> Thus, $(-2)^5 = -32$ since $(-2)^5 = (-2)(-2)(-2)(-2)(-2)$.

7.1.   RULE 1.   FINDING POWERS WHEN BASE IS POSITIVE

Find each power:

a) 3^2	*Ans.* 9	d) $(+5)^3$	*Ans.* 125	g) $\left(\frac{2}{3}\right)^2$	*Ans.* $\frac{4}{9}$
b) 2^4	16	e) $(+7)^2$	49	h) $\left(\frac{1}{2}\right)^3$	$\frac{1}{8}$
c) 10^3	1000	f) $(+1)^{10}$	1	i) $\left(\frac{1}{10}\right)^4$	$\frac{1}{10,000}$

7.2.   RULES 2 AND 3.   FINDING POWERS WHEN BASE IS NEGATIVE

Find each power:

a) $(-3)^2$	*Ans.* 9	d) $(-1)^{105}$	*Ans.* -1	g) $\left(-\frac{1}{2}\right)^2$	*Ans.* $\frac{1}{4}$
b) $(-3)^3$	-27	e) $(-0.5)^2$	0.25	h) $\left(-\frac{1}{3}\right)^3$	$-\frac{1}{27}$
c) $(-1)^{100}$	1	f) $(-0.2)^3$	-0.008	i) $\left(-\frac{2}{5}\right)^3$	$-\frac{8}{125}$

7.3 RULES 1 TO 3. FINDING POWERS OF SIGNED NUMBERS

Find each power:

a) 3^2	Ans. 9	d) $(\frac{1}{2})^3$	Ans. $\frac{1}{8}$	g) $(-1)^{100}$	Ans. 1
b) -3^2	-9	e) $(-\frac{1}{2})^3$	$-\frac{1}{8}$	h) -1^{100}	-1
c) $(-3)^2$	9	f) $-(\frac{1}{2})^3$	$-\frac{1}{8}$	i) $(-1)^{111}$	-1

7.4. FINDING BASES, EXPONENTS, OR POWERS

Complete each:

a) $(-3)^? = 81$	Ans. 4	c) $(?)^5 = -32$	Ans. -2	e) $(-\frac{3}{4})^3 = ?$	Ans. $-\frac{27}{64}$
b) $2^? = 128$	7	d) $(?)^{121} = -1$	-1	f) $-0.2^4 = ?$	-0.0016

8. DIVIDING SIGNED NUMBERS

Rules for Dividing Signed Numbers

Rule 1. To divide two signed numbers with like signs, divide the absolute value of the first by that of the second and make the quotient positive.

$$\text{Thus, } \frac{+8}{+2} = +4 \text{ and } \frac{-8}{-2} = +4.$$

Rule 2. To divide two signed numbers with unlike signs, divide the absolute value of the first by that of the second and make the quotient negative.

$$\text{Thus, } \frac{+12}{-4} = -3 \text{ and } \frac{-12}{+4} = -3.$$

Rule 3. Zero divided by any signed number is zero.

$$\text{Thus, } \frac{0}{+17} = 0 \text{ and } \frac{0}{-17} = 0.$$

Rule 4. Dividing a signed number by zero is an impossible operation.

Thus, $(+18) \div 0$ or $(-18) \div 0$ are impossible.

Combining Multiplying and Dividing of Signed Numbers

To multiply and divide signed numbers at the same time:

(1)   Isolate their absolute values to obtain the absolute value of the answer.
(2)   Find the sign of the answer, as follows:

Rule 5. **Make the sign of the answer positive** if all the numbers are positive or if there is an even number of negatives.

$$\text{Thus, } \frac{(+12)(+5)}{(+3)(+2)}, \frac{(+12)(+5)}{(-3)(-2)}, \text{ and } \frac{(-12)(-5)}{(-3)(-2)} \text{ equal } +10.$$

Rule 6. **Make the sign of the answer negative** if there is an odd number of negatives.

$$\text{Thus, } \frac{(+12)(+5)}{(+3)(-2)}, \frac{(+12)(-5)}{(-3)(-2)}, \text{ and } \frac{(-12)(-5)}{(+3)(-2)} \text{ equal } -10.$$

Rule 7. **Make the answer zero** if one of the numbers in the dividend is 0.

$$\text{Thus, } \frac{(+53)(0)}{(-17)(-84)} = 0.$$

Rule 8. A zero divisor makes the operation impossible.

Thus, $\dfrac{(+53)(-17)}{(+84)(0)}$ is meaningless.

8.1. RULE 1. DIVIDING SIGNED NUMBERS WITH LIKE SIGNS

Divide: a) $+12$ by $+6$ b) -12 by -4 c) -24.6 by -3

Procedure: **Solutions:**

	a) $\dfrac{+12}{+6}$	b) $\dfrac{-12}{-4}$	c) $\dfrac{-24.6}{-3}$
1. Divide absolute values:	$\dfrac{12}{6} = 2$	$\dfrac{12}{4} = 3$	$\dfrac{24.6}{3} = 8.2$
2. Make quotient positive:	**Ans.** $+2$	**Ans.** $+3$	**Ans.** $+8.2$

8.2. RULE 2. DIVIDING SIGNED NUMBERS WITH UNLIKE SIGNS

Divide   a) $+20$ by -5 b) -24 by $+8$ c) $+8$ by -16

Procedure: **Solutions:**

	a) $\dfrac{+20}{-5}$	b) $\dfrac{-24}{+8}$	c) $\dfrac{+8}{-16}$
1. Divide absolute values:	$\dfrac{20}{5} = 4$	$\dfrac{24}{8} = 3$	$\dfrac{8}{16} = \dfrac{1}{2}$
2. Make quotient negative:	**Ans.** -4	**Ans.** -3	**Ans.** $-\dfrac{1}{2}$

8.3. RULES 1 TO 4. DIVIDING SIGNED NUMBERS

Divide $+9$ by   a) $+3$ b) 0 c) -10 Divide each by -12:   d) $+36$ e) 0 f) -3

Ans.   a) $\dfrac{+9}{+3} = +3$ **(Rule 1)**   d) $\dfrac{+36}{-12} = -3$ **(Rule 2)**

b) $\dfrac{+9}{0}$ is meaningless   **(Rule 4)**   e) $\dfrac{0}{-12} = 0$ **(Rule 3)**

c) $\dfrac{+9}{10} = -\dfrac{9}{10}$ **(Rule 2)**   f) $\dfrac{-3}{12} = +\dfrac{1}{4}$ **(Rule 1)**

8.4. RULES 5 AND 6: COMBINING, MULTIPLYING, AND DIVIDING SIGNED NUMBERS

Solve: a) $\dfrac{(+12)(+8)}{(-3)(-4)}$ b) $\dfrac{(-4)(-2)(-3)}{(-5)(-10)}$

Procedure: **Solutions:**

1. Isolate absolute values:	$\dfrac{(12)(8)}{(3)(4)} = 8$	$\dfrac{(4)(2)(3)}{(5)(10)} = \dfrac{12}{25}$
2. Find the sign of the answer:	$\dfrac{(+)(+)}{(-)(-)} = +$	$\dfrac{(-)(-)(-)}{(-)(-)} = -$
	Ans. $+8$	**Ans.** $-\dfrac{12}{25}$

8.5.   RULES 7 AND 8.   ZERO IN DIVIDEND OR DIVISOR

Solve: a) $\dfrac{(+27)(0)}{(-3)(+15)}$ *Ans.* 0 **(Rule 7)**   b) $\dfrac{(+27)(-3)}{(+15)(0)}$ *Ans.* meaningless **(Rule 8)**

9.   EVALUATING EXPRESSIONS HAVING SIGNED NUMBERS

To evaluate an expression having signed numbers:

(1)   Substitute the values given for the letters, enclosing them in parentheses.
(2)   Perform the operations in the correct order, doing the power operations first.

9.1.   EVALUATING EXPRESSIONS HAVING ONE LETTER
Evaluate if $y = -2$:

a) $2y + 5$ b) $20 - 3y$ c) $4y^2$ d) $2y^3$ e) $20 - y^5$
$\quad$ $2(-2) + 5$ $\quad 20 - 3(-2)$ $\quad 4(-2)^2$ $\quad 2(-2)^3$ $\quad 20 - (-2)^5$
$\quad$ $-4 + 5$ $\quad 20 + 6$ $\quad 4(4)$ $\quad 2(-8)$ $\quad 20 - (-32)$

Ans.   1   *Ans.*   26   *Ans.*   16   *Ans.* -16 *Ans.*   52

9.2.   EVALUATING EXPRESSIONS HAVING TWO LETTERS
Evaluate if $x = -3$ and $y = +2$:

a) $2x - 5y$ b) $20 - 2xy$ c) $3xy^2$ d) $2x^2 \quad - y^2$
$\quad$ $2(-3) - 5(+2)$ $\quad 20 - 2(-3)(+2)$ $\quad 3(-3)(+2)^2$ $\quad 2(-3)^2 -(+2)^2$
$\quad$ $-6 - 10$ $\quad 20 + 12$ $\quad 3(-3)(4)$ $\quad 2(+9) \; - (+4)$

Ans. -16 *Ans.*   32   *Ans.* -36 *Ans.*   14

9.3.   EVALUATING EXPRESSIONS HAVING THREE LETTERS
Evaluate if $x = +1$, $y = -2$, $z = -3$:

a) $4xy^2 + z$ b) $x^2 + y^2 - z^2$ c) $\dfrac{3x - 5y}{2z}$ d) $\dfrac{2y^2}{x - z}$

$\quad$ $4(+1)(-2)^2 + (-3)$ $\quad (+1)^2 + (-2)^2 - (-3)^2$ $\dfrac{3(+1) - 5(-2)}{2(-3)}$ $\dfrac{2(-2)^2}{+1 - (-3)}$

$\quad$ $4(1)(4) - 3$ $\quad 1 + 4 - 9$ $\dfrac{3 + 10}{-6}$ $\dfrac{2 \cdot 4}{1 + 3}$

Ans.   13   *Ans.* -4 *Ans.* $-2\frac{1}{6}$ *Ans.*   2

Summary of Rules for Computations Involving Signed Numbers:

1. **To add** two signed numbers, add their absolute values if the numbers have the same signs; if the numbers have unlike signs, subtract the smaller absolute value from the larger, and to this result, prefix the sign of the number with the larger absolute value.
2. **To subtract** a signed number from a positive number, add its opposite negative; to subtract a signed number from a negative number, add its opposite positive.
3. **To multiply** two signed numbers, multiply their absolute values and prefix the result with a positive sign if the numbers are both positive or both negative; prefix the result with a negative sign if one of the numbers is positive and the other is negative.

4. **To find powers of signed numbers,** the power is positive if the exponent is even; the power is negative if the exponent is odd.
5. **To divide** one signed number by another, divide their absolute values and obey the rule of signs for multiplication to two signed numbers.

Supplementary Problems

3.1. State the quantity represented by each signed number: **(1.2)**

a) by $+12$, if -12 means 12 yd lost   d) by -32, if $+32$ means \$32 earned
b) by -7, if $+7$ means 7 flights up   e) by $+15$, if -15 means \$15 withdrawn
c) by $+25$, if -25 means 25 mi westward   f) by -13, if $+13$ means 13 steps forward

Ans.   a) 12 yd gained   c) 25 mi eastward   e) \$15 deposited
         b) 7 flights down   d) \$32 spent   f) 13 steps backward

3.2. State (*1*) the absolute value of each pair of numbers and (*2*) the differences of these absolute values:
a) $+40$ and -32 b) -3.7 and $+1.5$ c) $-40\frac{1}{2}$ and $-10\frac{1}{2}$ **(1.3)**

Ans.   a) (*1*) 40, 32; (*2*) 8   b) (*1*) 3.7, 1.5; (*2*) 2.2   c) (*1*) $40\frac{1}{2}$. $10\frac{1}{2}$; (*2*) 30

3.3. On a temperature scale, state the meanings of: a) -17 b) $+15$ **(2.1)**

Ans.   a) -17 means (*1*) 17° below zero or (*2*) a drop of 17° from any temperature.
         b) $+15$ means (*1*) 15° above zero or (*2*) a rise of 15° from any temperature.

3.4. Which is greater?   a) $-\frac{1}{2}$ or 0   b) $-\frac{1}{2}$ or -30 c) $+\frac{1}{2}$ or $-2\frac{1}{2}$ d) -110 or -120 **(2.2)**

Ans.   a) 0   b) $-\frac{1}{2}$ c) $+\frac{1}{2}$ d) -110

3.5. Beginning with 5° above zero, the temperature rose 2°, then dropped 8°, and finally rose 4°. Find the change in temperature and the final temperature after the changes.   **(2.4)**

Ans.   Change in temperature was a drop of 2° from the initial temperature of $+5$°. Final temperature after the changes was 3° above zero.

3.6. A plane starting from a point 200 km east of its base flew directly west to a point 300 km west of the base. How far did it travel?   **(2.6)**

Ans.   500 km westward

3.7. Using signed numbers, find each sum: **(3.1)**

a) 20 kg gained plus 7 kg lost   d) \$5 gained plus \$15 spent
b) 17 lb lost plus 3 lb gained   e) 8° rise plus 20° rise
c) \$5 spent plus \$15 spent   f) 8° drop plus 8° rise

Ans.   a) $+20 + (-7) = +13$ or 13 kg gained   d) $+5 + (-15) = -10$ or \$10 spent
         b) $-17 + (+3) = -14$ or 14 lb lost   e) $+8 + (+20) = +28$ or 28° rise
         c) $-5 + (-15) = -20$ or \$20 spent   f) $-8 + (+8) = 0$ or no change

3.8. Add: **(3.2)**

a) $+10\frac{1}{2}, +7\frac{1}{2}$ b) $-1.4, -2.5$ c) $+5, +3, +12$ d) $-1, -2, -3$

Ans.   a) $+18$ b) -3.9 c) $+20$ d) -6

3.9. Add: **(3.3)**

a) $+6\frac{1}{4}, -3\frac{1}{2}$ b) $-0.23, +0.18$ c) $+23, -13, +12$ d) $+5, -8, +17$

Ans.   a) $+2\frac{3}{4}$ b) -0.05 c) $+22$ d) $+14$

3.10. Add: (3.4)

a) $+25.7, -25.7$ b) $+120, -120, +19$ c) $+28, -16, -28, +16$

Ans. a) 0 b) $+19$ c) 0

3.11. Add -32 to: To $+11$, add (3.5)

a) -25 *Ans.* -57 c) $+32$ *Ans.* 0 e) -29 *Ans.* -18 g) $+117$ *Ans.* $+128$
b) $+25$ -7 d) $+100$ $+68$ f) -11 0 h) -108 -97

3.12. Express in simplified vertical form and add: (4.1)

a) $+11 + (+3) + (-9) + (-8)$ b) $-42 + (+42) + (-85)$ c) $+13.7 + (+2.4) - 8.9$

Ans. a)
$$\begin{array}{r} 11 \\ 3 \\ -9 \\ -8 \\ \hline \text{Sum} = -3 \end{array}$$
b)
$$\begin{array}{r} -42 \\ +42 \\ -85 \\ \hline \text{Sum} = -85 \end{array}$$
c)
$$\begin{array}{r} 13.7 \\ 2.4 \\ -8.9 \\ \hline \text{Sum} = 7.2 \end{array}$$

3.13. Express in simplified horizontal form and add: (4.1, 4.2)

a) $-14, -7, +22, -35$ b) $+3\frac{1}{4}, +8\frac{1}{2}, -40\frac{3}{4}$ c) $-1.78, -3.22, +16$

Ans. a) $-14 - 7 + 22 - 35$ b) $3\frac{1}{4} + 8\frac{1}{2} - 40\frac{3}{4}$ c) $-1.78 - 3.22 + 16$
  Sum $= -34$ Sum $= -29$ Sum $= 11$

3.14. In a football game, a team gained 8 yd on the first play, gained 1 yd on the second play, lost 12 yd on the third play, and lost 6 yd on the fourth play. Find the change in position due to these changes, using *a*) signed numbers and *b*) a number scale: (4.3)

a) $+8 + (+1) + (-12) + (-6)$ b)
$8 + 1 - 12 - 6$
-9

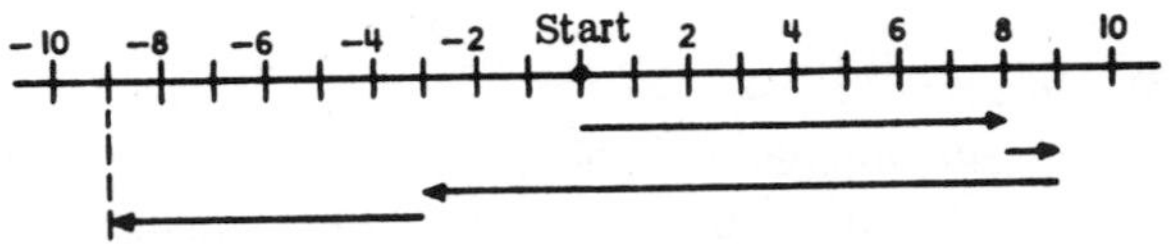

Fig. 3-14

Ans. Total change is 9-yd loss. See Fig. 3-14.

3.15. Subtract: (5.1)

a) $+11$ from $+16$ c) $+3.5$ from $+7.2$ e) $+3\frac{1}{3}$ from $+11\frac{2}{3}$
b) $+47$ from -47 d) $+7.2$ from -3.5 f) $+17\frac{2}{3}$ from -8

Ans. a) $+5$ b) -94 c) $+3.7$ d) -10.7 e) $+8\frac{1}{3}$ f) $-25\frac{2}{3}$

3.16. Subtract: (5.2)

a) -11 from $+16$ c) -0.81 from $+0.92$ e) $-7\frac{1}{5}$ from $+2\frac{4}{5}$
b) -47 from $+47$ d) -0.23 from -0.27 f) $-3\frac{5}{6}$ from -4

Ans. a) $+27$ b) $+94$ c) $+1.73$ d) -0.04 e) $+10$ f) $-\frac{1}{6}$

3.17. Subtract the lower number from the upper: (5.3)

a) $+22$ b) $+17$ c) -30.7 d) -0.123 e) $-13\frac{2}{3}$ f) $-27\frac{1}{2}$
$+17$ $+22$ $+30.7$ $+0.265$ $-10\frac{1}{3}$ $+3\frac{3}{4}$
Ans. a) $+5$ b) -5 c) -61.4 d) -0.388 e) $-3\frac{1}{3}$ f) $-31\frac{1}{4}$

3.18. Subtract $+4$ from: a) $+8\frac{1}{2}$ b) $+4$ c) $+2.3$ and d) -25 (5.4)
From -10 subtract e) -23 f) -8 g) -3.9 and h) $+20\frac{1}{4}$

Ans. a) $+4\frac{1}{2}$ b) 0 c) -1.7 d) -29 e) $+13$ f) -2 g) -6.1 h) $-30\frac{1}{4}$

3.19. Combine: **(5.5)**

a) $+11 + (+7) - (+24)$	*d*) $+25 - (-6) - (-22) + (+40)$
b) $-11 - (-5) - (-14)$	*e*) $-3.7 + (-2.4) - (+7.8) + (-11.4)$
c) $+0.13 - (+0.07) - (+0.32)$	*f*) $+2\frac{1}{2} - (-1\frac{1}{4}) - (+5\frac{3}{4}) - (-7)$

Ans.   *a*) -6 *b*) $+8$ *c*) -0.26 *d*) $+93$ *e*) -25.3 *f*) $+5$

3.20. Find the distance from 500 m above sea level to: *a*) 1200 m above sea level,   *b*) sea level,
c) 2000 m below sea level.   **(5.7)**

Ans.   *a*) 700 m up   *b*) 500 m down   *c*) 2500 m down

3.21. On a Monday, the hourly temperatures from 1 P.M. to 6 P.M. were: **(5.8)**

1 P.M.	2 P.M.	3 P.M.	4 P.M.	5 P.M.	6 P.M.
$-8°$	$-5°$	$0°$	$4°$	$-4°$	$-20°$

Find each hourly temperature change.

Ans.   *a*) 3° rise from 1 to 2 P.M.   *b*) 5° rise from 2 to 3 P.M.   *c*) 4° rise from 3 to 4 P.M.
d) 8° drop from 4 to 5 P.M.   *e*) 16° drop from 5 to 6 P.M.

3.22. Multiply: **(6.1)**

a) $(+3)(+22)$	*c*) $(-5)(-4)$	*e*) $(+8)(+2\frac{1}{2})$	*g*) $(-2\frac{1}{2})(-1\frac{1}{2})$
b) $(+0.3)(+2.2)$	*d*) $(-0.05)(-0.04)$	*f*) $(-35)(-2\frac{1}{5})$	*h*) $(+\frac{5}{4})(+\frac{15}{2})$

Ans.   *a*) $+66$ *b*) $+0.66$ *c*) $+20$ *d*) $+0.002$ *e*) $+20$ *f*) $+77$ *g*) $+3\frac{3}{4}$ *h*) $+9\frac{3}{8}$

3.23. Multiply: **(6.2)**

a) $(-20)(+6)$	*c*) $(-0.8)(+0.11)$	*e*) $(+6)(-7\frac{2}{3})$	*g*) $(-\frac{7}{2})(+\frac{11}{4})$
b) $(+13)(-8)$	*d*) $(+3.4)(-21)$	*f*) $(-2\frac{1}{7})(+21)$	*h*) $(+2\frac{1}{5})(-3\frac{2}{5})$

Ans.   *a*) -120 *b*) -104 *c*) -0.088 *d*) -71.4 *e*) -46 *f*) -45 *g*) $-\frac{77}{8}$ *h*) $-7\frac{12}{25}$

3.24. Multiply $+8$ by   *a*) $+7$ *b*) 0 *c*) -4.   Multiply -12 by   *d*) $+0.9$ *e*) $-1\frac{1}{2}$ *f*) -8.21.   **(6.3)**

Ans.   *a*) $+56$ *b*) 0 *c*) -32 *Ans.*   *d*) -10.8 *e*) $+18$ *f*) $+98.52$

3.25. Multiply: **(6.4)**

a) $(+3)(+4)(+12)$	*d*) $(-1)(-1)(+1)(-1)$	*g*) $(+\frac{1}{2})(+\frac{1}{2})(+\frac{1}{2})(-\frac{1}{2})$
b) $(+0.3)(+0.4)(+1.2)$	*e*) $(-2)(-2)(-2)(-2)$	*h*) $(-\frac{1}{2})(+8)(-\frac{1}{4})(+16)$
c) $(+0.3)(-4)(-0.12)$	*f*) $(-2)(+5)(-5)(+4)$	*i*) $(-1\frac{1}{2})(+2\frac{1}{2})(+3\frac{1}{2})$

Ans.   *a*) $+144$ *b*) $+0.144$ *c*) $+0.144$ *d*) -1 *e*) $+16$ *f*) $+200$ *g*) $-\dfrac{1}{16}$ *h*) $+16$ *i*) $-13\frac{1}{8}$

3.26. Complete each statement. In each case, indicate how signed numbers may be used to obtain the
answer.   **(6.5)**

a) If George withdraws $10 each week, then after 5 weeks his bank balance will be (   ).
b) If Lisa has been earning $50 a day, then in 3 days she will have (   ).
c) If the temperature has risen 8° each day, then 5 days ago it was (   ).
d) If a car has been traveling east at 40 mi/h, then 3 h ago it was (   ).
e) If a school decreases in the register 20 pupils per day, then 12 days ago the register was (   ).

Ans.   *a*) $150 less   *b*) $150 more   *c*) 40° less   *d*) 120 mi farther west
          $(-10)(+5) = -50$ $(+50)(+3) = +150$ $(+8)(-5) = -40$ $(+40)(-3) = -120$
e) 240 pupils more
          $(-20)(-12) = +240$

3.27. Find each power: $a)$ 2^3 $b)$ $(+4)^2$ $c)$ 10^4 $d)$ 0.3^2 $e)$ $(+0.2)^3$ $f)$ $(\frac{1}{4})^2$ $g)$ $(\frac{1}{10})^3$ $h)$ $(\frac{2}{5})^2$ **(7.1)**

$Ans.$ $a)$ 8 $b)$ 16 $c)$ 10,000 $d)$ 0.09 $e)$ 0.008 $f)$ $\frac{1}{16}$ $g)$ $\frac{1}{1000}$ $h)$ $\frac{4}{25}$

3.28. Find each power: **(7.1)**

$a)$ $(-1)^5$ $b)$ $(-1)^{82}$ $c)$ $(-0.3)^3$ $d)$ $(-0.1)^4$ $e)$ $(-0.12)^2$ $f)$ $(-\frac{2}{5})^3$ $g)$ $(-\frac{1}{3})^4$

$Ans.$ $a)$ -1 $b)$ $+1$ $c)$ -0.027 $d)$ $+0.0001$ $e)$ $+0.0144$ $f)$ $-\frac{8}{125}$ $g)$ $+\frac{1}{81}$

3.29. Complete each: **(7.4)**

$a)$ $(-5)^? = -125$ $c)$ $(-?)^4 = +0.0001$ $e)$ $(+\frac{2}{3})^4 = ?$ $g)$ $(?)^{171} = -1$
$b)$ $(+10)^? = 10,000$ $d)$ $(-?)^3 = -0.343$ $f)$ $(-\frac{1}{2})^5 = ?$ $h)$ $(?)^{242} = +1$

$Ans.$ $a)$ 3 $b)$ 4 $c)$ -0.1 $d)$ -0.7 $e)$ $\frac{16}{81}$ $f)$ $-\frac{1}{32}$ $g)$ -1 $h)$ $+$ or -1

3.30. Divide: **(8.1)**

$a)$ $+24$ by $+3$ $c)$ -49 by -7 $e)$ $+4.8$ by $+0.2$ $g)$ -18 by -4
$b)$ $+88$ by $+8$ $d)$ -78 by -6 $f)$ -95 by -0.5 $h)$ $+8$ by $+12$

$Ans.$ $a)$ $+8$ $b)$ $+11$ $c)$ $+7$ $d)$ $+13$ $e)$ $+24$ $f)$ $+190$ $g)$ $+4\frac{1}{2}$ $h)$ $+\frac{2}{3}$

3.31. Divide: **(8.2)**

$a)$ $+30$ by -5 $c)$ $+13$ by -2 $e)$ $+0.2$ by -0.04 $g)$ -100 by $+500$
$b)$ -30 by $+5$ $d)$ -36 by $+8$ $f)$ -30 by $+0.1$ $h)$ $+100$ by -3

$Ans.$ $a)$ -6 $b)$ -6 $c)$ $-6\frac{1}{2}$ $d)$ $-4\frac{1}{2}$ $e)$ -5 $f)$ -300 $g)$ $-\frac{1}{5}$ $h)$ $-33\frac{1}{3}$

3.32. Divide $+12$ by: $a)$ $+24$ $b)$ $+12$ $c)$ $+4$ $d)$ -1 $e)$ -3 $f)$ -48 **(8.2)**
Divide each by -20: $g)$ $+60$ $h)$ $+20$ $i)$ $+5$ $j)$ -225 $k)$ -10 $l)$ -100

$Ans.$ $a)$ $+\frac{1}{2}$ $b)$ $+1$ $c)$ $+3$ $d)$ -12 $e)$ -4 $f)$ $-\frac{1}{4}$ $g)$ -3 $h)$ -1 $i)$ $-\frac{1}{4}$ $j)$ $+11\frac{1}{4}$ $k)$ $+\frac{1}{2}$ $l)$ $+5$

3.33. Divide: **(8.3)**

$a)$ $\dfrac{+25}{+5}$ $b)$ $\dfrac{-25}{+5}$ $c)$ $\dfrac{-2.5}{-0.5}$ $d)$ $\dfrac{-0.25}{+0.5}$ $e)$ $\dfrac{-25}{-0.5}$ $f)$ $\dfrac{-0.025}{+0.5}$

$Ans.$ $a)$ $+5$ $b)$ -5 $c)$ $+5$ $d)$ $-\frac{1}{2}$ $e)$ $+50$ $f)$ -0.05

3.34. Multiply and divide as indicated: **(8.4, 8.5)**

$a)$ $\dfrac{(+20)(+12)}{(+3)(+5)}$ $c)$ $\dfrac{(+3)(+6)(+10)}{(+12)(-3)}$ $e)$ $\dfrac{0}{(-17)(-24)}$ $g)$ $\dfrac{(+1)(+2)(+3)}{(-4)(-6)}$

$b)$ $\dfrac{(+20)(-12)}{(-3)(-5)}$ $d)$ $\dfrac{(-3)(-6)(+18)}{(+12)(+3)}$ $f)$ $\dfrac{(-120)(+31)}{(-5)(-8)(-9)}$ $h)$ $\dfrac{(+0.1)(+0.2)(-30)}{(-0.4)(-0.1)}$

$Ans.$ $a)$ $+16$ $b)$ -16 $c)$ -5 $d)$ $+9$ $e)$ 0 $f)$ 0 $g)$ $+\frac{1}{4}$ $h)$ -15

3.35. Evaluate if $y = -3$:   **(9.1)**

a) $3y + 1$ c) $2y^2$ e) $(-y)^2$ g) $\dfrac{7y}{3}$ i) $\dfrac{3y + 15}{y}$

b) $20 - 4y$ d) $3y^3$ f) $2 - y^3$ h) $y(y^2 - 2)$ j) $2y^2 - 5y + 27$

Ans. a) -8 b) 32   c) 18   d) -81 e) 9   f) 29   g) -7 h) -21,   i) -2 j) 60

3.36. Evaluate if $x = -1$ and $y = +3$:   **(9.2)**

a) $x + y$ c) $x^2 + y^2$ e) $4xy - x^2$ g) $3xy^2$ i) $\dfrac{y^2}{6x}$

b) $y - 2x$ d) $3xy$ f) $x^2y + 10$ h) $x^3 + 10y$ j) $\dfrac{y + x}{y - x}$

Ans. a) 2   b) 5   c) 10   d) -9 e) -13 f) 13   g) -27 h) 29   i) $-\dfrac{3}{2}$ j) $\dfrac{1}{2}$

3.37. Evaluate if $x = -2$, $y = -1$, and $z = +3$:   **(9.3)**

a) $x + y + z$ c) $x^2 + y^2 + z^2$ e) $2xyz$ g) $xy + z^2$ i) $\dfrac{2x - 3y}{4z}$

b) $2x + 2y - 2z$ d) $x^3 - y^2 + z$ f) $xy + yz$ h) $y^2 - 5xz$ j) $\dfrac{x^2 - y^2}{z^2}$

Ans. a) 0   b) -12 c) 14   d) -6 e) 12   f) -1 g) 11   h) 31   i) $-\dfrac{1}{12}$ j) $\dfrac{1}{3}$

CHAPTER 4

Introduction to Monomials and Polynomials

1. UNDERSTANDING MONOMIALS AND POLYNOMIALS

A **term** is a number or the product of numbers. Each of the numbers being multiplied is a **factor** of the term.

Thus, the term $-5xy$ consists of three factors: -5, x, and y.

In $-5xy$, xy is the literal coefficient and -5 the numerical coefficient.

An **expression** consists of one or more terms. Expressions may be monomials or polynomials.

1. A **monomial** is an expression of one term.
2. A **polynomial** is an expression of two or more terms.
   a) A **binomial** is a polynomial of two terms.
   b) A **trinomial** is a polynomial of three terms.

Thus, $3x^2$ is a monomial.

   The polynomial $3x^2 + 5x$ is a binomial.

   The polynomial $3x^2 + 5x - 2$ is a trinomial.

Like terms are terms having the same literal factors. Like terms have the same literal coefficient.

Thus, $5x^2y$ and $3x^2y$ are like terms with the same literal coefficient, x^2y.

Unlike terms are terms that do not have the same literal coefficient.

Thus, $5x^2y$ and $-3xy^2$ are unlike terms.

1.1. SELECTING LIKE TERMS

In each polynomial select like terms, if any:

a) $10 - 5x + 2$

b) $x + 3y - z + y$

c) $2ab - 3ac + 4bc - bc$

Ans.   a) 10 and $+2$

b) $3y$ and $+ y$

c) $+4bc$ and $-bc$

d) $2abc + 3acd + 5bcd + 7acd$ *d)* $3acd$ and $7acd$
e) $y^2 - 2x^2 + 4z^2 - 5y^2$ *e)* y^2 and $-5\,y^2$
f) $xy^2 + x^2y^2 - 3x^2y + 7x^2y$ *f)* $-3x^2y$ and $7x^2y$
g) $2ab^2c^2 - 5a^2b^2c + 8a^2bc^2$ *g)* no like terms
h) $3(x + y) - 5(x + y) + 2(x - y)$ *h)* $3(x + y)$ and $-5(x + y)$

2. ADDING MONOMIALS

To add like terms:

1. Add their numerical coefficients.
2. Keep the common literal coefficient.
   Thus, $3a^2 - 5a^2 + 4a^2 = (3 - 5 + 4)a^2$ or $2a^2$.

To simplify monomials being added:

(1) Parentheses may be omitted.
(2) + meaning "add" may be omitted.
(3) If the first monomial is positive, its plus sign may be omitted.
   Thus, to simplify $(+ 8x) + (+ 9x) + (- 10x)$, write $8x + 9x - 10x$.

2.1. ADDING LIKE TERMS

Add: *a)* $5a^2$ and $3a^2$ *b)* $13(x + y)$ and $-7(x + y)$

Procedure:

1. Add the numerical coefficients: *a)* $5 + 3 = 8$ *b)* $13 - 7 = 6$
2. Keep common literal coefficient: **Ans.** $8a^2$ **Ans.** $6(x + y)$

2.2. ADDING LIKE TERMS HORIZONTALLY

Simplify and add:

a) $+5a + (+2a) + (-4a)$ *b)* $-8x^2 + (-12x^2)$ *c)* $-3wy + (+11wy)$ *d)* $+ 16r^2s + (-13r^2s)$

Solutions: (First simplify.)

a) $+5a + (+2a) + (-4a)$ *b)* $-8x^2 + (-12x^2)$ *c)* $-3wy + (+11wy)$
   $5a + 2a - 4a$ $-8x^2 - 12x^2$ $-3wy + 11wy$
Ans. $3a$ **Ans.** $-20x^2$ **Ans.** $8wy$

d) $+16r^2s + (-13r^2s)$
   $16r^2s - 13r^2s$
Ans. $3r^2s$

2.3. ADDING LIKE TERMS VERTICALLY

Add:	*a)*	*b)*	*c)*	*d)*	*e)*
	$8xy$	$-2abc$	$12c^3d^2$	$15(x - y)$	$-3(a^2 + b^2)$
	$-3xy$	$5abc$	$-10c^3d^2$	$-18(x - y)$	$-(a^2 + b^2)$
	$12xy$	$-8abc$	$-c^3d^2$	$14(x - y)$	$-8(a^2 + b^2)$
Ans.	$17xy$	$-5abc$	c^3d^2	$11(x - y)$	$-12(a^2 + b^2)$

2.4. ADDING LIKE AND UNLIKE TERMS

Simplify and add: (Combine only like terms.)

a) $+10x + (+ 5x) + (-7)$ b) $+3a^2 + (-4a^2) + (-3b^2)$ c) $-8a^2c + (-6ac^2) + (-3ac^2)$

$10x + 5x - 7$ $3a^2 - 4a^2 - 3b^2$ $-8a^2c - 6ac^2 - 3ac^2$

Ans. $15x - 7$ Ans. $-a^2 - 3b^2$ Ans. $-8a^2c - 9ac^2$

3.   ARRANGING AND ADDING POLYNOMIALS

Arranging the Terms of a Polynomial in Descending or Ascending Order

A polynomial may be arranged as follows:

1. In **descending order**, by having the exponents of the same letter decrease in successive terms.

2. In **ascending order**, by having the exponents of the same letter increase in successive terms.

Thus, $2x^2 + 3x^3 - 5x + 8$ becomes

$3x^3 + 2x^2 - 5x + 8$ in descending order

or $8 - 5x + 2x^2 + 3x^3$ in ascending order.

To Add Polynomials

Add: $5x - 2y$ and $y + 3x$

Procedure:

1. Arrange polynomials in order,
    placing like terms in same column:

2. Add like terms:

Solution:

1. $5x - 2y$
    $\underline{3x + \ y}$

2. $8x - \ y$ **Ans.**

To check addition of polynomials, substitute any convenient value, except 1 or 0, for each of the letters.

NOTE: If 1 is substituted for a letter, an incorrect exponent will not be detected.
      If $x = 1$, then $x^2 = 1$, $x^3 = 1$, $x^4 = 1$, etc.

In examples 3.4 and 3.5, checking is shown.

3.1. ARRANGING POLYNOMIALS AND COMBINING LIKE TERMS

Rearrange in descending order and combine:

a) $3a^2 + 2a^2 - 10a - 5a$ b) $5x + 8 - 6x + 10x^2$ c) $-5y^2 + 8x^2 + 5y^2 + x$

a) No rearrangement needed   b) $10x^2 + 5x - 6x + 8$ c) $8x^2 + x - 5y^2 + 5y^2$

Ans. $5a^2 - 15a$ Ans. $10x^2 - x + 8$ Ans. $8x^2 + x$

3.2. ADDING ARRANGED POLYNOMIALS

Add: a) $3x - 10$ b) $5x^2 + 5x$ c) $5x^3 + 7x^2 - \ 4$

      $\underline{2x + \ 4}$ $\underline{-x^2 - \ x}$ $\underline{-6x^3 \qquad - 10}$

Ans.   a) $5x - \ 6$ Ans. b) $4x^2 + 4x$ Ans. c) $-x^3 + 7x^2 - 14$

      d) $10x + 3y$

      $\underline{7x - 6y + 1}$

Ans.   d) $17x - 3y + 1$

3.3. ADDING POLYNOMIALS

Add: *a*) $3a + 5b$, $6b - 2a$, and $10b - 25$ *b*) $x^2 + x^3 - 3x$, $4 - 5x^2 + 3x^3$, and $10 - 8x^2 - 5x$

Procedure: | **Solutions:**

1. Rearrange polynomials with like terms in the same column:

a)
$$3a + 5b$$
$$-2a + 6b$$
$$\underline{\qquad\qquad 10b - 25}$$

b)
$$x^3 \ + \ x^2 - 3x$$
$$3x^3 \ - \ 5x^2 \qquad + 4$$
$$\underline{\qquad\quad - \ 8x^2 - 5x + 10}$$

2. Add like terms:

Ans. $a + 21b - 25$ **Ans.** $4x^3 - 12x^2 - 8x + 14$

3.4 and 3.5. CHECKING THE ADDITION OF POLYNOMIALS

3.4.   Check by letting $x = 2$ and $y = 3$: $25x - 10y + (5x - 2y) = 30x - 12y$
Check: Let $x = 2$ and $y = 3$. $25x - 10y \rightarrow 50 - 30 = 20$
$$\underline{5x - \ 2y \rightarrow 10 - \ 6 = \ 4}$$
$$30x - 12y \rightarrow 60 - 36 = 24$$

By adding vertically and horizontally, the sum is 24.

3.5.   Check by letting $a = 2$, $b = 4$, and $c = 3$: $a^2 + b^2 - c^2 + (3a^2 - b^2 + 4c^2) = 4a^2 + 3c^2$

Check: Let $a = 2$, $b = 4$, and $c = 3$. $a^2 + b^2 - c^2 \rightarrow \ 4 + 16 - 9 \ = 11$
$$\underline{3a^2 - b^2 + 4c^2 \rightarrow 12 - 16 + 36 = 32}$$
$$4a^2 \qquad + 3c^2 \rightarrow 16 \qquad + 27 = 43$$

By adding vertically and horizontally, the sum is 43.

4. SUBTRACTING MONOMIALS

To **subtract** a term, **add** its opposite. (Opposite terms differ only in sign.)
    Thus, to subtract $-3x$, add $+3x$; or to subtract $+5x^2$, add $-5x^2$.

4.1. SUBTRACTING LIKE TERMS

Subtract: *a*) $+2a - (-5a)$ *b*) $-3x^2 - (+5x^2)$ *c*) $-15cd - (-10cd)$
Solutions:

a) $+2a - (-5a)$ *b*) $-3x^2 - (+5x^2)$ *c*) $-15cd - (-10cd)$
$(+2a) + (+5a)$ $(-3x^2) + (-5x^2)$ $(-15cd) + (+10cd)$
$2a + 5a$ $-3x^2 - 5x^2$ $-15cd + 10cd$

Ans. $7a$ *Ans.* $-8x^2$ *Ans.* $-5cd$

d) $+3a^2b - (+8a^2b)$
d) $+3a^2b - (+8a^2b)$
$(+3a^2b) + (-8a^2b)$
$3a^2b - 8a^2b$

Ans. $-5a^2b$

4.2. DIFFERENT SUBTRACTION FORMS

> *a*) Subtract $-5a$ from $+2a$.
> *b*) From $-3x^2$ take $+5x^2$.
>
> *c*) Reduce $+7mn$ by $-2mn$.
> *d*) How much does $-8cd$ exceed $+3cd$?

Solutions:

> *a*) $+2a - (-5a)$
> $+2a + (+5a)$
> $2a + 5a$
>
> **Ans.** $7a$

> *b*) $-3x^2 - (+5x^2)$
> $-3x^2 + (-5x^2)$
> $-3x^2 - 5x^2$
>
> **Ans.** $-8x^2$

> *c*) $+7mn - (-2mn)$
> $+7mn + (+2mn)$
> $7mn + 2mn$
>
> **Ans.** $9mn$

> *d*) $-8cd - (+3cd)$
> $-8cd + (-3cd)$
> $-8cd - 3cd$
>
> **Ans.** $-11cd$

4.3. COMBINING ADDING AND SUBTRACTING OF LIKE TERMS

Combine:

> *a*) $+6x + (+8x) - (-3x)$ *b*) $-3a^2 - (-a^2) - (+10a^2)$ *c*) $+13ab - (-ab) - (-2ab) + (-3ab)$

Solutions:

> *a*) $+6x + (+8x) - (-3x)$
> $+6x + (+8x) + (+3x)$
> $6x + 8x + 3x$
>
> **Ans.** $17x$

> *b*) $-3a^2 - (-a^2) - (+10a^2)$
> $-3a^2 + (+a^2) + (-10a^2)$
> $-3a^2 + a^2 - 10a^2$
>
> **Ans.** $-12a^2$

> *c*) $+13ab - (-ab) - (-2ab) + (-3ab)$
> $+13ab + (+ab) + (+2ab) + (-3ab)$
> $13ab + ab + 2ab - 3ab$
>
> **Ans.** $13ab$

5. SUBTRACTING POLYNOMIALS

To Subtract Polynomials

Subtract $5x - 2y$ from $8x - 4y$.

Procedure:

1. Arrange polynomials in order, placing like terms in same column:
2. Subtract like terms:

Solution:

$(8x - 4y) - (5x - 2y)$

> **1.** $8x - 4y$ Minuend
>   **(S)** $\underline{5x - 2y}$ Subtrahend
>
> **2.** $3x - 2y$ Difference (**Ans.**)

> **NOTES:**   1) Use **(S)** to indicate subtraction.
>   2) To subtract a polynomial, change each of its signs **mentally**, then add.

To check subtraction, add the difference obtained to the subtrahend.
The sum should be the minuend.

5.1. SUBTRACTING ARRANGED POLYNOMIALS

Subtract:

> *a*) $5b + 7$
> **(S)** $\underline{3b - 2}$
> **Ans.** $2b + 9$

> *b*) $y^2 - 3y$
> **(S)** $\underline{-2y^2 - 4y}$
> $3y^2 + y$

> *c*) $8x^3 - 2x^2 + 5x$
> **(S)** $\underline{6x^3 + + 9x}$
> $2x^3 - 2x^2 - 4x$

> *d*) $3x^2y + 4xy^2$
> **(S)** $\underline{x^2y + xy^2 + y^3}$
> $2x^2y + 3xy^2 - y^3$

5.2. ARRANGING POLYNOMIALS AND SUBTRACTING

Arrange and subtract:

a) $2x - y + 8$ from $13x + 4y + 9$ c) $x^3 - 10$ from $25x^2 + 3x^3$
b) $10 + 3a - 5b$ from $7b + 2a - 8$ d) $m^2 - 18m$ from $12 + 3m^2$

Solutions:

a) $13x + 4y + 9$ b) $2a + 7b - 8$ c) $3x^3 + 25x^2$ d) $3m^2 \qquad + 12$
(S) $\underline{2x - y + 8}$ (S) $\underline{3a - 5b + 10}$ (S) $\underline{x^3 \qquad - 10}$ (S) $\underline{m^2 - 18m}$

Ans. $11x + 5y + 1$ $-a + 12b - 18$ $2x^3 + 25x^2 + 10$ $2m^2 + 18m + 12$

5.3. DIFFERENT SUBTRACTION FORMS

a) From $9 + x^2$ take $-3x^2 - 5x + 12$.   c) Find $8x^2 - 15x$ less $20x - 7$.
b) Reduce $17ab$ by $-8ab + 15$.   d) Subtract $10 - 4x + x^2$ from 0.

Solutions:

a) $x^2 \qquad + 9$ b) $17ab$ c) $8x^2 - 15x$ d) 0
(S) $\underline{-3x^2 - 5x + 12}$ (S) $\underline{-8ab + 15}$ (S) $\underline{\qquad 20x - 7}$ (S) $\underline{10 - 4x + x^2}$

Ans. $4x^2 + 5x - 3$ $25ab - 15$ $8x^2 - 35x + 7$ $-10 + 4x - x^2$

5.4. CHECKING THE SUBTRACTION OF POLYNOMIALS

Check each subtraction, using addition:

a) $8c - 3$ b) $5x^2 + 20$ c) $3y^2 + 5y$
(S) $\underline{5c - 8}$ (S) $\underline{x^2 + 15}$ (S) $\underline{y^2 - 8y + 5}$

      $3c + 5(?)$ $4x^2 + 5(?)$ $2y^2 + 13y + 5(?)$

Checking: (difference + subtrahend = minuend)

a) $3c + 5$ b) $4x^2 + 5$ c) $2y^2 + 13y + 5$
(A) $\underline{5c - 8}$ (A) $\underline{x^2 + 15}$ (A) $\underline{y^2 - 8y + 5}$

      $8c - 3$ $5x^2 + 20$ $3y^2 + 5y + 10$

(Correct)   (Correct)   (Incorrect)

Correct difference $= 2y^2 + 13y - 5$

6. USING PARENTHESES AND OTHER GROUPING SYMBOLS TO ADD OR SUBTRACT POLYNOMIALS

Symbols of grouping include:

1. **Parentheses**, ( ), as in $\frac{1}{2}(8 - 4x)$ or $5 - (10x - 4)$.
2. **Brackets**, [ ], to include parentheses as in $8 - [5 + (x - 2)]$.
3. **Braces**, { }, to include brackets as in $x - \{3 + [x - (y + 4)]\}$.
4. **Bar**, ——, as in the fraction $\dfrac{8 - 4x}{2}$.

Symbols of grouping may be used to show the addition or subtraction of polynomials. Thus:

1. To add $3a + 4b$ and $5a - b$, write $(3a + 4b) + (5a - b)$.
2. To subtact $2x - 5$ from $x^2 + 10x$, write $(x^2 + 10x) - (2x - 5)$.
3. To subtract the sum of $8x^2 + 9$ and $6 - x^2$ from $3x^2 - 12$, write $(3x^2 - 12) - [(8x^2 + 9) + (6 - x^2)]$.

Rules for Removing Parentheses and Grouping Symbols

Rule 1.   When removing parentheses preceded by a plus sign, **do not change** the signs of the enclosed terms.

Thus, $3a + (+5a - 10)$ $3x^2 + (-x^2 - 5x + 8)$

$3a + 5a - 10$ $3x^2 - x^2 - 5x + 8$

$8a - 10$ **Ans.** $2x^2 - 5x + 8$ **Ans.**

Rule 2.   When removing parentheses preceded by a minus sign, **change** the sign of each enclosed term.

Thus, $3a - (+5a - 10)$ $3x^2 - (-x^2 - 5x + 8)$

$3a - 5a + 10$ $3x^2 + x^2 + 5x - 8$

$-2a + 10$ **Ans.** $4x^2 + 5x - 8$ **Ans.**

Rule 3.   When more than one set of grouping symbols is used, remove one set at a time, beginning with the innermost one.

Thus, $2 + [r - (3 - r)]$ $6s - \{5 - [3 + (7s - 8)\}$

$2 + [r - 3 + r]$ $6s - \{5 - [3 + 7s - 8]\}$

$2 + r - 3 + r$ $6s - \{5 - 3 - 7s + 8\}$

$2r - 1$ **Ans.** $6s - 5 + 3 + 7s - 8$

$13s - 10$ **Ans.**

6.1.   RULE 1.   REMOVING PARENTHESES PRECEDED BY A PLUS SIGN

Simplify: (Do not change the sign of the enclosed terms.)

     a) $13x + (5x - 2)$ *b*) $(15a^2 - 3a) + (-7 - 2a)$ *c*) $3 + (6 - 5x) - 2x^2$

         $13x + 5x - 2$ $15a^2 - 3a - 7 - 2a$ $3 + 6 - 5x - 2x^2$

Ans. *a*) $18x - 2$ **Ans.** *b*) $15a^2 - 5a - 7$ **Ans.** *c*) $9 - 5x - 2x^2$

6.2.   RULE 2.   REMOVING PARENTHESES PRECEDED BY A MINUS SIGN

Simplify: (Change the sign of each enclosed term.)

     a) $13x - (5x - 2)$ *b*) $-(15a^2 - 3a) - (-7 - 2a)$ *c*) $3 - (5 - 5x) - 2x^2$

         $13x - 5x + 2$ $-15a^2 + 3a + 7 + 2a$ $3 - 5 + 5x - 2x^2$

Ans. *a*) $8x + 2$ **Ans.** *b*) $-15a^2 + 5a + 7$ **Ans.** *c*) $-2 + 5x - 2x^2$

6.3.   RULE 3.   BRACKETS CONTAINING PARENTHESES

Simplify:

Procedure:	**Solution:**	*a*) $2 - [r - (3 - r)]$	*b*) $5 - [(8 + 4x) - (3x - 2)]$
1. Remove ( ):		$2 - [r - 3 + r]$	$5 - [8 + 4x - 3x + 2]$
2. Remove [ ]:		$2 - r + 3 - r$	$5 - 8 - 4x + 3x - 2$
3. Combine:	**Ans.** *a*) $-2r + 5$	**Ans.** *b*) $-x - 5$	

6.4. and 6.5.   USING GROUPING SYMBOLS TO ADD OR SUBTRACT POLYNOMIALS

   6.4.   Subtract $2x - 5$ from the sum of $15x + 10$ and $3x - 5$.

Solution Using Parentheses   **Solution without Parentheses**

$$[(15x + 10) + (3x - 5)] - (2x - 5)$$

$$15x + 10 \ + 3x - 5 - 2x + 5 \qquad \textbf{(A)}$$

		$15x + 10$		$\longrightarrow$ $18x + 5$
	$\textbf{(A)}$	$\underline{3x - 5}$	$\textbf{(S)}$	$\underline{2x - 5}$
		$18x + 5$ $\longrightarrow$	**Ans.**	$16x + 10$

Ans. $16x + 10$

6.5. From the sum of $3a - 5b$ and $8a + 10$ subtract the sum of $2b - 8$ and $7a + 9$.

Solution Using ( ) and [ ]   **Solution without ( ) and [ ]**

$$[(3a - 5b) + (8a + 10)] - [(2b - 8) + (7a + 9)]$$

$$[3a - 5b + 8a + 10] - [2b - 8 + 7a + 9]$$

$$[11a - 5b + 10] - [7a + 2b + 1]$$

$$11a - 5b + 10 - 7a - 2b - 1$$

Ans. $4a - 7b + 9$

	$3a - 5b$			$2b$	$- 8$
$\textbf{(A)}$	$\underline{8a \qquad + 10}$	$\textbf{(A)}$		$\underline{7a \qquad + 9}$	
	$11a - 5b + 10$			$7a + 2b + 1$	
	$11a - 5b + 10$				
$\textbf{(S)}$	$\underline{7a \ + 2b + \ 1}$				
Ans.	$4a - 7b + 9$				

6.6. REMOVING SYMBOLS OF GROUPING

Simplify: (Remove innermost symbols first.)

Procedure:

1. Remove (   ): $5x - \{8x - [7 - (4x - 8 + 2x)]\} - 5$
2. Remove [   ]: $5x - \{8x - [7 - 4x - 8 + 2x]\} - 5$
3. Remove {   }: $5x - \{8x - 7 + 4x - 8 + 2x\} - 5$
4. Combine: $5x - 8x + 7 - 4x + 8 - 2x - 5$
                        Ans. $-9x + 10$

---

7. MULTIPLYING MONOMIALS AND POWERS OF THE SAME BASE

Rule 1. To multiply the powers of the same base, keep the base and add the exponents.

Thus, $x^4 \cdot x^3 = x^7 \qquad\qquad a^5 \cdot a \cdot b^2 \cdot b^3 = a^6 b^5$

               $3^{10} \cdot 3^2 = 3^{12} \qquad\quad 2^4 \cdot 2^5 \cdot 10^2 \cdot 10^4 = 2^9 \cdot 10^6$

RULE: $\boxed{x^a x^b = x^{a+b}}$

Rule 2. To find the power of a power of a base, keep the base and multiply the exponents.

Thus, $(x^4)^3 = x^{12}$ since $(x^4)^3 = (x^4)(x^4)(x^4)$

               $(5^2)^4 = 5^8$ since $(5^2)^4 = (5^2)(5^2)(5^2)(5^2)$

RULE: $\boxed{(x^a)^b = x^{ab}}$

Rule 3. Changing the order of factors does not change their product.

(This fundamental law is known as the **commutative law of multiplication.**)

Thus, $2 \cdot 3 \cdot 5 = 2 \cdot 5 \cdot 3 = 5 \cdot 2 \cdot 3 = 30$

          $3x \cdot 4x^2 = 3 \cdot 4 \cdot xx^2 = 12x^3$

To Multiply Monomials

Multiply $2x$ by $-3x^2$.

Procedure:	**Solution:**
	$(2x)(-3x^2)$
1. Multiply numerical coefficients:	1. $(2)(-3) = -6$
2. Multiply literal coefficients:	2. $(x)(x^2) = x^3$
3. Multiply results:	3. $-6x^3$ **Ans.**

7.1.   Rule 1.   Multiplying Powers of the Same Base

Multiply: (Keep base and add exponents.)

a) $b \cdot b^2$	Ans. b^3	d) $x^2 \cdot x^3 \cdot y$	Ans. $x^5 y$	g) $4^5 \cdot 4^2$	Ans. 4^7
b) $x^2 \cdot x^3 \cdot x$	x^6	e) $c^4 \cdot c \cdot d^5 \cdot d$	$c^5 d^6$	h) $2^3 \cdot 2 \cdot 5$	$2^4 \cdot 5$
c) $x^b x^c$	x^{b+c}	f) $x^3 y^2 x^4 y$	$x^7 y^3$	i) $3^2 3^3 y^4 y^5$	$3^5 y^9$

7.2.   Rule 2.   Finding the Power of a Power of a Base

Raise to a power: (Keep base and multiply exponents.)

a) $(a^4)^2$	Ans. a^8	d) $(3^5)^4$	Ans. 3^{20}	g) $(x^2)^3(y^3)^2$	Ans. $x^6 y^6$
b) $(b^3)^5$	b^{15}	e) $(10^4)^{10}$	10^{40}	h) $(4^2)^3(x^3)^4$	$4^6 x^{12}$
c) $(x^d)^e$	x^{de}	f) $(5^m)^n$	5^{mn}	i) $(y^3)^4 y^3 y^4$	y^{19}

7.3.   Multiplying Monomials

Multiply:

a) $2(-3b)$	Ans. $-6b$	f) $(10r^4)\left(-\frac{2}{5}rs\right)$	Ans. $-4r^5 s$
b) $(-2b)(5b)$	$-10b^2$	g) $(3a)(2b)(-10c)$	$-60abc$
c) $(4x)(-7y)$	$-28xy$	h) $(2x^2)(3x^3)(4x^4)$	$24x^9$
d) $(4x^2)(5x^3)$	$20x^5$	i) $(-3)(-r^4)(-s^5)$	$-3r^4 s^5$
e) $\left(-\frac{1}{2}y^3\right)(-10y^5)$	$5y^8$		

8.   MULTIPLYING A POLYNOMIAL BY A MONOMIAL

RULE: To multiply a polynomial by a monomial, multiply each term of the polynomial by the monomial.

(This fundamental law is known as the **distributive law**.)

Thus, $4(a + b - c) = 4a + 4b - 4c$

8.1.   Multiplying a Polynomial by a Monomial Horizontally

Multiply:

a) $10(a + b)$	Ans. $10a + 10b$	f) $\frac{2}{3}(9k - 30m)$	Ans. $6k - 20m$
b) $x(y - 7)$	$xy - 7x$	g) $3(x^2 - 2x + 8)$	$3x^2 - 6x + 24$
c) $-c(d - 5e)$	$-cd + 5ce$	h) $a(a^2 + a + 5)$	$a^3 + a^2 + 5a$
d) $\pi r(r + h)$	$\pi r^2 + \pi rh$	i) $-x^2(3 - 2x + x^2)$	$-3x^2 + 2x^3 - x^4$
e) $-ab(a + b)$	$-a^2 b - ab^2$		

8.2. Multiplying a Polynomial by a Monomial Vertically

Multiply:

a)
$$y^2 + 8y - 7$$
$$\underline{3y}$$
Ans. $3y^3 + 24y^2 - 21y$

b)
$$d^3 - 2d^2 - 20$$
$$\underline{-d^2}$$
Ans. $-d^5 + 2d^4 + 20d^2$

c)
$$5a + 2b - 3c + 8$$
$$\underline{-2abc}$$
Ans. $-10a^2bc - 4ab^2c + 6abc^2 - 16abc$

8.3. Removing Parentheses

Simplify:

a) $3x + 2(x - 3)$
     $3x + 2x - 6$

Ans. $5x - 6$

b) $3x^2 - x(5 - x)$
     $3x^2 - 5x + x^2$

Ans. $4x^2 - 5x$

c) $a(4a - 5) - 8(a^2 - 10)$
     $4a^2 - 5a - 8a^2 + 80$

Ans. $-4a^2 - 5a + 80$

8.4. Removing Brackets Containing Parentheses

Simplify: (Remove parentheses first.)

a) $3[5x - 2(x - 4)]$
     $3[5x - 2x + 8]$
     $3[3x + 8]$

Ans. $9x + 24$

b) $3 + 5[2 - 4(a - 6)]$
     $3 + 5[2 - 4a + 24]$
     $3 + 5[26 - 4a]$

Ans. $133 - 20a$

c) $a[ab - a(b - c)]$
     $a[ab - ab + ac]$
     $a[ac]$

Ans. a^2c

8.5 Removing Symbols of Grouping

Simplify: (Remove innermost symbols first.)

$$120y - 2\{y + 8[-7y - 5(y - 3y - 4)]\} + 320$$

1. Remove (   ): $120y - 2\{y + 8[-7y - 5y + 15y + 20]\} + 320$

2. Combine: $120y - 2\{y + 8[3y + 20]\} + 320$

3. Remove [   ]: $120y - 2\{y + 24y + 160\} + 320$

4. Remove {   }: $120y - 2y - 48y - 320 + 320$

5. Combine: $70y$ Ans.

9. MULTIPLYING POLYNOMIALS

To Multiply Polynomials

Multiply $3x + 4$ by $1 + 2x$.

Procedure:

1. Arrange each polynomial in order:

2. Multiply each term of one polynomial
by each term of the other:

3. Add like terms:

Solution:

$$3x + 4$$
$$\underline{2x + 1}$$

$$6x^2 + 8x$$
$$\underline{+ 3x + 4}$$
Ans. $6x^2 + 11x + 4$

To check multiplication of polynomials, interchange the polynomials and multiply again, or substitute any convenient values for the letters, except 1 or 0.

9.1. **MULTIPLYING POLYNOMIALS**

Multiply:

a) $3x + 4$
       $2x - 1$

       $6x^2 + 8x$
          $- 3x - 4$

Ans. $6x^2 + 5x - 4$

b) $8 + c$
       $3 - 2c$

       $24 + 3c$
          $- 16c - 2c^2$

Ans. $24 - 13c - 2c^2$

c) $4r + 7s$
       $4r - 7s$

       $16r^2 + 28rs$
              $- 28rs - 49s^2$

Ans. $16r^2 \qquad -49s^2$

d) $a^2 - 3a + 5$
          $5a - 2$

       $5a^3 - 15a^2 + 25a$
          $- 2a^2 + 6a - 10$

Ans. $5a^3 - 17a^2 + 31a - 10$

9.2. **CHECKING MULTIPLICATION**

Multiply and check: $(5r - 8)(3r - 2)$

Solution:

       $5r - 8$
       $3r - 2$

       $15r^2 - 24r$
              $- 10r + 16$

Ans. $15r^2 - 34r + 16$

Check by multiplication:

       $3r - 2$
       $5r - 8$

       $15r^2 - 10r$
              $- 24r + 16$

       $15r^2 - 34r + 16$

Check by substitution:

Let $r = 10$

$5r - 8 \rightarrow 42$

$3r - 2 \rightarrow 28$

$15r^2 - 34r + 16 \rightarrow 1500 - 340 + 16 = 1176$

Correct since $1176 = 42 \times 28$

9.3. **EXTENDED MULTIPLICATION**

Multiply $5p + 2$, $2p - 1$, and $p + 1$ and check.

Solution:

$5p + 2$
$2p - 1$

$10p^2 + 4p$
      $- 5p - 2$

$10p^2 - p - 2$

$10p^2 - p - 2$
       $p + 1$

$10p^3 - p^2 - 2p$
       $+ 10p^2 - p - 2$

$10p^3 + 9p^2 - 3p - 2$
                    Ans.

Check by substitution:

Let $p = 10$.

$5p + 2 \rightarrow 50 + 2 = 52$

$2p - 1 \rightarrow 20 - 1 = 19$

$p + 1 \rightarrow 10 + 1 = 11$

$10p^3 + 9p^2 - 3p - 2 \rightarrow 10{,}000 + 900 - 30 - 2 = 10{,}868$

Correct since $10{,}868 = 52 \times 19 \times 11$

10.   DIVIDING POWERS AND MONOMIALS

When you are dividing powers having the same base, arrange them into a fraction and apply the following rules:

Rule 1.   If the exponent of the numerator is **larger than** the exponent of the denominator, keep the base and subtract the smaller exponent from the larger.

Thus, $\dfrac{x^7}{x^4} = x^3$

RULE: | If a is larger than b, $\quad \dfrac{x^a}{x^b} = x^{a-b}$

Rule 2.   If the exponents **are equal**, then we have a number divided by itself; the quotient is 1.

Thus, $\dfrac{x^4}{x^4} = 1$

RULE: $\boxed{\dfrac{x^a}{x^a} = 1}$

Rule 3.   If the exponent of the denominator is larger, make the numerator of the quotient 1, and to obtain its denominator, keep the base and subtract the smaller exponent from the larger.

Thus, $\dfrac{x^4}{x^7} = \dfrac{1}{x^3}$

RULE: $\boxed{\text{If } a \text{ is smaller than } b, \quad \dfrac{x^a}{x^b} = \dfrac{1}{x^{b-a}}}$

To Divide Monomials

Divide $21ab^2$ by $-7a^2b$.

Procedure:	**Solutions:**
1. Arrange in fractional form:	**1.** $\dfrac{21ab^2}{-7a^2b}$ $\begin{array}{l}\text{dividend}\\[2pt]\text{divisor}\end{array}$
2. Divide numerical coefficients:	**2.** $\dfrac{21}{-7} = -3$
3. Divide literal coefficients:	**3.** $\dfrac{ab^2}{a^2b} = \dfrac{b}{a}$
4. Multiply the results:	**Ans.** **4.** $-\dfrac{3b}{a}$ quotient

To Check Division of Monomials

Multiply the quotient by the divisor. The result should be the dividend.

Thus, to check $\dfrac{15ab}{3a} = 5b$, multiply $3a$ by $5b$ to obtain $15ab$.

10.1.   **RULES 1 TO 3. DIVIDING POWERS OF THE SAME BASE**

Divide:

(Rule 1)	**(Rule 2)**	**(Rule 3)**
$a)$ $\dfrac{x^8}{x^2}$ *Ans.* x^6	$e)$ $\dfrac{x^2}{x^2}$ *Ans.* 1	$i)$ $\dfrac{x^2}{x^8}$ *Ans.* $\dfrac{1}{x^6}$
$b)$ $\dfrac{8^7}{8^3}$ 8^4	$f)$ $\dfrac{-8^3}{8^3}$ -1	$j)$ $\dfrac{8^3}{8^7}$ $\dfrac{1}{8^4}$
$c)$ $\dfrac{a^3b^4}{a^2b^2}$ ab^2	$g)$ $\dfrac{a^2b^2}{-a^2b^2}$ -1	$k)$ $\dfrac{a^2b^2}{a^3b^4}$ $\dfrac{1}{ab^2}$
$d)$ $\dfrac{x^{2a}}{x^a}$ x^a	$h)$ $\dfrac{-x^a}{-x^a}$ 1	$l)$ $\dfrac{x^a}{x^{2a}}$ $\dfrac{1}{x^a}$

10.2. DIVIDING MONOMIALS

Divide:

a) $\dfrac{24b^2}{3b}$ *Ans.* $8b$

d) $\dfrac{28x^2y^4}{28x^2y^2}$ *Ans.* y^2

g) $\dfrac{-14abc}{7abcd}$ *Ans.* $-\dfrac{2}{d}$

b) $\dfrac{24b}{8b^2}$ $\dfrac{3}{b}$

e) $\dfrac{28x^2y^2}{-28x^2y^2}$ -1

h) $\dfrac{-7a^2bc}{14ab^2c^2}$ $-\dfrac{a}{2bc}$

c) $\dfrac{-24b^2}{-24b^2}$ 1

f) $\dfrac{-25u^5w}{-25uw^5}$ $\dfrac{u^4}{w^4}$

i) $\dfrac{a^3b^4c^5}{a^3b^7c^2}$ $\dfrac{c^3}{b^3}$

10.3. CHECKING DIVISION

Check each division, using multiplication:

a) Does $\dfrac{24b^2}{3b} = 8b$?

b) Does $\dfrac{28x^2y^4}{28x^2y^2} = y^2$?

c) Does $\dfrac{a^3b^4c^6}{-ab^2c^2} = -a^2b^2c^3$?

Check: (Quotient $\times$ Divisor = Dividend)

a) Multiply:
$(8b)(3b) = 24b^2$
Correct.

b) Multiply:
$y^2(28x^2y^2) = 28x^2y^4$
Correct.

c) Multiply:
$(-a^2b^2c^3)(-a^2c^2) = a^3b^4c^5$
Incorrect.   The quotient should be $-a^2b^2c^4$.

11.   DIVIDING A POLYNOMIAL BY A MONOMIAL

To divide a polynomial by a monomial, divide each term of the polynomial by the monomial.

Thus, $\dfrac{10x + 15}{5} = \dfrac{10x}{5} + \dfrac{15}{5} = 2x + 3$.   Also, $\dfrac{ax + bx}{x} = \dfrac{ax}{x} + \dfrac{bx}{x} = a + b$.

To check the division, multiply the quotient by the divisor. The result should be the dividend.

Thus, to check $\dfrac{10x + 15}{5} = 2x + 3$, multiply $2x + 3$ by 5 to obtain $10x + 15$.

11.1. DIVIDING A POLYNOMIAL BY A MONOMIAL

Divide:

a) $\dfrac{3a + 6b}{3}$ *Ans.* $a + 2b$

f) $\dfrac{9x^2y - 36xy^2}{9xy}$ *Ans.* $x - 4y$

b) $\dfrac{r - rt}{r}$ $1 - t$

g) $\dfrac{7x - 14y + 56}{7}$ $x - 2y + 8$

c) $\dfrac{pq + pr}{-p}$ $-q - r$

h) $\dfrac{x^3 + 2x^2 + 5x}{x}$ $x^2 + 2x + 5$

d) $\dfrac{2\pi r + 2\pi R}{2\pi}$ $r + R$

i) $\dfrac{3x^5 - x^3 + 5x^2}{-x^2}$ $-3x^3 + x - 5$

e) $\dfrac{ab - abc}{ab}$ $1 - c$

11.2. CHECKING DIVISION

Check each, using multiplication:

a) Does $\dfrac{r - rt}{r} = 1 - t$?

b) Does $\dfrac{x^2y - 2xy^2}{xy} = x - y$?

c) Does $\dfrac{2x - 4y + 10}{-2} = -x + 2y - 5$?

Check: (Quotient × Divisor = Dividend)

a) Multiply:

$r(1 - t) = r - rt$

Correct.

b) Multiply:

$xy(x - y) = x^2y - xy^2$

Incorrect.   The quotient should be $x - 2y$.

c) Multiply:

$-2(-x + 2y - 5) = 2x - 4y + 10$

Correct.

11.3.   MULTIPLYING AND DIVIDING POLYNOMIALS BY A MONOMIAL

Simplify:

a) $5(x - 2) + \dfrac{3x - 12}{3}$

$5x - 10 + x - 4$

Ans. $6x - 14$

b) $\dfrac{x^2 - 5x}{x} - x(3 - x)$

$x - 5 - 3x + x^2$

Ans. $x^2 - 2x - 5$

c) $\dfrac{a^3 - a^2bc}{a} - b(b - ac)$

$a^2 - abc - b^2 + abc$

Ans. $a^2 - b^2$

12.   DIVIDING A POLYNOMIAL BY A POLYNOMIAL

To Divide Polynomial

Procedure:

1. Set up as a form of long division in which the polynomials are arranged in descending order, leaving space for missing terms:

2. Divide the first term of the divisor into the first term of the dividend to obtain the first term of the quotient:

3. Multiply the first term of the quotient by each term of the divisor:

4. Subtract like terms and bring down one or more terms as needed:

5. Repeat Steps 2 to 4 using the remainder as the new dividend: that is, divide, multiply, subtract and bring down:

6. Continue repeating Steps 2 to 4 as long as it is possible.

Solution: (By steps)

1. $x - 2 \overline{)x^2 - 5x + 6}$

2. $x - 2 \overline{)\overset{x}{x^2 - 5x + 6}}$

3. $x - 2 \overline{)\overset{x}{x^2 - 5x + 6}}$
$\underline{x^2 - 2x}$

4.
$-3x + 6$

5. $x - 2 \overline{)\overset{-3}{}}$
$-3x + 6$
$\underline{-3x + 6}$

6. No further steps needed

Full Solution:

$$\begin{array}{r} x - 3 \\ x - 2 \overline{)x^2 - 5x + 6} \\ \underline{x^2 - 2x} \\ -3x + 6 \\ \underline{-3x + 6} \end{array}$$

Ans. $x - 3$

To Check the Division

1. If no final remainder exists, then multiply the quotient by the divisor. The result should equal the dividend.

   Thus, to check $\dfrac{18}{3} = 6$, multiply 3 by 6 to obtain 18.

2. If there is a final remainder, add this to the product of the quotient and divisor. The result should equal the dividend.

   Thus, to check $\dfrac{19}{3} = 6\frac{1}{3}$, multiply 3 by 6 and then add 1 to obtain 19.

12.1.   DIVIDING A POLYNOMIAL BY A POLYNOMIAL (NO REMAINDER)

Divide: a) $x^2 - 9x + 14$ by $x - 7$ b) $x^3 - 6x^2 + 11x - 6$ by $x - 3$ and check.

Solutions:

$a)$

$$\begin{array}{r} x - 2 \\ x - 7\overline{\smash{\big)}\,x^2 - 9x + 14} \\ \underline{x^2 - 7x} \\ -2x + 14 \\ \underline{-2x + 14} \end{array}$$

Check:

$$\begin{array}{r} x - 2 \\ \underline{x - 7} \\ x^2 - 2x \\ \underline{-7x + 14} \\ x^2 - 9x + 14 \end{array}$$

Ans. $x - 2$ (dividend)

$b)$

$$\begin{array}{r} x^2 - 3x + 2 \\ x - 3\overline{\smash{\big)}\,x^3 - 6x^2 + 11x - 6} \\ \underline{x^3 - 3x^2} \\ -3x^2 + 11x \\ \underline{-3x^2 + 9x} \\ 2x - 6 \end{array}$$

Check:

$$\begin{array}{r} x^2 - 3x + 2 \\ \underline{x - 3} \\ x^3 - 3x^2 + 2x \\ \underline{-3x^2 + 9x - 6} \\ x^3 - 6x^2 + 11x - 6 \end{array}$$

Ans. $x^2 - 3x + 2$ $\underline{2x - 6}$ (dividend)

12.2. ARRANGING POLYNOMIALS AND DIVIDING

Divide $20a^2 - 3b^2 + 7ab$ by $-b + 4a$ and check:

Solution: Arrange both dividend and divisor in descending powers of a:

$$\begin{array}{r} 5a + 3b \\ 4a - b\overline{\smash{\big)}\,20a^2 + 7ab - 3b^2} \\ \underline{20a^2 - 5ab} \\ 12ab - 3b^2 \end{array}$$

Check:

$$\begin{array}{r} 5a + 3b \\ \underline{4a - b} \\ 20a^2 + 12ab \\ \underline{-5ab - 3b^2} \\ 20a^2 + 7ab - 3b^2 \end{array}$$ (dividend)

Ans. $5a + 3b$ $\underline{12ab - 3b^2}$

12.3. TERMS MISSING IN DIVIDEND

Divide $x^3 - 64$ by $x - 4$ and check:

Solution: Leave spaces for missing x^2 and x terms:

$$\begin{array}{r} x^2 + 4x + 16 \\ x - 4\overline{\smash{\big)}\,x^3 \qquad\qquad - 64} \\ \underline{x^3 - 4x^2} \\ 4x^2 \\ \underline{4x^2 - 16x} \\ 16x - 64 \end{array}$$

Check:

$$\begin{array}{r} x^2 + 4x + 16 \\ \underline{x - 4} \\ x^3 + 4x^2 + 16x \\ \underline{-4x^2 - 16x - 64} \\ x^3 \qquad\qquad - 64 \end{array}$$ (dividend)

Ans. $x^2 + 4x + 16$ $\underline{16x - 64}$

12.4. DIVIDING POLYNOMIALS (WITH REMAINDER)

Divide $8x^2 - 10x + 8$ by $2x - 4$ and check:

Solution:

$$4x + 3 + \frac{20}{2x - 4}$$

$$2x - 4 \overline{\smash{)}8x^2 - 10x + 8}$$

$$\underline{8x^2 - 16x}$$

$$+ \ 6x + \ 8$$

$$\underline{+ \ 6x - 12}$$

Remainder: $\qquad\qquad\qquad 20$

Complete quotient by adding $\dfrac{20}{2x - 4}$

Ans. $\qquad 4x + 3 + \dfrac{20}{2x - 4}$

Check: $\qquad 4x + 3$

$$\underline{2x - 4}$$

$$8x^2 + \ 6x$$

$$\underline{- 16x - 12}$$

$$8x^2 - 10x - 12$$

Add remainder: $\qquad\qquad + 20$

$$\overline{8x^2 - 10x + \ 8} \quad \text{(dividend)}$$

Supplementary Problems

4.1. In each polynomial, select like terms, if any: $\qquad\qquad\qquad\qquad\qquad\qquad$ **(1.1)**

a) $16 - 4y + 3y - y^2$

b) $16y^2 - 4y + 20y^2 + 8$

c) $5x^2 - 8xy + 7y^2 - 2$

d) $7a - 3b + 12ab + 10$

e) $a^2b + ab^2 - 2ab + 3ab^2$

f) $4(x + y) - 5(x + y) + 2(x^2 + y^2)$

g) $3xy + xz - 3x - 5z$

h) $3x^2y^2 + 2(x^2 + y^2) - 4(x^2 + y^2)$

Ans. a) $-4y$ and $+3y$

b) $16y^2$ and $20y^2$

c) No like terms

d) No like terms

e) $+ab^2$ and $+3ab^2$

f) $4(x + y)$ and $-5(x + y)$

g) No like terms

h) $2(x^2 + y^2)$ and $-4(x^2 + y^2)$

4.2. Add: $\qquad\qquad\qquad\qquad\qquad\qquad\qquad\qquad\qquad\qquad\qquad\qquad\qquad\qquad\qquad\qquad$ **(2.1 to 2.3)**

a) $+5b + (+ 16b)$

b) $-10y^3 + (-7y^3)$

c) $+7rs + (-10rs)$

d) $-20abc + (-abc)$

e) $+ \ 8x^2y^2$
$\quad +11x^2y^2$
$\quad \underline{+30x^2y^2}$

f) $+3.2h$
$\quad -2.2h$
$\quad \underline{+7.5h}$

g) $-3(a + b)$
$\quad +5(a + b)$
$\quad \underline{+ 2(a + b)}$

Ans. a) $21b$ b) $-17y^3$ c) $-3rs$ d) $-21abc$ e) $49x^2y^2$ f) $8.5h$ g) $4(a + b)$

4.3 Simplify and add: $\qquad\qquad\qquad\qquad\qquad\qquad\qquad\qquad\qquad\qquad\qquad\qquad\qquad\qquad$ **(2.2)**

a) $+13a + (-2a) + (-a)$

b) $- 2x^2 + (-8x^2) + (-15x^2)$

c) $+a^3 + (+3a^3) + (-7a^3)$

d) $+ xy^2 + (+xy^2) + (-xy^2)$

e) $+2.3ab + (+7.1ab) + (-3.7ab)$

f) $+\frac{3}{4}r^2s + (-\frac{1}{2}r^2s) + (1\frac{1}{8}r^2s)$

Ans. a) $13a - 2a - a = 10a$

b) $-2x^2 - 8x^2 - 15x^2 = - 25x^2$

c) $a^3 + 3a^3 - 7a^3 = - 3a^3$

d) $xy^2 + xy^2 - xy^2 = xy^2$

e) $2.3ab + 7.1ab - 3.7ab = 5.7ab$

f) $\frac{3}{4}r^2s - \frac{1}{2}r^2s + 1\frac{1}{8}r^2s = + 1\frac{3}{8}r^2s$

4.4. Simplify and add: $\qquad\qquad\qquad\qquad\qquad\qquad\qquad\qquad\qquad\qquad\qquad\qquad\qquad\qquad$ **(2.4)**

a) $+12a + (-3a) + (+10)$

b) $+12 + (-3a) + (+10a)$

c) $+12b + (-3) + (+10b)$

d) $- 10x^2 + (-3x^2) + (-5x)$

e) $-10x^2 + (-3x) + (-5x^2)$

f) $+5r^2s + (-2r^2s) + (+rs^2)$

g) $+5rs^2 + (-2r^2s) + (+rs^2)$

h) $+5rs^2 + (-2rs^2) + (+rs^2)$

Ans. a) $12a - 3a + 10 = 9a + 10$

b) $12 - 3a + 10a = 12 + 7a$

c) $12b - 3 + 10b = 22b - 3$

d) $-10x^2 - 3x^2 - 5x = -13x^2 - 5x$

e) $-10x^2 - 3x - 5x^2 = -15x^2 - 3x$

f) $5r^2s - 2r^2s + rs^2 = 3r^2s + rs^2$

g) $5rs^2 - 2r^2s + rs^2 = 6rs^2 - 2r^2s$

h) $5rs^2 - 2rs^2 + rs^2 = 4rs^2$

4.5. Rearrange in descending order and combine: **(3.1)**

a) $-6y + 2y^2 + y^3 + 3y^2$
$\quad y^3 + 2y^2 + 3y^2 - 6y$

b) $2x^3 + 3x^4 - x^3 - 10 + 5x^2$
$\quad 3x^4 + 2x^3 - x^3 + 5x^2 - 10$

Ans. $y^3 + 5y^2 - 6y$

Ans. $3x^4 + x^3 + 5x^2 - 10$

c) $6y^2 - x^2 + 10xy + 8x^2$
$\quad$ (order in terms of x)
$\quad 8x^2 - x^2 + 10xy + 6y^2$

Ans. $7x^2 + 10xy + 6y^2$

4.6. Add: **(3.2)**

a) $\quad 5y + 12$
$\quad -3y - 10$

b) $\quad 6x - 2y$
$\quad 3x + 10y$

c) $\quad 8x^3 + 5x^2 - 10x$
$\quad\ x^3 - 6x^2 + 11x$

d) $\quad\ x^2 - 3x + 25$
$\quad -3x^2 - 10x - 30$

Ans. $\quad 2y + 2$

$\quad 9x + 8y$

$\quad 9x^3 - x^2 + \ x$

$\quad -2x^2 - 13x - 5$

4.7. Add: **(3.3)**

a) $6x^2 - 7x$ and $-2x^2 - x$

c) $x^2 - x + 1$ and $7x - 4x^2 + 9$

b) $5a + 2, 3a - 7,$ and $-6a + 4$

d) $5y - 3x + 6$ and $-8 - 4x + y$

Ans. *a)* $\quad 6x^2 - 7x$
$\quad\ -2x^2 - \ x$
$\quad\ \ \ \ 4x^2 - 8x$

b) $\quad 5a + 2$
$\quad 3a - 7$
$\quad -6a + 4$
$\quad\ \ 2a - 1$

c) $\quad\ x^2 - \ x + \ 1$
$\quad -4x^2 + 7x + \ 9$
$\quad -3x^2 + 6x + 10$

d) $\quad -3x + 5y + 6$
$\quad -4x + \ y - 8$
$\quad -7x + 6y - 2$

4.8. Check, letting $x = 2, y = 3,$ and $z = 4$: **(3.4, 3.5)**

a) $\quad 5x - 3y + \ \ z$
$\quad 2x - 2y - 10z$
$\quad 7x - 5y - \ \ 9z$

Check: *a)* $5x - 3y + \ \ z \to 10 - 9 + \ 4 = \quad 5$
$\quad\ 2x - 2y - 10z \to 4 - 6 - 40 = -42$
$\quad\ 7x - 5y - \ \ 9z \to 14 - 15 - 36 = -37$

By adding vertically and horizontally, the sum is -37.

b) $\quad x^2 + 2y^2 - z^2$
$\quad x^2 - \ y^2 + 2z^2$
$\quad 2x^2 + y^2 + \ z^2$

c) $\quad x^2 + 2y^2 - \ z^2 \to 4 + 18 - 16 = 6$
$\quad x^2 - \ y^2 + 2z^2 \to 4 - \ 9 + 32 = 27$
$\quad 2x^2 + \ y^2 + \ z^2 \to 8 + 9 + 16 = 33$

By adding vertically and horizontally, the sum is 33.

4.9. Subtract: **(4.1)**

a) $+5c - (-3c)$

c) $-3rs - (+2rs)$

e) $+2x^2y - (+12x^2y)$

b) $-x^2 - (-4x^2)$

d) $+cd^2 - (+7cd^2)$

f) $+5abc - (-7abc)$

Ans. *a)* $8c$ *b)* $3x^2$ *c)* $-5rs$ *d)* $-6cd^2$ *e)* $-10x^2y$ *f)* $12abc$

4.10. Subtract: (4.1)

	a) $+3cd$	b) $+5x^2$	c) $-xy^2$	d) $-2pqr$	e) $-4(m-n)$
(S)	$+\ cd$	$-7x^2$	$-4xy^2$	$+7pqr$	$-14(m-n)$
Ans.	$2cd$	$12x^2$	$3xy^2$	$-9pqr$	$10(m-n)$

4.11. a) From $5y^2$ take $-2y^2$.   Ans. $5y^2 - (-2y^2) = 7y^2$ (4.1)
b) Reduce $-2ab$ by $-5ab$. $-2ab - (-5ab) = 3ab$
c) How much less than $+x^2$ is $+3x^2$? $x^2 - (+3x^2) = -2x^2$
d) By how much does $-17y$ exceed $-30y$? $-17y - (-30y) = 13y$

4.12. Combine: (4.3)

a) $+4b + (-2b) - (-3b)$ c) $+2cd - (-3cd) + (-10cd) - (+5cd)$
b) $-5x^2 - (-x^2) - (+3x^2)$ d) $-4abc^2 - (-abc^2) - (+3abc^2) - (-12abc^2)$

Ans. a) $4b - 2b + 3b = 5b$ c) $2cd + 3cd - 10cd - 5cd = -10cd$
b) $-5x^2 + x^2 - 3x^2 = -7x^2$ d) $-4abc^2 + abc^2 - 3abc^2 + 12abc^2 = 6abc^2$

4.13. Subtract: (5.1)

	a) $3k - 5$	b) $k^2 - 2k$	c) $-2a^2 - 5a + 12$	d) $3r^2 - 5rt$
(S)	$k - 8$	$-k^2 - 8k$	$-a^2\qquad -\ 7$	$-10r^2 -\ rt + 7t^2$
Ans.	$2k + 3$	$2k^2 + 6k$	$-a^2 - 5a + 19$	$13r^2 - 4rt - 7t^2$

4.14. Arrange and subtract: (5.2)

a) $5k - 3l - 2$ from $4l - 7 - k$ c) $3x + 5$ from $x^2 - 4x$
b) $3x + 4y - 5$ from $16 - 2x - y$ d) $m^3 - 18m$ from $8m^2 + 10m$

Ans.	a) $-k + 4l - 7$	b) $-2x - y + 16$	c) $x^2 - 4x$	d) $8m^2 + 10m$
(S)	$5k - 3l - 2$	$3x + 4y - 5$	$3x + 5$	$m^3\qquad -\ 18m$
	$-6k + 7l - 5$	$-5x - 5y + 21$	$x^2 - 7x - 5$	$-m^3 + 8m^2 + 28m$

4.15. a) Reduce $8x^2$ by $3x^2 + 5$.   Ans. $5x^2 - 5$ (5.3)

b) From $2x - 3y$ take $4y - 7x$.   Ans. $9x - 7y$
c) Subtract $a^3 - 2a^2 + 5$ from 0.   Ans. $-a^3 + 2a^2 - 5$
d) What is $5h^3 - 12h$ less $h^2 + 5h$? Ans. $5h^3 - h^2 - 17h$

4.16. Check each subtraction, using addition: (5.4)

	a) $3x - 2$	b) $8a^2 - 1$	c) $5y^2 - 3y$	d) $2h^2\qquad + 5$
(S)	$x - 5$	$-a^2 + 7$	$8y^2\qquad + 15$	$-3h^2 - h$
	$2x + 3(?)$	$9a^2 - 8(?)$	$-3y^2 + 3y - 15(?)$	$5h^2 - h + 5(?)$

Check by addition:

	a) $2x + 3$	b) $9a^2 - 8$	c) $-3y^2 + 3y - 15$	d) $5h^2 - h + 5$
(A)	$x - 5$	$-a^2 + 7$	$8y^2\qquad +15$	$-3h^2 - h$
	$3x - 2$	$8a^2\quad 1$	$5y^2 + 3y$	$2h^2 - 2h + 5$
	(Correct)	(Correct)	(Incorrect)	(Incorrect)
			Difference should be	Difference should be
			$-3y^2 - 3y - 15$	$5h^2 + h + 5$

4.17. Simplify: (6.1)

a) $2r + (3 - 7r)$ b) $7x^2 + 4 + (10 - 2x^2)$ c) $8y + (7y - 20) + (40 - 3y)$

Ans. a) $-5r + 3$ b) $5x^2 + 14$ c) $12y + 20$

4.18. Simplify: (6.2)

a) $17 - (3a + 4)$ c) $(5x^2 - 2) - (3 - 4x^2)$ e) $24h - (3h - 2) - 6(h - 1)$
b) $17a - (6 - 2a)$ d) $3 - (7y + 10) - (18 - y)$ f) $80 - (30 - 5y) - (3y - y^2)$

Ans. a) $13 - 3a$ c) $9x^2 - 5$ e) $15h + 8$
b) $19a - 6$ d) $-6y - 25$ f) $y^2 + 2y + 50$

4.19. Simplify: (6.3, 6.6)

a) $5a - [2a - (3a + 4)]$ b) $12x + 4 - [(3 - x) - (5x + 7)]$

Ans. $6a + 4$ $18x + 8$

c) $20 - \{[-(d - 1) + 3d] - 5d\}$
      $3d + 19$

4.20. a)   Subtract $3b - 8$ from the sum of $5b - 2$ and $-6b + 5$. (6.4, 6.5)
b)   From the sum of $8x^2 + 5$ and $7x^2 - 2$, subtract the sum of $20x - 8$ and $-x^2 + 5x$.

Ans. a) $[(5b - 2) + (-6b + 5)] - (3b - 8) = -4b + 11$
b) $[(8x^2 + 5) + (7x^2 - 2)] - [(20x - 8) + (-x^2 + 5x)] = 16x^2 - 25x + 11$

4.21. Multiply: (7.1)

a) $c^2 \cdot c \cdot c$ d) $x^2 \cdot y^3 \cdot x^5$ g) $7^2 7$
b) $y^3 \cdot y^4 \cdot y$ e) $m^2 p m^8 p$ h) $8^3 \cdot 8 \cdot 10^2$
c) $a^x a^y$ f) $a^3 b a b^4$ i) $3^5 3^4 x^2 x^9$

Ans. a) c^4 b) y^8 c) a^{x+y} d) $x^7 y^3$ e) $m^{10} p^2$ f) $a^4 b^5$ g) 7^3 h) $8^4 10^2$ i) $3^9 x^{11}$

4.22. Raise to the indicated power: (7.2)

a) $(x^2)^3$ c) $(b^2)^6$ e) $(3^3)^3$ g) $(r^3)^6 (t^2)^2$
b) $(x^3)^2$ d) $(a^x)^y$ f) $(5^2)^5$ h) $(4^7)^2 (10^2)^4$

Ans. a) x^6 b) x^6 c) b^{12} d) a^{xy} e) 3^9 f) 5^{10} g) $r^{18} t^4$ h) $4^{14} 10^8$

4.23. Multiply: (7.3)

a) $3(-5a)$ d) $5y(-y)(-y)$ g) $(-7ab)(-3a)(-b)$
b) $(-6x)(-x)$ e) $(0.2x^2)(0.3^3)$ h) $(3r^3)(5s^5)$
c) $(-4y)(+3y^2)$ f) $(\frac{2}{3}x)(3x)(-5y)$ i) $(-5xy^2)^2$

Ans. a) $-15a$ d) $5y^3$ g) $-21a^2 b^2$
b) $6x^2$ e) $0.06x^5$ h) $15r^3 s^5$
c) $-12y^3$ f) $-10x^2 y$ i) $25x^2 y^4$

4.24. Multiply: (8.1)

a) $2(a - b)$ d) $3a(b - 2c)$ g) $\frac{1}{2}(a^2 + 6b^2)$ j) $-5(x^2 + y^2 - z^2)$
b) $-3(x - 5)$ e) $-5x(x - 4)$ h) $ab(a - 2b)$ k) $-a(a^3 + a^2 - 5a)$
c) $a(c - 3d)$ f) $\frac{1}{2}(6y - 8z)$ i) $-4r^2(r^3 + 2r^2)$ l) $4x(5 - x - 10x^2)$

Ans. a) $2a - 2b$ d) $3ab - 6ac$ g) $\dfrac{a^2}{2} + 3b^2$ j) $-5x^2 - 5y^2 + 5z^2$

b) $-3x + 15$ e) $-5x^2 + 20x$ h) $a^2 b - 2ab^2$ k) $-a^4 - a^3 + 5a^2$
c) $ac - 3ad$ f) $3y - 4z$ i) $-4r^5 - 8r^4$ l) $20x - 4x^2 - 40x^3$

4.25. Multiply: (8.2)

$$
\begin{array}{lll}
a) \quad s^2 - s - 1 & b) \quad x^3 - 3x^2 + 11 & c) \quad 7h + 5k - 3m + 5 \\
\qquad\quad\; \underline{-3s} & \qquad\qquad\; \underline{7x^2} & \qquad\qquad\qquad \underline{2hkm}
\end{array}
$$

Ans. a) $-3s^3 + 3s^2 + 3s$ b) $7x^5 - 21x^4 + 77x^2$ c) $14h^2km + 10hk^2m - 6hkm^2 + 10hkm$

4.26. Simplify: (8.3)

a) $5 + 3(y - 2)$ c) $3x + 10(2x - 1)$ e) $-a(a - 2) + 6(a^2 - 5)$
b) $x + 2(3 - 4x)$ d) $-4(a - 3) - 2(5 - a)$ f) $\frac{1}{2}(6c - 10d) - \frac{2}{3}(9d - 30c)$

Ans. a) $3y - 1$ c) $23x - 10$ e) $5a^2 + 2a - 30$
b) $-7x + 6$ d) $-2a + 2$ f) $23c - 11d$

4.27. Simplify: (8.4, 8.5)

a) $2[3b + 5(b - 2)]$ c) $4(5 - c) - 3[8 - 7(c - 2)]$
b) $-a[2(a - 3) + 6a]$ d) $10\{25 - [5(y + 4) - 3(y - 1)]\}$

Ans. a) $16b - 20$ b) $-8a^2 + 6a$ c) $17c - 46$ d) $-20y + 20$

4.28. Simplify: (9.1)

a) $(5y + 2)(3y + 1)$ d) $(a + 2b)(a - b)$ g) $(x^2 + x + 1)(x + 2)$
b) $(x + 7)(x - 4)$ e) $(x - 4y)(x - y)$ h) $(r^2 - 2r + 3)(r - 1)$
c) $(1 + y)(2 - y)$ f) $(ab - 5)(ab + 2)$ i) $(t^2 + 5t - 6)(2t + 1)$

Ans. a) $15y^2 + 11y + 2$ d) $a^2 + ab - 2b^2$ g) $x^3 + 3x^2 + 3x + 2$
b) $x^2 + 3x - 28$ e) $x^2 - 5xy + 4y^2$ h) $r^3 - 3r^2 + 5r - 3$
c) $2 + y - y^2$ f) $a^2b^2 - 3ab - 10$ i) $2t^3 + 11t^2 - 7t - 6$

4.29. Check each multiplication, letting $x = 10$.   (9.2)

a) $3x + 1$
(M) $\underline{2x + 2}$

  $6x^2 + 8x + 2$

Check: a) $3x + 1 \rightarrow 30 + 1 = 31$
  $\underline{2x + 2} \rightarrow 20 + 2 = 22$

  $6x^2 + 8x + 2 \rightarrow 600 + 80 + 2 = 682$
  Correct since $682 = 31 \times 22$

b) $x^2 + x - 1$
(M) $\underline{\quad 3x - 2}$

  $3x^3 + x^2 - 5x + 2$

Check: b) $x^2 + x - 1 \rightarrow 100 + 10 - 1 = 109$
  $\underline{\quad 3x - 2} \rightarrow \quad\; 30 - 2 = 28$

  $3x^3 + x^2 - 5x + 2 \rightarrow 3000 + 100 - 50 + 2 = 3052$
  Correct since $3052 = 109 \times 28$

4.30. Multiply: (9.3)

a) $(x + 2)(x + 3)(x + 4)$ b) $(y - 1)(y - 2)(y + 5)$ c) $(2a + 3)(3a - 4)(a - 5)$

Ans. a) $(x^2 + 5x + 6)(x + 4)$ b) $(y^2 - 3y + 2)(y + 5)$ c) $(6a^2 + a - 12)(a - 5)$
  $x^3 + 9x^2 + 26x + 24$ $y^3 + 2y^2 - 13y + 10$ $6a^3 - 29a^2 - 17a + 60$

4.31. Divide: **(10.1)**

a) $\dfrac{x^5}{x^3}$ b) $\dfrac{x^3}{x^3}$ c) $\dfrac{x^3}{x^5}$ d) $\dfrac{a^4b^5}{ab^2}$ e) $\dfrac{ab^2}{a^4b^5}$ f) $\dfrac{10^5}{10^9}$ g) $\dfrac{10^9}{10^5}$ h) $\dfrac{10^5}{10^5}$

Ans. a) x^2 b) 1 c) $\dfrac{1}{x^2}$ d) a^3b^3 e) $\dfrac{1}{a^3b^3}$ f) $\dfrac{1}{10^4}$ g) 10^4 h) 1

4.32. Divide: **(10.2)**

a) $\dfrac{32a}{4a}$ b) $\dfrac{4a}{32a}$ c) $\dfrac{32a^4}{-8a^3}$ d) $\dfrac{-a^2b^2c^2}{-ab^3}$ e) $\dfrac{7xy}{-7xy}$ f) $\dfrac{-56c^4d}{7cd^4}$

Ans. a) 8 b) $\frac{1}{8}$ c) $-4a$ d) $\dfrac{ac^2}{b}$ e) -1 f) $-\dfrac{8c^3}{d^3}$

4.33. Divide: **(10.1, 10.2)**

a) r^{10} by r^3 d) u^2 by u^6 g) $-8x^2y^2 \div (-2xy)$ j) $a^5 \div a^b$
b) s^5 by s e) $2ab \div ab^2$ h) $10r^4s^4 \div 4r^4s^4$ k) $3^a \div 3^5$
c) $-t^2$ by t^2 f) $-a^2b \div ab^2$ i) $x^a \div x^2$ l) $x^a \div x^b$

Ans. a) r^7 b) s^4 c) -1 d) $\dfrac{1}{u^4}$ e) $\dfrac{2}{b}$ f) $-\dfrac{a}{b}$ g) $4xy$ h) $2\frac{1}{2}$ i) x^{a-2}

j) a^{5-b} k) 3^{a-5} l) x^{a-b}

4.34. Check each division, using multiplication: **(10.3)**

a) Does $\dfrac{18x^3}{3x} = 6x^2$? b) Does $\dfrac{6a^2b^2}{-2ab} = -3ab$? c) Does $\dfrac{-27a^2}{-9a^6} = \dfrac{3}{a^3}$?

Check, using multiplication:

a) $(3x)(6x^2) = 18x^3$ b) $(-2ab)(-3ab) = 6a^2b^2$ c) $(-9a^6)\left(\dfrac{3}{a^3}\right) = -27a^3$

  Correct Correct Incorrect. Quotient should be $\dfrac{3}{a^4}$.

4.35. Divide: **(11.1)**

a) $\dfrac{5a + 15c}{5}$ c) $\dfrac{xy^2 + x^2y}{xy}$ e) $\dfrac{P + Prt}{P}$ g) $\dfrac{20r + 30s - 40}{-10}$

b) $\dfrac{xy + x}{x}$ d) $\dfrac{\pi R^2 - \pi r^2}{\pi}$ f) $\dfrac{5x^3 - 5x}{5x}$ h) $\dfrac{y^5 + 2y^4 - y^3}{y^3}$

Ans. a) $a + 3c$ b) $y + 1$ c) $y + x$ d) $R^2 - r^2$ e) $1 + rt$ f) $x^2 - 1$
g) $-2r - 3s + 4$ h) $y^2 + 2y - 1$

4.36. Divide: **(11.1)**

a) $2x - 4$ by 2 d) $(45x^2 - 15) \div 15$ g) $4\,\overline{)8a^2 - 4a + 12}$

b) $3a - 6$ by -3 e) $(15x^2 - 45x) \div 15x$ h) $-x\,\overline{)x^3 - 7x^2 + 10x}$

c) $-7x^3 + 7x^2$ by $-7x$ f) $(27xy - 18y^2) \div 9y$ i) $3a^2\,\overline{)30a^4 - 33a^3 + 3a^2}$

Ans. a) $x - 2$ b) $-a + 2$ c) $x^2 - x$ d) $3x^2 - 1$ e) $x - 3$ f) $3x - 2y$
g) $2a^2 - a + 3$ h) $-x^2 + 7x - 10$ i) $10a^2 - 11a + 1$

4.37. Simplify: **(11.3)**

a) $2(x + 4) + \dfrac{6x - 8}{2}$ b) $\dfrac{r^2 - 10r}{r} + r(3 + r)$ c) $\dfrac{a^4 - a^3 + a^2}{a} - 2a(a^2 - 1)$

a) $2x + 8 + 3x - 4$ b) $r - 10 + 3r + r^2$ c) $a^3 - a^2 + a - 2a^3 + 2a$

Ans. $5x + 4$ $r^2 + 4r - 10$ $-a^3 - a^2 + 3a$

4.38. Divide: **(12.1a, 12.2)**

a) $d^2 + 9d + 14$ by $d + 7$ g) $x^2 + xy - 20y^2$ by $x - 4y$
b) $d^2 - 8d + 7$ by $d - 7$ h) $45 - 14b + b^2$ by $9 - b$
c) $r^2 - 13r + 30$ by $r - 10$ i) $15a^2b^2 - 8ab + 1$ by $3ab - 1$
d) $x^2 - 3x - 10$ by $x - 5$ j) $3x^2 - 40 - 2x$ by $3x + 10$
e) $2r^2 + 7r + 6$ by $r + 2$ k) $3 + 7y^2 - 22y$ by $3 - y$
f) $6y^2 - 29y + 28$ by $2y - 7$ l) $x^4 + 4x^2 - 45$ by $x^2 + 9$

Ans. a) $d + 2$ c) $r - 3$ e) $2r + 3$ g) $x + 5y$ i) $5ab - 1$ k) $1 - 7y$
       b) $d - 1$ d) $x + 2$ f) $3y - 4$ h) $5 - b$ j) $x - 4$ l) $x^2 - 5$

4.39. Divide: **(12.1b)**

a) $x^3 + 5x^2 + 7x + 2$ by $x + 2$ d) $6a^3 + 7a^2 + 12a - 5$ by $3a - 1$
b) $x^3 - 2x^2 - 5x + 6$ by $x - 1$ e) $6a^3 + 17a^2 + 27a + 20$ by $3a + 4$
c) $x^3 - 2x^2 - 5x + 6$ by $x - 3$ f) $21y^3 - 38y^2 + 29y - 40$ by $3y - 5$

Ans. a) $x^2 + 3x + 1$ b) $x^2 - x - 6$ c) $x^2 + x - 2$ d) $2a^2 + 3a + 5$

       e) $2a^2 + 3a + 5$ f) $7y^2 - y + 8$

4.40. Divide: **(12.3)**

a) $x^2 - 9$ by $x - 3$ d) $x^3 - 8$ by $x - 2$
b) $x^3 - 4x$ by $x + 2$ e) $x^4 - 1$ by $x + 1$
c) $7x^2 - 63$ by $x + 3$ f) $2x^3 - 30x - 8$ by $x - 4$

Ans. a) $x + 3$ b) $x^2 - 2x$ c) $7x - 21$ d) $x^2 + 2x + 4$ e) $x^3 - x^2 + x - 1$
       f) $2x^2 + 8x + 2$

4.41. Divide: **(12.4)**

a) $x^2 - 5x + 2$ by $x - 3$ c) $x^2 + 1$ by $x + 1$ e) $x^3 - 5x^2 + 6x + 5$ by $x - 3$
b) $2x^2 + 5x - 15$ by $x - 2$ d) $x^2 + 9$ by $x + 3$ f) $2x^3 + 3x^2 - x - 2$ by $2x - 3$

Ans. a) $x - 2 + \dfrac{-4}{x - 3}$ c) $x - 1 + \dfrac{2}{x + 1}$ e) $x^2 - 2x + \dfrac{5}{x - 3}$

       b) $2x + 9 + \dfrac{3}{x - 2}$ d) $x - 3 + \dfrac{18}{x + 3}$ f) $x^2 + 3x + 4 + \dfrac{10}{2x - 3}$

4.42. Check each division, using multiplication: **(12.2, 12.4)**

a) Does $\dfrac{21x^2 + 11xy - 2y^2}{3x + 2y} = 7x - y$?   b) Does $\dfrac{x^3 - 6}{x - 2} = x^2 + 2x + 4 + \dfrac{2}{x - 2}$?

Check, using multiplication:

a) $3x + 2y$ b) $x^2 + 2x + 4$
            $\underline{7x - y}$ $\underline{x - 2}$
Correct $21x^2 + 11xy - 2y^2$ $x^3 \qquad\quad - 8$
                                                            Add remainder: $\underline{\qquad\qquad + 2}$
                                                            Correct $x^3 \qquad\quad - 6$

CHAPTER 5

First-Degree Equations

1. REVIEWING THE SOLUTION OF FIRST-DEGREE EQUATIONS HAVING POSITIVE ROOTS

A first-degree equation in one unknown is one that contains only one unknown and where the unknown has the exponent 1.

Thus, $2x + 5 = 9$ is a first-degree equation in one unknown, but $x^2 + 5 = 9$ is not.

In Chap. 2, simple equations were solved by using the equality rules and inverse operations. In this chapter, these methods of solution are extended to equations having signed numbers and to more difficult equations.

Inverse Operations

Since addition and subtraction are inverse operations,

1. Use subtraction (**S**) to undo addition (**A**).
2. Use addition (**A**) to undo subtraction (**S**).

Since multiplication and division are inverse operations,

3. Use division (**D**) to undo multiplication (**M**).
4. Use multiplication (**M**) to undo division (**D**).

Order of Inverse Operations

Generally, undo addition and subtraction before undoing multiplication and division.

1.1. USING SUBTRACTION (S) TO UNDO ADDITION (A)

Solve:

a)	b)	c)	d)
$n + 2 = 8$	$n + 5 = 2n$	$2n + 6 = n + 10$	$n^2 + 8 = n^2 + n$
$\mathbf{S}_2 \quad \underline{\quad 2 = 2}$	$\mathbf{S}_n \quad \underline{n \quad = n}$	$\mathbf{S}_{n+6} \quad \underline{n + 6 = n + 6}$	$\mathbf{S}_n \quad \underline{n^2 \quad = n^2}$
Ans. $n \quad = 6$	*Ans.* $5 = n$	*Ans.* $n \quad = \quad 4$	*Ans.* $8 = \quad n$

1.2. USING ADDITION (A) TO UNDO SUBTRACTION (S)

Solve:

a)	b)	c)	d)
$x - 2 = 8$	$5 - x = 0$	$5 - n = -7$	$8 - n^2 = n - n^2$
$\mathbf{A}_2 \quad \underline{\quad 2 = 2}$	$\mathbf{A}_x \quad \underline{\quad x = x}$	$\mathbf{A}_{7+n} \quad \underline{7 + n = \quad 7 + n}$	$\mathbf{A}_{n^2} \quad \underline{\quad n^2 = \quad n^2}$
Ans. $x \quad = 10$	*Ans.* $5 \quad = x$	*Ans.* $12 \quad = \quad n$	*Ans.* $8 \quad = n$

1.3. Using Division (D) to Undo Multiplication (M)

Solve:

a) $2y = 10$　　b) $10w = 5$　　c) $0.7a = 77$　　d) $3\frac{1}{4}h = 32\frac{1}{2}$

$\mathbf{D_2}\quad \dfrac{2y}{2} = \dfrac{10}{2}$　$\mathbf{D_{10}}\quad \dfrac{10w}{10} = \dfrac{5}{10}$　$\mathbf{D_{0.7}}\quad \dfrac{0.7a}{0.7} = \dfrac{77}{0.7}$　$\mathbf{D_{3.25}}\quad \dfrac{3.25b}{3.25} = \dfrac{32.5}{3.25}$

Ans.　$y = 5$　Ans.　$w = \frac{1}{2}$　Ans.　$a = 110$　Ans.　$b = 10$

1.4. Using Multiplication (M) to Undo Division (D)

Solve:

a) $\dfrac{n}{3} = 6$　　b) $\dfrac{6}{n} = 1$　　c) $7 = \dfrac{1}{8}z$　　d) $\dfrac{b}{0.5} = 100$

$\mathbf{M_3}\quad 3\left(\dfrac{n}{3}\right) = 6\cdot 3$　$\mathbf{M_n}\quad n\left(\dfrac{6}{n}\right) = 1\cdot n$　$\mathbf{M_8}\quad 8\cdot 7 = 8\left(\dfrac{1}{8}z\right)$　$\mathbf{M_{0.5}}\quad 0.5\left(\dfrac{b}{0.5}\right) = 100(0.5)$

Ans.　$n = 18$　Ans.　$6 = n$　Ans.　$56 = z$　Ans.　$b = 50$

1.5. Using Two Operations to Solve Equations

Solve:

a) $5x + 2 = 17$　　b) $\dfrac{x}{6} - 2 = 8$　　c) $12 - x = 8$　　d) $\dfrac{12}{x} = 8$

$\mathbf{S_2}\quad \underline{\hspace{0.5cm}2 = \hspace{0.3cm}2}$　$\mathbf{A_2}\quad \underline{\hspace{1cm}2 = \hspace{0.3cm}2}$　$\mathbf{A_x}\quad \underline{\hspace{0.5cm}x = \hspace{0.3cm}+\,x}$　$\mathbf{M_x}\quad x\left(\dfrac{12}{x}\right) = 8x$

$\phantom{\mathbf{S_2}}\quad 5x \hspace{0.5cm}= 15$　$\phantom{\mathbf{A_2}}\quad \dfrac{x}{6} \hspace{0.5cm}= 10$　$\phantom{\mathbf{A_x}}\quad 12 \hspace{0.5cm}= 8 + x$　$\phantom{\mathbf{M_x}}\quad 12 = 8x$

　　$\mathbf{S_8}\quad \underline{\hspace{0.5cm}8 = 8}$　$\mathbf{D_8}\quad \dfrac{12}{8} = \dfrac{8x}{8}$

$\mathbf{D_5}\quad \dfrac{5x}{5} = \dfrac{15}{5}$　$\mathbf{M_6}\quad 6\left(\dfrac{x}{6}\right) = 10(6)$　Ans.　$4 = x$

Ans.　$x = 3$　Ans.　$x = 60$　　Ans.　$1\frac{1}{2} = x$

1.6. Equations Containing More Than One Term of the Unknown

Solve:

a) $8y = 3y + 35$　　b) $3y + y + 70 = 11y$　　c) $5y - 2 = 2y + 22$

$\mathbf{S_{3y}}\quad \dfrac{3y = 3y}{5y = \hspace{0.6cm}35}$

Combine:　$4y + 70 = 11y$　　$\mathbf{S_{2y}}\quad \dfrac{2y \hspace{0.5cm}= 2y}{3y - 2 = \hspace{0.6cm}22}$

$\mathbf{S_{4y}}\quad \underline{4y \hspace{1cm}= \hspace{0.3cm}4y}$

 $70 = 7y$

$\mathbf{A_2}\quad \underline{\hspace{1cm}2 = \hspace{0.3cm}2}$

$\mathbf{D_5}\quad \dfrac{5y}{5} = \dfrac{35}{5}$　$\mathbf{D_7}\quad \dfrac{70}{7} = \dfrac{7y}{7}$　$\phantom{\mathbf{A_2}}\quad 3y \hspace{0.5cm}= \hspace{0.6cm}24$

Ans.　$y = 7$　Ans.　$10 = y$　$\mathbf{D_3}\quad \dfrac{3y}{3} = \dfrac{24}{3}$

　　Ans.　$y = 8$

1.7. Equations Having Fractional Coefficients

Solve:

a) $\frac{1}{8}a = 3$　　　　b) $\frac{3}{8}b = 12$

$\mathbf{M_8}\quad 8(\frac{1}{8}a) = 3(8)$　　$\mathbf{M_{8/3}}\quad \frac{8}{3}(\frac{3}{8}b) = 12(\frac{8}{3})$

Ans.　$a = 24$　　Ans.　$b = 32$

$c)\quad \frac{2}{3}c - 3 = 7$

$\mathbf{A}_3\qquad\qquad \underline{3 = 3}$

$\qquad\qquad\quad \frac{2}{3}c = 10$

$\mathbf{M}_{3/2}\qquad \frac{3}{2}(\frac{2}{3}c) = \frac{3}{2}\cdot 10$

Ans.$\qquad\qquad c = 15$

$d)\quad \frac{8}{5}d + 12 = 36$

$\mathbf{S}_{12}\qquad\qquad \underline{12 = 12}$

$\qquad\qquad\quad \frac{8}{5}d \quad = 24$

$\mathbf{M}_{5/8}\qquad \frac{5}{8}(\frac{8}{5}d) = 24(\frac{5}{8})$

Ans.$\qquad\qquad d = 15$

1.8. NUMBER PROBLEMS LEADING TO FIRST-DEGREE EQUATIONS

a) Twice a number is equal to 81 less than 5 times the same number. Find the number.

b) Three times a number decreased by 8 equals the number increased by 12. Find the number.

Solutions:

a) Let n = the number

$\mathbf{A}_{81}\qquad\quad 2n = 5n - 81$

$\mathbf{S}_{2n}\qquad 2n + 81 = 5n$

$\mathbf{D}_3\qquad\qquad 81 = 3n$

$\qquad\qquad\quad 27 = n$

Ans. The number is 27.

b) Let n = the number

$\mathbf{A}_8\qquad 3n - 8 = n + 12$

$\mathbf{S}_n\qquad\quad 3n = n + 20$

$\mathbf{D}_2\qquad\quad 2n = 20$

$\qquad\qquad n = 10$

Ans. The number is 10.

2. SOLVING FIRST-DEGREE EQUATIONS HAVING NEGATIVE SOLUTIONS

To change $-x$ to $+x$, multiply by -1 or divide by -1.

Thus, if $\qquad -x = +5$

$\mathbf{M}_{-1}\ (-1)(-x) = (-1)(+5)$

$\qquad\qquad +x = -5$

2.1. USING ADDITION OR SUBTRACTION TO SOLVE EQUATIONS

Solve:

a)$\qquad n + 8 = \quad 2$

$\mathbf{S}_8\qquad\qquad \underline{8 = \quad 8}$

Ans.$\qquad\quad n = -6$

b)$\qquad n - 8 = -13$

$\mathbf{A}_8\qquad\qquad \underline{8 = \quad 8}$

Ans.$\quad n\qquad = -5$

c)$\qquad n - 15 = 2n - 9$

$\mathbf{S}_{n-9}\ \ \underline{n - 9 = \quad n - 9}$

Ans.$\qquad\quad -6 = \quad n$

d)$\qquad 14 = -n + 10$

$\mathbf{A}_{n-14}\quad \underline{n - 14 = \qquad n - 14}$

Ans.$\quad n\qquad = \qquad -4$

2.2. USING MULTIPLICATION OR DIVISION TO SOLVE EQUATIONS

Solve:

a)$\quad 2x = -8$

$\mathbf{D}_2\qquad \frac{2x}{2} = \frac{-8}{2}$

Ans.$\quad x = -4$

b)$\quad \frac{x}{2} = -8$

$\mathbf{M}_2\quad 2\left(\frac{x}{2}\right) = (-8)2$

Ans.$\quad x = -16$

c)$\quad -12x = 3$

$\mathbf{D}_{-12}\quad \frac{-12x}{-12} = \frac{3}{-12}$

Ans.$\quad x = -\frac{1}{4}$

d)$\quad -81 = 2.7x$

$\mathbf{D}_{2.7}\quad \frac{-81}{2.7} = \frac{2.7x}{2.7}$

Ans.$\quad -30 = x$

2.3. USING TWO OPERATIONS TO SOLVE EQUATIONS

Solve:

$a)$ $3y + 20 = 11$ $b)$ $8 - 3y = 29$ $c)$ $\dfrac{18}{y} = -6$ $d)$ $20 + 5y = 2y$

S_{20} $\dfrac{20 = 20}{3y = -9}$ S_8 $\dfrac{8 = 8}{-3y = 21}$ M_y $\dfrac{18}{y} \cdot y = -6y$ S_{5y} $\dfrac{5y = 5y}{20 = -3y}$

D_3 $\dfrac{3y}{3} = \dfrac{-9}{3}$ D_{-3} $\dfrac{-3y}{-3} = \dfrac{21}{-3}$ $18 = -6y$ D_{-3} $\dfrac{20}{-3} = \dfrac{-3y}{-3}$

Ans. $y = -3$ *Ans.* $y = -7$ D_{-6} $\dfrac{18}{-6} = \dfrac{-6y}{-6}$ *Ans.* $-6\frac{2}{3} = y$

                                        Ans. $-3 = y$

2.4. EQUATIONS HAVING FRACTIONAL COEFFICIENTS

Solve:

$a)$ $5 + \dfrac{y}{3} = 3$ $b)$ $\dfrac{3y}{4} - 14 = -23$ $c)$ $5\frac{3}{4}y = -10 + 5\frac{1}{2}y$

S_5 $\dfrac{5 = 5}{\dfrac{y}{3} = -2}$ A_{14} $\dfrac{ 14 = 14}{\dfrac{3y}{4} = -9}$ $S_{5(1/2)y}$ $\dfrac{5\frac{1}{2}y = 5\frac{1}{2}y}{\frac{1}{4}y = -10}$

M_3 $3\left(\dfrac{y}{3}\right) = (-2)3$ $M_{4/3}$ $\dfrac{4}{3}\left(\dfrac{3y}{4}\right) = (-9)\left(\dfrac{4}{3}\right)$ M_4 $4(\frac{1}{4}y) = (-10)\,4$

Ans. $y = -6$ *Ans.* $y = -12$ *Ans.* $y = -40$

2.5. MORE DIFFICULT EQUATIONS

Solve:

$a)$ $5y - 2y + 11 = 8y + 26$ $b)$ $4\frac{3}{4}a + 11 = 3a - 24$

Combine: $3y + 11 = 8y + 26$ $S_{3a + 11}$ $\dfrac{3a + 11 = 3a + 11}{1\frac{3}{4}a = -35}$

$S_{3y + 26}$ $\dfrac{3y + 26 = 3y + 26}{-15 = 5y}$ $M_{4/7}$ $\frac{4}{7}(\frac{7}{4}a) = (-35)(\frac{4}{7})$

D_5 $\dfrac{-15}{5} = \dfrac{5y}{5}$ *Ans.* $a = -20$

Ans. $-3 = y$

2.6. PROBLEMS HAVING NEGATIVE ROOTS

$a)$ The sum of three numbers, represented by n, $n - 4$, and $3n + 7$, is -2. Find the numbers.

$b)$ A merchant's profits in dollars on three articles are represented by p, $p - 3$, and $2p - 4$. If the total profit is \$1, find the profit on each.

Solutions:

$a)$ $n + (n - 4) + (3n + 7) = -2$
      $n + n - 4 + 3n + 7 = -2$
S_3 $5n + 3 = -2$
D_5 $5n = -5$
      $n = -1\begin{cases} n - 4 = -5 \\ 3n + 7 = 4 \end{cases}$

$b)$ $p + (p - 3) + (2p - 4) = 1$
           $p + p - 3 + 2p - 4 = 1$
A_7 $4p - 7 = 1$
D_4 $4p = 8$
      $p = 2\begin{cases} p - 3 = -1 \\ 2p - 4 = 0 \end{cases}$

Ans. Numbers are, -1, -5, and 4.   *Ans.* The merchant made \$2 on the first article, lost \$1 on the second, and made no profit on the third.

3. SOLVING EQUATIONS BY TRANSPOSING

Rule of Transposition

To transpose a term from one side of an equation to another, change its sign.

Thus, by transposing, $3x = 24 - x$ is changed into $3x + x = 24$.

In solving equations, **Tr** will indicate "transpose" or by "transposing."

Opposites are terms differing only in signs. Thus, $4xy$ and $-4xy$ are opposites.

The rule of transposition is based on the fact that a term may be eliminated from one side of an equation if its opposite is added to the other.

Thus,

$$13x = 24 + 5x$$
$$\underline{-5x = \quad\;\; -5x}$$
$$13x - 5x = 24$$

$\left\{\begin{array}{l} \text{Note how } 5x \text{ has been eliminated from the right} \\ \text{side and its opposite } -5x \text{ added to the left.} \end{array}\right.$

By using transposition, the second step becomes unnecessary.

To Solve Equations by Transposing Terms

Solve: $8a - 55 = 3a - 40$

Procedure:

1. Transpose (**Tr**) so that like terms will be on the same side of equation:
   (Change signs of transposed terms.)
2. Combine like terms:

3. Divide by the coefficient of the unknown:

Solutions:

1. $\quad 8a - 55 = 3a - 40$
    Tr $\;8a - 3a = 55 - 40$

2. $\qquad\qquad 5a = 15$

3. $\textbf{D}_5 \qquad \dfrac{5a}{5} = \dfrac{15}{5}$

Ans. $\qquad\qquad a = 3$

3.1. TRANSPOSING TERMS IN AN EQUATION

Solve, using transposition:

a) $\quad 31 - 6x = 25$
Tr $\;\; 31 - 25 = 6x$
$\textbf{D}_6 \qquad\quad 6 = 6x$
Ans. $\qquad\;\; 1 = x$

b) $\quad 2y - 28 = 42 - 5y$
Tr $\;\; 2y + 5y = 42 + 28$
$\textbf{D}_7 \qquad\;\; 7y = 70$
Ans. $\qquad\quad y = 10$

c) $\quad 11z + 3 - 4z = 5z$
Tr $\;\; 11z - 4z - 5z = -3$
$\textbf{D}_2 \qquad\qquad 2z = -3$
Ans. $\qquad\qquad z = -1\frac{1}{2}$

3.2. MORE DIFFICULT EQUATIONS

Solve and check:

a) $\qquad\quad 5a - 7 + 4a = 2a - 28 + 4a$
Tr $\;\; 5a + 4a - 4a - 2a = 7 - 28$
$\textbf{D}_3 \qquad\qquad\qquad 3a = -21$
$\qquad\qquad\qquad\quad a = -7$

b) $\qquad\quad n^2 + 3n - 8 = n^2 - 5n - 32$
Tr $\;\; n^2 - n^2 + 3n + 5n = -32 + 8$
$\textbf{D}_8 \qquad\qquad\qquad 8n = -24$
$\qquad\qquad\qquad\quad n = -3$

Check:

$$5a - 7 + 4a = 2a - 28 + 4a$$
$$5(-7) - 7 + 4(-7) \overset{?}{=} 2(-7) - 28 + 4(-7)$$
$$-35 - 7 - 28 \overset{?}{=} -14 - 28 - 28$$
$$-70 = -70$$

Ans. $a = -7$

Check:

$$n^2 + 3n - 8 = n^2 - 5n - 32$$
$$(-3)^2 + 3(-3) - 8 \overset{?}{=} (-3)^2 - 5(-3) - 32$$
$$9 - 9 - 8 \overset{?}{=} 9 + 15 - 32$$
$$-8 = -8$$

Ans. $n = -3$

4. SOLVING EQUATIONS CONTAINING PARENTHESES

To Solve Equations by Removing Parentheses

Solve: $8 + 2(x - 5) = 14$

Procedure:

1. Remove parentheses:

2. Solve the resulting equation:

Solutions:

1. $8 + 2(x - 5) = 14$
$$8 + 2x - 10 = 14$$

2. $2x - 2 = 14$

Ans. $x = 8$

4.1. REMOVING PARENTHESES TO SOLVE EQUATIONS

Remove parentheses, then solve:

a) Remove (   ):$3(n + 2) = 30 + n$
Tr $3n + 6 = 30 + n$
$$3n - n = 30 - 6$$
$$2n = 24$$

Ans. $n = 12$

b) Remove (   ):$-18 = 10 - 2(3 - x)$
$$-18 = 10 - 6 + 2x$$
$$-18 = 4 + 2x$$
$$-22 = 2x$$

Ans. $-11 = x$

4.2. MORE DIFFICULT EQUATIONS WITH PARENTHESES

Solve and check:

a) Remove(   ): $6y(y - 2) = 3y(2y - 1) + 27$

Tr $6y^2 - 12y = 6y^2 - 3y + 27$
Combine: $6y^2 - 6y^2 - 12y + 3y = 27$
$$-9y = 27$$
$$y = -3$$

Check:

$$6y(y - 2) = 3y(2y - 1) + 27$$
$$6(-3)(-5) \overset{?}{=} 3(-3)(-7) + 27$$
$$90 \overset{?}{=} 63 + 27$$
$$90 = 90$$

Ans. $y = -3$

b) Remove(   ): $(a + 1)(a - 5) + 7 = 44 - a(7 - a)$
Tr $a^2 - 4a - 5 + 7 = 44 - 7a + a^2$
Combine: $a^2 - a^2 - 4a + 7a = 44 + 5 - 7$
$$3a = 42$$
$$a = 14$$

Check:

$$(a + 1)(a - 5) + 7 = 44 - a(7 - a)$$
$$(15)(9) + 7 \overset{?}{=} 44 - 14(-7)$$
$$135 + 7 \overset{?}{=} 44 + 98$$
$$142 = 142$$

Ans. $a = 14$

4.3. and 4.4.   Problems Leading to Equations with Parentheses

4.3.   *a)* Find a number if twice the sum of the number and 7 equals 3 times the difference of the number and 10.

b) Find a number if 20 minus twice the number equals 3 times the sum of twice the number and 20.

Solutions:

a)   Let n = number.

Remove ( ): $2(n + 7) = 3(n - 10)$

Tr $2n + 14 = 3n - 30$

$14 + 30 = 3n - 2n$

$44 = n$

Ans.   Number is 44.

b)   Let n = number.

Remove( ): $20 - 2n = 3(2n + 20)$

Tr $20 - 2n = 6n + 60$

$20 - 60 = 6n + 2n$

$-40 = 8n$ or $n = -5$

Ans.   Number is -5.

4.4.   Marsha, José, and Gretchen earned \$120 together. José earned \$20 less than Marsha, and Gretchen earned twice as much as José. Find the earnings of each.

Solution:

$$\text{Let } n = \text{Marsha's earnings in dollars}$$
$$n - 20 = \text{José's earnings in dollars}$$
$$2(n - 20) = \text{Gretchen's earnings in dollars}$$

Remove ( ): $n + (n - 20) + 2(n - 20) = 120$

$n + n - 20 + 2n - 40 = 120$

$4n - 60 = 120$

$4n = 180$

$n = 45 \quad \begin{cases} n - 20 = 25 \\ 2(n - 20) = 50 \end{cases}$

Ans.   Marsha, José, and Gretchen earned \$45, \$25, and \$50, respectively.

5.   SOLVING EQUATIONS CONTAINING ONE FRACTION OR FRACTIONS HAVING THE SAME DENOMINATOR

To Solve Equations Having Same Denominator by Clearing of Fractions

Solve: *a)* $\dfrac{x}{3} + 5 = 2x$ *b)* $\dfrac{x}{5} + 6 = \dfrac{7x}{5}$

Procedure:

1. Clear of fractions by multiplying both sides of the equation by the denominator:

2. Solve the resulting equation:

Solutions:

1. $\dfrac{x}{3} + 5 = 2x$

Multiply by denominator 3:

$\mathbf{M_3} \quad 3\left(\dfrac{x}{3} + 5\right) = 3(2x)$

$x + 15 = 6x$

Ans. $3 = x$

1. $\dfrac{x}{5} + 6 = \dfrac{7x}{5}$

Multiply by denominator 5:

$\mathbf{M_5} \quad 5\left(\dfrac{x}{5} + 6\right) = 5\left(\dfrac{7x}{5}\right)$

$x + 30 = 7x$

Ans. $5 = x$

5.1. Fractional Equations Having the Same Denominator

Solve:

a) $\quad \dfrac{3x}{7} - 2 = \dfrac{x}{7}$

$\mathbf{M_7} \quad 7\left(\dfrac{3x}{7} - 2\right) = 7\left(\dfrac{x}{7}\right)$

$\quad\quad 3x - 14 = x$

$\quad\quad\quad 2x = 14$

Ans. $\quad\quad x = 7$

b) $\quad \dfrac{4}{y} = 5 - \dfrac{1}{y}$

$\mathbf{M_y} \quad y\left(\dfrac{4}{y}\right) = y\left(5 - \dfrac{1}{y}\right)$

$\quad\quad 4 = 5y - 1$

$\quad\quad 5 = 5y$

Ans. $\quad 1 = y$

c) $\quad \dfrac{5x}{4} - \dfrac{3}{4} = -12$

$\mathbf{M_4} \quad 4\left(\dfrac{5x}{4} - \dfrac{3}{4}\right) = 4(-12)$

$\quad\quad 5x - 3 = -48$

$\quad\quad\quad 5x = -45$

Ans. $\quad\quad x = -9$

5.2. FRACTIONAL EQUATIONS HAVING BINOMIAL NUMERATOR OR DENOMINATOR

Solve:

a) $\quad \dfrac{2x + 7}{4} = x - \dfrac{3}{4}$

$\mathbf{M_4} \quad 4\left(\dfrac{2x + 7}{4}\right) = 4\left(x - \dfrac{3}{4}\right)$

$\quad\quad 2x + 7 = 4x - 3$

$\quad\quad\quad 10 = 2x$

Ans. $\quad\quad 5 = x$

b) $\quad 10 = \dfrac{5x}{2x - 3}$

$\mathbf{M_{(2x-3)}} \quad 10(2x - 3) = \left(\dfrac{5x}{2x - 3}\right)(2x - 3)$

$\quad\quad 20x - 30 = 5x$

$\quad\quad 15x = 30$

Ans. $\quad\quad x = 2$

c) $\quad \dfrac{x - 3}{x} - 5 = \dfrac{x + 7}{x}$

$\mathbf{M_x} \quad x\left(\dfrac{x - 3}{x} - 5\right) = \left(\dfrac{x + 7}{x}\right)x$

$\quad\quad x - 3 - 5x = x + 7$

$\quad\quad\quad -5x = 10$

$\quad\quad\quad\quad x = -2$

6. SOLVING EQUATIONS CONTAINING FRACTIONS HAVING DIFFERENT DENOMINATORS: LOWEST COMMON DENOMINATOR

The **lowest common denominator (LCD)** of two or more fractions is the smallest number divisible by their denominators without a remainder.

Thus, in $\frac{1}{2} + x/3 = \frac{7}{4}$, 12 is the **LCD**, since 12 is the smallest number divisible by 2, 3, and 4 without a remainder. Larger common denominators of 2, 3, and 4 are 24, 36, 48, etc.

To Solve Equations Having Different Denominators by Clearing of Fractions

Solve: $\quad \dfrac{x}{2} + \dfrac{x}{3} = 20$

Procedure:

1. Clear of fractions by multiplying both sides of the equation by the LCD:

2. Solve the resulting equation:

Solutions:

1. $\quad\quad\quad$ LCD $= 6$

$\mathbf{M_6} \quad 6\left(\dfrac{x}{2} + \dfrac{x}{3}\right) = 6(20)$

2. $\quad\quad 3x + 2x = 120$

$\quad\quad\quad\quad 5x = 120$

Ans. $\quad\quad\quad x = 24$

6.1. FRACTIONAL EQUATIONS HAVING NUMERICAL DENOMINATORS

Solve:

a) $\dfrac{x}{2} - \dfrac{x}{3} = 5$ b) $\dfrac{a}{2} - \dfrac{a}{3} - \dfrac{a}{5} = 2$ c) $\dfrac{3y}{4} - \dfrac{y}{3} = 10$

LCD = 6   LCD = 30   LCD = 12

$\mathbf{M}_6 \quad 6\left(\dfrac{x}{2} - \dfrac{x}{3}\right) = 6(5)$ $\mathbf{M}_{30} \quad 30\left(\dfrac{a}{2} - \dfrac{a}{3} - \dfrac{a}{5}\right) = 30(2)$ $\mathbf{M}_{12} \quad 12\left(\dfrac{3y}{4} - \dfrac{y}{3}\right) = 12(10)$

$\qquad\qquad 3x - 2x = 30$ $\qquad\qquad 15a - 10a - 6a = 60$ $\qquad\qquad 9y - 4y = 120$

$\qquad\qquad\qquad\qquad\qquad\qquad\qquad\qquad\qquad\qquad\qquad\qquad\qquad\qquad\qquad\qquad\qquad 5y = 120$

Ans. $\qquad x = 30$ *Ans.* $\qquad a = -60$ *Ans.* $\qquad y = 24$

6.2. FRACTIONAL EQUATIONS HAVING LITERAL DENOMINATORS

Solve:

a) $\dfrac{10}{x} = \dfrac{25}{3x} - \dfrac{1}{3}$ b) $\dfrac{8}{a} - 3 = \dfrac{7}{2a}$ c) $\dfrac{5}{6} = \dfrac{7}{3x} + 1$

LCD = $3x$ LCD = $2a$ LCD = $6x$

$\mathbf{M}_{3x} \quad 3x\left(\dfrac{10}{x}\right) = 3x\left(\dfrac{25}{3x} - \dfrac{1}{3}\right)$ $\mathbf{M}_{2a} \quad 2a\left(\dfrac{8}{a} - 3\right) = 2a\left(\dfrac{7}{2a}\right)$ $\mathbf{M}_{6x} \quad 6x\left(\dfrac{5}{6}\right) = 6x\left(\dfrac{7}{3x} + 1\right)$

$\qquad\qquad 30 = 25 - x$ $\qquad\qquad 16 - 6a = 7$ $\qquad\qquad 5x = 14 + 6x$

Ans. $\qquad x = -5$ *Ans.* $\qquad a = 1\frac{1}{2}$ *Ans.* $\qquad x = -14$

6.3. FRACTIONAL EQUATIONS HAVING BINOMIAL NUMERATORS

Solve:

a) $\dfrac{6x + 13}{2} + \dfrac{x + 3}{3} = \dfrac{5}{6}$ b) $\dfrac{5y + 4}{9} = 2 + \dfrac{2y + 4}{6}$

LCD = 6   LCD = 18

$\mathbf{M}_6 \quad 6\left(\dfrac{6x + 13}{2}\right) + 6\left(\dfrac{x + 3}{3}\right) = 6\left(\dfrac{5}{6}\right)$ $\mathbf{M}_{18} \quad 18\left(\dfrac{5y + 4}{9}\right) = 18(2) + 18\left(\dfrac{2y + 4}{6}\right)$

$\qquad\qquad 18x + 39 + 2x + 6 = 5$ $\qquad\qquad 10y + 8 = 36 + 6y + 12$

Ans. $\qquad\qquad x = -2$ *Ans.* $\qquad y = 10$

6.4. FRACTIONAL EQUATIONS HAVING BINOMIAL DENOMINATORS

Solve:

a)

$\qquad\qquad\qquad\qquad\qquad\qquad\qquad \dfrac{3}{8} = \dfrac{6}{5 - y}$

LCD = $8(5 - y)$

$\mathbf{M}_{\text{LCD}}$

$\qquad\qquad\qquad\qquad\qquad\qquad \left(\dfrac{3}{8}\right)8(5 - y) = 8(5 - y)\left(\dfrac{6}{5 - y}\right)$

$\qquad\qquad\qquad\qquad\qquad\qquad\qquad 3(5 - y) = 48$

Ans. $\qquad\qquad\qquad\qquad\qquad\qquad y = -11$

b)

$\qquad\qquad\qquad\qquad\qquad\qquad \dfrac{6}{5} + \dfrac{3}{w - 3} = \dfrac{9}{5(w - 3)}$

$$\text{LCD} = 5(w - 3)$$

$$\mathbf{M}_{\text{LCD}} \quad \left(\frac{6}{5}\right)(5)(w - 3) + 5(w - 3)\left(\frac{3}{w - 3}\right) = \frac{9}{5(w - 3)}(5)(w - 3)$$

$$6(w - 3) + 15 = 9$$

Ans. $\qquad\qquad\qquad\qquad\qquad\qquad w = 2$

7. SOLVING EQUATIONS CONTAINING DECIMALS

A **decimal** may be written as a fraction whose denominator is 10, 100, or a power of 10. Thus, the denominator of 0.003 or $\frac{3}{1000}$ is 1000.

To Solve an Equation Having Decimals

Solve: $0.15x + 7 = 0.5x$

Procedure:

1. Clear of decimals by multiplying both sides of the equation by the denominator of the decimal having the greatest number of decimal places:
2. Solve the resulting equation:

Solutions:

1. $0.15x$ has more decimal places than $0.5x$. The denominator of $0.15x$ is 100.

$\mathbf{M}_{100} \quad 100(0.15x + 7) = 100(0.5x)$

2. $\qquad\qquad 15x + 700 = 50x$

Ans. $\qquad\qquad\quad x = 20$

In some cases, it may be better not to clear an equation of decimals. Thus, if $3a = 0.54$, simply divide by 3 to obtain $a = 0.18$. Or, if $2a - 0.28 = 0.44$, then $2a = 0.72$ and $a = 0.36$. These are cases where the coefficient of the unknown is not a decimal.

7.1. EQUATIONS WITH ONE DECIMAL

First clear each equation of decimals, then solve:

a) $\qquad 0.3a = 6$
$\mathbf{M}_{10} \quad 10(0.3a) = 60$
$\qquad\qquad 3a = 60$
Ans. $\qquad a = 20$

b) $\qquad\quad 8 = 0.05b$
$\mathbf{M}_{100} \quad 100(8) = 100(0.05b)$
$\qquad\qquad 800 = 5b$
Ans. $\quad 160 = b$

c) $\qquad 2.8c = 54 + c$
$\mathbf{M}_{10} \quad 10(2.8c) = 10(54 + c)$
$\qquad\qquad 28c = 540 + 10c$
Ans. $\qquad c = 30$

7.2. SOLVING EQUATIONS WITHOUT CLEARING OF DECIMALS

Solve, without clearing of decimals:

a) $\quad 3a = 0.6$
$\mathbf{D}_3 \quad \dfrac{3a}{3} = \dfrac{0.6}{3}$
Ans. $\quad a = 0.2$

b) $\; 5r - 5 = 0.05$
$\mathbf{D}_5 \quad \dfrac{5r}{5} = \dfrac{5.05}{5}$
Ans. $\quad r = 1.01$

c) $\quad \dfrac{x}{4} = 0.28$
$\mathbf{M}_4 \quad 4\left(\dfrac{x}{4}\right) = 4(0.28)$
Ans. $\quad x = 1.12$

d) $\quad \dfrac{x}{5} + 2 = 3.5$
$\mathbf{M}_5 \quad 5\left(\dfrac{x}{5}\right) = (1.5)5$
Ans. $\quad x = 7.5$

7.3. EQUATIONS WITH TWO OR MORE DECIMALS

Solve. (Multiply by the denominator of the decimal with the largest number of decimal places.)

$a)$	$0.05x = 2.5$	$b)$	$-0.9 = 0.003y$	$c)$	$0.2a = a - 0.8$
$\mathbf{M}_{100}$	$5x = 250$	$\mathbf{M}_{1000}$	$-900 = 3y$	$\mathbf{M}_{10}$	$2a = 10a - 8$
Ans.	$x = 50$	*Ans.*	$-300 = y$	*Ans.*	$a = 1$

$d)$	$0.5a - 3.5 = 0.75$
$\mathbf{M}_{100}$	$50a - 350 = 75$
Ans.	$a = 8.5$

7.4. EQUATIONS CONTAINING PERCENTS

Solve:

$a)$	25% of $x = 10$	$b)$	$x + 40\%x = 56$	$c)$	$x - 16\%x = 420$
$(25\% = \frac{1}{4})$		$(40\% = 0.4)$		$(16\% = 0.16)$	
	$\dfrac{x}{4} = 10$		$x + 0.4x = 56$		$x - 0.16x = 420$
Ans.	$x = 40$	*Ans.*	$x = 40$	*Ans.*	$x = 500$

7.5. EQUATIONS CONTAINING DECIMALS AND PARENTHESES

Solve:

$a)$	$0.3(50 - x) = 6$	$b)$	$0.8 = 0.02(x - 35)$	$c)$	$5(x + 0.8) = -16$
$\mathbf{M}_{10}$	$3(50 - x) = 60$	$\mathbf{M}_{100}$	$80 = 2(x - 35)$	$\mathbf{M}_{10}$	$50(x + 0.8) = -160$
	$150 - 3x = 60$		$80 = 2x - 70$		$50x + 40 = -160$
Ans.	$x = 30$	*Ans.*	$75 = x$	*Ans.*	$x = -4$

7.6. MORE DIFFICULT DECIMAL EQUATIONS

Solve:

$a)$	$0.04x + 0.03(5000 - x) = 190$	$b)$	$0.3(x - 200) + 0.03(1000 - x) = 105$
$\mathbf{M}_{100}$	$4x + 3(5000 - x) = 19{,}000$	$\mathbf{M}_{100}$	$30(x - 200) + 3(1000 - x) = 10{,}500$
	$4x + 15{,}000 - 3x = 19{,}000$		$30x - 6000 + 3000 - 3x = 10{,}500$
Ans.	$x = 4000$	*Ans.*	$x = 500$

8. SOLVING LITERAL EQUATIONS

Literal equations contain two or more letters.

Thus, $x + y = 20$, $5x = 15a$, $D = RT$, and $2x + 3y - 5z = 12$ are literal equations.

To solve a literal equation for any letter, follow the same procedure used in solving any equation for an unknown.

Thus, to solve for x in $5x = 25a$, divide both sides by 5 to obtain $x = 5a$.

Note that all formulas are examples of literal equations.

For example, $D = RT$ and $P = 2L + 2W$ are literal equations.

8.1. SOLVING LITERAL EQUATIONS BY USING ONE OPERATION

Solve for x:

$a)$ $x - y = 8$	$b)$ $x + 10 = h$	$c)$ $ax = b$	$d)$ $\dfrac{x}{a} = b$
$\mathbf{A}_y$ $\underline{\quad y = y \quad}$	$\mathbf{S}_{10}$ $\underline{\quad 10 = \quad 10}$	$\mathbf{D}_a$ $\dfrac{ax}{a} = \dfrac{b}{a}$	$\mathbf{M}_a$ $a\left(\dfrac{x}{a}\right) = a(b)$
Ans. $x = y + 8$	*Ans.* $x = h - 10$	*Ans.* $x = \dfrac{b}{a}$	*Ans.* $x = ab$

8.2. **SOLVING FOR ONE OF THE LETTERS IN A FORMULA**

Solve for the letter indicated:

$a)$ $RT = D$ $b)$ $S = C + P$ $c)$ $C = \dfrac{5}{9}(F \ \ 32)$ $d)$ $A - \dfrac{1}{2}bh$

Solve for R: Solve for C: Solve for $F(\mathbf{M}_{9/5}$ first):   Solve for $b(\mathbf{M}_2$ first):

$\mathbf{D}_T$ $\dfrac{RT}{T} = \dfrac{D}{T}$ $\mathbf{Tr}$ $S - P = C$ $\mathbf{Tr}$ $\dfrac{9}{5}C = F - 32$ $\mathbf{D}_h$ $2A = bh$

Ans. $R = \dfrac{D}{T}$ *Ans.* $S - P = C$ *Ans.* $\dfrac{9}{5}C + 32 = F$ *Ans.* $\dfrac{2A}{h} = b$

8.3. **SOLVING AND CHECKING LITERAL EQUATIONS**

Solve for y and check:

$a)$ $2y - 4a = 8a$ $b)$ $2y - 24a = 8y$ $c)$ $4b + 3y = 12b + y$

         $2y = 8a + 4a$ $2y - 8y = 24a$ $3y - y = 12b - 4b$

            $2y = 12a$ $-6y = 24a$ $2y = 8b$

             $y = 6a$ $y = -4a$ $y = 4b$

Check for $y = 6a$:   Check for $y = -4a$:   Check for $y = 4b$:

     $2y - 4a = 8a$ $2y - 24a = 8y$ $4b + 3y = 12b + y$

$2(6a) - 4a \overset{?}{=} 8a$ $2(-4a) - 24a \overset{?}{=} 8(-4a)$ $4b + 3(4b) \overset{?}{=} 12b + 4b$

   $12a - 4a \overset{?}{=} 8a$ $-8a - 24a \overset{?}{=} -32a$ $4b + 12b \overset{?}{=} 16b$

        $8a = 8a$ $-32a = -32a$ $16b = 16b$

8.4. **SOLVING A LITERAL EQUATION FOR EACH LETTER**

Solve for the letter indicated:

$a)$ $2x = 3y - 4z$ $b)$ $2x = 3y - 4z$ $c)$ $2x = 3y - 4z$

Solve for x: Solve for y, transposing first: Solve for z, transposing first:

$\mathbf{D}_2$ $2x = 3y - 4z$ $\mathbf{D}_3$ $2x + 4z = 3y$ $\mathbf{D}_4$ $4z - 3y - 2x$

Ans. $x = \dfrac{3y - 4z}{2}$ *Ans.* $\dfrac{2x + 4z}{3} = y$ *Ans.* $z = \dfrac{3y - 2x}{4}$

8.5. **SOLVING MORE DIFFICULT LITERAL EQUATIONS**

Solve for x or y:

$a)$ $3(x - 2b) = 9a - 15b$ $b)$ $\dfrac{y}{5} - h = f$ $c)$ $\dfrac{x}{a} - \dfrac{b}{5} = \dfrac{c}{10}$

      $3x - 6b = 9a - 15b$

$\mathbf{D}_3$ $3x = 9a - 9b$ Transpose first: LCD $= 10a$

Ans. $x = 3a - 3b$ $\mathbf{M}_5$ $\dfrac{y}{5} = f + h$ $\mathbf{M}_{10a}$ $10a\left(\dfrac{x}{a} - \dfrac{b}{5}\right) = \left(\dfrac{c}{10}\right)10a$

                                Ans. $y = 5f + 5h$ $\mathbf{Tr}$ $10x - 2ab = ac$

                                                    $\mathbf{D}_{10}$ $10x = 2ab + ac$

                                                    Ans. $x = \dfrac{2ab + ac}{10}$

9.   THE GRAPHING CALCULATOR

9.1. **GRAPHING CALCULATORS ARE USED TO SOLVE EQUATIONS.**
See Appendix B for further information pertaining to this topic.

Supplementary Problems

5.1. Solve: **(1.1)**

a) $n + 3 = 11$ *Ans.* $n = 8$ g) $n + 1\frac{1}{2} = 4\frac{1}{2}$ *Ans.* $n = 3$

b) $n + 7 = 20$ $n = 13$ h) $n + 2\frac{1}{8} = 5\frac{1}{2}$ $n = 3\frac{3}{8}$

c) $12 + n = 30$ $n = 29$ i) $2n + 5 = n + 12$ $n = 7$

d) $42 = n + 13$ $n = 29$ j) $n + 18 = 2n + 10$ $n = 8$

e) $n + 0.5 = 0.9$ $n = 0.4$ k) $n^2 + n = n^2 + 15$ $n = 15$

f) $n + 3.4 = 5.1$ $n = 1.7$ l) $2n^2 + 34 = 2n^2 + n$ $n = 34$

5.2. Solve: **(1.2)**

a) $x - 9 = 15$ *Ans.* $x = 24$ g) $x - 1\frac{1}{3} = 3\frac{2}{3}$ *Ans.* $x = 5$

b) $x - 20 = 50$ $x = 70$ h) $2\frac{1}{4} = x - 7\frac{1}{4}$ $x = 9\frac{1}{2}$

c) $17 = x - 13$ $x = 30$ i) $8 - x = 0$ $x = 8$

d) $100 = x - 41$ $x = 141$ j) $10 - x = 3$ $x = 7$

e) $x - 0.3 = 1.7$ $x = 2$ k) $10 - x = -4$ $x = 14$

f) $x - 5 = 8.3$ $x = 13.3$ l) $12 - x^2 = x - x^2$ $x = 12$

5.3. Solve: **(1.3)**

a) $3y = 15$ *Ans.* $y = 5$ e) $5y = 6.5$ *Ans.* $y = 1.3$ i) $1\frac{1}{2}y = 1\frac{1}{2}$ *Ans.* $y = 1$

b) $4y = 30$ $y = 7\frac{1}{2}$ f) $0.3y = 9$ $y = 30$ j) $2\frac{1}{2}y = 10$ $y = 4$

c) $36 = 12y$ $y = 3$ g) $1.1y = 8.8$ $y = 8$ k) $15 = 1\frac{1}{4}y$ $y = 12$

d) $6y = 3$ $y = \frac{1}{2}$ h) $4 = 0.4y$ $y = 10$ l) $14 = 2\frac{1}{3}y$ $y = 6$

5.4. Solve: **(1.4)**

a) $\dfrac{a}{4} = 3$ *Ans.* $a = 12$ e) $\frac{1}{3}a = 20$ *Ans.* $a = 60$ i) $\dfrac{5}{a} = 1$ *Ans.* $a = 5$

b) $\dfrac{a}{5} = 1$ $a = 5$ f) $\frac{1}{7}a = 12$ $a = 84$ j) $1 = \dfrac{78}{a}$ $a = 78$

c) $12 = \dfrac{a}{2}$ $a = 24$ g) $30 = \frac{1}{8}a$ $a = 240$ k) $\dfrac{a}{0.3} = 50$ $a = 15$

d) $25 = \dfrac{a}{10}$ $a = 250$ h) $5.7 = \frac{1}{10}a$ $a = 57$ l) $200 = \dfrac{a}{0.7}$ $a = 140$

5.5. Solve: **(1.5)**

a) $3x + 1 = 13$ *Ans.* $x = 4$ g) $3.4 = 1.4 + 2x$ *Ans.* $x = 1$

b) $5x + 3 = 33$ $x = 6$ h) $7.9 = 3.1 + 4x$ $x = 1.2$

c) $8 + 4x = 44$ $x = 9$ i) $8x + 3\frac{1}{4} = 19\frac{1}{4}$ $x = 2$

d) $11 + 7x = 88$ $x = 11$ j) $20x + 5\frac{7}{8} = 15\frac{7}{8}$ $x = \frac{1}{2}$

e) $76 = 10x + 6$ $x = 7$ k) $3x + 2\frac{1}{4} = 6$ $x = 1\frac{1}{4}$

f) $45 = 15 + 12x$ $x = 2\frac{1}{2}$ l) $4x + 3 = 5\frac{2}{3}$ $x = \frac{2}{3}$

5.6. Solve (use **A** and **D** or **A** and **M**):   **(1.5)**

a) $5b - 2 = 33$ *Ans.* $b = 7$ g) $7b - 2\frac{1}{2} = 4\frac{1}{2}$ *Ans.* $b = 1$

b) $6b - 8 = 25$ $b = 5\frac{1}{2}$ h) $2b - \frac{2}{5} = \frac{4}{5}$ $b = \frac{3}{5}$

c) $89 = 9b - 1$ $b = 10$ i) $\dfrac{b}{3} - 5 = 9$ $b = 42$

d) $42 = 10b - 3$ $b = 4\frac{1}{2}$ j) $\dfrac{b}{6} - 5 = 10$ $b = 90$

$e)$ $3b - 0.4 = 0.8$ $Ans.$ $b = 0.4$ $k)$ $\frac{1}{4}b - 9 = 12$ $Ans.$ $b = 84$

$f)$ $11b - 0.34 = 0.21$ $b = 0.05$ $l)$ $20 - \frac{1}{8}b - 11$ $b = 248$

5.7.   Solve (use **M** and **D** or **A** and **S**):   **(1.5)**

$a)$ $\dfrac{21}{r} = 3$ $Ans.$ $r = 7$ $e)$ $\dfrac{5.4}{r} = 3$ $Ans.$ $r = 1.8$ $i)$ $30 - r = 17$ $Ans.$ $r = 13$

$b)$ $\dfrac{14}{r} = 4$ $r = 3\frac{1}{2}$ $f)$ $0.4 = \dfrac{10}{r}$ $r = 25$ $j)$ $8.4 - r = 5.7$ $r = 2.7$

$c)$ $8 = \dfrac{24}{r}$ $r = 3$ $g)$ $\dfrac{5}{r} = 2\frac{1}{2}$ $r = 2$ $k)$ $5.24 = 8.29 - r$ $r = 3.05$

$d)$ $12 = \dfrac{3}{r}$ $r = \frac{1}{4}$ $h)$ $2\frac{1}{4} = \dfrac{9}{r}$ $r = 4$ $l)$ $17\frac{3}{4} = 20\frac{1}{4} - r$ $r = 2\frac{1}{2}$

5.8.   Solve: **(1.6)**

$a)$ $6y = 2y + 16$ $Ans.$ $y = 4$ $g)$ $21 = 10y - 4y$ $Ans.$ $y = 3\frac{1}{2}$
$b)$ $10y = 30 + 5y$ $y = 6$ $h)$ $24y - 21y = 14$ $y = 4\frac{2}{3}$
$c)$ $18 + 2y = 11y$ $y = 2$ $i)$ $13y - 4 = 10y + 2$ $y = 2$
$d)$ $3y + 40 = 8y$ $y = 8$ $j)$ $20y + 16 = 30y - 44$ $y = 6$
$e)$ $2y + 7y - 72$ $y = 8$ $k)$ $2y - 1.7 = y + 1.4$ $y = 3.1$
$f)$ $21 = 10y - 4y$ $y = 3\frac{1}{2}$ $l)$ $7y - 0.8 = 4y - 0.2$ $y = 0.2$

5.9.   Solve: **(1.7)**

$a)$ $\frac{1}{5}t = 14$ $Ans.$ $t = 70$ $e)$ $\frac{1}{3}t + 9 = 14$ $Ans.$ $t = 15$ $i)$ $24 + \frac{1}{8}t = 31$ $Ans.$ $t = 56$
$b)$ $20 = \frac{1}{10}t$ $t = 200$ $f)$ $\frac{1}{4}t - 6 = 8$ $t = 56$ $j)$ $17 + \frac{3}{7}t = 29$ $t = 28$
$c)$ $\frac{2}{5}t = 16$ $t = 40$ $g)$ $\frac{2}{5}t + 7 = 13$ $t = 15$ $k)$ $40 = \frac{3}{2}t - 11$ $t = 34$
$d)$ $30 - \frac{6}{7}t$ $t = 35$ $h)$ $\frac{3}{7}t 6 = 6$ $t 28$ $l)$ $\frac{8}{7}t + 9 = \frac{3}{7}t + 39$ $t = 42$

5.10.   Four times a number increased by 45 equals 7 times the number. Find the number.   **(1.8)**

$Ans.$ 15

5.11.   Ten times a number decreased by 7 equals 8 times the number increased by 21.   **(1.8)**
Find the number. $Ans.$ 14

5.12.   Two-thirds of a number increased by 10 equals 24. Find the number. $Ans.$ 21   **(1.8)**

5.13.   Solve: **(2.1)**

$a)$ $n + 11 = 3$ $Ans.$ $n = -8$ $g)$ $n - 3\frac{1}{3} = -8\frac{1}{3}$ $Ans.$ $n = -5$
$b)$ $n + 25 = -5$ $n = -30$ $h)$ $n - \frac{4}{5} = -8$ $n = -7\frac{1}{5}$
$c)$ $30 + n = 24$ $n = -6$ $i)$ $n - 20 = 2n - 3$ $n = -17$
$d)$ $40 = n + 70$ $n = -30$ $j)$ $2n + 6 = n - 10$ $n = -16$
$e)$ $n - 5 = -12$ $n = -7$ $k)$ $-n + 2 = 5$ $n = -3$
$f)$ $n - 0.9 = -2$ $n = -1.1$ $l)$ $0.34 = 0.29 - n$ $n = -0.05$

5.14.   Solve: **(2.2)**

$a)$ $3x = -66$ $Ans.$ $x = -22$ $g)$ $-4.8 = \dfrac{x}{10}$ $Ans.$ $x = -48$

$b)$ $13x = -130$ $x = -10$ $h)$ $-0.13 = \dfrac{x}{6}$ $x = -0.78$

$c)$ $-50 = 15x$ $x = -3\frac{1}{3}$ $i)$ $-7x = 35$ $x = -5$

d) $-7 = 21x$　　　Ans.　$x = -\frac{1}{3}$　　　j) $-\dfrac{x}{4} = 120$　　　Ans.　$x = -480$

e) $\dfrac{x}{5} = -4$　　　　　$x = -20$　　　k) $3.1x = -31$　　　　　$x = -10$

f) $\dfrac{x}{2} = -3\frac{1}{2}$　　　　　$x = -7$　　　l) $-3\frac{1}{4}x = 130$　　　　$x = -40$

5.15. Solve:　　　　　　　　　　　　　　　　　　　　　　　　　　　　**(2.3)**

a) $2y + 16 = 2$　　　Ans.　$y = -7$　　　g) $13y = 6y - 84$　　　Ans.　$y = -12$

b) $10y + 42 = 37$　　　$y = -\frac{1}{2}$　　　h) $3y = 9y + 78$　　　$y = -13$

c) $3y - 5 = -17$　　　$y = -4$　　　i) $\dfrac{40}{y} = -5$　　　$y = -8$

d) $20 + 11y = 9$　　　$y = -1$　　　j) $-3 = \dfrac{18}{y}$　　　$y = -6$

e) $6y + 35 = y$　　　$y = -7$　　　k) $-\dfrac{20}{y} = 4$　　　$y = -5$

f) $8y - 20 = 10y$　　　$y = -10$　　　l) $6 = -\dfrac{3}{y}$　　　$y = -\frac{1}{2}$

5.16. Solve:　　　　　　　　　　　　　　　　　　　　　　　　　　　　**(2.4)**

a) $\dfrac{y}{4} + 6 = 5$　　　Ans.　$y = -4$　　　f) $24 = \dfrac{7y}{5} + 31$　　　Ans.　$y = -5$

b) $8 + \dfrac{y}{5} = -1$　　　$y = -45$　　　g) $y - \frac{1}{2}y = -20$　　　$y = -40$

c) $30 = 25 - \dfrac{y}{3}$　　　$y = -15$　　　h) $y + \frac{2}{3}y = -45$　　　$y = -27$

d) $\dfrac{2y}{3} + 7 = -7$　　　$y = -21$　　　i) $8\frac{1}{2}y + 6 = 7\frac{3}{4}y$　　　$y = -8$

e) $12 + \dfrac{2y}{5} = -8$　　　$y = -50$

5.17. Solve:　　　　　　　　　　　　　　　　　　　　　　　　　　　　**(2.5)**

a) $4y - 9y + 22 = 3y + 30$　　　Ans.　$y = -1$　　　c) $2\frac{1}{2}a + 10 = 4\frac{1}{4}a + 52$　　　Ans.　$a = -24$

b) $12y - 10 = 8 + 3y - 36$　　　$y = -2$　　　d) $5.4b - 14 = 8b + 38$　　　$b = -20$

5.18. The sum of two numbers represented by n and $2n + 8$ is -7. Find the numbers.　　　**(2.6)**

　Ans. -5 and -2

5.19. The sum of three numbers represented by x, $3x$, and $3 - 2x$ is 1. Find the numbers.　　　**(2.6)**

　Ans. -1, -3, and 5

5.20. Solve, using transposition:　　　　　　　　　　　　　　　　　　**(3.1)**

a) $7 - 2r = 3$　　　Ans.　$r = 2$　　　f) $40 - 9s = 3s + 64$　　　Ans.　$s = -2$

b) $27 = 30 - 6r$　　　　　$r = \frac{1}{2}$　　　g) $6t + t = 10 + 11t$　　　$t = -2\frac{1}{2}$

c) $10r + 37 = -23$　　　$r = -6$　　　h) $4t + 40 - 65 = -t$　　　$t = 5$

d) $4s - 8 = 16 - 2s$　　　$s = 4$　　　i) $20 + 8t = 40 - 22$　　　$t = -\frac{1}{4}$

e) $12 + s = 6s + 7$　　　$s = 1$

5.21. Solve:　　　　　　　　　　　　　　　　　　　　　　　　　　　　**(3.2)**

a) $8a + 1 + 3a = 7 + 9a - 12$　　　Ans.　$a = -3$　　　b) $n^2 - 6n + 1 = n^2 - 8n - 9$　　　Ans.　$n = -5$

5.22. Solve: **(4.1)**

a) $4(x + 1) - 20$	*Ans.* $x - 4$	*g)* $20 + 8(2 - y) - 44$	*Ans.*	$y - -1$
b) $3(x - 2) = -6$	$x = 0$	*h)* $12y - 3 = 5(2y + 1)$		$y = 4$
c) $5(7 - x) = 25$	$x = 2$	*i)* $3(z + 1) = 4(6 - z)$		$z = 3$
d) $42 = 7(2x - 1)$	$x = 3\frac{1}{2}$	*j)* $10(2 - z) = 4(z - 9)$		$z = 4$
e) $6(y - 1) = 7y - 12$	$y = 6$	*k)* $6(3z - 1) = -7(8 + z)$		$z = -2$
f) $30 - 2(y - 1) = 38$	$y = -3$	*l)* $2(z + 1) - 3(4z - 2) = 6z$		$z = \frac{1}{2}$

5.23. Solve: **(4.2)**

a) $3r(2r + 4) = 2r(3r + 8) - 12$ *Ans.* $r = 3$

b) $(s + 3)(s + 5) + s(10 - s) = 11s + 1$ *Ans.* $s = -2$

5.24. Find a number if twice the sum of the number and 4 equals 11 more than the number.   **(4.3)**

Ans. 3

5.25. Find a number such that 3 times the sum of the number and 2 equals 4 times the number decreased by 3.   **(4.3)**

Ans. 9

5.26. Find a number if 25 minus 3 times the number equals 8 times the difference obtained when 1 is subtracted from the number.   **(4.3)**

Ans. 3

5.27. Three boys earned \$60 together. Henry earned \$2 less than Ed, and Jack earned twice as much as Henry. Find their earnings.   **(4.4)**

Ans. Henry, Ed, and Jack earned \$14.50, \$16.50, and \$29, respectively.

5.28. Solve: **(5.1)**

a) $\dfrac{3x}{4} = 9$	*Ans.* $x = 12$	*f)* $\dfrac{10}{x} - 2 = 18$	*Ans.* $x = \frac{1}{2}$
b) $\dfrac{2x}{5} + 8 = 6$	$x = -5$	*g)* $\dfrac{y}{3} + 10 = y$	$y = 15$
c) $\dfrac{x}{3} - 5 = 5$	$x = 30$	*h)* $20 - \dfrac{3y}{5} = y - 12$	$y = 20$
d) $\dfrac{12}{x} = -3$	$x = -4$	*i)* $\dfrac{15}{3y} + 3 = 18$	$y = \frac{1}{3}$
e) $7 = \dfrac{84}{x}$	$x = 12$		

5.29. Solve: **(5.1)**

a) $\dfrac{2x}{5} + 6 = \dfrac{x}{5}$	*Ans.* $x = -30$	*d)* $\dfrac{6h}{5} + 6 = \dfrac{2h}{5}$	*Ans.* $h = -7\frac{1}{2}$
b) $10 - \dfrac{x}{7} = \dfrac{4x}{7}$	$x = 14$	*e)* $\dfrac{3}{r} = 2 - \dfrac{7}{r}$	$r = 5$
c) $\dfrac{4h}{3} - \dfrac{5h}{3} = -2$	$h = 6$	*f)* $\dfrac{1}{r} + 3 = \dfrac{9}{r} - \dfrac{2}{r}$	$r = 2$

5.30. Solve: **(5.2)**

a) $\dfrac{3x - 1}{7} = 2x + 3$ *b)* $\dfrac{y}{2y - 9} = 2$ *c)* $\dfrac{z - 6}{5} + z = \dfrac{4z + 16}{5}$

Ans.   *a)* $x = -2$ *b)* $y = 6$ *c)* $z = 11$

5.31. Solve: **(6.1)**

a) $\dfrac{x}{2} - \dfrac{x}{3} = 7$ *Ans.* $x = 42$ f) $10 + \dfrac{y}{6} = \dfrac{y}{3} - 4$ *Ans.* $y = 84$

b) $\dfrac{x}{5} + \dfrac{x}{6} = 11$ $x = 30$ g) $\dfrac{3x}{4} - \dfrac{2x}{3} = \dfrac{3}{4}$ $x = 9$

c) $\dfrac{x}{2} = 12 - \dfrac{x}{4}$ $x = 16$ h) $\dfrac{x}{2} = \dfrac{3x}{7} - 5$ $x = -70$

d) $\dfrac{y}{4} + \dfrac{y}{3} + \dfrac{y}{2} = 26$ $y = 24$ i) $\dfrac{5x}{2} - \dfrac{2x}{3} = -\dfrac{11}{6}$ $x = -1$

e) $\dfrac{y}{5} + \dfrac{y}{3} - \dfrac{y}{2} = 3$ $y = 90$

5.32. Solve: **(6.2)**

a) $\dfrac{5}{x} - \dfrac{2}{x} = 3$ *Ans.* $x = 1$ c) $\dfrac{1}{x} + \dfrac{1}{2} = \dfrac{5}{x}$ *Ans.* $x = 8$ e) $\dfrac{2}{3x} + \dfrac{1}{x} = 5$ *Ans.* $x = \dfrac{1}{3}$

b) $\dfrac{7}{a} = 2 + \dfrac{1}{a}$ $a = 3$ d) $\dfrac{3}{b} + \dfrac{1}{4} = \dfrac{2}{b}$ $b = -4$ f) $\dfrac{3}{4c} = \dfrac{1}{c} - \dfrac{1}{4}$ $c = 1$

5.33. Solve: **(6.1, 6.2)**

a) $\dfrac{r}{6} = \dfrac{1}{2}$ *Ans.* $r = 3$ f) $\dfrac{2b - 12}{b} = \dfrac{10}{7}$ *Ans.* $b = 21$

b) $\dfrac{r}{12} = \dfrac{3}{4}$ $r = 9$ g) $\dfrac{2c + 4}{12} = \dfrac{c + 4}{7}$ $c = 10$

c) $\dfrac{8}{r} = \dfrac{4}{3}$ $r = 6$ h) $\dfrac{6c + 3}{11} = \dfrac{3c}{5}$ $c = 5$

d) $\dfrac{b + 3}{b} = \dfrac{2}{5}$ $b = -5$ i) $\dfrac{c}{10} = \dfrac{c - 12}{6}$ $c = 30$

e) $\dfrac{b + 6}{b} = \dfrac{7}{5}$ $b = 15$

5.34. Solve: **(6.3)**

a) $\dfrac{x - 2}{3} - \dfrac{x + 1}{4} = 4$ *Ans.* $x = 59$ c) $\dfrac{w - 2}{4} - \dfrac{w + 4}{3} = -\dfrac{5}{6}$ *Ans.* $w = -12$

b) $\dfrac{y - 3}{5} - 1 = \dfrac{y - 5}{4}$ $y = -7$ d) $1 - \dfrac{2m - 5}{3} = \dfrac{m + 3}{2}$ $m = 1$

5.35. Solve: **(6.4)**

a) $\dfrac{d + 8}{d - 2} = \dfrac{9}{4}$ *Ans.* $d = 10$ c) $\dfrac{8}{y - 2} - \dfrac{13}{2} = \dfrac{3}{2y - 4}$ *Ans.* $y = 3$

b) $\dfrac{4}{x - 4} = \dfrac{7}{x + 2}$ $x = 12$ d) $\dfrac{10}{r - 3} + \dfrac{4}{3 - r} = 6$ $r = 4$

5.36. Solve: **(7.1, 7.3)**

a) $0.5d = 3.5$ *Ans.* $d = 7$ f) $8.6m + 3 = 7.1m$ *Ans.* $m = -2$

b) $0.05e = 4$ $e = 80$ g) $x - 4.2x = 0.8x - 12$ $x = 3$

c) $60 = 0.3f$ $f = 200$ h) $x + 0.4x + 8 = -20$ $x = -20$

$d)$ $3.1c = 0.42 + c$ *Ans.* $c = 0.2$ $i)$ $x - 0.125x - 1.2 = 19.8$ *Ans.* $x = 24$

$e)$ $6d - 10 = 3.5d$ $d = 4$

5.37. Solve: (7.2)

$a)$ $6a = 3.3$ *Ans.* $a = 0.55$ $f)$ $\dfrac{b}{5} + 0.05 = 1.03$ *Ans.* $b = 4.9$

$b)$ $7a = 7.217$ $a = 1.031$ $g)$ $4c - 6.6 = c - 0.18$ $c = 2.14$

$c)$ $\dfrac{a}{2} = 12.45$ $a = 24.9$ $h)$ $3c - 2.6 = c + 5$ $c = 3.8$

$d)$ $\tfrac{2}{3}b = 7.8$ $b = 11.7$ $i)$ $4c + 0.8 = c + 0.44$ $c = -0.12$

$e)$ $8b - 4 = 0.32$ $b = 0.54$

5.38. Solve: (7.4)

$a)$ $33\tfrac{1}{3}\%$ of $x = 24$ *Ans.* $x = 72$ $f)$ $x = 40.5 - 35\%x$ *Ans.* $x = 30$
$b)$ $16\tfrac{2}{3}\%$ of $x = 3.2$ $x = 19.2$ $g)$ $x - 75\%x = 70$ $x = 280$
$c)$ 70% of $x = 140$ $x = 200$ $h)$ $2x - 50\%x = -9$ $x = -6$
$d)$ $x + 20\%x = 30$ $x = 25$ $i)$ $x + 10 = 4 + 87\tfrac{1}{2}\%x$ $x = -48$
$e)$ $2x + 10\%x = 7$ $x = 3\tfrac{1}{3}$

5.39. Solve: (7.5, 7.6)

$a)$ $0.2(x + 5) = 10$ *Ans.* $x = 45$
$b)$ $4(x - 0.3) = 12$ $x = 3.3$
$c)$ $0.03(x + 200) = 45$ $x = 1300$
$d)$ $50 - 0.05(x - 100) = 20$ $x = 700$
$e)$ $0.03y + 0.02(5000 - y) = 140$ $y = 4000$
$f)$ $0.05y - 0.03(600 - y) = 14$ $y = 400$
$g)$ $0.1(1000 - x) + 0.07(2000 - x) = 104$ $x = 800$
$h)$ $250 - 0.3(x + 100) = 0.5(600 - x) - 40$ $x = 200$

5.40. Solve for x: (8.1)

$a)$ $x - b = 3b$ $d)$ $ax = -10a$ $g)$ $\dfrac{x}{a} = 10b$

$b)$ $x - 5a = 20$ $e)$ $bx = b^3$ $h)$ $\dfrac{x}{4b} = \dfrac{b}{2}$

$c)$ $x + 10c = c + 8$ $f)$ $2cx = -8cd$ $i)$ $a + b = \dfrac{x}{3}$

Ans. $a)$ $x = 4b$ $d)$ $x = -10$ $g)$ $x = 10ab$
      $b)$ $x = 5a + 20$ $e)$ $x = b^2$ $h)$ $x = 2b^2$
      $c)$ $x = -9c + 8$ $f)$ $x = -4d$ $i)$ $x = 3a + 3b$

5.41. Solve for the letter indicated: (8.2)

$a)$ $LW = A$ for L $d)$ $A = \tfrac{1}{2}bh$ for h $g)$ $A = \tfrac{1}{2}h(b + b')$ for h
$b)$ $RP = I$ for P $e)$ $V = \tfrac{1}{3}Bh$ for B $h)$ $s = \tfrac{1}{2}at^2$ for a
$c)$ $P = S - C$ for S $f)$ $S = 2\pi rh$ for r $i)$ $F = \tfrac{9}{5}C + 32$ for C

Ans. $a)$ $L = \dfrac{A}{W}$ $d)$ $h = \dfrac{2A}{b}$ $g)$ $h = \dfrac{2A}{b + b'}$

      $b)$ $P = \dfrac{I}{R}$ $e)$ $B = \dfrac{3V}{h}$ $h)$ $a = \dfrac{2s}{t^2}$

      $c)$ $S = P + C$ $f)$ $r = \dfrac{S}{2\pi h}$ $i)$ $C = \tfrac{5}{9}(F - 32)$

5.42   Solve for x or y: (8.3)

         a) $2x = 6a + 22a$ *d*) $ax - 3a = 5a$ *g*) $ay - 6b = 3ay$

         b) $3y - a = -10a$ *e*) $\dfrac{x}{3} + b = 7b$ *h*) $\dfrac{y}{m} + 3m = 4m$

         c) $\dfrac{x}{4} + b = -b$ *f*) $\dfrac{2x}{3} - 4c = 10c$ *i*) $\dfrac{y}{r} - 5 = r + 2$

   Ans. *a*) $x = 14a$ *d*) $x = 8$ *g*) $y = -\dfrac{3b}{a}$

         b) $y = -3a$ *e*) $x = 18b$ *h*) $y = m^2$

         c) $x = -8b$ *f*) $x = 21c$ *i*) $y = r^2 + 7r$

5.43.   Solve for the letter indicated: (8.4)

         a) $x - 10 = y$ for x *d*) $2x + 3y = 10$ for y *g*) $x + y = z$ for x

         b) $2y = 6x + 8$ for y *e*) $2x + 3y = 12$ for x *h*) $\dfrac{x}{3} - 2y = 3z$ for x

         c) $x + 2y = 20$ for x *f*) $\dfrac{x}{2} + 8 = y$ for x *i*) $ax - by = cz$ for x

   Ans. *a*) $x = y + 10$ *d*) $y = \dfrac{10 - 2x}{3}$ *g*) $x = z - y$

         b) $y = 3x + 4$ *e*) $x = \dfrac{12 - 3y}{2}$ *h*) $x = 6y + 9z$

         c) $x = 20 - 2y$ *f*) $x = 2y - 16$ *i*) $x = \dfrac{by + cz}{a}$

5.44.   Solve for x or y: (8.5)

    a) $5(x + a) = 10(x - 2a)$ *Ans.* $x = 5a$ *d*) $\dfrac{x}{6} - \dfrac{a}{3} = \dfrac{b}{2}$ *Ans.* $x = 2a + 3b$

    b) $\dfrac{x}{3} + b = c - 4b$ $x = 3c - 15b$ *e*) $3(5 - y) = 7(y - b)$ $y = \dfrac{7b + 15}{10}$

    c) $\dfrac{x}{12} = \dfrac{a}{6} + b$ $x = 2a + 12b$ *f*) $\dfrac{y}{3} = \dfrac{y}{4} + c$ $y = 12c$

5.45.   Use a graphing calculator to solve 5.43 *a*) to 5.43 *f*).   (9.1)

CHAPTER 6

Formulas

1. POINTS AND LINES

The word *geometry* is derived from the Greek words *geos* (meaning earth) and *metron* (meaning measure). The ancient Egyptians, Chinese, Babylonians, Romans, and Greeks used geometry for surveying, navigation, astronomy, and other practical occupations.

The Greeks sought to systematize the geometric facts they knew by establishing logical reasons for them and relationships among them. The work of such men as Thales (600 B.C.), Pythagoras (540 B.C.), Plato (390 B.C.), and Aristotle (350 B.C.) in systematizing geometric facts and principles culminated in the geometry text *Elements*, written about 325 B.C. by Euclid. This most remarkable text has been in use for over 2000 years.

1.1. Undefined Terms of Geometry: Point, Line, and Plane

These undefined terms underlie the definitions of all geometric terms. They can be given meanings by way of descriptions. However, these descriptions, which follow, are not to be thought of as definitions.

1.2. Point

A point has position only. It has no length, width, or thickness.

A point is represented by a dot. Keep in mind, however, that the dot *represents* a point but *is not* a point, just as a dot on a map may represent a locality but is not the locality. A dot, unlike a point, has size.

A point is designated by a capital letter next to the dot, thus: $A \cdot$ or $\cdot \, ^{\cdot}P$.

1.3. Line

A line has length but has no width or thickness.

A line may be represented by the path of a piece of chalk on the blackboard or by a stretched elastic band.

A line is designated by the capital letters of any two of its points or by a small letter, thus: $\overleftrightarrow{A \ B}$, $C \ D$, $^a\!\!/$, or $\overrightarrow{AB}$.

A line may be straight, curved, or a combination of these. To understand how lines differ, think of a line as being generated by a moving point. A *straight line*, such as ↔, is generated by a point moving always in the same direction. A *curved line*, such as ⟩, is generated by a point moving in a changing direction.

A straight line is unlimited in extent. It may be extended in either direction indefinitely.

A *ray* is the part of a straight line beginning at a given point and extending limitlessly in one direction: $\overrightarrow{AB}$ and $\overleftarrow{A \; B}$ designate rays.

In this book, the word *line* will mean straight line unless you are told otherwise.

Two straight lines, if they intersect, do so at a single point.

1.4.   Planes

A plane has length with width but no thickness. It may be represented by a blackboard or a side of a box; remember, however, that these are representations of a plane but are not planes.

A plane surface (or plane) is a surface such that a straight line connecting any two of its points lies entirely in it. A plane is a flat surface.

Plane geometry is the geometry of plane figures—those that may be drawn on a plane. Unless you are told otherwise, the word *figure* will mean plane figure in this book.

1.5.   Line Segments

A straight line segment is the part of a straight line between two of its points, including the two points. It is designated by the capital letters of these points or by a small letter. Thus $\overline{AB}$ or r represents the straight line segment $\overset{r}{A \; B}$ between A and B.

The expression *straight line segment* may be shortened to *line segment* or *segment*, if the meaning is clear. Thus, $\overline{AB}$ and *segment AB* both mean "the straight line segment AB."

1.6.   Dividing a Line Segment into Parts

If a line segment is divided into parts:

1. The length of the whole line segment equals the sum of the lengths of its parts. Note that the length of $\overline{AB}$ is designated AB.
2. The length of the whole line segment is greater than the length of any part.

Suppose $\overline{AB}$ is divided into three parts of lengths a, b, and c thus: $\underset{A \quad\quad a \quad\; b \quad\; c \quad\quad B}{\rule{4cm}{0.4pt}}$. Then $AB = a + b + c$. Also, AB is greater than a; this may be written $AB > a$.

1.7.   Congruent Segments

Two line segments having the same length are said to be *congruent*. Thus, if $AB = CD$, then $\overline{AB}$ is congruent to $\overline{CD}$, written $\overline{AB} \cong \overline{CD}$.

2.   UNDERSTANDING POLYGONS, CIRCLES, AND SOLIDS

A.   *Understanding Polygons in General*

A **polygon** is a closed figure in a plane (flat surface) bounded by straight lines.

Names of Polygons According to the Number of Sides

Sides	Polygon	Sides	Polygon
3	Triangle	8	Octagon
4	Quadrilateral	10	Decagon
5	Pentagon	12	Dodecagon
6	Hexagon	n	n-gon

An **equilateral polygon** is a polygon having congruent sides.

Thus, a square is an equilateral polygon.

An **equiangular polygon** is a polygon having congruent angles.

Thus, a rectangle is an equiangular polygon.

A **regular polygon** is an equilateral and equiangular polygon.

Thus, the polygon in Fig. 6-1 is a five-sided equilateral and equiangular polygon.

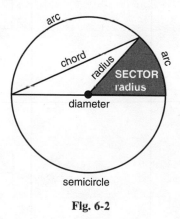

Regular Pentagon

Fig. 6-1

Symbols for Congruent Sides and Congruent Angles

Using the same letter for sides indicates they are congruent. Congruent angles may be shown by using arcs crossed by the same number of strokes, as in Fig. 6-1. Two angles are congruent if their measures are equal. Thus, two 60° angles are congruent.

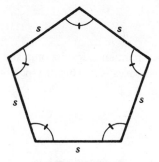

Fig. 6-2

B.   Understanding Circles

The **circumference** is the distance around a circle.

A **radius** is a line joining the center to a point on the circumference. All radii of a circle are equal.

A chord is a line joining any two points on the circumference.

A diameter is a chord through the center. A diameter is twice a radius. It is the largest chord.

An arc is a part of the circumference of a circle.

A sector is a part of the area of a circle bounded by an arc and two radii.

See Fig. 6-2.

C. *Understanding Triangles*

1. An **equilateral triangle** has three congruent sides. It also has three equal angles, each 60°.

2. An **isosceles triangle** has at least two congruent sides. It also has at least two congruent angles. The congruent angles shown lie along the base *b* and are called the **base angles**.

3. A **scalene triangle** has no congruent sides.

4. A **right triangle** has one right angle. Its **hypotenuse** is opposite the right angle. **Its legs** (or **arms**) are the other two sides. The symbol for the right angle is a square corner.

5. An **obtuse triangle** has one obtuse angle (more than 90° and less than 180°).

6. An **acute triangle** has all acute angles (less than 90°).

See Fig. 6-3.

D. *Understanding Quadrilaterals*

1. A **parallelogram** has two pairs of parallel sides. Its opposite sides and its opposite angles are congruent. The distance between the two bases is *h*. This distance is at right angles to both bases.

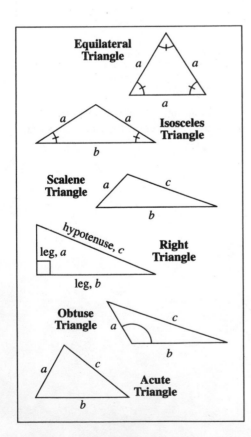

Fig. 6-3

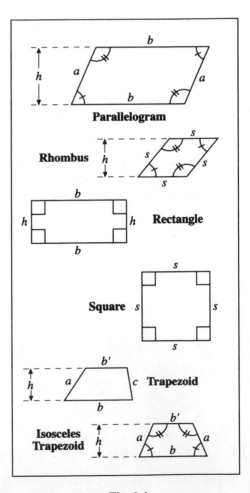

Fig. 6-4

2. A **rhombus** has four congruent sides. Its opposite angles are congruent. It is an equilateral parallelogram.

3. A **rectangle** has four right angles. Its bases b and its heights h are congruent. It is an equiangular parallelogram.

4. A **square** has four congruent sides and four right angles. It is an equilateral and equiangular parallelogram. A **square unit** is a square whose side is 1 unit. Thus, 1 square foot ($1\ \text{ft}^2$) is a square whose side is 1 ft.

5. A **trapezoid** has one and only one pair of parallel sides. The bases are represented by b and b'.

6. An **isosceles trapezoid** has two congruent legs (nonparallel sides). Note the congruent angles.

See Fig. 6-4.

E. *Understanding Solids*

1. A **solid** is an enclosed portion of space bounded by plane and curved surfaces.

Thus, the **pyramid**, the **cube**, the **cone**, the **cylinder**, and the **sphere** are solids.

2. A **polyhedron** is a solid bounded by plane (flat) surfaces only. Thus, the pyramid and cube are polyhedrons. The cone, cylinder, and sphere are not polyhedrons since each has a curved surface. The faces of a polyhedron are its bounding polygons. The edges of a polyhedron are the sides of its faces.

3. A **prism** is a polyhedron, two of whose faces are parallel polygons and whose remaining faces are parallelograms. The bases of a prism are its parallel polygons. These may have any number of sides. The lateral (side) faces are its parallelograms. The distance between the two bases is h. This line is at right angles to each base. See Fig. 6-5.

Fig. 6-5   Fig. 6-6

A **right prism** is a prism whose lateral faces are rectangles. The distance h is the height of any of the lateral faces. See Fig. 6-6.

4. A **rectangular solid** (box) is a prism bounded by six rectangles. The rectangular solid can be formed from the pattern of six rectangles folded along the dotted lines. The length l, the width w, and the height h are its dimensions. See Fig. 6-7.

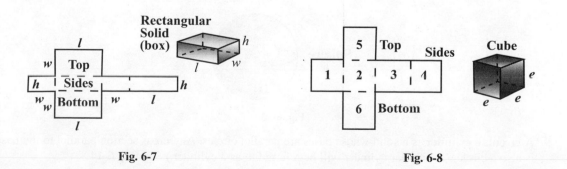

Fig. 6-7   Fig. 6-8

5. A **cube** is a rectangular solid bounded by six squares. The cube can be formed from the pattern of six squares folded along the dotted lines. Each equal dimension is represented by e in the diagram. See Fig. 6-8.

   A **cubic unit** is a cube whose edge measures 1 unit. Thus, 1 cubic inch (1 in^3) is a cube whose edge measures 1 in.

6. A **pyramid** is a polyhedron whose base is a polygon and whose other faces meet at a point, its vertex. The base B may have any number of sides. However, the other faces must be triangles. The distance from the vertex to the base is equal to the altitude or height h, a line from the vertex at right angles to the base. See Fig. 6-9.

   A **regular pyramid** is a pyramid whose base is a regular polygon and whose altitude joins the vertex and the center of the base. See Fig. 6-10.

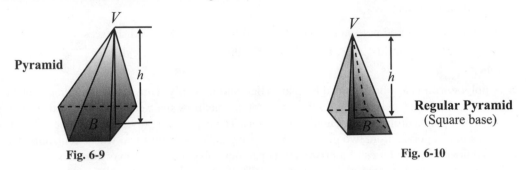

Fig. 6-9

Fig. 6-10

7. A **circular cone** is a solid whose base is a circle and whose lateral surface comes to a point. (A circular cone will be referred to as a cone.) See Fig. 6-11.

   A **cone of revolution** is formed by revolving a right triangle about one of its legs. This leg becomes the altitude or height h of the cone, and the other becomes the radius r of the base. See Fig. 6-12.

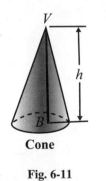

Cone

Fig. 6-11

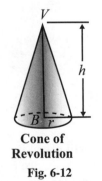

Cone of Revolution

Fig. 6-12

8. A **sphere** is a solid such that every point on its surface is an equal distance from the same point, its center.

   A sphere is formed by revolving a semicircle about its diameter. See Fig. 6-13.

Sphere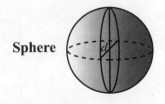

Fig. 6-13

9. A **circular cylinder** is a solid whose bases are parallel circles. Any cross section parallel to the bases is also a circle. (A circular cylinder will be referred to as a cylinder.) See Fig. 6-14.

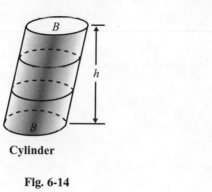

Cylinder

Fig. 6-14

**Cylinder of
Revolution**

Fig. 6-15

A **cylinder of revolution** is formed by revolving a rectangle about one of its two dimensions. This dimension becomes the height h of the cylinder, and the other becomes the radius r of the base. See Fig. 6-15.

3. FORMULAS FOR PERIMETERS AND CIRCUMFERENCES: LINEAR MEASURE

A formula is an equality expressing in mathematical symbols a numerical rule or relationship among quantities.

Thus, $S = C + P$ is a formula for the rule: **selling price equals cost plus profit**.

In perimeter formulas, the perimeter is in the same unit as the dimensions. Thus, if the side of a square is 3 m, the perimeter is 12 m.

The **perimeter of a polygon** is the distance around it. Thus, the perimeter p of the triangle shown is the sum of the lengths of its three sides; that is,

$$p = a + b + c$$

See Fig. 6-16.

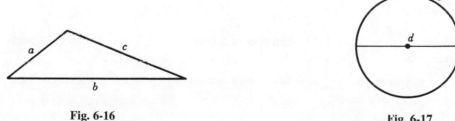

Fig. 6-16   **Fig. 6-17**

The **circumference of a circle** is the distance around it. For any circle, the circumference c is π times the diameter d; that is, $c = \pi d$. See Fig. 6-17. The value of π, using five digits, is 3.1416. If less accuracy is needed, 3.142, 3.14, or $\frac{22}{7}$ may be used.

3.1. PERIMETER FORMULAS FOR TRIANGLES AND QUADRILATERALS

State the formula for the perimeter p of each, using the letters shown in Figs. 6-3 and 6-4.

a) Equilateral triangle	*Ans.* $p = 3a$	d) Square	*Ans.* $p = 4s$
b) Isosceles triangle	$p = 2a + b$	e) Rectangle	$p = 2b + 2h$
c) Scalene triangle	$p = a + b + c$	f) Obtuse triangle	$p = a + b + c$

3.2. PERIMETER FORMULAS FOR POLYGONS

State the name of the equilateral polygon to which each perimeter formula applies:

a) $p = 3s$	*Ans.* equilateral triangle	c) $p = 10s$	*Ans.* equilateral decagon
b) $p = 4s$	square or rhombus	d) $p = 12s$	equilateral dodecagon

3.3. FINDING A SIDE OF AN EQUILATERAL POLYGON

If each of the following polygons has a perimeter of 36 in, find a side:

a) square   b) equilateral triangle   c) equilateral hexagon   d) equilateral decagon

<u>Solutions:</u>

a) $p = 4s$ b) $p = 3s$ c) $p = 6s$ d) $p = 10s$
    $36 = 4s$ $36 = 3s$ $36 = 6s$ $36 = 10s$
     $9 = s$ $12 = s$ $6 = s$ $3.6 = s$

Ans.   9 in   *Ans.*   12 in   *Ans.*   6 in   *Ans.*   3.6 in

3.4. FINDING PERIMETERS OF REGULAR POLYGONS

Find the perimeter of:

a)   an equilateral triangle   b)   a square with a side of   c)   a regular hexagon with a
    with a side of 5 m $4\frac{1}{2}$ ft   side of 2 yd 1 ft

<u>Solutions:</u>

a) See Fig. 6-18.   b) See Fig. 6-19.   c) See Fig. 6-20.

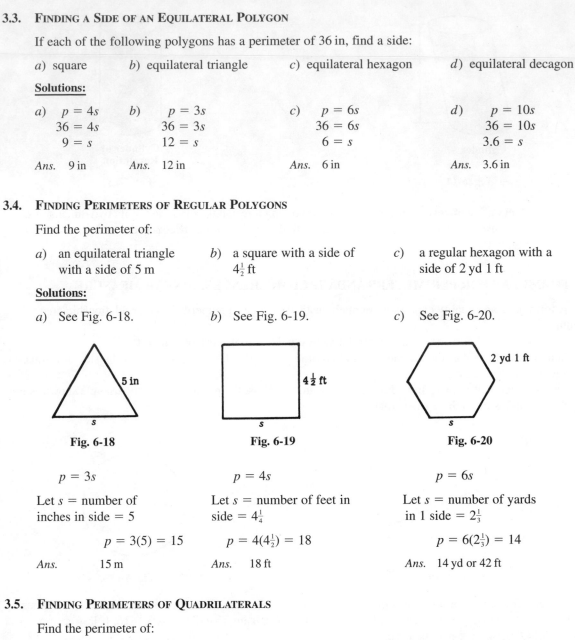

Fig. 6-18   Fig. 6-19   Fig. 6-20

$p = 3s$ $p = 4s$ $p = 6s$

Let s = number of   Let s = number of feet in   Let s = number of yards
inches in side = 5   side = $4\frac{1}{4}$ in 1 side = $2\frac{1}{3}$

$p = 3(5) = 15$ $p = 4(4\frac{1}{2}) = 18$ $p = 6(2\frac{1}{3}) = 14$

Ans.   15 m   *Ans.*   18 ft   *Ans.*   14 yd or 42 ft

3.5. FINDING PERIMETERS OF QUADRILATERALS

Find the perimeter of:

a) a rectangle with sides of 4 and $1\frac{1}{2}$ m   b) a parallelogram with sides of 4 yd and 2 ft

<u>Solutions:</u>

a) See Fig. 6-21.   b) See Fig. 6-22.

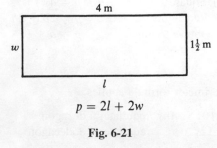

$p = 2l + 2w$

Fig. 6-21

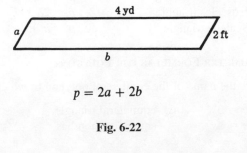

$p = 2a + 2b$

Fig. 6-22

Lct l = number of meters in length = 4   Let a = number of yards in 1 side = $\frac{2}{3}$
and w = number of meters in width = $1\frac{1}{2}$ and b = number of yards in other side = 4

$$p = 2(4) + 2(1\tfrac{1}{2}) = 11$$ $$p = 2(\tfrac{2}{3}) + 2(4) = 9\tfrac{1}{3}$$

Ans.   11 m   *Ans.* $9\frac{1}{3}$ yd or 28 ft

3.6. FINDING THE PERIMETER OF A RECTANGLE

If l, w, and p are in centimeters, find the perimeter of the rectangle shown in Fig. 6-23 if:

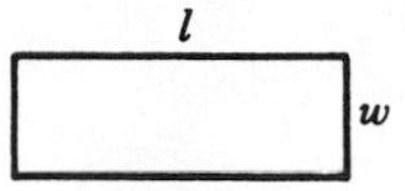

Fig. 6-23

a) $l = 10$, $w = 3$ *b)* $l = 3\frac{1}{2}$, $w = 4\frac{1}{4}$ *c)* $l = 3.1$, $w = 2.6$

Solutions:

a) $p = 2l + 2w$ *b)* $p = 2l + 2w$ *c)* $p = 2l + 2w$
      $p = 2(10) + 2(3)$ $p = 2(3\tfrac{1}{2}) + 2(4\tfrac{1}{4})$ $p = 2(3.1) + 2(2.6)$
      $p = 26$ $p = 15\tfrac{1}{2}$ $p = 11.4$

Ans. 26 cm   *Ans.* $15\frac{1}{2}$ cm   *Ans.* 11.4 cm

3.7. FINDING THE LENGTH OR WIDTH OF A RECTANGLE

For a rectangle, if l, w, and p are in meters, find:

a) the length if $p = 20$ and $w = 3$ *c)* the width if $p = 40$ and $l = w + 5$
b) the width if $p = 27$ and $l = 5\frac{1}{2}$ *d)* the length if $p = 30$ and $w = 3l$

Solutions:

a) $p = 2l + 2w$ *b)* $p = 2l + 2w$ *c)* $p = 2l + 2w$ *d)* $p = 2l + 2w$
       $20 = 2l + 6$ $27 = 2(5\tfrac{1}{2}) + 2w$ $40 = 2(w + 5) + 2w$ $30 = 2l + 2(3l)$
       $7 = l$ $8 = w$ $7\tfrac{1}{2} = w$ $3\tfrac{3}{4} = l$

Ans.   7 m   *Ans.* 8 m   *Ans.* $7\frac{1}{2}$ m   *Ans.* $3\frac{3}{4}$ m

3.8. PERIMETER OF AN ISOCELES TRIANGLE

For the isosceles triangle shown in Fig. 6-24, if a, b, and p are in inches, find:

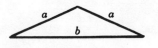

Fig. 6-24

a) the perimeter if $a = 10$ and $b = 12$ *c)* the base if $p = 33$ and $a = b - 3$
b) the base if $p = 25$ and $a = 8$ *d)* an equal side if $p = 35$ and $b = 1.5a$

Solutions:

a) $p = 2a + b$ *b)* $p = 2a + b$ *c)* $p = 2a + b$ *d)* $p = 2a + b$

          $p = 2(10) + 12$ $25 = 2(8) + b$ $33 = 2(b - 3) + b$ $35 = 2a + 1.5a$

          $p = 32$ $9 = b$ $13 = b$ $10 = a$

Ans. 32 in   *Ans.* 9 in   *Ans.* 13 in   *Ans.* 10 in

3.9. CIRCUMFERENCE AND ARC FORMULAS

For any circle (see Fig. 6-25), state a formula which relates:

a) the diameter d and radius r *Ans.* $d = 2r$

b) the circumference c and the diameter d $c = \pi d$

c) the circumference c and the radius r $c = 2\pi r$

d) the circumference c and a 90° arc a $c = 4a$

e) a 60° arc a' and the radius r (Read a' as "a prime.") $a' = \dfrac{\pi r}{3}$

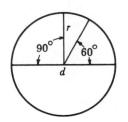

Fig. 6-25

3.10. CIRCUMFERENCE OF A CIRCLE

For a circle (see Fig. 6-26), if r, d, and c are in feet, find, using $\pi = 3.14$,

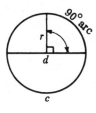

Fig. 6-26

a) the circumference if $r = 5$ *c)* the radius if $c = 942$

b) a 90° arc if $d = 8$ *d)* the diameter if $c = 1570$

Solutions:

a) $c = 2\pi r$ *b)*   90° arc $= \dfrac{c}{4} = \dfrac{\pi d}{4}$ *c)* $c = 2\pi r$ *d)* $c = \pi d$

       $= 2(3.14)5$ $942 = 2(3.14)r$ $1570 = 3.14d$

       $= 31.4$ $= \dfrac{(3.14)8}{4}$ $150 = r$ $500 = d$

Ans. 31.4 ft   *Ans.* 150 ft   *Ans.* 500 ft

                                   $= 6.28$

                           Ans. 6.28 ft

3.11. PERIMETERS OF COMBINED FIGURES

State the formula for the perimeter p for each of the regions in Figs. 6 27 to 6 32.

(sq. = square, each curved figure is a semicircle)

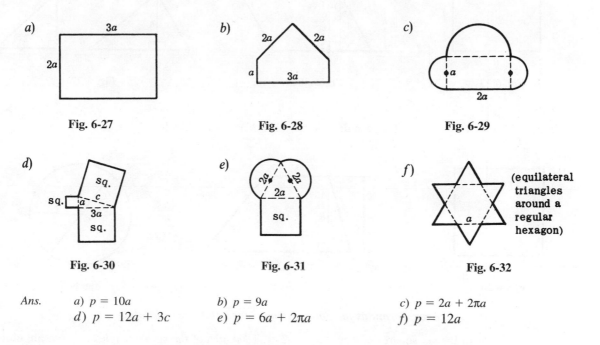

a)   b)   c)

Fig. 6-27   Fig. 6-28   Fig. 6-29

d)   e)   f)

Fig. 6-30   Fig. 6-31   Fig. 6-32

(equilateral triangles around a regular hexagon)

Ans.   a) $p = 10a$ b) $p = 9a$ c) $p = 2a + 2\pi a$
            d) $p = 12a + 3c$ e) $p = 6a + 2\pi a$ f) $p = 12a$

4. FORMULAS FOR AREAS: SQUARE MEASURE

A **square unit** is a square whose side is 1 unit. Thus, a square inch (1 in^2) is a square whose side is 1 in. See Fig. 6-33.

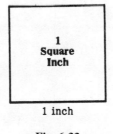

**1
Square
Inch**

1 inch

Fig. 6-33

The **area of a polygon or circle** is the number of square units contained in its surface. Thus, the area of a rectangle 5 units long and 4 units wide contains 20 square units.

In area formulas, the area is in **square units**, the unit being the same as that used for the dimensions. Thus, if the side of a square is 3 yards, its area is 9 square yards. See Fig. 6-34.

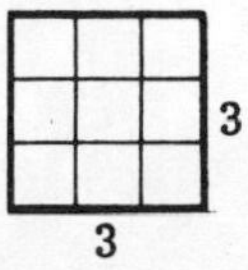

3

3

Fig. 6-34

Area Formulas Using A for Area of Figure

1. Rectangle: $A = bh$ **2. Parallelogram:** $A = bh$ **3. Triangle:** $A = \dfrac{bh}{2}$

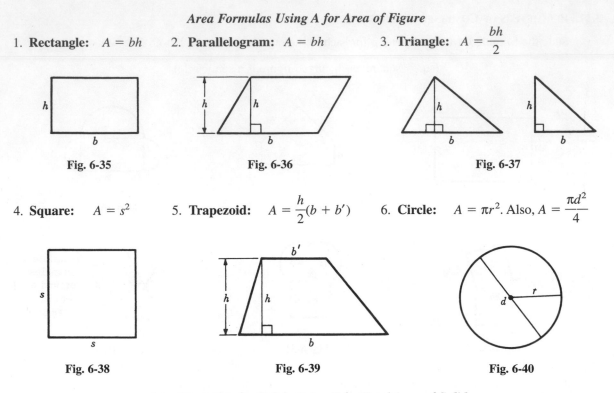

Fig. 6-35   Fig. 6-36   Fig. 6-37

4. Square: $A = s^2$ **5. Trapezoid:** $A = \dfrac{h}{2}(b + b')$ **6. Circle:** $A = \pi r^2$. Also, $A = \dfrac{\pi d^2}{4}$

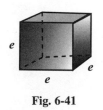

Fig. 6-38   Fig. 6-39   Fig. 6-40

Area Formulas for Solids Using T for Total Area of Solid

1. Total area of the six squares of a **cube:**

$$T = 6e^2$$

   See Fig. 6-41.

2. Total area of the six rectangles of a **rectangular solid:**

$$T = 2lw + 2lh + 2wh$$

   See Fig. 6-42.

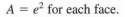

Fig. 6-41

$A = e^2$ for each face.

Fig. 6-42

$A = lw$ for top or bottom faces
$A = lh$ for front or back faces
$A = wh$ for left or right faces

3. Total area of a **sphere:**
$A = 4\pi r^2$

   See Fig. 6-43.

4. Total area of a **cylinder of revolution:**
$T = 2\pi rh + 2\pi r^2$
$T = 2\pi r(r + h)$

   See Fig. 6-44.

Fig. 6-43

Fig. 6-44

4.1. RELATIONS AMONG SQUARE UNITS

Find the area A of (*Hint:* The square unit in Fig. 6-45 is a square whose side is 1 unit.):

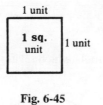

Fig. 6-45

a) a square foot in square inches
b) a square yard in square feet
c) a square meter in square centimeters

<u>Solutions:</u>

a) $A = s^2$
Since 1 ft = 12 in,
$A = 12^2 = 144$

Ans. $1\,\text{ft}^2 = 144\,\text{in}^2$

b) $A = s^2$
Since 1 yd = 3 ft
$A = 3^2 = 9$

Ans. $1\,\text{yd}^2 = 9\,\text{ft}^2$

c) $A = s^2$
Since 1 m = 100 cm,
$A = 100^2 = 10,000$

Ans. $1\,\text{m}^2 = 10,000\,\text{cm}^2$

4.2. FINDING THE AREA OF SQUARES

Find the area of a square A whose side is:

a) 6 in   *b*) 6 m   *c*) 6 yd

<u>Solutions:</u>

a) $A = s^2$
Since 6 in = $\frac{1}{2}$ ft
$A = \left(\frac{1}{2}\right)^2 = \frac{1}{4}$

Ans. $\frac{1}{4}\,\text{ft}^2$

b) $A = s^2$
$= 6^2$
$= 36$

Ans. $36\,\text{m}^2$

c) $A = s^2$
Since 6 yd = 18 ft,
$A = 18^2 = 324$

Ans. $324\,\text{ft}^2$

4.3. FINDING AREAS
Find the area of:

a) a rectangle with sides of 4 and $2\frac{1}{2}$ cm (Fig. 6-46)
b) a parallelogram with a base of 5.8 in and a height of 2.3 in (Fig. 6-47)
c) a triangle with a base of 4 ft and an altitude to the base of 3 ft 6 in (Fig. 6-48)

<u>Solutions:</u>

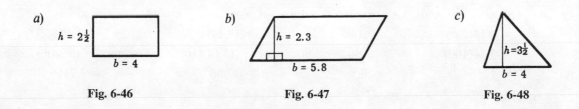

a)   *b*)   *c*)

Fig. 6-46   Fig. 6-47   Fig. 6-48

$$A = bh$$
$$= 4(2\tfrac{1}{2}) = 10$$

$$A = bh$$
$$= (5.8)(2.3) = 13.34$$

$$A = \tfrac{1}{2}bh$$
$$= \tfrac{1}{2}(4)(3\tfrac{1}{2}) = 7$$

Ans. $10\,\text{cm}^2$ *Ans.* $13.34\,\text{in}^2$ *Ans.* $7\,\text{ft}^2$

4.4. AREA-OF-CIRCLE FORMULAS

For a circle (see Fig. 6-49), state a formula which relates:

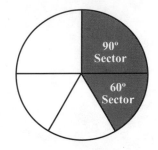

Fig. 6-49

a) the area A and the radius r *Ans.* $A = \pi r^2$

b) the area A and the diameter d $A = \dfrac{\pi d^2}{4}$

c) the area A and a sector S of $90°$ $A = 4S$

d) a $60°$ sector S' and the area A $S' = \dfrac{A}{6}$

e) a $40°$ sector S'' and the radius r (Read S'' as "S double prime.") $S'' = \dfrac{\pi r^2}{9}$

4.5. AREA OF A CIRCLE

For a circle (see Fig. 6-50), if r and d are in centimeters (cm), find, *to the nearest integer*, the area of:

Fig. 6-50

a) the circle if $r = 20$ *c*) a $90°$ sector if $r = 4$
b) the circle if $d = 10$ *d*) a $30°$ sector if $d = 12.2$

Solutions:

a) $A = \pi r^2$ *b*) $A = \pi r^2$ *c*) $A = \pi r^2$ *d*) $A = \pi r^2$
       $= 3.14(20^2)$ $= 3.14(5^2)$ $= (3.14)4^2$ $= (3.14)(6.1^2)$
       $= 3.14(400)$ $= 3.14(25)$ $90°$ sector $= \tfrac{1}{4}(3.14)(16)$ $30°$ sector $= \tfrac{1}{12}(3.14)(37.21)$
       $= 1256$ $= 78.5$ $= 12.56$ $= 9.7366$

Ans. $1256\,\text{cm}^2$ *Ans.* $79\,\text{cm}^2$ *Ans.* $13\,\text{cm}^2$ *Ans.* $10\,\text{cm}^2$

4.6. FORMULAS FOR COMBINED AREAS

State the formula for the area A of each shaded region in Figs. 6-51 to 6-54.

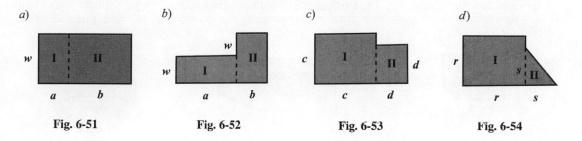

a)   b)   c)   d)

Fig. 6-51   Fig. 6-52   Fig. 6-53   Fig. 6-54

Solution:

(Abbreviations used are rect. for rectangle, sq. for square, and $\triangle$ for triangle.)

a) A = rect. I + rect. II   b) A = rect. I + rect. II   c) A = sq. I + sq. II   d) A = sq. I + $\triangle$II

Ans. $A = aw + bw$ Ans. $A = aw + 2bw$ Ans. $A = c^2 + d^2$ Ans. $A = r^2 + \dfrac{s^2}{2}$

4.7. FORMULAS FOR REDUCED AREAS

State the formula for the area A of each shaded region in Figs. 6-55 to 6-58.

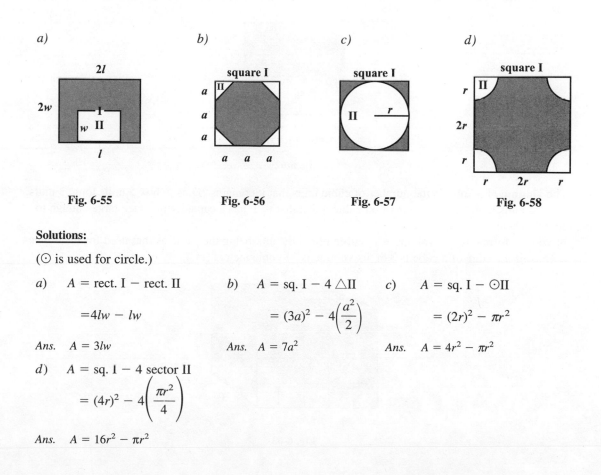

a)   b)   c)   d)

Fig. 6-55   Fig. 6-56   Fig. 6-57   Fig. 6-58

Solutions:

($\odot$ is used for circle.)

a) A = rect. I − rect. II   b) A = sq. I − 4 $\triangle$II   c) A = sq. I − $\odot$II

    $= 4lw - lw$ $= (3a)^2 - 4\left(\dfrac{a^2}{2}\right)$ $= (2r)^2 - \pi r^2$

Ans. $A = 3lw$ Ans. $A = 7a^2$ Ans. $A = 4r^2 - \pi r^2$

d) A = sq. 1 − 4 sector II

    $= (4r)^2 - 4\left(\dfrac{\pi r^2}{4}\right)$

Ans. $A = 16r^2 - \pi r^2$

4.8. FINDING TOTAL AREAS OF SOLIDS

Find, to the nearest integer, the total area of (see Figs. 6-59 to 6-61):

a) a cube with an edge   *b*) a rectangular solid with dimensions   *c*) a sphere with a radius
of 5 in   of 10, 7, and $4\frac{1}{2}$ ft   of 1.1 m

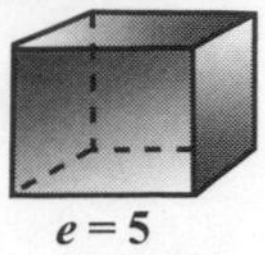

$e = 5$

Fig. 6-59

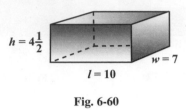

$h = 4\frac{1}{2}$ $w = 7$

$l = 10$

Fig. 6-60

$r = 1.1$

Fig. 6-61

<u>Solutions:</u>

a) $T = 6e^2$ *b*) $T = 2lw + 2lh + 2wh$ *c*) $T = 4\pi r^2$

$= 6(5^2)$ $= 2(10)(7) + 2(10)(4\frac{1}{2}) + 2(7)(4\frac{1}{2})$ $= 4(3.14)(1.1^2)$

$= 150$ $= 293$ $= 15.1976$

Ans.   150 in^2 *Ans.*   293 ft^2 *Ans.*   15 m^2

5. FORMULAS FOR VOLUMES: CUBIC MEASURE

A **cubic unit** is a cube whose edge is 1 unit. Thus, a cubic inch (1 in^3) is a cube whose side is 1 in.

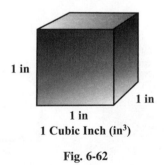

1 in

1 in

1 in

1 Cubic Inch (in^3)

Fig. 6-62

The **volume of a solid** is the number of cubic units that it contains. Thus, a box 5 units long, 3 units wide, and 4 units high has a volume of 60 cubic units; that is, it has a capacity or space large enough to contain 60 cubes, 1 unit on a side.

In volume formulas, the volume is in **cubic units**, the unit being the same as that used for the dimensions. Thus, if the edge of a cube is 3 m, its volume is 27 cubic meters (m^3).

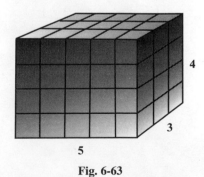

4

3

5

Fig. 6-63

Volume Formulas

Here we use V for the volume of the solid, B for the area of a base, and h for the distance between the bases or between the vertex and a base.

1. **Rectangular solid:** $V = lwh$. See Fig. 6-64.

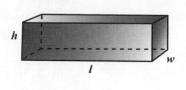

Fig. 6-64

2. **Prism:** $V = Bh$. See Fig. 6-65.

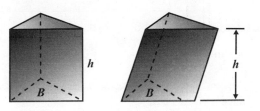

Fig. 6-65

3. **Cylinder:** $V = Bh$ or $V = \pi r^2 h$. See Fig. 6-66.

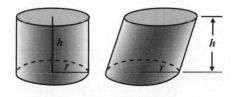

Fig. 6-66

4. **Cube:** $V = e^3$. See Fig. 6-67.

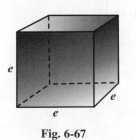

Fig. 6-67

5. **Pyramid:** $V = \frac{1}{3}Bh$. See Fig. 6-68.

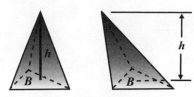

Fig. 6-68

6. **Cone:** $V = \frac{1}{3}Bh$ or $V = \frac{1}{3}\pi r^2 h$.

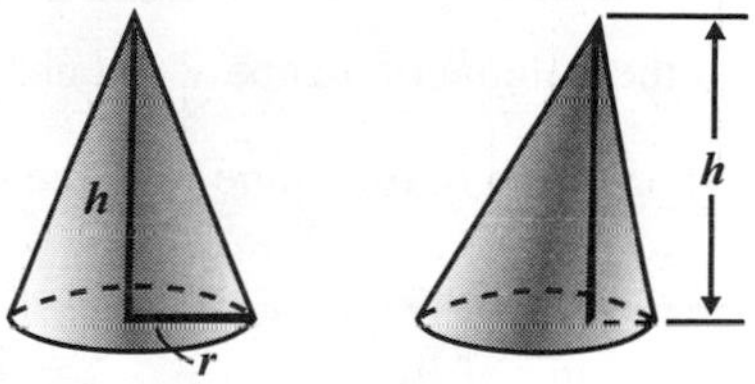

Fig. 6-69

7. **Sphere:** $V = \frac{4}{3}\pi r^3$. See Fig. 6-70.

Fig. 6-70

5.1.   RELATIONS AMONG CUBIC UNITS

Find the volume V of:

a) a cubic foot in cubic inches
b) a cubic yard in cubic feet
c) a liter (cubic decimeter) in cubic centimeters
(*Hint*: A cubic unit is a cube whose edge is 1 unit. See Fig. 6-71.)

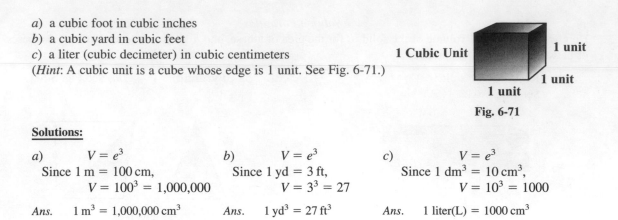

Fig. 6-71

Solutions:

a)　　　$V = e^3$
Since 1 m = 100 cm,
　　　$V = 100^3 = 1,000,000$

Ans.　1 m^3 = 1,000,000 cm^3

b)　　　$V = e^3$
Since 1 yd = 3 ft,
　　　$V = 3^3 = 27$

Ans.　1 yd^3 = 27 ft^3

c)　　　$V = e^3$
Since 1 dm^3 = 10 cm^3,
　　　$V = 10^3 = 1000$

Ans.　1 liter(L) = 1000 cm^3

5.2. FINDING VOLUMES OF CUBES

Find the volume of a cube *V* in cubic feet whose edge is *a*) 4 in, *b*) 4 ft, *c*) 4 yd:

Solutions: (To find volume in cubic feet, express side in feet.)

a)　　　$V = e^3$
Since 4 in = $\frac{1}{3}$ ft,
　　　$V = (\frac{1}{3})^3 = \frac{1}{27}$

Ans.　$\frac{1}{27}$ ft^3

b)　$V = e^3$
　　$= 4^3 = 64$

Ans.　64 ft^3

c)　　　$V = e^3$
Since 4 yd = 12 ft,
　　　$V = 12^3 = 1728$

Ans.　1728 ft^3

5.3. FINDING VOLUMES OF RECTANGULAR SOLID, PRISM, AND PYRAMID

Find the volume of:

a) a rectangular solid having a length of 6 cm, a width of 4 cm, and a height of 1 cm
b) a prism having a height of 15 yd and a triangular base of 120 ft^2
c) a pyramid having a height of 8 km and a square base whose side is $4\frac{1}{2}$ km

Solutions:

a)　$V = lwh$

　　　$= 6(4)(1) = 24$

Ans. 24 cm^3

b)　$V = Bh$

　　　$= 120(45) = 5400$

Ans. 5400 ft^3 or 200 yd^3

c)　$V = \dfrac{1}{3} Bh$

　　　$= \dfrac{1}{3}\left(\dfrac{9}{2}\right)^2 (8) = 54$

Ans. 54 km^3

5.4. FINDING VOLUMES OF SPHERE, CYLINDER, AND CONE

Find the volume, to the nearest integer, of:

a) a sphere with a radius of 10 in
b) a cylinder with a height of 4 yd and a base whose radius is 2 ft
c) a cone with a height of 2 ft and a base whose radius is 2 yd

Solutions:

(Let $\pi = 3.14$)

a) $V = \frac{4}{3}\pi r^3$
    $= \frac{4}{3}(3.14)10^3$
    $= 4186\frac{2}{3}$

Ans. 4187 in^3

b) $V = \pi r^2 h$
    $= (3.14)(2^2)12$
    $= 150.72$

Ans. 151 ft^3

c) $V = \frac{1}{3}\pi r^2 h$
    $= \frac{1}{3}(3.14)(6^2)(2)$
    $= 75.36$

Ans. 75 ft^3

5.5. DERIVING FORMULAS FROM $V = Bh$

From $V = Bh$, the volume formula for a prism or cylinder, derive the volume formulas for the solids in Figs. 6-72 to 6-75.

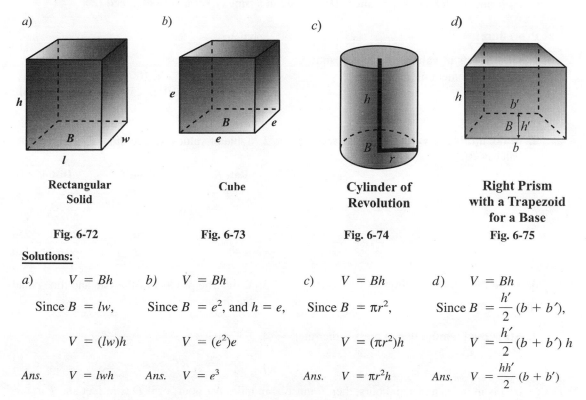

a)   b)   c)   d)

Rectangular   Cube   Cylinder of   Right Prism
Solid   Revolution   with a Trapezoid
                                                                         for a Base

Fig. 6-72   Fig. 6-73   Fig. 6-74   Fig. 6-75

Solutions:

a) $V = Bh$

Since $B = lw$,

$V = (lw)h$

Ans. $V = lwh$

b) $V = Bh$

Since $B = e^2$, and $h = e$,

$V = (e^2)e$

Ans. $V = e^3$

c) $V = Bh$

Since $B = \pi r^2$,

$V = (\pi r^2)h$

Ans. $V = \pi r^2 h$

d) $V = Bh$

Since $B = \dfrac{h'}{2}(b + b')$,

$V = \dfrac{h'}{2}(b + b')h$

Ans. $V = \dfrac{hh'}{2}(b + b')$

5.6. FORMULAS FOR COMBINED VOLUMES

State the formula for the volume of each solid in Figs. 6-76, 6-77, and 6-78.

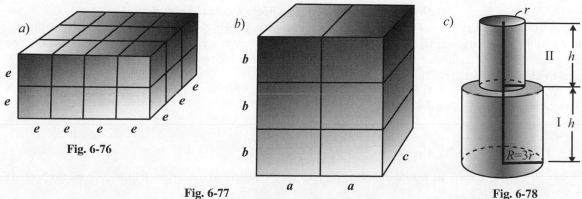

a)

Fig. 6-76

b)

Fig. 6-77

c)

II h

I h

$R = 3r$

Fig. 6-78

Solutions:

a) $V = lwh$
Now, $l = 4e$, $w = 3e$, $h = 2e$
Hence, $V = (4e)(3e)(2e)$

Ans. $V = 24e^3$

b) $V = lwh$
Now, $l = 2a$, $w = c$, $h - 3b$
Hence, $V = (2a)(c)(3b)$

Ans. $V = 6abc$

c) $V = $ cyl. I $+$ cyl. II
$- \pi R^2 h + \pi r^2 h$
Now, $R = 3r$.
Hence, $V = \pi(3r)^2 h + \pi r^2 h$

Ans. $V = 10\pi r^2 h$

6. DERIVING FORMULAS

To Derive a Formula for Related Quantities

Derive a formula relating the **distance** (D) traveled in a **time** (T) at a **rate of speed** (R).

Procedure:

1. **Obtain sets of values** for these quantities, using convenient numbers:

2. **Make a table of values** for those sets of values:

3. **State the rule** that follows:

4. **State the formula** that expresses the rule:

Solution:

1. Sets of values:
   At 50 mi/h for 2 h, 100 mi will be traveled.
   At 25 mi/h for 10 h, 250 mi will be traveled.
   At 40 mi/h for 3 h, 120 mi will be traveled.

2. Table of values: (Place units above quantities.)

Rate R (mi/h)	Time T (h)	Distance D (mi)
50	2	$50 \cdot 2 = 100$
25	10	$25 \cdot 10 = 250$
40	3	$40 \cdot 3 = 120$

3. **Rule**: The product of the rate and time equals the distance.

4. Formula:

$$RT = D$$

NOTE: If D is in miles and T in hours, then R must be in miles per hour; or if D is in feet and T in seconds, R must be in feet per second. **Rate** *must be in* **distance** *units* **per time** *unit*.

Obtaining Formulas from a More General Formula

A formula such as $RT = D$ relates **three** quantities: time, rate, and distance. Each of these quantities may vary in value: that is, they may have many values. However, in a problem, situation, or discussion, one of these quantities may have a fixed or unchanging value. When such is the case, this constant value may be used to obtain a formula relating the other **two** quantities.

Thus, $D = RT$ leads to $D = 30T$ if the rate of speed is fixed at 30 mi/h or 30 ft/min. Or $D = RT$ leads to $D = 3R$ when the time of travel is fixed at 3 h or 3 min.

6.1. DERIVING A COIN FORMULA

Derive a formula for the number of nickels n equivalent to q quarters (*equivalent* means "equal in value"):

Solutions:

1. Sets of values: **2.**   Table of values:

	No. of Quarters q	No. of Nickels n
2 quarters equal 10 nickels	2	$5 \cdot 2$ or 10
4 quarters equal 20 nickels	4	$5 \cdot 4$ or 20
10 quarters equal 50 nickels	10	$5 \cdot 10$ or 50
q quarters equal $5q$ nickels	q	$5q$

3. Rule: The number of nickels equivalent to a number of quarters is five times that number.

4. Formula: $n = 5q$ *Ans.*

6.2. DERIVING A COIN FORMULA

Derive a formula for the value in cents c of d dimes and n nickels:

Solution:

1.   Sets of values: **2.**   Table of values:

	No. of Dimes d	No. of Nickels n	Value of Dimes & Nickels c (cents)
In 3 dimes and 4 nickels, there is 50¢.	3	4	$10 \cdot 3 + 5 \cdot 4$ or 50
In 4 dimes and 2 nickels, there is 50¢.	4	2	$10 \cdot 4 + 5 \cdot 2$ or 50
In 5 dimes and 3 nickels, there is 65¢.	5	3	$10 \cdot 5 + 5 \cdot 3$ or 65
In d dimes and n nickels, there is $(10d + 5n)$¢.	d	n	$10d + 5n$

3.   Rule: The value in cents of dimes and nickels is 10 times the number of dimes plus 5 times the number of nickels.

4.   Formula: $c = 10d + 5n$ *Ans.*

6.3. DERIVING COIN FORMULAS

Derive a formula for each relationship:

a) The number of pennies p equivalent to q quarters   *Ans.* $p = 25q$
b) The number of nickels n equivalent to d dimes $n = 2d$
c) The number of quarters q equivalent to D dollars $q = 4D$
d) The number of pennies p equivalent to n nickels and q quarters $p = 5n + 25q$
e) The number of nickels n equivalent to q quarters and d dimes $n = 5q + 2d$

6.4. DERIVING TIME FORMULAS

Derive a formula for each relationship:

a) The number of seconds s in m min   *Ans.* $s = 60m$
b) The number of hours h in d days $h = 24d$
c) The number of weeks w in d days $w = \dfrac{d}{7}$
d) The number of days d in w weeks and 5 days $d = 7w + 5$
e) The number of minutes m in h h and 30s $m = 60h + \frac{1}{2}$

6.5. Deriving Length Formulas

Derive a formula for each relationship:

a) The number of feet *f* in *y* yd   *Ans.* $f = 3y$

b) The number of yards *y* in *i* in $y = \dfrac{i}{36}$

c) The number of meters *m* in 5 km $m = 5000\, m$

6.6. Obtaining Formulas from $D = RT$

From $D = RT$, derive a formula for each relationship:

a) The distance in miles and time in hours when the rate is 35 mi/h   *Ans.* $D = 35T$

b) The distance in feet and the time in seconds when sound travels $D = 1100T$
     at 1100 ft/s

c) The distance in miles and the time in seconds when light travels $D = 186,000T$
     at 186,000 mi/s

d) The distance in miles and the rate in miles per hour when the time $D = 1\frac{1}{2}R$
     of travel is 1 h 30 min

e) The rate in miles per hour and the time in hours when the distance $125 = RT$
     traveled is 125 mi

7. TRANSFORMING FORMULAS

The **subject of a formula** is the letter that has been isolated and expressed in terms of the other letters. Thus, in $p = 4s$, p is the subject of the formula.

Transforming a formula is the process of changing the subject.

Thus, $p = 4s$ becomes $p/4 = s$ when both sides are divided by 4. In the transforming of the formula, the subject has changed from p to s.

In solving a formula for a letter, the formula is transformed in order that the letter be made the subject. Thus, to solve $D = 5T$ for T, transform it into $T = D/5$.

Use Inverse Operations to Transform Formulas

1. Use **division** to undo **multiplication**.
    Thus, $c = 25q$ becomes $c/25 = q$ by division.
2. Use **multiplication** to undo **division**.
    Thus, $w = d/7$ becomes $7w = d$ by multiplication.
3. Use **subtraction** to undo **addition**.
    Thus, $S = P + C$ becomes $S - P = C$ by subtraction.
4. Use **addition** to undo **subtraction**.
    Thus, $S = C - L$ becomes $S + L = C$ by addition.

Formulas may be transformed by transposing terms. In transposing a term, change its sign. Thus, $a + b = 180$ becomes $a = 180 - b$ when $+b$ is transposed. Actually, $+ b$ has been subtracted from both sides to undo addition.

7.1. Transformations Requiring Division

Solve for the letter indicated:

a) $D = RT$ for R *b*) $D = RT$ for T *c*) $V = LWH$ for L

d) $c = 10d$ for d *e*) $C = 2\pi r$ for r

Solutions:

a) $D = RT$	b) $D = RT$	c) $V = LWH$	d) $c = 10d$	e) $C = 2\pi r$
$\mathbf{D}_T$ $\dfrac{D}{T} = \dfrac{RT}{T}$	$\mathbf{D}_R$ $\dfrac{D}{R} = \dfrac{RT}{R}$	$\mathbf{D}_{WH}$ $\dfrac{V}{WH} = \dfrac{LWH}{WH}$	$\mathbf{D}_{10}$ $\dfrac{c}{10} = \dfrac{10d}{10}$	$\mathbf{D}_{2\pi}$ $\dfrac{C}{2\pi} = \dfrac{2\pi r}{2\pi}$
Ans. $\dfrac{D}{T} = R$	Ans. $\dfrac{D}{R} = T$	Ans. $\dfrac{V}{WH} = L$	Ans. $\dfrac{c}{10} = d$	Ans. $\dfrac{C}{2\pi} = r$

7.2. TRANSFORMATIONS REQUIRING MULTIPLICATION

Solve for the letter indicated:

a) $\dfrac{i}{12} = f$ for i b) $f = \dfrac{n}{d}$ for n c) $\dfrac{V}{LW} = H$ for V d) $\dfrac{b}{2} = \dfrac{A}{h}$ for A

Solutions:

a) $\dfrac{i}{12} = f$	b) $f = \dfrac{n}{d}$	c) $\dfrac{V}{LW} = H$	d) $\dfrac{b}{2} = \dfrac{A}{h}$
$\mathbf{M}_{12}$ $12\left(\dfrac{i}{12}\right) = 12f$	$\mathbf{M}_d$ $fd = d\left(\dfrac{n}{d}\right)$	$\mathbf{M}_{LW}$ $LW\left(\dfrac{V}{LW}\right) = LWH$	$\mathbf{M}_h$ $\left(\dfrac{b}{2}\right)h = \left(\dfrac{A}{h}\right)h$
Ans. $i = 12f$	Ans. $fd = n$	Ans. $V = LWH$	Ans. $\dfrac{bh}{2} = A$

7.3. TRANSFORMATIONS REQUIRING ADDITION OR SUBTRACTION

(Transposing is the result of adding or subtracting.) Solve for the letter indicated:

a) $a + b = 90$ for a b) $a = b - 180$ for b c) $a + c = b + 100d$ for b
d) $a - b - 25 = c$ for a

Solutions:

a) $a + b = 90$	b) $a = b - 180$	c) $a + c = b + 100d$	d) $a - b - 25 = c$
Transpose b:	Transpose -180:	Transpose $100d$:	Transpose $-b - 25$:
Ans. $a = 90 - b$	Ans. $a + 180 = b$	Ans. $a + c - 100d = b$	Ans. $a = b + c + 25$

7.4. TRANSFORMATIONS REQUIRING TWO OPERATIONS

Solve each formula for the letter indicated:

a) $P = 2a + b$ for a b) $c = 10d + 25q$ for q c) $F = \frac{9}{5}C + 32$ for C
d) $V = \frac{1}{3}Bh$ for B

Solutions:

a) $P = 2a + b$	b) $c = 10d + 25q$	c) $F = \frac{9}{5}C + 32$	d) $V = \frac{1}{3}Bh$
Transpose b:	Transpose $10d$:	Transpose 32:	$\mathbf{M}_3$ $3V = 3(\frac{1}{3}Bh)$
$\mathbf{D}_2$ $P - b = 2a$	$\mathbf{D}_{25}$ $c - 10d = 25q$	$\mathbf{M}_{5/9}$ $F - 32 = \frac{9}{5}C$	$\mathbf{D}_h$ $3V = Bh$
Ans. $\dfrac{P - b}{2} = a$	Ans. $\dfrac{c - 10d}{25} = q$	Ans. $\frac{5}{9}(F - 32) = C$	Ans. $\dfrac{3V}{h} = B$

7.5. MORE DIFFICULT TRANSFORMATIONS

Solve each formula for the letter indicated:

a) $A = \dfrac{h}{2}(b + b')$ for h b) $S = \dfrac{n}{2}(a + l)$ for a c) $l = a + (n - 1)d$ for n

Solutions:

a) $\qquad\qquad A = \dfrac{h}{2}(b + b')$ $\qquad$ b) $\quad$ $\mathbf{M_2}$ $\quad$ $S = \dfrac{n}{2}(a + l)$ $\qquad$ c) $\quad$ $l = a + (n-1)d$

$\mathbf{M_2}$ $\qquad$ $2A = 2\left(\dfrac{h}{2}\right)(b + b')$ $\qquad\qquad$ $\mathbf{D_n}$ $\quad$ $2S = n(a + l)$ $\qquad$ Transpose a:

$\mathbf{D_{b+b'}}$ $\dfrac{2A}{b + b'} = \dfrac{h(b + b')}{b + b'}$ $\qquad\qquad\qquad$ $\dfrac{2S}{n} = a + l$ $\qquad\qquad$ $\mathbf{D_d}$ $\quad$ $l - a = (n-1)d$

$\qquad\qquad\qquad\qquad\qquad\qquad\qquad\qquad$ Transpose $+l$: $\qquad\qquad\qquad$ $\dfrac{l - a}{d} = n - 1$

$\qquad\qquad\qquad\qquad\qquad\qquad\qquad\qquad\qquad\qquad\qquad\qquad\qquad\qquad$ Transpose -1:

Ans. $\quad \dfrac{2A}{b + b'} = h$ $\qquad\qquad$ Ans. $\quad \dfrac{2S}{n} - l = a$ $\qquad\qquad$ Ans. $\quad \dfrac{l - a}{d} + 1 = n$

8. FINDING THE VALUE OF AN UNKNOWN IN A FORMULA

To Find an Isolated Unknown, Substitute and Solve

The value of an unknown in a formula may be found if values are given for the other letters. By substitution replace the letters by their given values and then solve for the unknown.

Thus, in $A = bh$ if $b = 10$ and $h = 5$, then $A = (10)(5) = 50$.

To Find an Unisolated Unknown, Substitute First or Transform First

When the unknown is not isolated, its value may be found by one of two methods. The first method is to substitute first and then solve. The second method is to transform the formula first to isolate the unknown. After the transformation, substitution is then used.

Thus, in $p = 3s$ if $p = 27$, the value of s may be found.

(1) By substituting first: $27 = 3s$, $s = 9$.
(2) By transforming first: Transform $p = 3s$ into $s = \dfrac{p}{3}$. Then $s = \frac{27}{3}$ or 9.

8.1. FINDING THE VALUE OF AN ISOLATED UNKNOWN

Find each unknown:

a) V if $V = lwh$ and $l = 10$, $w = 2$, $h = 3.2$

b) S if $S = \dfrac{n}{2}(a + l)$ and $n = 8$, $a = 5$, $l = 12$

c) A if $A = p + prt$ and $p = 800$, $r = 0.04$, $t = 3$

d) S if $S = \frac{1}{2}gt^2$, $g = 32$, $t = 5$

Solutions:

a) $\quad V = lwh$ $\qquad$ b) $\quad S = \dfrac{n}{2}(a + l)$ $\qquad$ c) $\quad A = p + prt$ $\qquad\qquad$ d) $\quad S = \frac{1}{2}gt^2$

$\qquad\quad = 10(2)(3.2)$ $\qquad\qquad = \dfrac{8}{2}(5 + 12)$ $\qquad\qquad = 800 + 800(0.04)3$ $\qquad\qquad = \frac{1}{2} \cdot 32 \cdot 5^2$

Ans. $\quad V = 64$ $\qquad$ Ans. $\quad S = 68$ $\qquad$ Ans. $\quad A = 896$ $\qquad\qquad$ Ans. $\quad S = 400$

8.2. FINDING THE VALUE OF AN UNKNOWN THAT IS NOT ISOLATED

Find each unknown, using substitution first or transformation first:

a) Find h if $A = bh$, $b = 13$, $\qquad$ b) Find a if $p = 2a + b$, $\qquad$ c) Find h if $V = \frac{1}{3}Bh$,
and $A = 156$. $\qquad\qquad\qquad$ $b = 20$, and $p = 74$. $\qquad\qquad$ $B = 240$, and $V = 960$.

Solutions:

(1) By substitution first:

$$A = bh$$

$$156 = 13h$$

Ans. $12 = h$

(2) By transformation first:

$$A = bh$$

Transform: $\dfrac{A}{b} = h$

Substitute: $\dfrac{156}{13} = h$

Ans. $12 = h$

(1) By substitution first:

$$p = 2a + b$$

$$74 = 2a + 20$$

Ans. $27 = a$

(2) By transformation first:

$$p = 2a + b$$

Transform: $\dfrac{p - b}{2} = a$

Substitute: $\dfrac{74 - 20}{2} = a$

Ans. $27 = a$

(1) By substitution first:

$$V = \frac{1}{3}Bh$$

$$960 = \frac{1}{3}(240)h$$

Ans. $12 = h$

(2) By transformation first:

$$V = \frac{1}{3}Bh$$

Transform: $\dfrac{3V}{B} = h$

Substitute: $\dfrac{3(960)}{240} = h$

Ans. $12 = h$

8.3. MORE DIFFICULT EVALUATIONS USING TRANSFORMATIONS

Find each unknown:

a) Find h if $V = \pi r^2 h$, $\pi = 3.14$, $V = 9420$, and $r = 10$.
b) Find t if $A = p + prt$, $A = 864$, $p = 800$, and $r = 2$.

Solutions:

a) $V = \pi r^2 h$

Transform: $\dfrac{V}{\pi r^2} = h$

Substitute: $\dfrac{9420}{(3.14)(100)} = h$

                  $\dfrac{9420}{314} = h$

Ans. $30 = h$

b) $A = p + prt$

Transform: $\dfrac{A - p}{pr} = t$

Substitute: $\dfrac{864 - 800}{800(2)} = t$

                  $\dfrac{64}{1600} = t$

Ans. $0.04 = t$

8.4. FINDING AN UNKNOWN IN A PROBLEM

A train takes 3 h 15 min to go a distance of 247 mi. Find its average speed.

Solutions:

Here, $D = RT$, $T = 3\frac{1}{4}$ h, and $D = 247$. To find R:

(1) By substitution first:

$$247 = \frac{13}{4}R$$

$\mathbf{M}_{4/13}$ $\dfrac{4}{13} \cdot \overset{19}{\cancel{247}} = R$

$$76 = R$$

(2) By transposition first:

$$D = RT$$

Transform: $\dfrac{D}{T} = R$

$$247 \div \frac{13}{4} = R$$

$$247 \cdot \frac{4}{13} = R$$

Ans.   Average rate is 76 mi/h

Supplementary Problems

6.1. State the formula for the perimeter p of each: **(3.1)**

    a) right triangle   *Ans.* $p = a + b + c$ *d*) trapezoid   *Ans.* $p = a + b + b' + c$
    b) acute triangle $p = a + b + c$ *e*) rhombus $p = 4s$
    c) parallelogram $p = 2a + 2b$ *f*) isosceles trapezoid $p = 2a + b + b'$

6.2. State the name of the equilateral polygon to which each perimeter formula applies: **(3.2)**

    a) $p = 5s$ *Ans.*   equilateral pentagon   *c*) $p = 6s$ *Ans.*   equilateral hexagon
    b) $p = 8s$ equilateral octagon   *d*) $p = ns$ equilateral n-gon

6.3. Stating the formula used, find the perimeter of each equilateral polygon having a side of 6 in. Express each answer in inches, feet, and yards.   **(3.1, 3.4)**

    a) equilateral triangle   *c*) regular hexagon   *e*) equilateral centagon (100 sides)
    b) square or rhombus   *d*) equilateral decagon   *f*) regular n-gon

    Ans. *a*) $p = 3s$; 18 in, $1\frac{1}{2}$ ft, $\frac{1}{2}$ yd   *c*) $p = 6s$; 36 in, 3 ft, 1 yd   *e*) $p = 100s$; 600 in, 50 ft, $16\frac{2}{3}$ yd

         b) $p = 4s$; 24 in, 2 ft, $\frac{2}{3}$ yd   *d*) $p = 10s$; 60 in, 5 ft, $1\frac{2}{3}$ yd   *f*) $p = ns$; $6n$ in, $\dfrac{n}{2}$ ft, $\dfrac{n}{6}$ yd

6.4. A piece of wire 12 ft long is to be bent into the form of an equilateral polygon. Find the number of feet available for each side of: **(3.3)**

    a) an equilateral triangle   *c*) a regular pentagon   *e*) an equilateral dodecagon
    b) a square   *d*) an equilateral octagon   *f*) an equilateral centagon

    Ans. *a*) each of 3 sides = 4 ft   *c*) each of 5 sides = 2.4 ft   *e*) each of 12 sides = 1 ft
           b) each of 4 sides = 3 ft   *d*) each of 8 sides = $1\frac{1}{2}$ ft   *f*) each of 100 sides = 0.12 ft

6.5. Find the perimeter of: **(3.5)**

    a) a rectangle with sides of 3 and $5\frac{1}{2}$ m   *Ans.* *a*) 17 m
    b) a rhombus with side of 4 yd 1 ft   *b*) $17\frac{1}{3}$ yd
    c) a parallelogram with sides of 3.4 and 4.7 m   *c*) 16.2 m
    d) an isosceles trapezoid with bases of 13 yd 1 ft
          and 6 yd 1 ft and each remaining side 5 yd 2 ft   *d*) 31 yd

6.6. For a rectangle of width w and length l, in inches, find: **(3.6, 3.7)**

    a) the perimeter if $l = 12$ and $w = 7$ *d*) the length if $p = 30$ and $w = 9\frac{1}{2}$
    b) the perimeter if $l = 5\frac{1}{2}$ and $w = 2\frac{3}{8}$ *e*) the length if $p = 52$ and $w = l - 4$
    c) the perimeter if $l = 7.6$ and $w = 4.3$ *f*) the width if $p = 60$ and $l = 2w + 3$

    Ans. *a*) 38 in   *b*) $15\frac{3}{4}$ in   *c*) 23.8 in   *d*) $5\frac{1}{2}$ in
         e) Since $52 = 2l + 2(l - 4)$, length is 15 in   *f*) Since $60 = 2(2w + 3) + 2w$, width is 9 in

6.7. For an isosceles triangle with base b and congruent sides a, in yards, find: **(3.8)**

    a) the perimeter if $a = 12$ and $b = 16$ *d*) the base if $p = 20$ and $s = 7$
    b) the perimeter if $a = 3\frac{1}{4}$ and $b = 4\frac{3}{8}$ *e*) each equal side if $p = 32$ and $b = a + 5$
    c) the perimeter if $a = 1.35$ and $b = 2.04$ *f*) the base if $p = 81$ and $a = b - 3$

    Ans. *a*) 40 yd   *b*) $10\frac{7}{8}$ yd   *c*) 4.74 yd   *d*) Since $20 = 14 + b$, base is 6 yd.
         e) Since $32 = 2a + (a + 5)$, equal side is 9 yd   *f*) Since $81 = 2(b - 3) + b$, base is 29 yd.

6.8. For any circle, state a formula which relates   **(3.9)**

    a) the circumference c and a $45°$ arc a *Ans.* *a*) $c = 8a$
    b) the semicircumference s and the radius r *b*) $s = \pi r$

c) a 120° arc *a'* and the diameter *d*   *Ans.*   *c*) $a' = \dfrac{\pi d}{3}$

d) a 20° arc *a''* and semicircumference *s*   *d*) $a'' = \dfrac{s}{9}$

6.9.   For a circle (see Fig. 6-79), if *r*, *d*, and *c* (*c* = circumference) are in miles, find, using $\pi = 3.14$. **(3.10)**

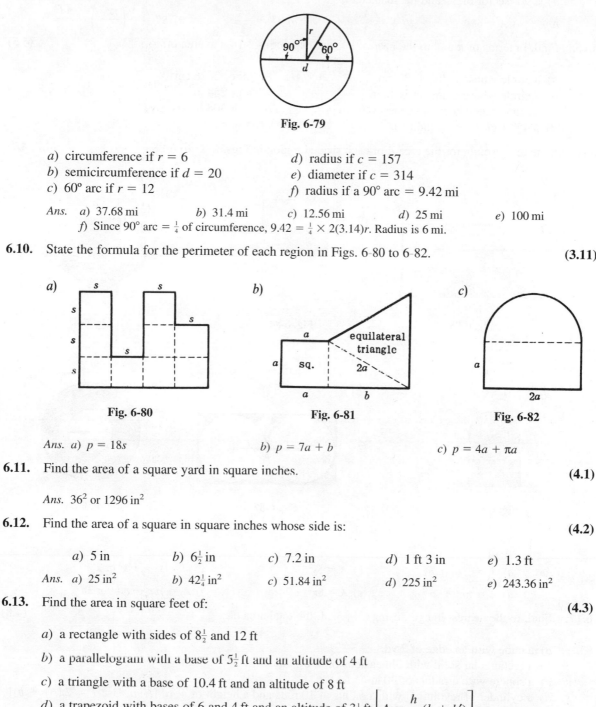

Fig. 6-79

a) circumference if $r = 6$ *d*) radius if $c = 157$

b) semicircumference if $d = 20$ *e*) diameter if $c = 314$

c) 60° arc if $r = 12$ *f*) radius if a 90° arc = 9.42 mi

Ans.   *a*) 37.68 mi   *b*) 31.4 mi   *c*) 12.56 mi   *d*) 25 mi   *e*) 100 mi

     f) Since 90° arc = $\frac{1}{4}$ of circumference, $9.42 = \frac{1}{4} \times 2(3.14)r$. Radius is 6 mi.

6.10.   State the formula for the perimeter of each region in Figs. 6-80 to 6-82.   **(3.11)**

Ans.   *a*) $p = 18s$ *b*) $p = 7a + b$ *c*) $p = 4a + \pi a$

6.11.   Find the area of a square yard in square inches.   **(4.1)**

Ans. 36^2 or 1296 in^2

6.12.   Find the area of a square in square inches whose side is: **(4.2)**

          a) 5 in   *b*) $6\frac{1}{2}$ in   *c*) 7.2 in   *d*) 1 ft 3 in   *e*) 1.3 ft

Ans.   *a*) 25 in^2 *b*) $42\frac{1}{4}$ in^2 *c*) 51.84 in^2 *d*) 225 in^2 *e*) 243.36 in^2

6.13.   Find the area in square feet of: **(4.3)**

a) a rectangle with sides of $8\frac{1}{2}$ and 12 ft

b) a parallelogram with a base of $5\frac{1}{2}$ ft and an altitude of 4 ft

c) a triangle with a base of 10.4 ft and an altitude of 8 ft

d) a trapezoid with bases of 6 and 4 ft and an altitude of $3\frac{1}{2}$ ft $\left[A = \dfrac{h}{2}(b + b') \right]$

Ans.   *a*) 102 ft^2 *b*) 22 ft^2 *c*) 41.6 ft^2 *d*) $17\frac{1}{2}$ ft^2

6.14. For a circle, state a formula which relates: **(4.4)**

a) the area of the circle A and a 60° sector S
b) a 120° sector of S' and the area of the circle A
c) a 90° sector of S'' and the radius r
d) a 90° sector of S'' and the diameter d

Ans. a) $A = 6S$ b) $S' = \dfrac{A}{3}$ c) $S'' = \dfrac{\pi r^2}{4}$ d) $S'' = \dfrac{1}{4}\left(\dfrac{\pi d^2}{4}\right) = \dfrac{\pi d^2}{16}$

6.15. Find, in terms of π and to the nearest integer, using $\pi = 3.14$, the area of: **(4.5)**

a) a circle whose radius is 30 cm   *Ans.* 900π or 2826 cm^2
b) a circle whose diameter is 18 in 81π or 254 in^2
c) a semicircle whose radius is 14 ft 98π or 308 ft^2
d) a 45° sector whose radius is 12 m 18π or 57 m^2

6.16. State the formula for the area A of each shaded region in Figs. 6-83 to 6-88.   **(4.6, 4.7)**

a)

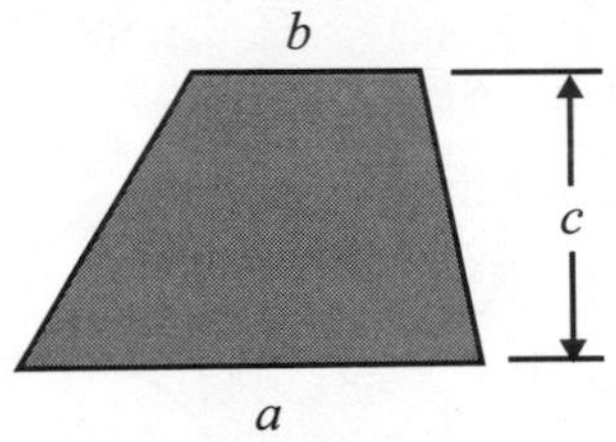

Fig. 6-83

b)

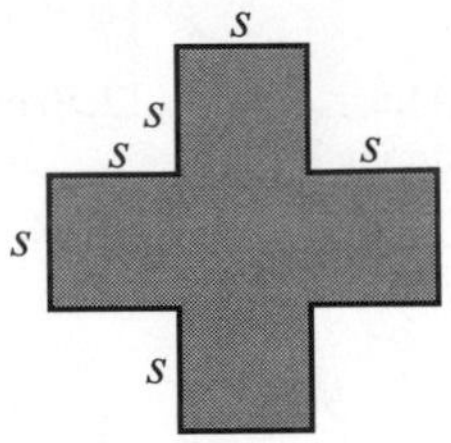

Fig. 6-84

c)

Fig. 6-85

d)

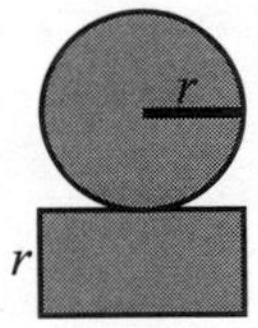

Fig. 6-86

e)

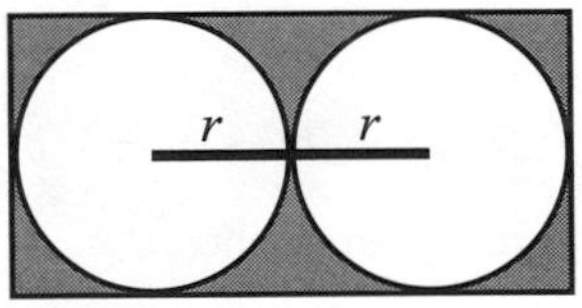

Fig. 6-87

f)

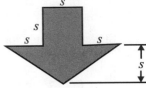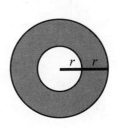

Fig. 6-88

Ans. a) $A = \dfrac{c}{2}(a + b)$ b) $A = 5s^2$ c) $A = s^2 + 1\tfrac{1}{2}s^2$ or $2\tfrac{1}{2}s^2$

d) $A = \pi r^2 + 2r^2$ e) $A = 8r^2 - 2\pi r^2$ f) $A = 4\pi r^2 - \pi r^2$ or $3\pi r^2$

6.17. Find, to the nearest integer, using $\pi = 3.14$, the total area of: **(4.8)**

a) a cube with an edge of 7 yd
b) a rectangular solid with dimensions of 8, $6\tfrac{1}{2}$, and 14 ft
c) a sphere with a radius of 30 in
d) a cylinder of revolution with a radius of 10 rods and a height of $4\tfrac{1}{2}$ rd [*Hint.* Use $T = 2\pi r(r + h)$]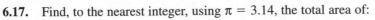

Ans. a) $6(7^2)$ or 294 yd^2 b) $2(8)(6\tfrac{1}{2}) + 2(8)(14) + 2(6\tfrac{1}{2})(14)$ or 510 ft^2

c) $4(3.14)30^2$ or 11,304 in^2 d) $2(3.14)(10)(10 + 4\tfrac{1}{2})$ or 911 rd^2

6.18. Find the volume of: **(5.1)**

    a) a cubic yard in cubic inches

    b) a cubic rod in cubic yards

    c) a cubic meter in cubic centimeters

    Ans. a) 36^3 or 46,656 in^3 *b)* $(5\frac{1}{2})^3$ or $166\frac{3}{8}$ yd^3 *c)* 100^3 or 1,000,000 cm^3

6.19. Find, to the nearest cubic inch, the volume of a cube whose edge is: **(5.2)**

    a) 3 in   *b)* $4\frac{1}{2}$ in   *c)* 7.5 in   *d)* 3 ft   *e)* 1 ft 2 in

  Ans.   *a)* 27 in^3 *b)* 91 in^3 *c)* 422 in^3 *d)* 46,656 in^3 *e)* 2744 in^3

6.20. Find, to the nearest integer, the volume of: **(5.3)**

    a) a rectangular solid whose length is 3 in, width $8\frac{1}{2}$ in, and height 8 in

    b) a prism having a height of 2 ft and a square base whose side is 3 yd

    c) a pyramid having a height of 2 yd and a base whose area is 6.4 ft^2

    Ans. a) $3(8\frac{1}{2})(8)$ or 204 in^3 *b)* $2(9)(9)$ or 162 ft^3 *c)* $\frac{1}{3}(6)(6.4)$ or 13 ft^3

6.21. Find, to the nearest integer, the volume of: **(5.4)**

    a) a sphere with a radius of 6 in

    b) a cylinder having a height of 10 ft and a base whose radius is 2 yd

    c) a cone having a height of 3 yd and a base whose radius is 1.4 ft

  a) $\frac{4}{3}(3.14)6^3$ or 904.32   *b)* $(3.14)(6^2)10$ or 1130.4   *c)* $\frac{1}{3}(3.14)(1.4^2)(9)$ or 18.4632

  Ans. 904 in^3 *Ans.* 1130 ft^3 *Ans.* 18 ft^3

6.22. From $V = \frac{1}{3}Bh$, the volume formula for a pyramid or cone, derive volume formulas for each of the solids in Figs. 6-89 through 6-92. **(5.5)**

a) 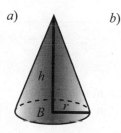   *b)*   *c)* 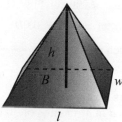   *d)*

        Cone   **Pyramid with a**   **Pyramid with a**   **Cone where**

                 Square Base   **Rectangular Base** $h = 2r$

       Fig. 6-89   **Fig. 6-90**   **Fig. 6-91**   **Fig. 6-92**

  a) $V = \frac{1}{3}Bh$ *b)* $V = \frac{1}{3}Bh$ *c)* $V = \frac{1}{3}Bh$ *d)* $V = \frac{1}{3}Bh$

  Since $B = \pi r^2$ Since $B = s^2$ Since $B = lw$ Since $B = \pi r^2$ and $h = 2r$

     $V = \frac{1}{3}(\pi r^2)h$ $V = \frac{1}{3}(s^2)h$ $V = \frac{1}{3}(lw)h$ $V = \frac{1}{3}(\pi r^2)(2r)$

  Ans. $V = \frac{1}{3}\pi r^2 h$ *Ans.* $V = \frac{1}{3}s^2 h$ *Ans.* $V = \frac{1}{3}lwh$ *Ans.* $V = \frac{2}{3}\pi r^3$

6.23. State a formula for the volume of each solid in Figs. 6-93 to 6-95.   **(5.6)**

a)

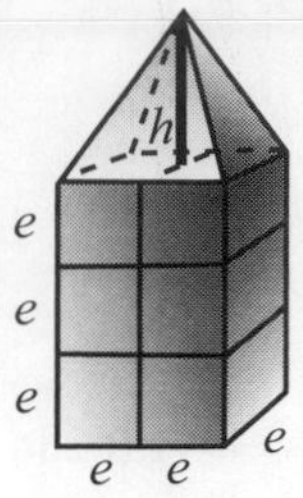

e
e
e
e e e

Fig. 6-93

b)

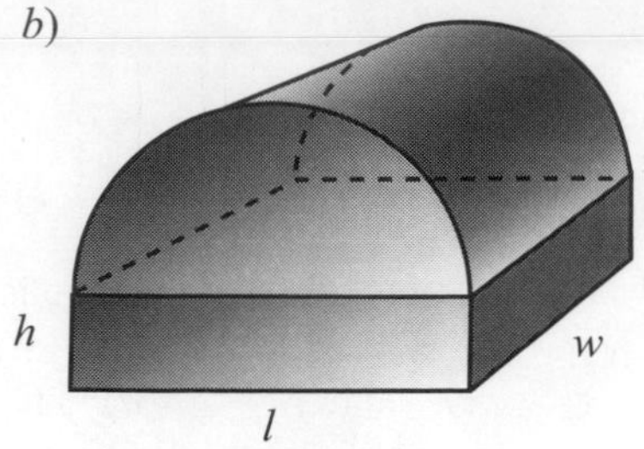

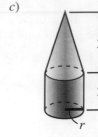

h
l w

Fig. 6-94

c)

$3r$
$2r$
r

Fig. 6-95

    a) $(2e)(3e)e + \frac{1}{3}(2e^2)h$ *b)* $lwh + \frac{1}{2} \cdot \pi\left(\frac{1}{2}\right)^2 w$ *c)* $\pi r^2(2r) + \frac{1}{3}\pi r^2(3r)$

Ans. *a)* $6e^3 + \dfrac{2e^2h}{3}$ *b)* $lwh + \dfrac{\pi l^2 w}{8}$ *c)* $3\pi r^3$

6.24. Derive a formula for each relationship: **(6.1)**

    a) The number of pennies p equivalent to d dimes
    b) The number of dimes d equivalent to p pennies
    c) The number of nickels n equivalent to D dollars
    d) The number of half-dollars h equivalent to q quarters
    e) The number of quarters q equivalent to d dimes

    Ans. *a)* $p = 10d$ *b)* $d = \dfrac{p}{10}$ *c)* $n = 20D$ *d)* $h = \frac{1}{2}q$ *e)* $q = \dfrac{2}{5}d$ or $\dfrac{2d}{5}$

6.25. Derive a formula for each relationship: **(6.2, 6.3)**

    a) The value in cents c of d dimes and q quarters
    b) The value in cents c of n nickels and D dollars
    c) The number of nickels n equivalent to d dimes and h half-dollars
    d) The number of dimes d equivalent to n nickels and p pennies
    e) The number of quarters q equivalent to D dollars and n nickels

    Ans. *a)* $c = 10d + 25q$ *b)* $c = 5n + 100D$ *c)* $n = 2d + 10h$ *d)* $d = \dfrac{n}{2} + \dfrac{p}{10}$ *e)* $q = 4D + \dfrac{n}{5}$

6.26. Derive a formula for each relationship: **(6.4)**

    a) The number of seconds s in h h   *c)* The number of hours h in w weeks
    b) The number of hours h in m min   *d)* The number of days d in M months of 30 days
    e) The number of days d in M months of 30 days, w weeks, and 5 days
    f) The number of minutes m in h h 30 s
    g) The number of days d in y years of 365 days and 3 weeks

    Ans. *a)* $s = 3600h$ *b)* $h = \dfrac{m}{60}$ *c)* $h = 168w$ *d)* $d = 30M$ *e)* $d = 30M + 7w + 5$

           f) $m = 60h + \frac{1}{2}$ *g)* $d = 365y + 21$

6.27. Derive a formula for each relationship: **(6.5)**

    a) The number of inches i in y yd   *d)* The number of miles m in f ft
    b) The number of yards y in f ft   *e)* The number of meters m in c cm
    c) The number of yards y in r rd   *f)* The number of centimeters c in d dm

    Ans. *a)* $i = 36y$ *b)* $y = \dfrac{f}{3}$ *c)* $y = 5\frac{1}{2}r$ *d)* $m = \dfrac{f}{5280}$ *e)* $m = \dfrac{c}{100}$ *f)* $c = 10d$

6.28. From $D = RT$, obtain a formula for each relationship: **(6.6)**

    a) Distance in miles and rate in miles per hour for 5 h
    b) Distance in miles and rate in miles per hour for 30 min
    c) Time in hours and rate in miles per hour for a distance of 25 mi
    d) Time in seconds and rate in feet per second for a distance of 100 yd
    e) Distance in feet and time in minutes for a rate of 20 ft/min
    f) Distance in feet and time in minutes for a rate of 20 ft/s

    Ans.   *a*) $D = 5R$ *b*) $D = \frac{1}{2}R$ *c*) $RT = 25$ *d*) $RT = 300$ *e*) $D = 20T$
           f) $D = 1200T$ (20 ft/s = 1200 ft/min)

6.29. Solve for the letter indicated: **(7.1)**

    a) $d = 2r$ for r *e*) $c = \pi d$ for d *i*) $V = LWH$ for H
    b) $p = 5s$ for s *f*) $c = \pi d$ for π *j*) $V = 2\pi r^2 h$ for h
    c) $D = 30T$ for T *g*) $NP = C$ for N *k*) $9C = 5(F - 32)$ for C
    d) $25W = A$ for W *h*) $I = PR$ for R *l*) $2A = h(b + b')$ for h

    Ans. *a*) $\dfrac{d}{2} = r$ *c*) $\dfrac{D}{30} = T$ *e*) $\dfrac{c}{\pi} = d$ *g*) $N = \dfrac{C}{P}$ *i*) $\dfrac{V}{LW} = H$ *k*) $C = \dfrac{5(F - 32)}{9}$

          b) $\dfrac{p}{5} = s$ *d*) $W = \dfrac{A}{25}$ *f*) $\pi = \dfrac{c}{d}$ *h*) $\dfrac{I}{P} = R$ *j*) $\dfrac{V}{2\pi r^2} = h$ *l*) $\dfrac{2A}{b + b'} = h$

6.30. Solve for the letter indicated: **(7.2)**

    a) $\dfrac{p}{10} = s$ for p *e*) $\pi = \dfrac{c}{2r}$ for c *i*) $\dfrac{V}{3LW} = H$ for V

    b) $R = \dfrac{D}{15}$ for D *f*) $\dfrac{M}{D} = F$ for M *j*) $\dfrac{T}{14RS} = \dfrac{1}{2}$ for T

    c) $W = \dfrac{A}{8}$ for A *g*) $P = \dfrac{A}{2F}$ for A *k*) $\dfrac{L}{KA} = V^2$ for L

    d) $w = \dfrac{d}{7}$ for d *h*) $\dfrac{T}{Q} = 5R$ for T *l*) $\dfrac{V}{\pi r^2} = \dfrac{h}{3}$ for V

    Ans. *a*) $p = 10s$ *c*) $8W = A$ *e*) $2\pi r = c$ *g*) $2PF = A$ *i*) $V = 3LWH$ *k*) $L = KAV^2$

          b) $15R = D$ *d*) $7w = d$ *f*) $M = FD$ *h*) $T = 5RQ$ *j*) $T = 7RS$ *l*) $V = \dfrac{\pi r^2 h}{3}$

6.31. Solve for the letter indicated: **(7.3)**

    a) $a + b = 60$ for a *d*) $3m = 4n + p$ for p *g*) $5a + b = c - d$ for b
    b) $3c + g = 85$ for g *e*) $10r = s - 5t$ for s *h*) $5a - 4c = 3e + f$ for f
    c) $h - 10r = l$ for h *f*) $4g + h - 12 = j$ for h *i*) $\dfrac{b}{2} - 10 + c = 100p$ for c

    Ans. *a*) $a = 60 - b$ *d*) $3m - 4n = p$ *g*) $b = c - d - 5a$
          b) $g = 85 - 3c$ *e*) $10r + 5t = s$ *h*) $5a - 4c - 3e = f$
          c) $h = l + 10r$ *f*) $h = j + 12 - 4g$ *i*) $c = 100p + 10 - \dfrac{b}{2}$

6.32. Solve for the letter indicated: **(7.4)**

    a) $4P - 3R = 40$ for P *d*) $A = \frac{1}{2}bh$ for b *g*) $\dfrac{R}{2} - 4S = T$ for R

    b) $36 - 5i = 12f$ for i *e*) $V = \frac{1}{3}\pi r^2 h$ for h *h*) $8h - \dfrac{k}{5} = 12$ for k

    c) $\dfrac{P}{2} + R = S$ for P *f*) $A = \frac{1}{2}h(b + b')$ for h *i*) $20p - \frac{2}{3}q = 8t$ for q

Ans. a) $P = \dfrac{3R + 40}{4}$　　　d) $\dfrac{2A}{h} = b$　　　g) $R = 8S + 2T$

b) $\dfrac{36 - 12f}{5} = i$　　　e) $\dfrac{3V}{\pi r^2} = h$　　　h) $5(8h - 12) = k$ or $40h - 60 = k$

c) $P = 2S - 2R$　　　f) $\dfrac{2A}{b + b'} = h$　　　i) $\frac{3}{2}(20p - 8t) = q$ or $30p - 12t = q$

6.33. Solve for the letter indicated:　　　　　　　　　　　　　**(7.5)**

　　　a) $l = a + (n - 1)d$ for d　　　b) $s = \dfrac{n}{2}(a + l)$ for l　　　c) $F = \frac{9}{5}C + 32$ for C

Ans. a) $\dfrac{l - a}{n - 1} = d$　　　b) $l = \dfrac{2s}{n} - a$ or $\dfrac{2s - an}{n}$　　　c) $\frac{5}{9}(F - 32) = C$

6.34. Find each unknown:　　　　　　　　　　　　　　　　**(8.1)**

a) Find I if $I = prt$ and $p = 3000, r = 0.05, t = 2$.　　　*Ans.*　a) 300

b) Find t if $t = \dfrac{I}{pr}$ and $I = 40, p = 2000, r = 0.01$.　　　b) 2

c) Find F if $F = \frac{9}{5}C + 32$ and $C = 55$.　　　c) 131

d) Find F if $F = \frac{9}{5}C + 32$ and $C = -40$.　　　d) -40

e) Find C if $C = \frac{5}{9}(F - 32)$ and $F = 212$.　　　e) 100

f) Find S if $S = \frac{1}{2}gt^2$ and $g = 32, t = 8$.　　　f) 1024

g) Find g if $g = \dfrac{2S}{t^2}$ and $S = 800, t = 10$.　　　g) 16

h) Find S if $S = \dfrac{a - lr}{1 - r}$ and $a = 5, l = 40, r = -1$.　　　h) $22\frac{1}{2}$

i) Find A if $A = p + prt$ and $p = 500, r = 0.04, t = 2\frac{1}{2}$.　　　i) 550

6.35. Find each unknown:　　　　　　　　　　　　　　　**(8.2, 8.3)**

a) Find R if $D = RT$ and $D = 30, T = 4$.　　　*Ans.*　a) $7\frac{1}{2}$

b) Find b if $A = bh$ and $A = 22, h = 2.2$.　　　b) 10

c) Find a if $P = 2a + b$ and $P = 12, b = 3$.　　　c) $4\frac{1}{2}$

d) Find w if $P = 2l + 2w$ and $P = 68, l = 21$.　　　d) 13

e) Find c if $p = a + 2b + c$ and $p = 33, a = 11, b = 3\frac{1}{2}$.　　　e) 15

f) Find h if $2A = h(b + b')$ and $A = 70, b = 3.3, b' = 6.7$.　　　f) 14

g) Find B if $V = \frac{1}{3}Bh$ and $V = 480, h = 12$.　　　g) 120

h) Find a if $l = a + (n - 1)d$ and $l = 140, n = 8, d = 3$.　　　h) 119

i) Find E if $l = \dfrac{E}{R + r}$ and $I = 40, R = 3, r = 1\frac{1}{2}$.　　　i) 180

j) Find E if $C = \dfrac{nE}{R + nr}$ if $C = 240, n = 8, R = 14, r = 2$.　　　j) 900

6.36. a) A train takes 5 h 20 min to go a distance of 304 mi. Find its average speed.　　　**(8.4)**

Ans. 57 mi/h $\left(R = \dfrac{D}{T} \right)$

b) A rectangle has a perimeter of 2 yd and a length of 22 in. Find its width.

Ans. 14 in

CHAPTER 7

Graphs of Linear Equations

1. UNDERSTANDING GRAPHS

Reviewing Number Scales

A **number scale** is a line on which distances from a point are numbered in equal units, positively in one direction and negatively in the other.

The **origin** is the zero point from which distances are numbered.

Note on the horizontal number scale in Fig. 7-1 how **positive numbers** are to the **right** of the origin, while on the vertical scale in Fig. 7-2 they are **above** the origin.

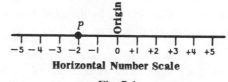

Horizontal Number Scale

Fig. 7-1

The position of any point on a number scale is found by determining its distance from 0 and attaching the appropriate sign. Thus, *P* on the horizontal scale is at position −2.

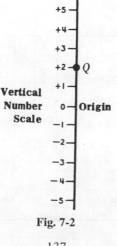

Fig. 7-2

Forming a Graph by Combining Number Scales

The graph shown in Fig. 7-3 is formed by combining two number scales at right angles to each other so that their zero points coincide.

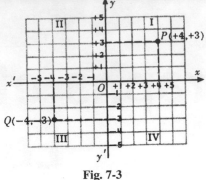

Fig. 7-3

The horizontal number scale is usually called the x **axis**. To show this, x and x' are placed at each end. (x' is read "x prime.")

The vertical number scale is called the y **axis**. Its end letters are y and y'.

The origin O is the point where the two scales cross each other.

To locate a point, determine its **distance from each axis**. Note on the graph that for P, the distances are $+4$ and $+3$.

The **coordinates of a point** are its distances from the axes, with appropriate signs attached.

  (1) The x **coordinate of a point**, its **abscissa**, is its distance from the y axis. This distance is measured along the x axis. For P, this is $+4$; for Q, -4.
  (2) The y **coordinate of a point**, its **ordinate**, is its distance from the x axis. This distance is measured along the y axis. For P, this is $+3$; for Q, -3.
  (3) In stating the coordinates of a point, the x coordinate precedes the y coordinate, just as in the alphabet x precedes y. Place the coordinates in parentheses. Thus, the coordinates of P are written ($+4$, $+3$) or (4, 3), those for Q, (-4, -3).

The **quadrants of a graph** are the four parts cut off by the axes. Note on the graph how these are numbered I, II, III, and IV in a counterclockwise direction.

Comparing a Map and a Graph

A map is a special kind of graph. Each position on a map, such as the map of Graphtown shown in Fig. 7-4, may be located using a street number and an avenue number. Each point on a graph is located by using

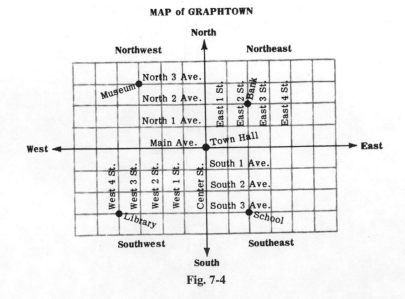

Fig. 7-4

an *x* number and a *y* number. The points *B, M, L, S,* and *T* on the graph correspond to the map position of the bank, museum, library, school, and town hall, respectively.

Note how, in Fig. 7-5, quadrant I corresponds to the northeast map section, quadrant II to the northwest, quadrant III to the southwest, and quadrant IV to the southeast.

Use the map and graph in Exercises **1.1** and **1.2**.

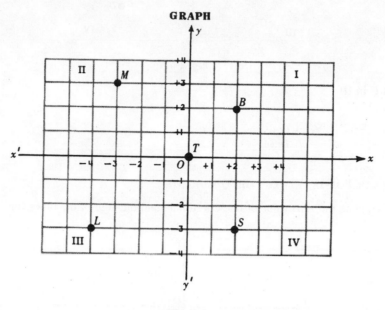

Fig. 7-5

1.1. LOCATING POINTS ON A GRAPH

On the graph shown in Fig. 7-5, locate each point, placing the *x* coordinate before the *y* coordinate:

	Point	Coordinates (Ans.)		Point	Coordinates (Ans.)
a)	B	(+2, +2) or (2, 2)	d)	L	(−4, −3)
b)	M	(−3, +3) or (−3, 3)	e)	S	(+2, −3) or (2, −3)
c)	T	(0, 0), the origin			

1.2. LOCATING POSITIONS ON A MAP

On the map of Graphtown, locate the position of each indicated building, placing the street before the avenue.

	Building	Position (Ans.)		Building	Position (Ans.)
a)	Bank	(E 2 St., N 2 Ave.)	d)	Library	(W 4 St., S 3 Ave.)
b)	Museum	(W 3 St., N 3 Ave.)	e)	School	(E 2 St., S 3 Ave.)
c)	Town hall	(Center St, Main Ave.)			

1.3. COORDINATES OF POINTS IN THE FOUR QUADRANTS

Using Fig. 7-6, state the signs of the coordinates of:

a) any point *A* in I

b) any point *B* in II

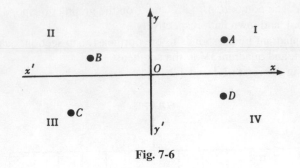

Fig. 7-6

c) any point C in III
d) any point D in IV

Ans.   a) A: (+, +)   c) C: (−, −)
       b) B: (−, +)   d) D: (+, −)

1.4.   COORDINATES OF POINTS BETWEEN THE QUADRANTS

Using Fig. 7-7, state the zero value of one coordinate and the sign of the other for:

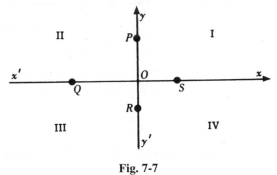

Fig. 7-7

a) any point P between I and II
b) any point Q between II and III
c) any point R between III and IV
d) any point S between IV and I

Ans.   a) P: (0, +)   c) R: (0, −)
       b) Q: (−, 0)   d) S: (+, 0)

1.5.   GRAPHING A QUADRILATERAL

If $A(3, 1)$, $B(−5, 1)$, $C(−5, −3)$, and $D(3, −3)$ are the vertices of the rectangle shown in Fig. 7-8, find its perimeter and area.

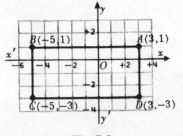

Fig. 7-8

Solution: The base and the height of rectangle *ABCD* are 8 and 4. Hence, the perimeter is 24 units, and the area is 32 square units.   *Ans.*   24, 32

1.6.   GRAPHING A TRIANGLE

If $A(4\frac{1}{2}, -2)$, $B(-2\frac{1}{2}, -2)$, and $C(1,5)$ are the vertices of the triangle shown in Fig. 7-9, find its area.

Solution: The base $BA = 7$. The height $CD = 7$. Since $A = \frac{1}{2}bh$, $A = \frac{1}{2}(7)(7) = 24\frac{1}{2}$.   *Ans.* $24\frac{1}{2}$

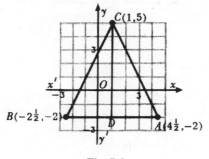

Fig. 7-9

2.   GRAPHING LINEAR EQUATIONS

A linear equation is an equation whose graph is a straight line. (*Note:* See Appendix B for an introduction to graphing lines using a calculator.)

To Graph a Linear Equation

Graph: $y = x + 4$ (see Fig. 7-10)

Procedure:

1. **Make a table of coordinates for three pairs of values as follows:**

   Let *x* have convenient values such as 2, 0, and −2. Substitute each of these for *x*, and find the corresponding value of *y*.

2. **Plot the points and draw the straight line joining them:**

   NOTE: If correct, two points determine a line. The third point serves as a checkpoint to ensure correctness.

Solution:

1. Table of coordinate values:
   Since $y = x + 4$,

   (1) If $x = 2$, $y = 2 + 4 = 6 \rightarrow$
   (2) If $x = 0$, $y = 0 + 4 = 4 \rightarrow$
   (3) If $x = -2$, $y = -2 + 4 = 2 \rightarrow$

Point	Coordinates (x, y)
A	(2, 6)
B	(0, 4)
C	(−2, 2)

2. **Join the plotted points:**

Fig. 7-10

An intercept of a graph is the distance from the origin to the point where the graph crosses either axis. See Fig. 7-11.

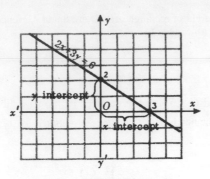

Fig. 7-11

(1) The *x* **intercept** of a graph is the value of *x* for the point where the graph crosses the *x* axis. At this point, $y = 0$.

Thus, for $2x + 3y = 6$, the *x* intercept $= 3$.

(2) The *y* **intercept** of a graph is the value of *y* for the point where the graph crosses the *y* axis. At this point, $x = 0$.

Thus, for $2x + 3y = 6$, the *y* intercept $= 2$.

To Graph a Linear Equation by Using Intercepts

Graph: $2x + 3y = 6$

Procedure

1. Make a table of pairs of values as follows:

   a) Let $x = 0$ to obtain the *y* intercept.
   b) Let $y = 0$ to obtain the *x* intercept.
   c) Obtain a third or checkpoint, using any convenient value for either unknown.

2. Plot the points and join them with a straight line. (See Fig. 7-12.)

Solution:

1. Table of coordinate values:

		Point	Coordinates (x, y)
a)	If $x = 0$, $y = 2$.	A	(0,2)
b)	If $y = 0$, $x = 3$.	B	(3,0)
c)	If $x = -3$, $y = 4$.	C	$(-3, 4)$

2. Join the plotted points:

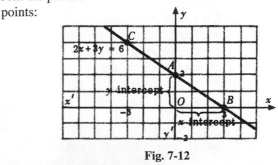

Fig. 7-12

Equations of the First Degree

(1) **An equation of the first degree in one unknown** is one which, after it has been simplified, contains only one unknown having the exponent 1.

Thus, $2x = 7$ is an equation of the first degree in one unknown.

(2) **An equation of the first degree in two unknowns** is one which, after it has been simplified, contains only two unknowns, each of them in a separate term and having the exponent 1.

Thus, $2x = y + 7$ is an equation of the first degree in two unknowns, but $2xy = 7$ is not, since *x* and *y* are not in separate terms.

Rule 1.   The graph of a first-degree equation in one or two unknowns is a straight line. Hence, such equations are **linear equations.**

Thus, the graphs of the equations $y = x + 4$ and $2x + 3y = 6$ are straight lines.

Rule 2.   The graph of a first-degree equation in only one unknown is the x axis, the y axis, or a line parallel to one of the axes. Thus, in Fig. 7-13.

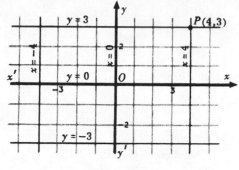

Fig. 7-13

(1) The graph of $y = 0$ is the x axis, and the graph of $x = 0$ is the y axis.
(2) The graphs of $y = 3$ and $y = -3$ are lines parallel to the x axis.
(3) The graphs of $x = 4$ and $x = -4$ are lines parallel to the y axis.

NOTE: The coordinates of any joint of intersection of two such lines are obtainable from the equations of these lines. Thus, $(4, 3)$ is the intersection of $x = 4$ and $y = 3$.

Rule 3.   If a point is **on the graph** of an equation, its coordinates **satisfy** the equation. Thus, in Fig. 7-14, $x = 1$, and $y = 4$, the coordinates of P on the graph of $y = x + 3$, satisfy the equation $y = x + 3$.

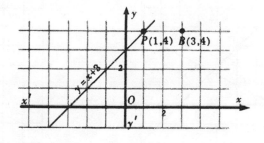

Fig. 7-14

Rule 4.   If a point is **not on the graph** of an equation, its coordinates **do not satisfy** the equation.

Thus, $x = 3$ and $y = 4$, the coordinates of B, which is not on the graph of $y = x + 3$, do not satisfy the equation $y = x + 3$.

2.1.   MAKING TABLES OF COORDINATE VALUES

Complete the table of values for each equation:

a) $y = 2x - 3$

	(x,y)
(1)	$(-2, ?)$
(2)	$(0, ?)$
(3)	$(2, ?)$

b) $x = 3y + 1$

	(x,y)
(1)	$(?, -2)$
(2)	$(?, 0)$
(3)	$(?, 2)$

c) $x + 2y = 10$

	(1)	*(2)*	*(3)*
x	?	?	?
y	-2	0	2

d) $y = 3$

	(1)	*(2)*	*(3)*
x	-2	0	2
y	?	?	?

(Vertically arranged tables)   (Horizontally arranged tables)

Solutions:

If one of two coordinate values is given, the corresponding value is found by substituting the given value in the equation and solving.

a)

	(x, y)
(1)	$(-2, -7)$
(2)	$(0, -3)$
(3)	$(2, 1)$

b)

	(x, y)
(1)	$(-5, -2)$
(2)	$(1, 0)$
(3)	$(7, 2)$

c)

	(1)	(2)	(3)
x	14	10	6
y	-2	0	2

d)

	(1)	(2)	(3)
x	-2	0	2
y	3	3	3

(*y* = 3 for all values of *x*.)

2.2.   RULE 1. GRAPHING LINEAR EQUATIONS

Using the same set of axes, graph:

$$a)\quad y = \frac{x}{2} \text{ and } y = 5 - x \qquad\qquad b)\quad y = \frac{x-2}{2} \text{ and } y = 3$$

Procedure:

1. Make a table of values:

Solutions: (See Fig. 7-15.)   (See Fig. 7-16.)

Point	(x, y)
A	$(-2, -1)$
B	$(0, 0)$
C	$(2, 1)$

Point	(x, y)
D	$(-2, 7)$
E	$(0, 5)$
F	$(2, 3)$

	G	H	I
x	-2	0	2
y	-2	-1	0

	J	K	L
x	-2	0	2
y	3	3	3

2. Join the plotted points:

(a)   (b)

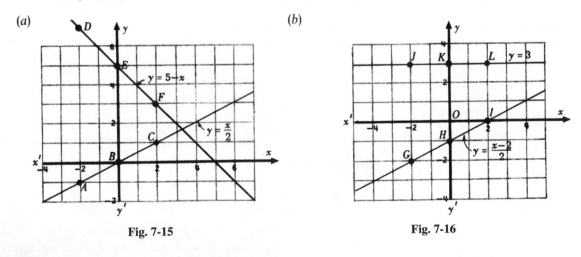

Fig. 7-15   Fig. 7-16

2.3.   GRAPHING LINEAR EQUATIONS, USING INTERCEPTS

Graph, using intercepts:

$$a)\ 2x + 5y = 10 \qquad\qquad b)\ 3x - 4y = 6$$

Procedure:

1. Make a table of values:

   a) Let $x = 0$ to find y intercept:
   b) Let $y = 0$ to find x intercept:
   c) Obtain a checkpoint:

Solutions: (See Fig. 7-17.)   (See Fig. 7-18.)

Point	(x, y)
A	$(0, 2)$
B	$(5, 0)$
C	$(2\frac{1}{2}, 1)$

Point	(x, y)
D	$(0, -1\frac{1}{2})$
E	$(2, 0)$
F	$(4, 1\frac{1}{2})$

2. Join the plotted points:

(a)   (b)

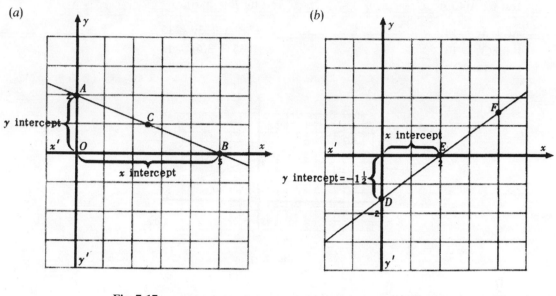

Fig. 7-17   Fig. 7-18

2.4.   RULE 2.   GRAPHING EQUATIONS OF FIRST DEGREE IN ONLY ONE UNKNOWN

Using a separate set of axes for each, graph:

a) $x = -4$, $x = 0$, $x = 2\frac{1}{2}$ (See Fig. 7-19.)   b) $y = 3\frac{1}{2}$, $y = 0$, $y = -4$ (See Fig. 7-20.)

Solutions: (Each graph is an axis or a line parallel to an axis.)

a)   Graphs of $x = -4$ and $x = 2\frac{1}{2}$ are parallel   b)   Graphs of $y = 3\frac{1}{2}$ and $y = -4$ are parallel to x
     to y axis. Graph of $x = 0$ is the y axis.   axis. Graph of $y = 0$ is the x axis.

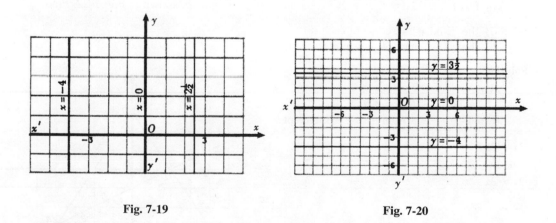

Fig. 7-19   Fig. 7-20

2.5.   RULE 3.   COORDINATE VALUES OF ANY POINT ON A LINE

Points $A(4, 3)$ and $B(-5, 6)$ are on the graph of $x + 3y = 13$. Show that their coordinates satisfy the equation. See Fig. 7-21.

Solutions: (Substitute to test each pair of coordinate values.)

Test for $A(4, 3)$:

$$x + 3y = 13$$
$$4 + 3(3) \stackrel{?}{=} 13$$
$$13 = 13$$

Test for $B(-5, 6)$:

$$x + 3y = 13$$
$$-5 + 3(6) \stackrel{?}{=} 13$$
$$13 = 13$$

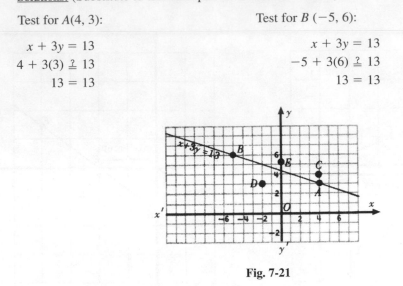

Fig. 7-21

2.6. RULE 4. COORDINATE VALUES OF ANY POINT NOT ON A LINE

Points $C(4, 4)$, $D(-2, 3)$, and $E(0, 5\frac{1}{3})$ are not on the graph of $x + 3y = 13$ used in **2.5**. Show that their coordinates do not satisfy the equation.

Solutions: (Substitute to test each pair of coordinate values.)

Test for $C(4, 4)$:

$$x + 3y = 13$$
$$4 + 3(4) \stackrel{?}{=} 13$$
$$4 + 12 \stackrel{?}{=} 13$$
$$16 \neq 13$$

Test for $D(-2, 3)$:

$$x + 3y = 13$$
$$-2 + 3(3) \stackrel{?}{=} 13$$
$$-2 + 9 \stackrel{?}{=} 13$$
$$7 \neq 13$$

Test for $E(0, 5\frac{1}{3})$:

$$x + 3y = 13$$
$$0 + 3(5\frac{1}{3}) \stackrel{?}{=} 13$$
$$16 \neq 13$$

2.7. INTERCEPTS AND POINTS OF INTERSECTION

The graphs of $3x + 2y = 6$, $x + 2y = -2$, and $2y - 3x = 6$ are shown in Fig. 7-22. Find:

a) the x and y intercepts of each line
b) the coordinates of their points of intersection

Solutions:

a)	(Let $y = 0$) x **Intercept**	(Let $x = 0$) y **Intercept**
$3x + 2y = 6$	2	3
$x + 2y = -2$	-2	-1
$2y - 3x = 6$	-2	3

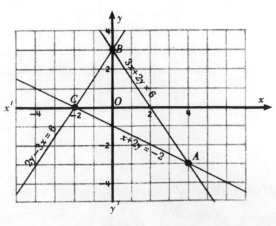

Fig. 7-22

b) $A(4, -3)$ is the point of intersection of $3x + 2y = 6$ and $x + 2y = -2$.
    $B(0, 3)$ is the point of intersection of $3x + 2y = 6$ and $2y - 3x = 6$.
    $C(-2, 0)$ is the point of intersection of $x + 2y = -2$ and $2y - 3x = 6$.

3. SOLVING A PAIR OF LINEAR EQUATIONS GRAPHICALLY

The common solution of two linear equations is the **one and only one** pair of values that satisfies both equations.

Thus, $x = 3$, $y = 7$ is the common solution of the equations $x + y = 10$ and $y = x + 4$. Note that the graphs of these equations meet at the point $x = 3$ and $y = 7$. Since two intersecting lines meet in one and only one point, this pair of values is the common solution.

Consistent, Inconsistent, and Dependent Equations

(1) Equations are **consistent** if one and only one pair of values satisfies both equations. Thus,
$x + y = 10$ and $y = x + 4$, shown in Fig. 7-23, are consistent equations.

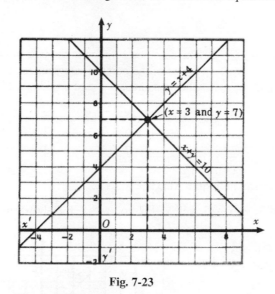

Fig. 7-23

(2) Equations are **inconsistent** if no pairs of values satisfies both equations. Thus, in Fig. 7-24,
$x + y = 3$ and $x + y = 5$ are inconsistent equations. Note that the graphs of these equations are
parallel lines and cannot meet. There is no common solution. Two numbers cannot have a sum of 3
and also 5.

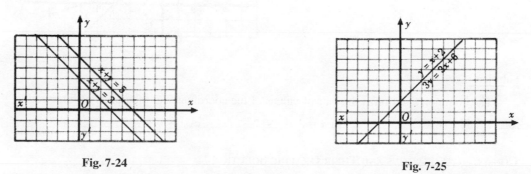

Fig. 7-24   Fig. 7-25

(3) Equations are **dependent** if any pair of values that satisfies one also satisfies the other. Thus, in Fig.
7-25, $y = x + 2$ and $3y = 3x + 6$ are dependent equations. Note that the same line is the graph of
either equation. Hence, any pair of values of a point on this line satisfies either equation.

If equations are dependent, one equation can be obtained from the other by performing the same operation with the same number on both sides. Thus, $3y = 3x + 6$ can be obtained from $y = x + 2$ by multiplying both sides of $y = x + 2$ by 3.

To Solve a Pair of Linear Equations Graphically

Solve $x + 2y = 7$ and $y = 2x + 1$ graphically and check their common solution.

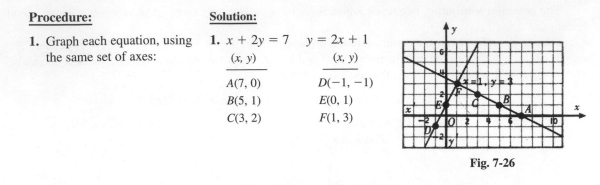

Procedure:	**Solution:**

1. Graph each equation, using the same set of axes:

1. $x + 2y = 7$ $y = 2x + 1$

(x, y)	(x, y)
$A(7, 0)$	$D(-1, -1)$
$B(5, 1)$	$E(0, 1)$
$C(3, 2)$	$F(1, 3)$

Fig. 7-26

2. Find the common solution:

2. The common solution is the pair of coordinate values of the point of intersection: $x = 1$, $y = 3$.

3. Check the values found:

3. Check, using $x = 1$ and $y = 3$:

$$x + 2y = 7 \qquad\qquad y = 2x + 1$$
$$1 + 6 \overset{?}{=} 7 \qquad\qquad 3 \overset{?}{=} 2 + 1$$
$$7 = 7 \qquad\qquad 3 = 3$$

3.1. FINDING COMMON SOLUTIONS GRAPHICALLY

From the graphs of $3y + x = 5$, $x + y = 3$, and $y = x + 3$ shown in Fig. 7-27, find the common solution of:

a) $3y + x = 5$ and $x + y = 3$
b) $3y + x = 5$ and $y = x + 3$
c) $x + y = 3$ and $y = x + 3$

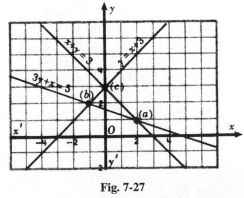

Fig. 7-27

Solutions:

The common solution is the pair of values of the coordinates at the points of intersection.

Ans.   *a)* $x = 2, y = 1$ *b)* $x = -1, y = 2$ *c)* $x = 0, y = 3$

3.2. CONSISTENT EQUATIONS AND THEIR GRAPHIC SOLUTION

Solve each pair of equations graphically:

a) $y = 3$ *b)* $y = 1$ *c)* $x + y = 8$
       $x = 4$ $y = x - 2$ $y = x + 6$

Solutions: See Fig. 7-28.   See Fig. 7-29.

a) Graph of $y = 3$ Graph of $x - 4$ *b)* Graph of $y = 1$ $y = x - 2$
      is a line parallel   is a line parallel   is a line parallel
      to x axis.   to y axis.   to x axis $\dfrac{(x, y)}{(-2, -4)}$

                                                                               $(0, -2)$
                                                                               $(2, 0)$

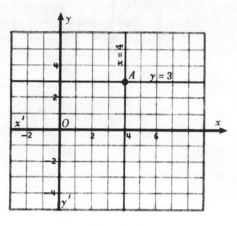

Fig. 7-28

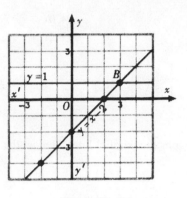

Fig. 7-29

Using intersection point A,   Using intersection point B,

   $x = 4, y = 3$ *Ans.* $x = 3, y = 1$ *Ans.*

c) See Fig. 7-30.

   $x + y = 8$ $y = x + 6$

   $\dfrac{(x, y)}{}$ $\dfrac{(x, y)}{}$

   $(0, 8)$ $(-2, 4)$
   $(2, 6)$ $(0, 6)$
   $(4, 4)$ $(2, 8)$

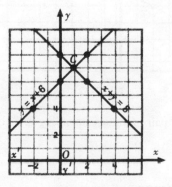

Fig. 7-30

Using intersection point C,

   $x = 1, y = 7$ *Ans.*

3.3. INCONSISTENT EQUATIONS

Show graphically that there is no common solution for the following pair of equations:

$$x + y = 4 \qquad 2x = 6 - 2y$$

Solution: See Fig. 7-31. $x + y = 4 \qquad 2x = 6 - 2y$

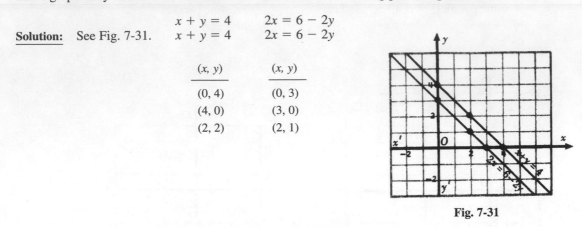

(x, y)	(x, y)
(0, 4)	(0, 3)
(4, 0)	(3, 0)
(2, 2)	(2, 1)

Fig. 7-31

The graphs are parallel lines. Hence, the equations are inconsistent, and there is no common solution.

3.4. DEPENDENT EQUATIONS

For the equations $x - 2y = 2$ and $2x = 4y + 4$, show graphically that any pair of values satisfying one equation satisfies the other equation also. See Fig. 7-32.

Solution:

$$x - 2y = 2 \qquad 2x = 4y + 4$$

(x, y)	(x, y)
(0, −1)	(−2, −2)
(2, 0)	(−4, −3)
(4, 1)	$(1, -\frac{1}{2})$

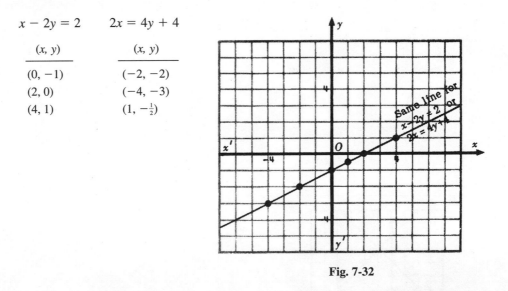

Fig. 7-32

The same line is the graph of the two equations. Hence, any pair of values satisfying one equation satisfies the other also.

4. DERIVING A LINEAR EQUATION FROM A TABLE OF VALUES

Deriving a Simple Linear Equation by Inspection

A simple linear equation such as $y = x + 3$ involves only one operation. Any y value equals 3 **added to** its corresponding x value. Note this in the following table of values for $y = x + 3$:

y	-7	0	3	4	6	13
x	-10	-3	0	1	3	10

A simple equation can be derived from a table of values by inspection. Once the single operation is found, the equation follows. For example, suppose we are given the table:

y	4	8	16	40
x	1	2	4	10

An inspection shows that any y value equals 4 times its corresponding x value. Hence, its equation is $y = 4x$.

Note also that in $y = x + 3$ and $y = 4x$, y is expressed in terms of x. In such cases, it is useful to place y values in a horizontally arranged table.

Deriving a Linear Equation by the Ratio Method

How can we derive a linear equation when two operations are involved, as in the case of $y = 4x + 2$? Here any y value equals 2 added to 4 times its corresponding x value.

Examine the following table of values for $y = 4x + 2$ and notice that as x increases 1, y increases 4; as x increases 2, y increases 8; and finally as x decreases 1, y decreases 4.

Change in y

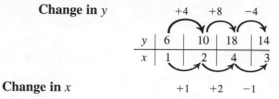

Change in x

Compare each change in y (above the table) with the corresponding change in x (below the table). Notice that the y change or y difference equals 4 times the x change or x difference. From this, we may conclude that the equation is of the form $y = 4x + b$. The value of b may now be found. To find b, substitute any tabular pair of values for x and y in $y = 4x + b$.

Thus, in $y = 4x + b$, substitute $x = 1$ and $y = 6$.

$$6 = 4(1) + b$$
$$2 = b \qquad \text{Since } b = 2, \text{ the equation is } y = 4x + 2.$$

RULE: If a linear equation has the form $y = mx + b$, the value of m can be obtained from a table by using the ratio of the y difference to the corresponding x difference; that is,

$$m = \frac{y \text{ difference}}{x \text{ difference}} = \frac{\text{difference of two } y \text{ values}}{\text{corresponding difference of two } x \text{ values}}$$

(Think of $\underline{m}$ as the $\underline{m}$ultiplier of x in $y = \underline{m}x + b$.)

4.1. DERIVING SIMPLE LINEAR EQUATIONS BY INSPECTION

Derive the linear equation for each table:

a)

y	3	4	10
x	1	2	8

b)

y	2	0	-4
x	5	3	-1

c)

y	7	21	70
x	1	3	10

d)

y	2	5	9
x	6	15	27

Solutions:

a) Since each y value is 2 more than its corresponding x value, $y = x + 2$.   *Ans.*

b) Since each y value is 3 less than its corresponding x value, $y = x - 3$.

c) Since each y value is 7 times its corresponding x value, $y = 7x$.

d) Since each y value is one-third of its corresponding x value, $y = \dfrac{x}{3}$.

4.2. DERIVING LINEAR EQUATION OF FORM $y = mx + b$ BY RATIO METHOD
Derive the linear equation for each table:

a)

y	1	4	10
x	0	1	3

b)

y	-12	-2	18
x	-2	0	4

Procedure:

Solutions:

1. Find m:

$$m = \frac{y \text{ difference}}{x \text{ difference}}$$

1.

$$
\begin{array}{c}
\overset{+3}{\frown}\quad\overset{+6}{\frown}
\end{array}
$$

y	1	4	10
x	0	1	3

$$
\begin{array}{c}
\underset{+1}{\smile}\quad\underset{+2}{\smile}
\end{array}
$$

$$m = \frac{+3}{+1} = \frac{+6}{+2} = 3$$

1.

$$
\begin{array}{c}
\overset{+10}{\frown}\quad\overset{+20}{\frown}
\end{array}
$$

y	-12	-2	18
x	-2	0	4

$$
\begin{array}{c}
\underset{+2}{\smile}\quad\underset{+4}{\smile}
\end{array}
$$

$$m = \frac{+10}{+2} = \frac{+20}{+4} = 5$$

2. Find b:

Substitute any tabular pairs of values in:

$$y = mx + b$$

2. Since $m = 3$,
$y = 3x + b$. Substitute
$x = 0$, $y = 1$
in $y = 3x + b$
$1 = 3(0) + b$
$1 = b$

2. Since $m = 5$,
$y = 5x + b$.   Substitute
$y = 18$, $x = 4$
in $y = 5x + b$
$18 = 5(4) + b$
$-2 = b$

3. Form equation:

$$y = mx + b$$

3. $y = 3x + 1$ *Ans.*

3. $y = 5x - 2$ *Ans.*

5. MIDPOINT OF A SEGMENT

The coordinates (x_m, y_m) of midpoint M of the line segment joining $P(x_1, y_1)$ to $Q(x_2, y_2)$ are:

$$x_m = \tfrac{1}{2}(x_1 + x_2) \quad \text{and} \quad y_m = \tfrac{1}{2}(y_1 + y_2)$$

In Fig. 7-33, segment y_m is the median of trapezoid $CPQD$, whose bases are y_1 and y_2. Since the length of a median is one-half the sum of the bases, $y_m = \tfrac{1}{2}(y_1 + y_2)$. Similarly, segment x_m is the median of trapezoid $ABQP$, whose bases are x_1 and x_2; hence, $x_m = \tfrac{1}{2}(x_1 + x_2)$.

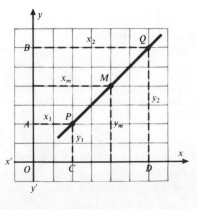

Fig. 7-33

5.1. APPLYING THE MIDPOINT FORMULA

If M is the midpoint of $\overline{PQ}$, find the coordinates of (a) M if the coordinates of P and Q are $P(3, 4)$ and $Q(5, 8)$; (b) Q if the coordinates of P and M are $P(1, 5)$ and $M(3, 4)$.

Solutions:

(a) $x_m = \frac{1}{2}(x_1 + x_2) = \frac{1}{2}(3 + 5) = 4$; $y_m = \frac{1}{2}(y_1 + y_2) = \frac{1}{2}(4 + 8) = 6$.

(b) $x_m = \frac{1}{2}(x_1 + x_2)$, so $3 = \frac{1}{2}(1 + x_2)$ and $x_2 = 5$; $y_m = \frac{1}{2}(y_1 + y_2)$, so $4 = \frac{1}{2}(5 + y_2)$ and $y_2 = 3$.

5.2. DETERMINING IF SEGMENTS BISECT EACH OTHER

The vertices of a quadrilateral are $A(0, 0)$, $B(0, 3)$, $C(4, 3)$, and $D(4, 0)$.

(a) Show that $ABCD$ is a rectangle.

(b) Show that the midpoint of $\overline{AC}$ is also the midpoint of $\overline{BD}$.

(c) Do the diagonals bisect each other? Why?

Solutions:

(a) From Fig. 7-34, $AB = CD = 3$ and $BC = AD = 4$; hence, $ABCD$ is a parallelogram. Since $\angle BAD$ is a right angle, $ABCD$ is a rectangle.

(b) Midpoint of $\overline{AC} = (2, \frac{3}{2}) = $ midpoint of $\overline{BD}$.

(c) Yes; they are congruent and have the same midpoint.

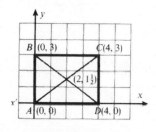

Fig. 7-34

6. DISTANCE BETWEEN TWO POINTS

Rule 1. The distance between two points having the same ordinate (or y value) is the absolute value of the difference of their abscissas. (Hence, the distance between two points must be positive.)

Thus, the distance between points $P(6, 1)$ and $Q(9, 1)$ is $9 - 6 = 3$.

Rule 2. The distance between two points having the same abscissa (or x value) is the absolute value of the difference of their ordinates.

Thus, the distance between points $P(2, 1)$ and $Q(2, 4)$ is $4 - 1 = 3$.

Rule 3. The distance d between points $P_1(x_1, y_1)$ and $P_2(x_2, y_2)$ is:

$$d = \sqrt{(x_2 - x_1)^2 + (y_2 - y_1)^2} \quad \text{or} \quad d = \sqrt{(\Delta x)^2 + (\Delta y)^2}$$

The difference $x_2 - x_1$ is denoted by the symbol Δx; the difference $y_2 - y_1$ is denoted by Δy. Delta (Δ) is the fourth letter of the Greek alphabet, corresponding to our d. The differences Δx and Δy may be positive or negative.

6.1. FINDING THE DISTANCE BETWEEN TWO POINTS BY USING THE DISTANCE FORMULA

Find the distance between the points (a) $(-3, 5)$ and $(1, 5)$; (b) $(3, -2)$ and $(3, 4)$; (c) $(3, 4)$ and $(6, 8)$; (d) $(-3, 2)$ and $(9, -3)$.

Solutions:

(a) Since both points have the same ordinate (or y value), $d = x_2 - x_1 = 1 - (-3) = 4$.

(b) Since both points have the same abscissa (or x value), $d = y_2 - y_1 = 4 - (-2) = 6$.

(c) $d = \sqrt{(x_2 - x_1)^2 + (y_2 - y_1)^2} = \sqrt{(6 - 3)^2 + (8 - 4)^2} = \sqrt{3^2 + 4^2} = 5$

(d) $d = \sqrt{(x_2 - x_1)^2 + (y_2 - y_1)^2} = \sqrt{[9 - (-3)]^2 + (-3 - 2)^2} = \sqrt{12^2 + (-5)^2} = 13$

6.2.   APPLYING THE DISTANCE FORMULA TO A TRIANGLE

(a) Find the lengths of the sides of a triangle whose vertices are $A(1, 1)$, $B(1, 4)$, and $C(5, 1)$.

(b) Show that the triangle whose vertices are $G(2, 10)$, $H(3, 2)$, and $J(6, 4)$ is a right triangle.

Solutions:

See Fig. 7-35.

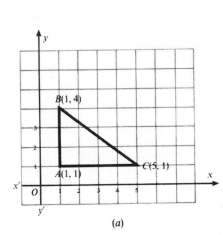

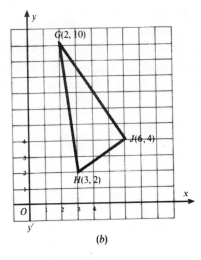

(a)   (b)

Fig. 7-35

(a) $AC = 5 - 1 = 4$ and $AB = 4 - 1 = 3$; $BC = \sqrt{(5 - 1)^2 + (1 - 4)^2} = \sqrt{4^2 + (-3)^2} = 5$.

(b) $(GJ)^2 = (6 - 2)^2 + (4 - 10)^2 = 52$; $(HJ)^2 = (6 - 3)^2 + (4 - 2)^2 = 13$; $(GH)^2 = (2 - 3)^2 + (10 - 2)^2 = 65$. Since $(GJ)^2 + (HJ)^2 = (GH)^2$, ΔGHJ is a right triangle.

6.3.   APPLYING THE DISTANCE FORMULA TO A PARALLELOGRAM

The coordinates of the vertices of a quadrilateral are $A(2, 2)$, $B(3, 5)$, $C(6, 7)$, and $D(5, 4)$. Show that $ABCD$ is a parallelogram.

Solution:

See Fig. 7-36, where we have:

$$AB = \sqrt{(3 - 2)^2 + (5 - 2)^2} = \sqrt{1^2 + 3^2} = \sqrt{10}$$

$$CD = \sqrt{(6 - 5)^2 + (7 - 4)^2} = \sqrt{1^2 + 3^2} = \sqrt{10}$$

$$BC = \sqrt{(6 - 3)^2 + (7 - 5)^2} = \sqrt{3^2 + 2^2} = \sqrt{13}$$

$$AD = \sqrt{(5 - 2)^2 + (4 - 2)^2} = \sqrt{3^2 + 2^2} = \sqrt{13}$$

Thus, $\overline{AB} \cong \overline{CD}$ and $\overline{BC} \cong \overline{AD}$. Since opposite sides are congruent, $ABCD$ is a parallelogram.

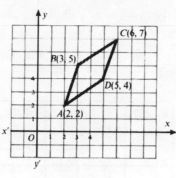

Fig. 7-36

Supplementary Problems

7.1. State the coordinates of each lettered point on the graph in Fig. 7-37.　　　　**(1.1)**

Ans.

A(3, 0)
B(4, 3)
C(3, 4)
D(0, 2)
E(−2, 4)
F(−4, 2)
G(−1, 0)
H(−3½, −2)
I(−2, −3)
J(0, −4)
K(1½, −2½)
L(4, −2½)

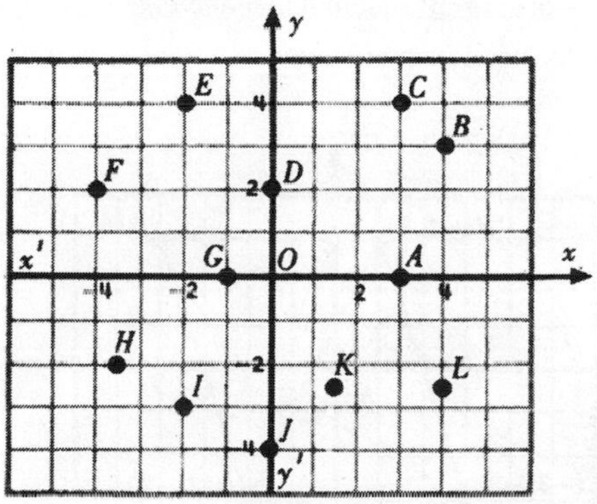

Fig. 7-37

7.2. Plot each point and locate it with reference to quadrants I to IV:　　　　**(1.3, 1.4)**

A(−2, −3)　C(0, −1)　E(3, −4)　G(0, 3)
B(−3, 2)　　D(−3, 0)　F(1½, 2½)　H(3½, 0)

Ans. See Fig. 7-38.
F is in I
B is in II
A is in III
E is in IV
G is between I and II
D is between II and III
C is between III and IV
H is between IV and I

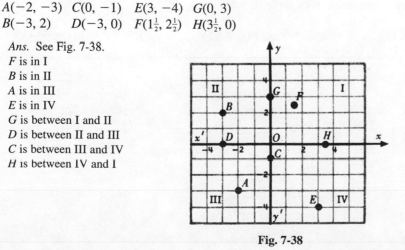

Fig. 7-38

7.3. Plot each point: $A(2, 3)$, $B(-3, 3)$, $C(-3, -2)$, $D(2, -2)$. Find the perimeter and area of square $ABCD$.   **(1.5)**

    Ans. Perimeter of square formed is 20 units, its area is 25 square units. See Fig. 7-39.

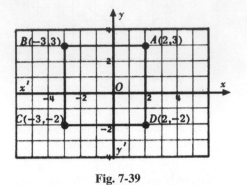

Fig. 7-39

7.4. Plot each point: $A(4, 3)$, $B(-1, 3)$, $C(-3, -3)$, $D(2, -3)$. Find the area of parallelogram $ABCD$ and triangle BCD.   **(1.5, 1.6)**

    Ans. Area of parallelogram $= bh = 5(6)$ or 30 square units. See Fig. 7-40.

        Area of $\triangle BCD = \frac{1}{2}bh = \frac{1}{2}(30) = 15$ square units

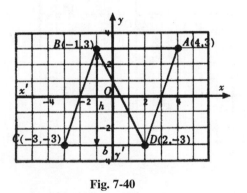

Fig. 7-40

7.5. Graph each equation after completing each table of values: **(2.1, 2.2)**

    a) $y = -4$ (See Fig. 7-41.)   *b*) $y = 2x$ (See Fig. 7-42.)   *c*) $x = 2y + 3$ (Fig. 7-43.)

	(x, y)
A	$(-2, ?)$
B	$(0, ?)$
C	$(2, ?)$

	(x, y)
D	$(-2, ?)$
E	$(0, ?)$
F	$(2, ?)$

	(x, y)
G	$(?, -1)$
H	$(?, 0)$
I	$(?, 1)$

Ans. *a*)

	(x, y)
A	$(-2, -4)$
B	$(0, -4)$
C	$(2, -4)$

Ans. (*b*)

	(x, y)
D	$(-2, -4)$
E	$(0, 0)$
F	$(2, 4)$

Ans. *c*)

	(x, y)
G	$(1, -1)$
H	$(3, 0)$
I	$(5, 1)$

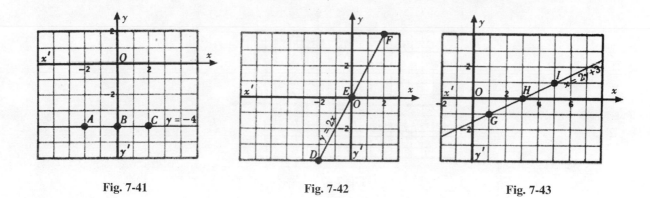

Fig. 7-41   Fig. 7-42   Fig. 7-43

7.6. Graph each equation, using intercepts: **(2.3)**

a) $3x + 2y = 6$ (See Fig. 7-44.)   *b)* $4y - 3x = 6$ (See Fig. 7-45.)   *c)* $\dfrac{x}{4} + \dfrac{y}{3} = 1$ (See Fig. 7-46.)

Ans. a)	(x, y)
A	$(0, 3)$
B	$(2, 0)$
C	$(4, -3)$

Ans. b)	(x, y)
D	$(0, 1\frac{1}{2})$
E	$(-2, 0)$
F	$(2, 3)$

Ans. c)	(x, y)
G	$(0, 3)$
H	$(4, 0)$
I	$(-4, 6)$

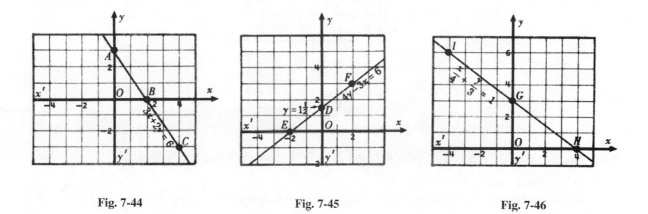

Fig. 7-44   Fig. 7-45   Fig. 7-46

7.7. Using one set of axes, graph each line and state the coordinates of their nine points of intersection:

$x = 1\frac{1}{2}, x = 0, x = -3, y = 5, y = 0, y = -2\frac{1}{2}.$ **(2.4)**

Ans.   The points of intersection are:

$A(1\frac{1}{2}, 0)$	$D(-3, 5)$	$G(0, -2\frac{1}{2})$
$B(1\frac{1}{2}, 5)$	$E(-3, 0)$	$H(1\frac{1}{2}, -2\frac{1}{2})$
$C(0, 5)$	$F(-3, -2\frac{1}{2})$	$I(0, 0)$

See Fig. 7-47.

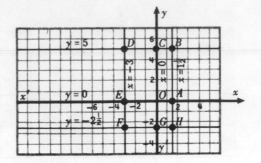

Fig. 7-47

7.8. Locate $A(1, 0)$ and $B(2, -4)$ on the graph of $y + 4x = 4$, and show that their coordinates satisfy the equation. Also show that $C(2, 2)$ and $D(-1, 3)$ are not on the graph of $y + 4x = 4$, and show that their coordinates do not satisfy the equation.   **(2.5, 2.6)**

Ans. Test $A(1, 0)$.   Test $B(2, -4)$.   Test $C(2, 2)$.   Test $D(-1, 3)$.
$$y + 4x = 4 \qquad\qquad y + 4x = 4 \qquad\qquad y + 4x = 4 \qquad\qquad y + 4x = 4$$
$$0 + 4(1) \overset{?}{=} 4 \qquad -4 + 4(2) \overset{?}{=} 4 \qquad 2 + 4(2) \overset{?}{=} 4 \qquad 3 + 4(-1) \overset{?}{=} 4$$
$$4 = 4 \qquad\qquad\quad 4 = 4 \qquad\qquad\quad 10 \neq 4 \qquad\qquad\quad -1 \neq 4$$

        A is on graph. B is on graph. C is not on graph. D is not on graph.

See Fig. 7-48.

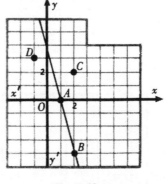

Fig. 7-48

7.9. The graphs of $2y - x = 6$, $2y - x = 2$, and $2y = 3x - 2$ are shown in Fig. 7-49. Find: **(2.7)**

    a) the x and y intercepts of each line

    b) the coordinates of any point of intersection common to two graphs

Ans. a) **Equation**	(Let $y = 0$.) x **Intercept**	(Let $x = 0$.) y **Intercept**
$2y - x = 6$	-6	3
$2y - x = 2$	-2	1
$2y = 3x - 2$	$\frac{2}{3}$	-1

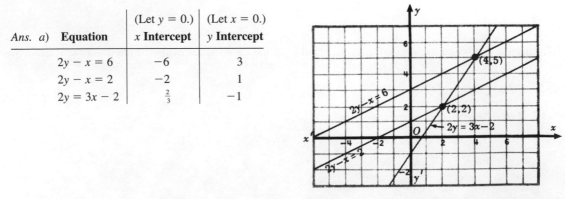

Fig. 7-49

b) (2, 2) is the point of intersection of $2y - x = 2$ and $2y = 3x - 2$.
      (4, 5) is the point of intersection of $2y - x = 6$ and $2y = 3x - 2$.
      Since $2y - x = 2$ and $2y - x = 6$ are parallel, they have no point of intersection.

7.10. From the graph of $3x + 4y = 12$, $4y + 7x = 12$, and $3x + 4y = -4$, find any solution to any two equations.   **(3.1)**

    Ans.   The common solution of $3x + 4y = 12$ and $4y + 7x = 12$ is $x = 0$, $y = 3$. The common solution of $3x + 4y = -4$ and $4y + 7x = 12$ is $x = 4$, $y = -4$. Since $3x + 4y = 12$ and $3x + 4y = -4$ are parallel, there is no common solution. See Fig. 7-50.

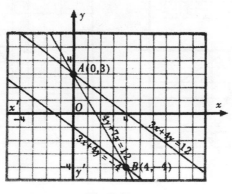

Fig. 7-50

7.11. Solve each pair of equations graphically: **(3.2)**

    a) $y = -2$ *b*) $y = 4$ *c*) $3x + 4y = -6$
        $x = 3\frac{1}{2}$ $x + y = 2$ $y = 3 - 3x$

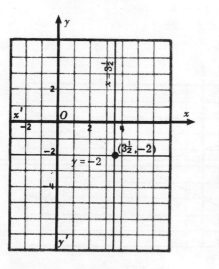
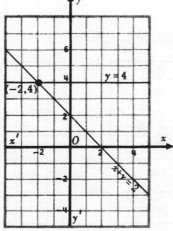
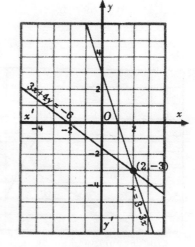

    Fig. 7-51   **Fig. 7-52**   **Fig. 7-53**

  Ans. $x = 3\frac{1}{2}$, $y = -2$ *Ans.* $x = -2$, $y = 4$ *Ans.* $x = 2$, $y = -3$
    See Fig. 7-51.   See Fig. 7-52.   See Fig. 7-53.

7.12. Graph each equation and determine which equations have no common solution: (3.3)

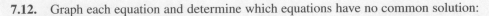

a) $2x + 3y = 6$
    $5x + 3y = 6$
    $2x + 3y = -3$

b) $y + 2x = 0$
    $2y - 3x = 6$
    $y = 3 - 2x$

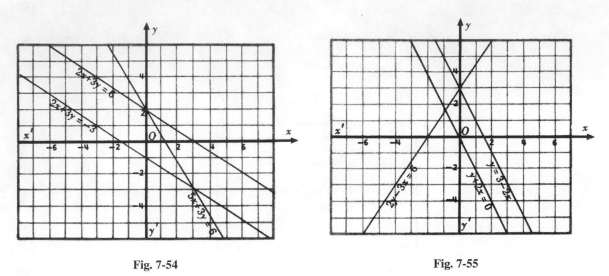

| Fig. 7-54 | Fig. 7-55 |

$2x + 3y = 6$ and $2x + 3y = -3$ have no common solution. See Fig. 7-54.

$y = 3 - 2x$ and $y + 2x = 0$ have no common solution. See Fig. 7-55.

7.13. Graph each equation and determine which equations are such that any pair of values satisfying one equation satisfies the other also: (3.4)

a) $2y - x = 2$
    $y + 2x = 6$
    $3x = 6y - 6$

b) $2y = x - 4$
    $x + 4y = 4$
    $8y = 8 - 2x$

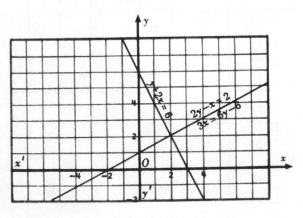

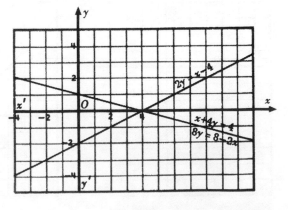

| Fig. 7-56 | Fig. 7-57 |

The graphs of $2y - x = 2$ and $3x = 6y - 6$ are the same line. A pair of values satisfying one equation satisfies the other. See Fig. 7-56.

The graphs of $x + 4y = 4$ and $8y = 8 - 2x$ are the same line. A pair of values satisfying one equation satisfies the other. See Fig. 7-57.

7.14. Derive the linear equation for each table:　　　　**(4.1)**

a)

y	5	8	9
x	2	5	6

b)

y	-4	-6	-10
x	-2	-3	-5

c)

y	-8	-7	-3
x	0	1	5

d)

y	-1	1	4
x	-3	3	12

e)

y	2	2	2
x	7	-7	20

f)

y	4	2	0
x	3	5	7

g)

p	6	9	15
q	8	12	20

h)

r	$4\frac{1}{2}$	$\frac{1}{2}$	-2
s	$1\frac{1}{2}$	$-2\frac{1}{2}$	5

Ans.　a)　$y = x + 3$　　b)　$y = 2x$　　　　c)　$y = x - 8$　　　　d)　$y = \dfrac{x}{3}$

e)　$y = 2$　　f)　$x + y = 7$　　　　g)　$p = \frac{3}{4}q$　　　　h)　$r = s + 3$

7.15. Derive the linear equation for each table:　　　　**(4.2)**

a)

y	-1	1	5
x	0	1	3

b)

y	-1	3	15
x	-1	0	3

c)

y	4	3	-2
x	-1	0	5

d)

p	8	0	-4
q	0	4	6

e)

y	2	8	14
x	-2	0	2

f)

y	-9	-4	6
x	-3	0	2

g)

y	6	0	-2
x	0	3	4

h)

s	1	3	7
t	-4	0	8

i)

y	7	16	28
x	1	4	8

j)

y	27	13	-1
x	5	3	1

k)

y	7	-2	-14
x	-1	2	6

l)

a	5	7	11
b	6	9	15

Ans.　a)　$y = 2x - 1$　　　　b)　$y = 4x + 3$　　　　c)　$y = -x + 3$　　　　d)　$p = -2q + 8$

e)　$y = 3x + 8$　　　　f)　$y = 5x - 4$　　　　g)　$y = -2x + 6$　　　　h)　$s = \frac{1}{2}t + 3$

i)　$y = 3x + 4$　　　　j)　$y = 7x - 8$　　　　k)　$y = -3x + 4$　　　　l)　$a = \frac{2}{3}b + 1$

7.16. Find the midpoint of the segment joining:　　　　**(5.1)**

a)　$(0, 0)$ and $(8, 6)$　　　　e)　$(-20, -5)$ and $(0, 0)$　　　　i)　$(3, 4)$ and $(7, 6)$

b)　$(0, 0)$ and $(5, 7)$　　　　f)　$(0, 4)$ and $(0, 16)$　　　　j)　$(-2, -8)$ and $(-4, -12)$

c)　$(0, 0)$ and $(-8, 12)$　　　　g)　$(8, 0)$ and $(0, -2)$　　　　k)　$(7, 9)$ and $(3, 3)$

d)　$(14, 10)$ and $(0, 0)$　　　　h)　$(-10, 0)$ and $(0, -5)$　　　　l)　$(2, -1)$ and $(-2, -5)$

Ans.　a) $(4, 3)$　　　b) $(2\frac{1}{2}, 3\frac{1}{2})$　　　c) $(-4, 6)$　　　d) $(7, 5)$　　　e) $(-10, -2\frac{1}{2})$　　　f) $(0, 10)$

g) $(4, -1)$　　　h) $(-5, -2\frac{1}{2})$　　　i) $(5, 5)$　　　j) $(-3, -10)$　　　k) $(5, 6)$　　　l) $(0, -3)$

7.17. Find the midpoints of the sides of a triangle whose vertices are:　　　　**(5.1)**

a)　$(0, 0), (8, 0), (0, 6)$　　　　d)　$(3, 5), (5, 7), (3, 11)$

b)　$(-6, 0), (0, 0), (0, 10)$　　　　e)　$(4, 0), (0, -6), (-4, 10)$

c)　$(12, 0), (0, -4), (0, 0)$　　　　f)　$(-1, -2), (0, 2), (1, -1)$

Ans.　a)　$(4, 0), (0, 3), (4, 3)$　　　　d)　$(4, 6), (4, 9), (3, 8)$

b)　$(-3, 0), (0, 5), (-3, 5)$　　　　e)　$(2, -3), (-2, 2), (0, 5)$

c)　$(6, -2), (0, -2), (6, 0)$　　　　f)　$(-\frac{1}{2}, 0), (\frac{1}{2}, \frac{1}{2}), (0, -\frac{3}{2})$

7.18. Find the midpoints of the sides of the quadrilateral whose successive vertices are:　　　　**(5.2)**

a)　$(0, 0), (0, 4), (2, 10), (6, 0)$　　　　c)　$(-2, 0), (0, 4), (6, 2), (0, -10)$

b)　$(-3, 5), (-1, 9), (7, 3), (5, -1)$　　　　d)　$(-3, -7), (-1, 5), (9, 0), (5, -8)$

Ans.　a)　$(0, 2), (1, 7), (4, 5), (3, 0)$　　　　c)　$(-1, 2), (3, 3), (3, -4), (-1, -5)$

b)　$(-2, 7), (3, 6), (6, 1), (1, 2)$　　　　d)　$(-2, -1), (4, \frac{5}{2}), (7, -4), (1, -\frac{15}{2})$

7.19. Find the distance between each of the following pairs of points: **(6.1)**

        a)   (0, 0) and (0, 5)   *d)* (−6, −1) and (−6, 11)   *g)* $(-3, -4\frac{1}{2})$ and $(-3, 4\frac{1}{2})$

        b)   (4, 0) and (−2, 0)   *e)* (5, 3) and (5, 8.4)   *h)* (*a*, *b*) and (2*a*, *b*)

        c)   (0, −3) and (0, 7)   *f)* (−1.5, 7) and (6, 7)

  Ans. *a)* 5   *b)* 6   *c)* 10   *d)* 12   *e)* 5.4   *f)* 7.5   *g)* 9   *h)* *a*

7.20. Find the distance between each of the following pairs of points: **(6.1)**

        a)   (0, 0) and (5, 12)   *e)* (−3, −6) and (3, 2)   *i)* (3, 4) and (4, 7)

        b)   (−3, −4) and (0, 0)   *f)* (2, 3) and (−10, 12)   *j)* (−1, −1) and (1, 3)

        c)   (0, −6) and (9, 6)   *g)* (2, 2) and (5, 5)   *k)* $(-3, 0)$ and $(0, \sqrt{7})$

        d)   (4, 1) and (7, 5)   *h)* (0, 5) and (−5, 0)   *l)* (*a*, 0) and (0, *a*)

  Ans. *a)* 13   *b)* 5   *c)* 15   *d)* 5   *e)* 10   *f)* 15   *g)* $3\sqrt{2}$ *h)* $5\sqrt{2}$ *i)* $\sqrt{10}$

       j) $2\sqrt{5}$ *k)* 4   *l)* $a\sqrt{2}$

7.21. Repeat Problems 7.10. and 7.11 using a graphing calculator. (See Appendix B for further details.)

CHAPTER 8

Introduction to Simultaneous Equations

1. SOLVING A PAIR OF EQUATIONS BY ADDITION OR SUBTRACTION

To Solve a Pair of Equations by Adding or Subtracting

Solve:
(1) $3x - 7 = y$
(2) $4x - 5y = 2$

Procedure:

1. **Arrange** so that like terms are in the same column:

2. **Multiply** so that the coefficients of one of the unknowns will have the same absolute value:

3. **To eliminate the unknown** whose coefficients have the same absolute value:
   a) **Add** if their signs are **unlike**.
   b) **Subtract** if their signs are **like**.

4. **Find the remaining unknown** by solving the resulting equation:

5. **Find the other unknown** by substituting the value found in any equation having both unknowns:

6. **Check** the common solution in each of the original equations:

Solution:

1. Arrange: (1) $3x - 7 = y \rightarrow 3x - y = 7$
             (2) $\qquad\qquad\quad 4x - 5y = 2$

2. Multiply:
   $$\mathbf{M_5} \quad 3x - y - 7 \rightarrow 15x - 5y = 35$$

3. Eliminate y:

$$15x - 5y = 35$$
$$\text{Subtract:} \quad \underline{4x - 5y = 2}$$
$$11x = 33$$

4. Find x: $\qquad \mathbf{D_{11}} \qquad 11x = 33$
   $$x = 3$$

5. Find y: $\qquad\qquad\qquad 4x - 5y = 2$
   Substitute 3 for x: $\quad 4(3) - 5y = 2$
   $$y = 2$$

6. Check for $x = 3$, $y = 2$:

   (1) $\quad 3x - 7 = y$ $\qquad$ (2) $\quad 4x - 5y = 2$
   $\quad 3(3) - 7 \stackrel{?}{=} 2$ $\qquad\qquad 4(3) - 5(2) \stackrel{?}{=} 2$
   $\qquad\qquad 2 = 2$ $\qquad\qquad\qquad\qquad\quad 2 = 2$

1.1.   USING ADDITION OR SUBTRACTION TO ELIMINATE ONE UNKNOWN

Add or subtract to eliminate one unknown; then find the value of remaining unknown:

$a)$ $5x + 3y = 19$ $b)$ $10x + 4y = 58$ $c)$ $10y = 38 - 6x$ $d)$ $3x + 10 = 5y$

        $x + 3y = 11$ $13x - 4y = 57$ $12y = 48 - 6x$ $7x + 20 = -5y$

Solutions:

$a)$ By subtraction, $4x = 8$ $b)$ By adding, $23x = 115$ $c)$ By subtraction, $-2y = -10$

               Ans. $x = 2$ *Ans.* $x = 5$ *Ans.* $y = 5$

$d)$ By adding, $10x + 30 = 0$

               Ans. $x = -3$

1.2.   SOLUTIONS NOT REQUIRING MULTIPLICATION

Solve by addition or subtraction and check:

$a)$ (1) $5x + 3y = 17$ $b)$ (1) $10x + 4y = 20$

    (2) $x + 3y = 1$ (2) $13x - 4y = -66$

Solutions:

$a)$ By subtracting, $4x = 16$ $b)$ By adding, $23x = -46$

                $x = 4$ $x = -2$

    Substitute 4 for x in $x + 3y = 1$ Substitute -2 for x in $10x + 4y = 20$

                $4 + 3y = 1$ $-20 + 4y = 20$

                     $y = -1$ $y = 10$

    Ans. $x = 4, y = -1$ *Ans.* $x = -2, y = 10$

    Check for $x = 4, y = -1$:   Check for $x = -2, y = 10$:

    (1) $5x + 3y = 17$ (2) $x + 3y = 1$ (1) $10x + 4y = 20$ (2) $13x - 4y = -66$

      $20 - 3 \stackrel{?}{=} 17$ $4 - 3 \stackrel{?}{=} 1$ $-20 + 40 \stackrel{?}{=} 20$ $-26 - 40 \stackrel{?}{=} -66$

            $17 = 17$ $1 = 1$ $20 = 20$ $-66 = -66$

1.3.   SOLUTIONS REQUIRING MULTIPLICATION

Solve by addition or subtraction:

$a)$ (1) $13x - 4y = 57$ $b)$ (1) $3r - 5s = 19$

    (2) $5x + 2y = 29$ (2) $2r - 4s = 16$

Solutions: [In (a) one multiplication is needed, in (b) two are needed.]

$a)$ (1) $13x - 4y = 57$ $b)$ (1) $\mathbf{M_2}$ $3r - 5s = 19 \rightarrow 6r - 10s = 38$

    (2) $\mathbf{M_2}$ $5x + 2y = 29 \rightarrow \underline{10x + 4y = 58}$ (2) $\mathbf{M_3}$ $2r - 4s = 16 \rightarrow \underline{6r - 12s = 48}$

        By adding, $23x\;\;\;\;\; = 115$ By subtracting, $2s = -10$

                   $x = 5$ $s = -5$

    Substitute 5 for x in $5x + 2y = 29$ Substitute -5 for s in $3r - 5s = 19$

             $25 + 2y = 29$ $3r + 25 = 19$

                   $y = 2$ $r = -2$

    Ans. $x = 5, y = 2$ *Ans.* $s = -5, r = -2$

1.4.   SOLUTIONS REQUIRING REARRANGEMENT OF TERMS

Rearrange, then solve by addition or subtraction:

$a)$ (1) $2x + y = 16$ $b)$ (1) $3(x - 2) = 8y + 1$
       (2) $y = 25 - 5x$ (2) $8(y + 1) = 4x - 4$

Procedure: **Solutions:**

1. Rearrange: (1) $2x + y = 16$ (1) $3x - 8y = \quad 7$
                          (2) $\underline{5x + y = 25}$ (2) $\underline{-4x + 8y = -12}$
2. Add or subtract: By subtraction $-3x = -9$ By adding $-x = -5$
                                         $x = 3$ $x = 5$
3. Substitute: Substitute 3 for x in   Substitute 5 for x in
                                   $2x + y = 16$ $3(x - 2) = 8y + 1$
                                    $6 + y = 16$ $9 = 8y + 1$
                                        $y = 10$ $1 = y$

            Ans. $x = 3, y = 10$ *Ans.* $x = 5, y = 1$

1.5.   FRACTIONAL PAIRS OF EQUATIONS

Solve by addition or subtraction:

$a)$ (1) $\frac{1}{2}x + \frac{1}{4}y = 5$ $b)$ (1) $\dfrac{a}{b} + 5 = -\dfrac{4}{b}$
       (2) $\frac{1}{2}x - \frac{3}{4}y = -3$ (2) $-2a - 3b = -6$

Solutions:

$a)$ (1) $\frac{1}{2}x + \frac{1}{4}y = \quad 5$ $b)$ (1) $\mathbf{M}_b$ $\dfrac{a}{b} + 5 = -\dfrac{4}{b}$
              (2) $\underline{\frac{1}{2}x - \frac{3}{4}y = -3}$

   By subtracting $y = 8$ $\mathbf{M}_2$ $a + 5b = -4$
   Substitute 8 for y in $\frac{1}{2}x + \frac{1}{4}y = 5$ $2a + 10b = -8$
                          $\frac{1}{2}x + 2 = 5$ (2) $\underline{-2a - 3b = -6}$
                               $x = 6$

                                         By adding $7b = -14$ $b = -2$
Ans. $x = 6, y = 8$ Substitute -2 for b in $-2a - 3b = -6$
                                                           $-2a + 6 = -6$

                                         Ans. $a = 6, b = -2$

1.6.   DECIMAL PAIRS OF EQUATIONS

Solve by addition or subtraction:

$a)$ (1) $0.3x = 0.6y$ $b)$ (1) $0.2x = 0.05y + 75$
       (2) $x + 6y = 8000$ (2) $3x - 5y = 700$

Solutions:

       (1) $\mathbf{M}_{10}$ $0.3x = 0.6y$ (1) $\mathbf{M}_{100}$ $0.2x = 0.05y + 75$
       Rearrange $3x = 6y$ Rearrange $20x = 5y + 7500$
                   $3x - 6y = 0$ $20x - 5y = 7500$
       (2) $\underline{x + 6y = 8000}$ (2) $\underline{3x - 5y = \quad 700}$

   By adding $4x = 8000$ $x = 2000$ By subtracting $17x = 6800$ $x = 400$
   Substitute 2000 for x in $x + 6y = 8000$ Substitute 400 for x in $3x - 5y = 700$
                          $2000 + 6y = 8000$ $1200 - 5y = 700$

Ans. $x = 2000, y = 1000$ *Ans.* $x = 400, y = 100$

2. SOLVING A PAIR OF EQUATIONS BY SUBSTITUTION

To Solve a Pair of Equations by Substitution

$$\text{Solve:} \quad (1) \quad x - 2y = 7$$
$$\quad\quad\quad (2) \quad 3x + y = 35$$

Procedure:

1. After studying the equations to determine which transformation provides the easier expression for one of the unknowns, **express one unknown in terms of the other** by transforming one of the equations:

2. In the other equation, **substitute** for the first unknown the expression containing the other unknown:

3. **Find the remaining unknown** by solving the resulting equation:

4. **Find the other unknown** by substituting the value found in any equation having both unknowns:

5. **Check** the common solution in each of the original equations:

Solution:

1. Express x in terms of y:
   (1) $x - 2y = 7 \rightarrow x = 2y + 7$

2. Substitute for x:
   Substitute $2y + 7$ for x in $3x + y = 35$
   $$3(2y + 7) + y = 35$$

3. Find y:
   $$6y + 21 + y = 35$$
   $$7y = 14$$
   $$y = 2$$

4. Find x: Substitute 2 for y in $x = 2y + 7$
   $$x = 4 + 7$$
   $$x = 11$$

5. Check for $x = 11$, $y = 2$:
   (1) $\quad x - 2y = 7 \quad$ (2) $\quad 3x + y = 35$
   $\quad 11 - 2(2) \stackrel{?}{=} 7 \quad\quad 3(11) + 2 \stackrel{?}{=} 35$
   $\quad\quad\quad 7 = 7 \quad\quad\quad\quad 35 = 35$

Supplementary Problems

8.1. Add or subtract to eliminate one unknown; then find the value of the remaining unknown: **(1.1)**

$\quad a)\quad 3x - y = 21$
$\quad\quad\quad 2x + y = 4$
$\quad b)\quad 7a + b = 22$
$\quad\quad\quad 5a + b = 14$

$\quad c)\quad m + 4n = 6$
$\quad\quad\quad m - 2n = 18$
$\quad d)\quad 3k + 5p = 9$
$\quad\quad\quad 3k - p = -9$

$\quad e)\quad 8r = 11 - 5s$
$\quad\quad\quad 5r = 15 + 5s$
$\quad f)\quad -10u = 2v + 30$
$\quad\quad\quad -u = -2v + 14$

Ans. $a)$ $x = 5$ (Add, $5x = 25$)
$\quad\quad\ b)$ $a = 4$ (Subtract, $2a = 8$)

$c)$ $n = -2$ (Subtract, $6n = -12$)
$d)$ $p = 3$ (Subtract, $6p = 18$)

$e)$ $r = 2$ (Add, $13r = 26$)
$f)$ $u = -4$ (Add, $-11u = 44$)

8.2. Solve by addition or subtraction: **(1.2)**

$\quad a)\quad 7a + t = 42$
$\quad\quad\quad 3a - t = 8$
$\quad b)\quad 4r - 2s = -14$
$\quad\quad\quad 4r - 5s = -32$

$\quad c)\quad 2a + 5s = 44$
$\quad\quad\quad 6a - 5s = -8$
$\quad d)\quad 7c - 3d = 23$
$\quad\quad\quad -2c - 3d = 5$

$\quad e)\quad 8r - 1 = 3t$
$\quad\quad\quad 8r + 1 = 9t$
$\quad f)\quad -7g = 31 - 2h$
$\quad\quad\quad -17g = 17 + 2h$

Ans. $a)$ $a = 5$, $t = 7$ (Add)

$\quad\quad\ b)$ $r = -\frac{1}{2}$, $s = 6$ (Subtract)

$c)$ $a = 4\frac{1}{2}$, $s = 7$ (Add)

$d)$ $c = 2$, $d = -3$ (Subtract)

$e)$ $r = \frac{1}{4}$, $t = \frac{1}{3}$ (Subtract)

$f)$ $g = -2$, $h = 8\frac{1}{2}$ (Add)

8.3. Solve by addition or subtraction. (One multiplication is needed before adding or subtracting.) **(1.3a)**

a) (1) $2x - y = 5$
(2) $3x + 3y = 21$

b) (1) $A + 4B = 18$
(2) $5A + 3B = 5$

c) (1) $3c + 5d = 11$
(2) $2c - d = 16$

d) (1) $7r = 5s + 15$
(2) $2r = s + 9$

e) (1) $3h - 4j = 13$
(2) $-6h + 3j = -21$

f) (1) $6t = 3v + 63$
(2) $5t = 9v + 85$

Ans. a) $x = 4, y = 3$ (**M₃** in equation 1 and add)

b) $A = -2, B = 5$ (**M₅** in equation 1 and subtract)

c) $c = 7, d = -2$ (**M₅** in equation 2 and add)

d) $r = 10, s = 11$ (**M₅** in equation 2 and subtract)

e) $h = 3, j = -1$ (**M₂** in equation 1 and add)

f) $t = 8, v = -5$ (**M₃** in equation 1 and subtract)

8.4. Solve by addition or subtraction. (Two multiplications are needed before adding or subtracting.) **(1.3b)**

a) (1) $3x + 5y = 11$
(2) $4x - 3y = 5$

b) (1) $5t - 2s = 17$
(2) $8t + 5s = 19$

c) (1) $8a + 3b = 13$
(2) $3a + 2b = 11$

d) (1) $6c + 12b = 7$
(2) $8c - 15b = -1$

e) (1) $-3m + 4p = -6$
(2) $5m - 6p = 8$

f) (1) $7r + 11s = 35$
(2) $6r - 12s = 30$

Ans. a) $x = 2, y = 1$ (**M₄** in equation 1, **M₃** in equation 2, and subtract)

b) $t = 3, s = -1$ (**M₅** in equation 1, **M₂** in equation 2, and add)

c) $a = -1, b = 7$ (**M₂** in equation 1, **M₃** in equation 2, and subtract)

d) $c = \frac{1}{2}, b = \frac{1}{3}$ (**M₄** in equation 1, **M₃** in equation 2, and subtract)

e) $m = -2, p = -3$ (**M₃** in equation 1, **M₂** in equation 2, and subtract)

f) $r = 5, s = 0$ (**M₆** in equation 1, **M₇** in equation 2, and subtract)

8.5. Rearrange; then solve by addition or subtraction: **(1.4)**

a) $3x - y = 17$
$y = 8 - 2x$

b) $2y - 7x = 2$
$3x = 14 - y$

c) $9 = 8p - 5q$
$10 + 7q = 6p$

d) $10C - 9D = 18$
$6D + 2C = 1$

e) $12r - 3s = 3$
$3(r - 1) = s - 4$

f) $4(W + 3) = 3Z + 7$
$2(Z - 5) = W + 5$

Ans. a) $x = 5, y = -2$
e) $r = 2, s = 7$

b) $x = 2, y = 8$
f) $W = 7; Z = 11$

c) $p = \frac{1}{2}, q = -1$

d) $C = \frac{3}{2}, D = -\frac{1}{3}$

8.6. Solve by addition or subtraction: **(1.5)**

a) $\frac{1}{2}c + \frac{1}{2}d = 4$

$\frac{1}{2}c - \frac{1}{2}d = -2$

b) $\frac{4}{5}r - \frac{1}{4}s = 11$

$\frac{3}{5}r - \frac{1}{4}s = 8$

c) $3K - \frac{1}{3}N = 11$

$2K + \frac{1}{3}N = 4$

d) $x - y = 17$

$\frac{4}{3}x + \frac{3}{2}y = 0$

e) $3r - 2p = 32$

$\frac{r}{5} + 3p = -1$

f) $\frac{3b}{a} - 4 = \frac{3}{a}$

$\frac{5a}{b} + 2 = \frac{25}{b}$

Ans. a) $c = 2, d = 6$
e) $r = 10, p = -1$

b) $r = 15, s = 4$
f) $a = 3, b = 5$

c) $K = 3, N = -6$

d) $x = 9, y = -8$

8.7. Solve by addition or subtraction: **(1.6)**

a) $3x - y = 500$
$0.7x + 0.2y = 550$

b) $a - 2b = 500$
$0.03a + 0.02b = 51$

c) $y = 4x - 100$
$0.06y = 0.05x + 32$

d) $0.03C + 0.04D = 44$
$0.04C + 0.02D = 42$

e) $0.8R - 0.7T = 140$
$0.03R + 0.05T = 51$

f) $0.05(W + 2000) = 0.03(Y + 3000)$
$W = \frac{Y}{2} + 500$

Ans.   a) $x = 500, y = 1000$ c) $y = 700, x = 200$ e) $R = 700, T = 600$
         b) $a = 1400, b = 450$ d) $C = 800, D = 500$ f) $W = 4000, Y = 7000$

8.8.   Substitute to eliminate one unknown, then find the value of the remaining unknown: **(2.1)**

a) $y = 2x$
     $7x - y = 35$

b) $a = b + 2$
     $3a + 4b = 20$

c) $R = \dfrac{S}{3}$
     $3R + 2S = 36$

d) $r = 4t - 1$
     $6t + r = 79$

e) $3p = 27 - q$
     $2q = 3p$

f) $5y - 9x = -24$
     $5y = 11x$

g) $a = 9 - 3b$
     $7b + 5a = 33$

h) $3d - 2g = 27$
     $d = 4 - g$

i) $s = \dfrac{t}{3} - 1$
     $6s + t = 21$

Ans.   a)   (Substitute $2x$ for y)
              $x = 7$

b)   (Substitute $b + 2$ for a)
     $b = 2$

c) $\left(\text{Substitute } \dfrac{S}{3} \text{ for } R\right)$
     $S = 12$

d)   (Substitute $4t - 1$ for r)
     $t = 8$

e)   (Substitute $2q$ for $3p$)
     $q = 9$

f)   (Substitute $11x$ for $5y$)
     $x = -12$

g)   (Substitue $9 - 3b$ for a)
     $b = 1\frac{1}{2}$

h)   (Substitute $4 - g$ for d)
     $g = -3$

i) $\left(\text{Substitute } \dfrac{t}{3} - 1 \text{ for } s\right)$
     $t = 9$

8.9.   Solve by substitution: **(2.2)**

a) $x - y = 12$
     $3x = 1 - 4y$

b) $5A - 8B = 8$

     $B + A = 12$

c) $r - 3s = 11$
     $5s + 30 = 4r$

d) $p = 2(r - 5)$

     $4p + 40 = r - 7$

e) $6a = 7c + 7$
     $7c - a = 28$

f) $h - 5 = \dfrac{d}{3}$

     $3h - 2d = -6$

Ans.   a)   (Substitute $y + 12$
              for x) $x = 7, y = -5$

b)   (Substitute $12 - B$ for A)

     $A = 8, B = 4$

c)   (Substitute $3s + 11$ for r)
     $r = 5, s = -2$

d)   (Substitute $2r - 10$ for p)

     $p = -12, r = -1$

e)   (Substitute $7c - 28$ for a)
     $a = 7, c = 5$

f) $\left(\text{Substitute } \dfrac{d}{3} + 5 \text{ for } h\right)$

     $h = 12, d = 21$

8.10.   Solve by substitution: **(2.3)**

a) $x - 9b = 0$

     $\dfrac{x}{3} = 2b + \dfrac{1}{3}$

b) $r + 5 = 2s$

     $\dfrac{4s + 1}{5} = 3r - 3$

c) $\dfrac{c}{2} + \dfrac{d}{3} = 9$

     $c = 4d + 4$

d) $h + 10m = 900$
     $0.4h = -2m + 300$

Ans.   a)   (Substitute $9b$ for x)
              $b = \frac{1}{3}, x = 3$

b)   (Substitute $2s - 5$ for r)
     $s = 3\frac{1}{2}, r = 2$

c)   (Substitute $4d + 4$ for c)
     $c = 16, d = 3$

d)   (Substitute $900 - 10m$ for h)
     $h = 600, m = 30$

8.11.   Solve the systems of equations in Problem 8.10 using a graphing calculator.

CHAPTER 9

Problem Solving and Mathematical Modeling

The **four steps of problem solving** are as follows:

1. **Representation** of unknowns by letters (preceded, obviously, by identification of all unknowns!).
2. **Translation** of relationships about unknowns into equations.
3. **Solution** of equations to find the values of the unknowns.
4. **Verification** or **check** of the values found to see if they **satisfy the original problem.**

1. NUMBER PROBLEMS HAVING ONE UNKNOWN: INTRODUCTION TO PROBLEM SOLVING

In number problems having one unknown, a single relationship about the unknown is needed. After the unknown is represented by a letter such as n or x, this relationship is used to obtain an equation.

The value of the unknown is found by solving the equation. However, the value found must be checked in the original problem. A check in any equation is insufficient since the equation may be incorrect.

To Solve a Number Problem Having One Unknown

Solve: Twice a certain number increased by 10
equals 32. Find the number.

Procedure:

1. **Represent** the unknown by a letter:

2. **Translate** the relationship about the unknown into an equation:

3. **Solve** the equation:

Solution:

1. Representation: Let n = number.

2. Translation:
Twice a certain number increased by 10 equals 32.

$$2n \qquad + 10 \qquad = \qquad 32$$

3. Solution: $2n = 22$ or $n = 11$.   *Ans.*   The number is 11.

4. Check the value found in the original problem: (**Do not check in just any equation!**)	**4.** Check: Does 11 satisfy the statement "twice the number increased by 10 equals 32"? If it does, then:

$$2(11) + 10 \overset{?}{=} 32$$
$$32 = 32$$

1.1. TRANSLATION OF STATEMENTS INTO EQUATIONS

Using n to represent the unknown, express each statement as an equation; then find the unknown.

		Equation	Value of n
a)	If a number is decreased by 5, the result is 28.	*Ans.* $n - 5 = 28$	$n = 33$
b)	Three times a number, increased by 8, equals 41.	$3n + 8 = 41$	$n = 11$
c)	Two-fifths of a number equals 18.	$\dfrac{2n}{5} = 18$	$n = 45$
d)	A number added to one-fourth of itself equals 45.	$n + \dfrac{n}{4} = 45$	$n = 36$
e)	When 8 times a number is diminished by 20, the remainder is 28.	$8n - 20 = 28$	$n = 6$
f)	Ten exceeds one-fourth of a number by 3.	$10 - \dfrac{n}{4} = 3$	$n = 28$

1.2. VERIFICATION OR CHECK IN ORIGINAL STATEMENT

Check each statement for the number indicated:

a) Does 20 check in "one-fifth of the number increased by 3 is 7"?   *Ans.*   Yes

    Check: $\frac{20}{5} + 3 \overset{?}{=} 7, 4 + 3 \overset{?}{=} 7, 7 = 7$

b) Does 3 check in "7 times a number less 12 is 9"?   *Ans.*   Yes

    Check: $7(3) - 12 \overset{?}{=} 9, 21 - 12 \overset{?}{=} 9, 9 = 9$

c) Does 24 check in "three-fourths of the number decreased by 10 equals 8"?   *Ans.*   Yes

    Check: $\frac{3}{4}(24) - 10 \overset{?}{=} 8, 18 - 10 \overset{?}{=} 8, 8 = 8$

d) Does 21 check in "5 times the sum of the number and 8 is 3 times the number less 2"? *Ans.*   No

    Check: $5(21 + 8) \overset{?}{=} 3(21) - 2, 5(29) \overset{?}{=} 63 - 2, 145 \neq 61$

1.3. COMPLETE SOLUTIONS OF NUMBER PROBLEMS

a) Find a number such that 3 times the number decreased by 5 equals 19.	b) What number added to 40 is the same as 5 times the number?	c) If 11 times a number is increased by 10, the result is 14 times the number less 5. Find the number.

Solutions:

a) Let n = number.	b) Let n = number.	c) Let n = number.
$3n - 5 = 19$	$n + 40 = 5n$	$11n + 10 \overset{?}{=} 14n - 5$
$3n = 24$	$40 = 4n$	$15 = 3n$
$n = 8$	$n = 10$	$n = 5$
Ans. The number is 8.	*Ans.* The number is 10.	*Ans.* The number is 5.

Check:

Does 8 check in "3 times the number decreased by 5 equals 19"?	Does 10 check in "the number added to 40 is 5 times the number"?	Does 5 check in "11 times the number increased by 10 is 14 times the number less 5"?

$$24 - 5 \overset{?}{=} 19 \qquad\qquad 40 + 10 \overset{?}{=} 50 \qquad\qquad 55 + 10 \overset{?}{=} 70 - 5$$
$$19 = 19 \qquad\qquad\qquad 50 = 50 \qquad\qquad\qquad 65 = 65$$

2. NUMBER PROBLEMS HAVING TWO UNKNOWNS

In number problems having two unknowns, two relationships concerning the unknowns are needed. Such problems may be solved by one of two methods:

Method 1. Using One Letter and Obtaining One Equation
> One of the relationships is used to represent the two unknowns in terms of one letter. The other relationship is then used to obtain a single equation.

Method 2. Using Two Letters and Obtaining Two Equations
> Each of the unknowns is represented by a different letter. Each of the two relationships is then used to obtain a separate equation.

The value found for the unknown by either method must be checked in the **original problem**.

> **A Check in Any Equations(s) Is Insufficient Since It (They) May Be Incorrect!**

To Solve a Number Problem Having Two Unknowns

Solve: One positive number is twice another. The larger is 10 more than the smaller. Find the numbers.

Procedure, Using One Letter:

1. **Represent** one of the unknowns by a letter. Represent the other unknown in terms of the letter, using one of the relationships.
2. **Translate** the other relationship into an equation:

3. **Solve** the equation:

4. **Check** the values found in the **original problem:**

Solution: Method 1

1. Representation: Let $s =$ smaller number.
   Then $2s =$ larger number, since the larger is twice the smaller.

2. Translation: The larger is 10 more than the smaller.
   $$2s \qquad = \qquad s + 10$$

3. Solution: $2s - s = 10$, $s = 10$, and $2s = 20$.

   Ans. The numbers are 20 and 10.

4. Check: Are the numbers 20 and 10?
   One number is twice another.   The larger is 10 more than the smaller.
   Hence, $20 \overset{?}{=} 2(10)$ Hence, $20 \overset{?}{=} 10 + 10$
   $\qquad\quad 20 = 20$ $\qquad\quad 20 = 20$

Procedure, Using Two Letters:

1. **Represent** each of the two unknowns by a different letter:
2. **Translate** each relationship into a separate equation:

3. **Solve** both equations:

4. **Check** in the **original problem:**

Solution: Method 2

1. Representation: Let $l =$ larger number
   $s =$ smaller number

2. Translation: One number is twice another.
   $$l \qquad = \qquad 2s$$
   The larger is 10 more than the smaller.
   $$l \qquad = \qquad s + 10$$

3. Solution: Substitute $2s$ for l in $l = s + 10$,
   $$2s = s + 10, s = 10.$$
   Since $l = 2s$, $l = 20$.

   Ans. The numbers are 20 and 10.

4. Check: (Same as check for method 1.)

2.1. REPRESENTING TWO UNKNOWNS BY USING ONE LETTER

If n represents a number, represent another number that is

a) 5 more than n	*Ans.*	$n + 5$
b) 10 less than n		$n - 10$
c) 5 times as large as n		$5n$
d) one-fifth of n		$\dfrac{n}{5}$
e) the sum of twice n and 8		$2n + 8$
f) the product of n and 15		$15n$
g) the quotient of n and 3		$\dfrac{n}{3}$
h) 3 more than twice n		$2n + 3$
i) 80 reduced by 6 times n		$80 - 6n$
j) 10 less than the product of n and 5		$5n - 10$

2.2. USING ONE EQUATION FOR TWO UNKNOWNS

Express each statement as an equation: then find the numbers.

a) Two numbers are represented by n and $n + 5$.   *Ans.* $n + (n + 5) = 3n - 3, n = 8$
   Their sum is 3 less than 3 times the smaller.   Numbers are 8 and 13.

b) Two numbers are represented by n and $20 - n$. $n = 3(20 - n), n = 15$
   The first is 3 times the second.   Numbers are 15 and 5.

c) Two numbers are represented by n and $3n - 2$. $(3n - 2) = 2n - 6, n = -4$
   The second number is twice the first number less 6.   Numbers are -4 and -14.

d) Two numbers are represented by n and $6n$. $10n - 6n = 18, n = 4\frac{1}{2}$
   Ten times the first exceeds the second by 18.   Numbers are $4\frac{1}{2}$ and 27.

2.3. USING TWO EQUATIONS FOR TWO UNKNOWNS

Using s for the smaller number and l for the larger, obtain two equations for each problem; then find l and s.

a) The sum of two numbers is 15.   *Ans.* a) $l + s = 15$ $l = 11\frac{1}{2}, s = 3\frac{1}{2}$
   Their difference is 8. $l - s = 8$

b) The sum of the larger and 8 is 3 times the smaller.   b) $l + 8 = 3s$ $l = 46, s = 18$
   The larger reduced by 10 equals twice the smaller. $l - 10 = 2s$

c) Separate 40 into two parts such that the larger   c) $l + s = 40$ $l = 44, s = -4$
   exceeds twice the smaller by 52. $l - 2s = 52$

2.4. COMPLETE SOLUTION OF NUMBER PROBLEM HAVING TWO UNKNOWNS

The larger of two numbers is 3 times the smaller. Their sum is 8 more than twice the smaller. Find the numbers.

Solution:

| **Method 1** | **Method 2** |

Method 1

1. Representation, using one letter:

   Let s = smaller number
   $3s$ = larger number, since larger is 3
        times as large

2. Translation, using one equation:
   Their sum is 8 more than twice the smaller.

   $$3s + s \quad = \quad\quad 2s + 8$$

3. Solution:
   $$4s = 2s + 8, \, 2s = 8$$
   $$s = 4 \text{ and } 3s = 12$$

   Ans.   Numbers are 12 and 4.

4. Check: (Do this in the original problem.)

Method 2

1. Representation, using two letters:

   Let s = smaller number
       l = larger number

2. Translation, using two equations:
   Larger is 3 times the smaller.

   (1) $l = \quad\quad 3s$

   Their sum is 8 more than twice the smaller.

   (2) $l + s \quad = \quad\quad 2s + 8$

3. Solution: Substitute $3s$ for l in (2):
   $$3s + s = 2s + 8, \, 2s = 8$$
   $$s = 4 \text{ and } l = 3s = 12$$

   Ans.   Numbers are 12 and 4.

4. Check: (Do this in the original problem.)

3. CONSECUTIVE-INTEGER PROBLEMS

An integer is a signed whole number. An integer may be a positive whole number such as 25, a negative whole number such as -15, or zero.

Each consecutive-integer problem involves a set of consecutive integers, a set of consecutive even integers, or a set of consecutive odd integers. Each such set involves integers arranged in **increasing order** from left to right.

NOTE: In the table in Fig. 9-1, n represents the first number of a set. However, n may be used to represent any other number in the set. Thus, a set of three consecutive integers may be represented by $n - 1$, n, and $n + 1$.

3.1. REPRESENTATION USING n FOR FIRST INTEGER

Using n for the first integer, represent:

Table of Integers

	Consecutive Integers	Consecutive Even Integers	Consecutive Odd Integers
Illustrations	4, 5, 6, 7 $-4, -3, -2, -1$	4, 6, 8, 10 $-4, -2, 0, 2$	5, 7, 9, 11 $-5, -3, -1, 1$
Kinds of integers	Odd or even	Even only	Odd only
Differ by	1	2	2

Representation
  of first consecutive number
  of second consecutive number
  of third consecutive number
	
	n
	$n + 1$
	$n + 2$
		
		n
		$n + 2$
		$n + 4$
			
			n
			$n + 2$
			$n + 4$

Fig. 9-1

		Representation	
a) Three consecutive integers and their sum	Ans.	a) $n, n + 1, n + 2$	Sum $= 3n + 3$
b) Three consecutive even integers and their sum		b) $n, n + 2, n + 4$	Sum $= 3n + 6$
c) Three consecutive odd integers and their sum		c) $n, n + 2, n + 4$	Sum $= 3n + 6$
d) Four consecutive integers and their sum		d) $n, n + 1, n + 2, n + 3$	Sum $= 4n + 6$

3.2.   REPRESENTATION USING n FOR MIDDLE INTEGER

Using n for the middle integer, represent:

		Representation	
a) Three consecutive integers and their sum	Ans.	a) $n - 1, n, n + 1$	Sum $= 3n$
b) Three consecutive even integers and their sum		b) $n - 2, n, n + 2$	Sum $= 3n$
c) Five consecutive odd integers and their sum		c) $n - 4, n - 2, n,$	
		$n + 2, n + 4$	Sum $= 5n$

3.3.   TRANSLATION IN CONSECUTIVE-INTEGER PROBLEMS

Using n, $n + 1$, and $n + 2$ for three consecutive integers, express each statement as an equation; then find the integers.

		Equations	Integers
a) Their sum is 21.	Ans.	a) $3n + 3 = 21$	6, 7, 8
b) The sum of the first two is 7 more than the third.		b) $2n + 1 = (n + 2) + 7$	8, 9, 10
c) The sum of the second and third is 2 less than 3 times the first.		c) $2n + 3 = 3n - 2$	5, 6, 7
d) The third added to twice the first is 12 more than twice the second.		d) $2n + (n + 2) = 2(n + 1) + 12$	12, 13, 14

3.4.   TRANSLATION IN CONSECUTIVE EVEN INTEGER PROBLEMS

Using n, $n + 2$, and $n + 4$ for three consecutive even integers, express each statement as an equation; then find the integers.

		Equations	Integers
a) Their sum is 42.	Ans.	a) $3n + 6 = 42$	12, 14, 16
b) The second is half the first.		b) $n + 2 = \dfrac{n}{2}$	$-4, -2, 0$
c) The first equals the sum of the second and third.		c) $n = 2n + 6$	$-6, -4, -2$

3.5.   COMPLETE SOLUTIONS OF AN INTEGER PROBLEM

Find five consecutive odd integers whose sum is 45.

<u>Solutions:</u>

Method 1	**Method 2**
Represent the five consecutive odd integers, using n, $n + 2$, $n + 4$, $n + 6$, and $n + 8$.	Represent the five consecutive odd integers, using $n - 4$, $n - 2$, n, $n + 2$, and $n + 4$.

Then, their sum $= 5n + 20 = 45$. | Then, their sum $= 5n = 45$.

$5n = 25, n = 5$ (the first) | $n = 9$ (the third)

Ans.   5, 7, 9, 11, 13 | *Ans.*   5, 7, 9, 11, 13

4. AGE PROBLEMS

Rule 1.   To find a person's future age in a number of years hence, **add** that number of years to her or his present age.
Thus, in 10 years, a person 17 years old will be $17 + 10$ or 27 years old.

Rule 2.   To find a person's past age a number of years ago, **subtract** that number of years from his or her present age.
Thus, 10 years ago, a person 17 years old was $17 - 10$ or 7 years old.

4.1. REPRESENTING AGES ON BASIS OF PRESENT AGE

Represent the age of a person, in years,

a) 10 years hence if present age is x years   *Ans.* a) $x + 10$
b) 10 years ago if present age is x years   b) $x - 10$
c) In y years if present age is 40 years   c) $40 + y$
d) y years ago if present age is 40 years   d) $40 - y$
e) y years ago if present age is p years   e) $p - y$

4.2. REPRESENTING AGES

Find or represent the age of a person (in years). (In the answer, the expression in parentheses is present age.)

a) 5 years hence if she was 20 years old 10 years ago   *Ans.* a) $(20 + 10) + 5 = 35$
b) y years hence if he was 30 years old 5 years ago   b) $(30 + 5) + y = y + 35$
c) 5 years ago if she will be 20 years old in y years   c) $(20 - y) - 5 = 15 - y$

4.3. USING ONE EQUATION FOR TWO UNKNOWNS IN AGE PROBLEMS

Using $4S$ and S to represent the present ages of a father and his son, express each statement as an equation; then find their present ages.

	Equations	**Ages Now**
a) In 14 years, the father will be twice as old as his son will be then.	*Ans.* a) $4S + 14 = 2(S + 14)$ $S = 7$	28, 7
b) In 5 years, the father will be 21 years older than twice the age of his son now.	b) $3S + 5 = 2S + 21$ $S = 8$	32, 8
c) Three years ago, the father was 5 times as old as his son was then.	c) $4S - 3 = 5(S - 3)$ $S = 12$	48, 12

4.4. USING TWO EQUATIONS FOR TWO UNKNOWNS IN AGE PROBLEMS

Obtain two equations for each problem, using M and D to represent the present ages of a mother and daughter, respectively; then find their present ages.

	Equations	Ages Now
a) The sum of their present ages is 40. The mother is 20 years older than the daughter.	*Ans.* *a)* (1) $M + D = 40$ (2) $M = D + 20$	30, 10
b) The sum of their present ages is 50. In 5 years, the mother will be 3 times as old as the daughter will be then.	*b)* (1) $M + D = 50$ (2) $M + 5 = 3(D + 5)$	40, 10
c) In 8 years the mother will be twice as old as her daughter will be then. Three years ago, the mother was 3 times as old as her daughter was then.	*c)* (1) $M + 8 = 2(D + 8)$ (2) $M - 3 = 3(D - 3)$	36, 14

4.5. COMPLETE SOLUTION OF AN AGE PROBLEM: TWO METHODS

A father is now 20 years older than his daughter. In 8 years, the father's age will be 5 years more than twice the daughter's age then. Find their present ages.

Solutions:

Method 1, using two letters.	**Method 2, using one letter.**

Let F = father's present age
D = daughter's present age
Then (1) $F = D + 20$
      (2) $F + 8 = 2(D + 8) + 5$
Substitute $D + 20$ for F in (2):
$(D + 20) + 8 = 2(D + 8) + 5, D = 7$
	Let D = daughter's present age
	and $D + 20$ = father's present age
	Then $(D + 20) + 8 = 2(D + 8) + 5$
	          $D + 28 = 2D + 16 + 5$
	                    $7 = D$
	
	Ans. 27 and 7 years

5. RATIO PROBLEMS

Ratios are used to compare quantities by division.

The **ratio of two quantities** expressed in the same unit is the first divided by the second.

Thus, the ratio of 10 ft to 5 ft is 10 ft ÷ 5 ft, which equals 2.

Ways of Expressing a Ratio

A ratio can be expressed in the following ways:

1. Using a colon: 3:4
2. Using *to*: 3 to 4
3. As a common fraction: $\frac{3}{4}$
4. As a decimal fraction: 0.75
5. As a percent: 75 percent

General Principles of Ratios

1. To find the ratios between quantities, the quantities must have the same unit.

     Thus, to find the ratio of 1 ft to 4 in, first change the foot to 12 in.

     Then, take the ratio of 12 in to 4 in. The result is a ratio of 3 to 1 = 3.

2. A ratio is an abstract number, that is, a number without a unit of measure.

     Thus, the ratio of \$7 to \$10 = 7:10 = 0.7. The common unit of dollars must be removed.

3. A ratio should be simplified by reducing to lowest terms and eliminating fractions contained in the ratio.

     Thus, the ratio of 20 to 30 = 2 to 3 = $\frac{2}{3}$.

     Also, the ratio of $2\frac{1}{2}$ to $\frac{1}{2}$ = 5 to 1 = 5.

4. The ratios of three or more quantities may be expressed as a **continued ratio**. This is simply an enlarged ratio statement.

Thus, the ratio of \$2 to \$3 to \$5 is the continued ratio 2:3:5. This enlarged ratio is a combination of three separate ratios. These are 2:3, 3:5, and 2:5, as shown:

5.1.   RATIO OF TWO QUANTITIES WITH SAME UNIT

Express each ratio in lowest terms:

a) \$15 to \$3   *Ans.*   a) $\dfrac{15}{3} = 5$ e) \$2.50 to \$1.50   *Ans.*   e) $\dfrac{2.50}{1.50} = \dfrac{5}{3}$

b) 15 lb to 3 lb   b) $\dfrac{15}{3} = 5$ f) \$1.25 to \$5   f) $\dfrac{1.25}{5} = \dfrac{1}{4}$

c) 3 oz to 15 oz   c) $\dfrac{3}{15} = \dfrac{1}{5}$ g) $2\frac{1}{2}$ days to 2 days   g) $2\frac{1}{2} \div 2 = \dfrac{5}{4}$

d) 24 s to 18 s   d) $\dfrac{24}{18} = \dfrac{4}{3}$ h) $2\frac{1}{4}$ years to $\frac{1}{4}$ year   h) $2\frac{1}{4} \div \dfrac{1}{4} = 9$

5.2.   RATIO OF TWO QUANTITIES WITH DIFFERENT UNITS

Express each ratio in lowest terms:

	Change to Same Unit	**Ratio**
a) 2 years to 3 months	a) 24 months to 3 months	$\frac{24}{3} = 8$ *Ans.*
b) 80¢ to \$3.20	b) 80¢ to 320¢	$\frac{80}{320} = \frac{1}{4}$ *Ans.*
c) $1\frac{2}{3}$ yd to 2 ft	c) 5 ft to 2 ft	$\frac{5}{2} = 2\frac{1}{2}$ *Ans.*
d) 50 mi/h to 1 mi/min	d) 50 mi/h to 60 mi/h	$\frac{50}{60} = \frac{5}{6}$ *Ans.*

5.3.   CONTINUED RATIO OF THREE QUANTITIES

Express each continued ratio in lowest terms:

	Change to Same Unit	**Continued Ratio**
a) 1 gal to 2 qt to 2 pt	a) 8 pt to 4 pt to 2 pt	$8:4:2 = 4:2:1$ *Ans.*
b) 100 km to 5 m	b) 100,000 m to 5 m	$100,000:5 = 20,000:1$
c) \$1 to 1 quarter to 2 dimes	c) 100¢ to 25¢ to 20¢	$100:25:20 = 20:5:4$
d) 30 s to 2 min to 1 h	d) 30 s to 120 s to 3600 s	$30:120:3600 = 1:4:120$

5.4.   NUMERICAL RATIOS

Express each ratio in lowest terms:

a) 50 to 60   *Ans.* $\dfrac{50}{60} = \dfrac{5}{6}$ e) $1\frac{3}{4}$ to 7   *Ans.* $\dfrac{7}{4} \div 7 = \dfrac{1}{4}$

b) 70 to 55 $\dfrac{70}{55} = \dfrac{14}{11}$ f) 12 to $\dfrac{3}{8}$ $12 \div \dfrac{3}{8} = 32$

c) 175 to 75 $\dfrac{175}{75} = \dfrac{7}{3}$ g) $7\frac{1}{6}$ to $1\frac{1}{3}$ $\dfrac{43}{6} \div \dfrac{4}{3} = \dfrac{43}{8}$

d) 6.3 to 0.9 $\dfrac{6.3}{0.9} = 7$ h) 80% to 30% $\dfrac{80\%}{30\%} = \dfrac{8}{3}$

5.5. ALGEBRAIC RATIOS

Express each ratio in lowest terms:

a) $2x$ to $5x$ *Ans.* $\dfrac{2x}{5x} = \dfrac{2}{5}$ d) $4ab$ to $3ab$ *Ans.* $\dfrac{4ab}{3ab} = \dfrac{4}{3}$

b) $3a$ to $6b$ $\dfrac{3a}{6b} = \dfrac{a}{2b}$ e) $5s^2$ to s^3 $\dfrac{5s^2}{s^3} = \dfrac{5}{s}$

c) πD to πd $\dfrac{\pi D}{\pi d} = \dfrac{D}{d}$ f) x to $5x$ to $7x$ $x : 5x : 7x = 1{:}5{:}7$

5.6. REPRESENTATION OF NUMBERS IN A FIXED RATIO

Using x as their common factor, represent the numbers and their sum if:

			Numbers	**Sum**
a) Two numbers have a ratio of 4 to 3	*Ans.*		$4x$ and $3x$	$7x$
b) Two numbers have a ratio of 7 to 1			$7x$ and x	$8x$
c) Three numbers have a ratio of $4 : 3 : 1$			$4x$, $3x$, and x	$8x$
d) Three numbers have a ratio of $2 : 5 : 8$			$2x$, $5x$, and $8x$	$15x$
e) Five numbers have a ratio of $8 : 5 : 3 : 2 : 1$			$8x$, $5x$, $3x$, $2x$, and x	$19x$

5.7. RATIO IN NUMBER PROBLEMS

If two numbers in the ratio of 5:3 are represented by $5x$ and $3x$, express each statement as an equation; then find x and the numbers.

		Equations	**Value of x**	**Numbers**
a) The sum of the numbers is 88.	*Ans.*	a) $8x = 88$	$x = 11$	55 and 33
b) The difference of the numbers is 4.		b) $2x = 4$	$x = 2$	10 and 6
c) Twice the larger added to 3 times the smaller is 57.		c) $10x + 9x = 57$	$x = 3$	15 and 9
d) Three times the smaller equals the larger increased by 20.		d) $9x = 5x + 20$	$x = 5$	25 and 15

5.8. RATIO IN A TRIANGLE PROBLEM

The lengths of the sides of the triangle in Fig. 9-2 are in the ratio of 2:3:4. If the perimeter of the triangle is 45 cm, find the length of each side.

Solution:

Let $2x$, $3x$, and $4x$ represent the lengths of the sides in centimeters.
Then the perimeter $9x = 45$;

$x = 5$, $2x = 10$, $3x = 15$, $4x = 20$.   *Ans.*   10 cm, 15 cm, 20 cm

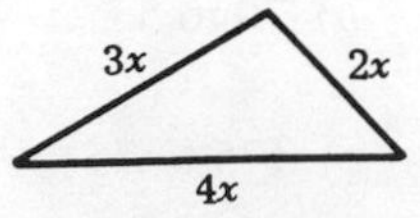

Fig. 9-2

5.9.   RATIO IN A MONEY PROBLEM

Sharon has $3\frac{1}{2}$ times as much money as Lester. If Sharon gives Lester a quarter, they will then have the same amount. How much did each have?

Solution:

Since the ratio of their money, $3\frac{1}{2}$ to 1, is 7:2,

let $7x =$ Sharon's money in cents and $2x =$ Lester's money in cents.

Then, $7x - 25 = 2x + 25$, and $x = 10$, $7x = 70$, $2x = 20$.

Ans.   70¢ and 20¢

5.10.   RATIO IN A WILL PROBLEM

In her will, a woman left her husband $20,000 and her son $12,000. Upon her death, her estate amounted to only $16,400. If the court divides the estate in the ratio of the bequests in the will, what should each receive?

Solution:

The ratio of $20,000 to $12,000 is 5:3.

Let $5x =$ the husband's share in dollars and $3x =$ the son's share in dollars.

Then $8x = 16,400$, and $x = 2050$, $5x = 10,250$, $3x = 6150$.

Ans. $10,250 and $6150

5.11.   RATIO IN A WAGE PROBLEM

Henry and George receive the same hourly wage. After Henry worked 4 h and George $3\frac{1}{2}$ h, Henry found he had $1.50 more than George. What did each earn?

Solution:

Since the ratio of 4 to $3\frac{1}{2}$ is 8:7,

let $8x =$ Henry's earnings in dollars and $7x =$ George's earnings in dollars,

Then $8x = 7x + 1.50$, and $x = 1.50$, $8x = 12$, $7x = 10.50$.

Ans. $12 and $10.50

6.   ANGLE PROBLEMS

Pairs of Angles:

(1) Adjacent Angles
Adjacent angles are two angles having the same vertex and a common side between them.

(2) Complementary Angles
Complementary angles are two angles the sum of whose measures equals 90° or a right angle.

(3) Supplementary Angles
Supplementary angles are two angles the sum of whose measures equals 180° or a straight angle.

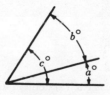

Fig. 9-3a

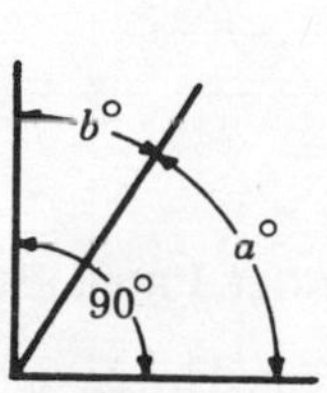

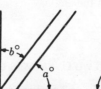

Fig. 9-4

Fig. 9-5

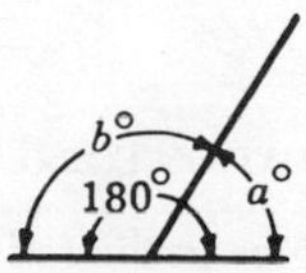

Fig. 9-6

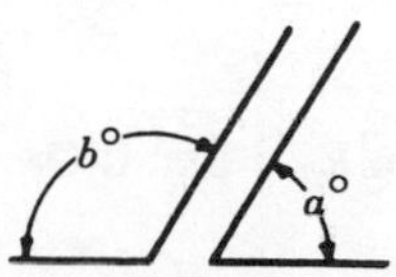

Fig. 9-7

Rule 1. If an angle of $c°$ consists of two adjacent angles of $a°$ and $b°$, as in Fig. 9-3a, then

$$a° + b° = c°$$

Thus, if $a° = 25$ and $b° = 32°$, then $c° = 57°$. See Fig. 9-3b.

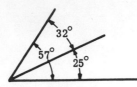

Fig. 9-3b

Rule 2. If two complementary angles contain $a°$ and $b°$, then

$$a° + b° = 90°$$

or

$$b° = 90° - a°$$

Thus, 70° and 20° are complementary angles.

Complementary angles may be adjacent, as in Fig. 9-4, or nonadjacent, as in Fig. 9-5.

Either of two complementary angles is the complement of the other. If $a°$ and $b°$ are complementary,

$b° =$ complement of $a°$

$a° = 90° - a°$

$a° =$ complement of $b°$

$b° = 90° - b°$

Rule 3. If two supplementary angles contain $a°$ and $b°$, then

$$a° + b° = 180°$$

or

$$b° = 180° - a°$$

Thus, 70° and 110° are supplementary angles.

Supplementary angles may be adjacent, as in Fig. 9-6, or nonadjacent, as in Fig. 9-7.

Either of two supplementary angles is the supplement of the other. If $a°$ and $b°$ are supplementary,

$b° =$ supplement of

$a° = 180° - a°$

$a° =$ supplement of

$b° = 180° - b°$

(4) Angles of Any Triangle

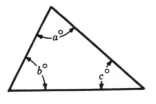

Fig. 9-8

(5) Angles of a Right Triangle

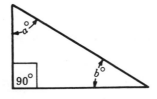

Fig. 9-9

(6) Angles of Isosceles Triangle

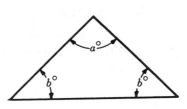

Fig. 9-10

Rule 4. The sum of the measures of the angles of any triangle equals 180°.

Thus, if the angles of a triangle contain $a°$, $b°$, and $c°$, as in Fig 9-8, then,

$$a° + b° + c° = 180°$$

Rule 5. The sum of the measure of the acute angles of a right triangle equals 90°.

Thus, if the acute angles of a right triangle, as in Fig. 9-9, contain $a°$ and $b°$, then,

$$a° + b° = 90°$$

Rule 6. The angles opposite the congruent sides of an isosceles triangle are congruent.

Thus, if the congruent angles of an isosceles triangle contain $b°$ and the third angle $a°$, as in Fig. 9-10, then,

$$a° + 2b° = 180°$$

NOTE: Henceforth, unless otherwise indicated, the word *angle* will mean "the number of degrees in the angle" or "the measure of the angle."

6.1.   USING ONE EQUATION FOR TWO UNKNOWNS IN ANGLE PROBLEMS

If two angles in the ratio of 3 to 2 are represented by $3x$ and $2x$, express each statement as an equation; then find x and the angles.

	Equation	x	Angles
a) The angles are adjacent and form an angle of 40°.	*Ans.* $3x + 2x - 40$	8	24°, 16°
b) The angles are complementary.	$3x + 2x = 90$	18	54°, 36°
c) The angles are supplementary.	$3x + 2x = 180$	36	108°, 72°
d) The larger angle is 28° less than twice the smaller.	$3x = 3x - 28$	28	84°, 56°
e) The smaller is 40° more than one-third of the larger.	$2x = \dfrac{3x}{3} + 40$	40	120°, 80°
f) The angles are two angles of a triangle whose third angle is 70°.	$3x + 2x + 70 = 180$	22	66°, 44°
g) The angles are the acute angles of a right triangle.	$3x + 2x = 90$	18	54°, 36°
h) The first angle is one of the two congruent angles of an isosceles triangle, and the other is the remaining angle.	$3x + 3x + 2x = 180$	$22\frac{1}{2}$	$67\frac{1}{2}°$, 45°

6.2. **USING TWO EQUATIONS FOR TWO UNKNOWNS IN ANGLE PROBLEMS**

If two angles are represented by a and b, obtain two equations for each problem; then find the angles.

	Equations	Angles
a) The angles are adjacent, forming an angle of 88°. One is 36° more than the other.	*Ans.* $a + b = 88$ $a = b + 36$	62°, 26°
b) The angles are complementary. One is twice as large as the other.	$a + b = 90$ $a = 2b$	60°, 30°
c) The angles are supplementary. One is 60° less than twice the other.	$a + b = 180$ $a = 2b - 60$	100°, 80°
d) The angles are two angles of a triangle whose third angle is 40°. The difference of the angles is 24°.	$a + b = 140$ $a - b = 24$	82°, 58°
e) The first angle is one of the two congruent angles of an isosceles triangle, and the other is the remaining angle. The second angle is 3 times the first.	$2a + b = 180$ $b = 3a$	36°, 108°

6.3. **THREE ANGLES HAVING A FIXED RATIO**

If three angles in the ratio of 4:3:2 are represented by $4x$, $3x$, and $2x$, express each statement as an equation; then find x and the angles.

	Equation	x	Angles
a) The first and second are adjacent and form an angle of 84°.	*Ans.* $4x + 3x = 84$	12	48°, 36°, 24°
b) The second and third are complementary.	$3x + 2x = 90$	18	72°, 54°, 36°
c) The first and third are supplementary.	$4x + 2x = 180$	30	120°, 90°, 60°
d) The angles are the three angles of a triangle.	$4x + 3x + 2x = 180$	20	80°, 60°, 40°

	Equation	x	Angles
e) The sum of the first and second is 27° more than twice the third.	*Ans.* $4x + 3x = 4x + 27$	9	36°, 27°, 18°
f) The first and third are acute angles of a right triangle.	$4x + 2x = 90$	15	60°, 45°, 30°

6.4. SUPPLEMENTARY-ANGLES PROBLEM

The numbers of degrees in each of two supplementary angles are consecutive odd integers. Find each angle.

Solution: See Fig. 9-11.
Let n and $n + 2$ = number of degrees in each of the angles.
Then $2n + 2 = 180, n = 89$.   *Ans.* 89°, 91°

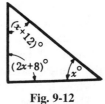

Fig. 9-11

6.5. SUM-OF-ANGLES-OF-A-TRIANGLE PROBLEM

In a triangle, one angle exceeds another by 12°. The third angle is 4° less than the sum of the other two. Find the angles.

Solution:

Let x = number of degrees in one angle
$x + 12$ = number of degrees in the second
$2x + 8$ = number of degrees in the third, since $x + (x + 12) - 4 = 2x + 8$. See Fig. 9-12.
Then their sum $4x + 20 = 180$, and,

$$x = 40, x + 12 = 52, 2x + 8 = 88.   Ans.   40°, 52°, 88°$$

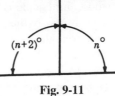

Fig. 9-12

7. PERIMETER PROBLEMS

NOTE.   Unless otherwise indicated, the word *side* means "the number of linear units in the side."

The **perimeter of a polygon** is the sum of its sides.
Thus, for a quadrilateral as shown in Fig. 9-13, if p = the perimeter,

$$p = a + b + c + d$$

The perimeter of a regular polygon equals the product of one side and the number of sides.
Thus, for the regular pentagon in Fig. 9-14, if p = the perimeter,

$$p = 5s$$

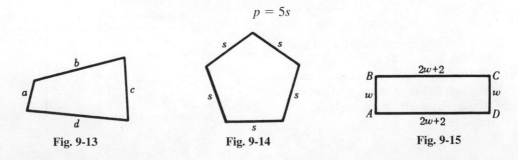

Fig. 9-13   **Fig. 9-14**   **Fig. 9-15**

7.1.   ONE-EQUATION METHOD: PERIMETER OF RECTANGLE

The width and length of a rectangle such as the one in Fig. 9-15 are represented by w and $2w + 2$, respectively. Express each statement as an equation; then find w and the dimensions of the rectangle.

		Equation	w	Width, Length
a)	The perimeter is 76 ft. *Ans.*	$6w + 4 = 76$	12	12 ft, 26 ft
b)	The semiperimeter (half-perimeter) is 56 in.	$3w + 2 = 56$	18	18 in, 38 in
c)	The sum of three sides not including $\overline{BC}$ is 82 yd.	$4w + 2 = 82$	20	20 yd, 42 yd
d)	A rectangle having each dimension 3 m less than those of the one shown has a perimeter of 28 m.	$2(2w - 1) + 2(w - 3) = 28$	6	6 m, 14 m

7.2.   TWO-EQUATION METHOD: PERIMETER OF TRIANGLE

The base of the triangle in Fig. 9-16 is 20. If x and y represent the remaining sides, obtain two equations for each problem; then find AB and BC.

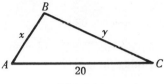

Fig. 9-16

		Equation	AB, BC
a)	The perimeter is 50. *Ans.*	$x + y + 20 = 50$	10, 20
	BC is twice AB.	$y - 2x$	
b)	The sum of AB and BC is 25.	$x + y = 25$	10, 15
	BC is 15 less than 3 times AB.	$y = 3x - 15$	
c)	AB is 4 more than half of BC.	$x = \dfrac{y}{2} + 4$	10, 12
	The difference of BC and AB is 2.	$y - x = 2$	

7.3.   ONE-EQUATION METHOD: PERIMETER OF TRAPEZOID

Figure 9-17 shows a trapezoid with congruent sides and with bases and sides with indicated lengths. Express each statement as an equation; then find x and the length of each base.

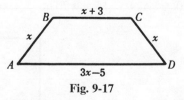

Fig. 9-17

		Equation	x	BC, AD
a)	The perimeter is 70 yd. *Ans.*	$6x - 2 = 70$	12	15 yd, 31 yd
b)	The combined length of the upper and lower bases is 36 in.	$4x - 2 = 36$	$9\frac{1}{2}$	$12\frac{1}{2}$ in, $23\frac{1}{2}$ in
c)	The sum of three sides not including the upper base is 65 ft.	$5x - 5 = 65$	14	17 ft, 37 ft

	Equation	x	BC, AD
d) The perimeter of the trapezoid exceeds the perimeter of a square having $\overline{BC}$ as a side by 10 cm.	$6x - 2 = 4(x + 3) + 10$	12	15 cm, 31 cm

7.4.   PERIMETER OF A QUADRILATERAL: RATIO OF SIDES

The sides of a quadrilateral in the ratio of $2 : 3 : 4 : 5$ are represented by $2x$, $3x$, $4x$, and $5x$, as shown in Fig. 9-18. Express each statement as an equation; then find x and AB.

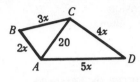

Fig. 9-18

		Equation	x	AB
a) The perimeter is 98 ft.	*Ans.*	$14x = 98$	7	14 ft
b) The largest side is 8 in less than the sum of the other three sides.		$5x = 9x - 8$	2	4 in
c) The sum of twice the smaller side and the other sides is 84 mi.		$2(2x) + 12x = 84$	$5\frac{1}{4}$	$10\frac{1}{2}$ mi
d) The perimeter of the quadrilateral is 22 m more than that of a square having a side 4 m less than BC.		$14x = 4(3x - 4) + 22$	3	6 cm
e) The perimeter of triangle ACD exceeds that of triangle ABC by 18 cm.		$9x + 20 = (5x + 20) + 18$	$4\frac{1}{2}$	9 cm

7.5.   PERIMETER-OF-AN-ISOSCELES-TRIANGLE PROBLEM

In an isosceles triangle, the ratio of one of the congruent sides to the base is $4 : 3$. A piece of wire 55 in long is to be used to make a wire model of it. Find the base and the side. See Fig. 9-19.

Solution: Let $4x$ = each congruent side in inches and $3x$ = base in inches.

Hence, $4x + 4x + 3x = 55$, $11x = 55$,

$x = 5, 3x = 15, 4x = 20.$ *Ans.*   15 and 20 in

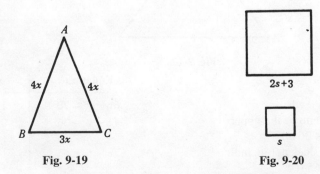

Fig. 9-19   **Fig. 9-20**

7.6.   PERIMETER-OF-A-SQUARE PROBLEM

The side of a larger square is 3 ft more than twice the side of a smaller square. The perimeter of the larger square exceeds that of the smaller square by 46 ft. Find the sides of the squares. See Fig. 9-20.

> **Solution:**   Let s = each side of the smaller square in feet
> $2s + 3$ = each side of the larger square in feet.
> Then $4(2s + 3) = 4s + 46$, $8s + 12 = 4s + 46$,
> $s = 8\frac{1}{2}$, $2s + 3 = 20$.   *Ans.* $8\frac{1}{2}$ and 20 ft

8. COIN OR STAMP PROBLEMS

The total value T of a number of coins or stamps of the same kind equals the number N of the coins or stamps multiplied by the value V of one of the coins or stamps.

$$T = NV$$

Thus, 8 nickels have a total value = $8(5) = 40¢$ and twenty 3-cent stamps have a total value = $20(3) = 60¢$.

NOTE: In $T = NV$, T and V must have the same unit of money: that is, if V is in dollars, then T must be in dollars.

8.1.   FINDING TOTAL VALUES

Find the total value *in cents* of each of the following:

a) 3 nickels and 5 dimes

b) q quarters and 7 nickels

c) 8 20-cent and t 29-cent stamps

d) x 4-cent and $(x + 5)$ 6-cent stamps

Ans.   *a*) $3(5) + 5(10) = 65$

b) $25q + 7(5) = 25q + 35$

c) $8(20) + 29t = 160 + 29t$

d) $4x + 6(x + 5) = 10x + 30$

8.2.   REPRESENTATION IN COIN OR STAMP PROBLEMS

Write the expression that represents the total value *in cents* of each; then simplify it.

		Representation	In Simplified Form
a) n dimes and $n + 3$ quarters	*Ans.*	*a*) $10n + 25(n + 3)$	$35n + 75$
b) n nickels and $2n - 3$ half-dollars		*b*) $5n + 50(2n - 3)$	$105n - 150$
c) n 20-cent and $2n$ 25-cent stamps		*c*) $20n + 25(2n)$	$70n$
d) n 6-cent and $(3n - 10)$ 12-cent stamps		*d*) $6n + 12(3n - 10)$	$42n - 120$

8.3.   USING ONE EQUATION IN COIN OR STAMP PROBLEMS

Express each statement as an equation; then find n and the number of each coin or stamp.

a) The total value of n dimes and $2n$ nickels is $2.40.

b) The total value of n nickels and $n - 3$ quarters is $7.65.

c) The value of $n + 5$ dimes is 3 times n nickels.

d) The value of n 20-cent and $(n + 3)$ 30-cent stamps is 20¢ more than $(n + 1)$ 60¢ stamps.

Ans.		Equations		n	Number of Each Coin or Stamp
	a)	$10n + 5(2n) = 240$		12	12 dimes and 24 nickels
	b)	$5n + 25(n - 3) = 765$		28	28 nickels and 25 quarters
	c)	$10(n + 5) = 3(5n)$		10	10 nickels and 15 dimes
	d)	$20n + 30(n + 3) = 60(n + 1) + 20$		1	1 20-cent, 4 30-cent, 2 60-cent

8.4.   USING TWO EQUATIONS IN COIN OR STAMP PROBLEMS

Using x for the value in cents of the first kind of coin and y for the value in cents of the second, obtain two equations for each problem; then find the kind of coins.

Equations

a) The value of 5 of the first kind and 6 of the second is 85¢. Each second coin has twice the value of each first.

Ans.   *a)* $5x + 6y = 85$
$y = 2x$
$x = 5 \quad y = 10$
nickels   dimes

b) The value of 4 of the first kind is 85¢ less than 5 of the second. The value of 2 of the first kind and 4 of the second is $1.20.

b) $\quad 4x = 5y - 85$
$2x + 4y = 120$
$x = 10 \quad y = 25$
dimes   quarters

8.5.   COMPLETE SOLUTION OF COIN PROBLEM

In her coin bank, Maria has three times as many quarters as nickels. If the value of the quarters is $5.60 more than the value of the nickels, how many of each kind does she have?

Solution: Let n = the number of nickels.

$$NV = T$$

	Number of Coins	Value of Each (¢)	Total Value (¢)
Nickels	n	5	$5n$
Quarters	$3n$	25	$75n$

The value of the quarters is $5.60 more than the value of the nickels.

$75n = 5n + 560, \quad n = 8, 3n = 24.$ *Ans.* 8 nickels, 24 quarters

8.6.   COMPLETE SOLUTION OF A CHANGE PROBLEM

Is it possible to change a $10 bill into an equal number of half-dollars, quarters, and dimes?

Solution: Let the number of half-dollars, the number of quarters, and the number of dimes each be represented by n. The combined value is $10.

$$\text{Then } 10n + 25n + 50n = 1000 \qquad 85n = 1000 \qquad n = 11\tfrac{13}{17}$$

Ans. Impossible, the number of coins must be a whole number.

8.7.   COMPLETE SOLUTION OF A FARES PROBLEM

On a trip, the fare for each adult was $5.00 and for each child $2.50. The number of passengers was 30, and the total paid was $122.50. How many adults and children were there?

Solution: Let $a =$ the number of adults and $c =$ the number of children.

$$\text{Then } 5a + 2.5c = 122.15$$
$$a + c = 30$$

Solving gives $a = 19, c = 11$.   *Ans.* 19 adults, 11 children

9. COST AND MIXTURE PROBLEMS

The total value T of a number of units of the same kind equals the number N of units multiplied by the value V of 1 unit.

$$T = NV$$

Thus, the total value of five books at 15¢ per book $= 5(15) = 75$¢.
The total value of six tickets at 50¢ each and five tickets at 25¢ each $= 6(50) + 5(25) = 425$¢.

Coin or Stamp Problems

The coin or stamp problems previously considered are special cases of mixture problems. Stamps and coins may be "mixed" as coffees, teas, nuts, and other items are mixed.

Cost Problems

When $T = NV$ is applied to a cost problem, the formula $C = NP$ should be used as follows:

$$C = NP \qquad \begin{aligned} C &= \text{total cost of number of units of a kind} \\ N &= \text{number of such units} \\ P &= \text{price of each unit purchased} \end{aligned}$$

Thus, the cost of five pencils at 8¢ per pencil $= 5(8) = 40$¢.

9.1. REPRESENTATION IN MIXTURE PROBLEMS: $T = NV$

Write the expression that represents the total value in cents; then simplify it:

	Representation	In Simplified Form
a) n lb of coffee valued at 90¢ a pound and $(n + 3)$ lb of coffee valued at $1.05 a pound	*Ans.* a) $90n + 105(n + 3)$	$195n + 315$
b) 3 lb of tea valued at $1.50 per pound and n lb of tea valued at $1.75 per pound	b) $3(150) + 175n$	$175n + 450$
c) n stamps valued at 35¢ each and $20 - n$ stamps valued at 50¢ each	c) $35n + 50(20 - n)$	$1000 - 15n$
d) d dozen pencils at $1.30 a dozen and 3 dozen pencils at 60¢ a dozen	d) $130d + 3(60)$	$130d + 180$

9.2. REPRESENTATION IN COST PROBLEMS

Write the expression that represents the total cost in dollars; then simplify it:

	Representation	In Simplified Form
a) 10 tables priced at $5 a piece and n tables at $7.50 a piece	*Ans.* a) $5(10) + 7.50n$	$7.5n + 50$
b) n lb of candy priced at 65¢ a pound and $(30 - n)$ lb of candy at $1.45 a pound	b) $0.65n + 1.45(30 - n)$	$43.5 - 0.8n$
c) q qt of cream at 35¢ a quart and $(2q + 3)$ qt of cream at 42¢ a quart	c) $0.35q + 0.42(2q + 3)$	$1.19q + 1.26$

9.3.   **TRANSLATION IN VALUE AND COST PROBLEMS**

Express each statement as an equation; then find n and the number of each kind:

	Equation

a) The cost of n lb of coffee at 95¢ a pound and   *Ans.* $95n + 110(25 - n) = 2600$, $n = 10$
       $25 - n$ lb of coffee at $1.10 a pound is   10 lb at 95¢ and 15 lb at $1.10
       $26.00.

b) The value of 40 tickets at 75¢ each and n $40(75) + 135n = 4080$, $n = 8$
       tickets at $1.35 each is $40.80.   8 tickets at $1.35

c) The cost of n lb of cookies at 80¢ a pound $80n + 160(2n - 3) = 1520$, $n = 5$
       and $(2n - 3)$ lb of cookies at $1.60 a pound   5 lb at 80¢ and 7 lb at $1.60
       is $15.20.

d) The value of n dollar bills, $n + 10$ five-   In dollars:
       dollar bills, and $3n - 2$ ten-dollar bills is $n + 5(n + 10) + 10(3n - 2) = 174$, $n = 4$
       $174.   4 dollar bills, 14 five-dollar bills,
                                                               10 ten-dollar bills

9.4.   **COMPLETE SOLUTION: BLENDING-COFFEE PROBLEM**

A coffee merchant blended coffee worth $0.93 a pound with coffee worth $1.20 a pound. The mixture of 30 lb was valued by her at $1.02 a pound. How many pounds of each grade of coffee did she use?

Solutions:

Method 1, using one letter

	Value per Pound (¢)	No. of Pounds	Total Value (¢)
Cheaper	93	x	$93x$
Better	120	$30 - x$	$120(30 - x)$
Mixture	102	30	3060

$93x + 120(30 - x) = 3060$

$93x + 3600 - 120x = 3060$

$x = 20$, $30 - x = 10$

Ans. 20 lb of $0.93 coffee and 10 lb of $1.20 coffee

Method 2, using two letters

	Value per Pound (¢)	No. of Pounds	Total Value (¢)
Cheaper	93	x	$93x$
Better	120	y	$120y$
Mixture	102	30	3060

$93x + 120y = 3060$

$\mathbf{M_{93}}$ $x + y = 30$

$$93x + 120y = 3060$$
$$\underline{93x + \;\;93y = 2790}$$
Subtract: $27y = \;\;270$

$y = 10$, $x = 20$

9.5.   **COMPLETE SOLUTION: SELLING TICKETS PROBLEM**

At a game, tickets were bought at $3, $5, and $7.50 each. The number of $5 tickets was 3 times the number of $3 tickets and 10 less than the number at $7.50. The receipts amounted to $885.00. How many of each ticket were sold?

Solution:

Let n = number of tickets at $3 each
        $3n$ = number of tickets at $5 each
$3n + 10$ = number of tickets at $7.50 each.
Hence, $3n + 5(3n) + 7.5(3n + 10) = 885$
                $18n + 22.5n + 75 = 885$
                        $n = 20$, $3n = 60$, $3n + 10 = 70$

Ans.   20 at $3, 60 at $5, 70 at $7.50

10. INVESTMENT OR INTEREST PROBLEMS

Annual interest I equals the principal P multiplied by the rate of interest R per year.

$$I = PR$$

Thus, the annual interest from $200 at 6 percent per year is (200)(0.06), or $12.

NOTE: Unless otherwise stated, the rate of interest is the rate per year; that is, 5 percent means 5 percent per year.

10.1. REPRESENTATION IN INTEREST PROBLEMS: $I = PR$

Write the expression that represents the annual interest in dollars earned by each principal; then simplify it:

		Representation	In Simplified Form
a)	$2000 at 5 percent and $$(P + 200)$ at 6 percent	*Ans.* $0.05(2000) + 0.06(P + 200)$	$0.06P + 112$
b)	$$P$ at 5 percent and $2P$ at $5\frac{1}{2}$ percent	$0.05P + 0.05\frac{1}{2}(2P)$	$0.16P$
c)	$$P$ at 3 percent and $$(2P - 400)$ at 6 percent	$0.03P + 0.06(2P - 400)$	$0.15P - 24$
d)	$$P$ at 7 percent, $2P$ at 5 percent, and $3P$ at 3 percent	$0.07P + 0.05(2P) + 0.03(3P)$	$0.26P$

10.2. TRANSLATION IN AN INTEREST PROBLEM

Express each statement as an equation; then solve and state each principal.

a) The total annual income from $500 at 4 percent and $$P$ at 5 percent is $55.

b) The total annual income from $$P$ at 3 percent and $$(5000 - P)$ at 2 percent is $120.

c) The annual interest from $$P$ at 3 percent equals that from $$(P - 2000)$ at 5 percent.

d) The annual interest from $2P$ at 5 percent exceeds that from $3P$ at 3 percent by $25.

e) The total annual interest from $$P$ at 6 percent, $2P$ at 5 percent, and $$(2P - 300)$ at 4 percent is $180.

Ans.   a) $0.04(500) + 0.05P - 55$, $P=700$ $700 at 5 percent

b) $0.03P + 0.02(5000 - P) - 120$, $P = 2000$ $2000 at 3 percent and $3000 at 2 percent

c) $0.03P = 0.05(P - 2000)$, $P = 5000$, $5000 at 3 percent and $3000 at 5 percent

d) $0.05(2P) = 0.03(3P) + 25$, $P = 2500$, $5000 at 5 percent and $7500 at 3 percent

e) $0.06P + 0.05(2P) + 0.04(2P - 300) = 180$, $P = 800$, $800 at 6 percent, $1600 at 5 percent, $1300 at 4 percent

10.3. COMPLETE SOLUTION: RATIO IN AN INTEREST PROBLEM

Mr. Wong invested two sums of money in the ratio of 5:3. The first sum was invested at 4 percent and the second at 2 percent. The annual interest from the first exceeds that of the second by $112. How much was each investment?

Solution: Since the ratio is 5:3, let $5x$ = first investment in dollars
and $3x$ = second investment in dollars

	Principal ($)	Rate of Interest	Annual Interest ($)
First investment	$5x$	0.08	$0.08(5x)$
Second investment	$3x$	0.04	$0.04(3x)$

$$0.08(5x) = 0.04(3x) + 112$$

$$0.4x = 0.12x + 112$$

Ans. $2000 at 8 percent, $1200 at 4 percent

10.4.   ALTERNATE METHODS IN AN INVESTMENT PROBLEM

Mr. Wong invested $8000, part at 5 percent and the rest at 2 percent. The yearly income from the 5 percent investment exceeded that from the 2 percent investment by $85. Find the investment at each rate.

Solution:

Method 1, using one letter

	Principal ($)	Rate of Interest	Annual Interest ($)
First	x	0.05	$0.05x$
Second	$8000 - x$	0.02	$0.02(8000 - x)$

$$M_{100} \quad 0.05x - 0.02(8000 - x) = 85$$
$$5x - 2(8000 - x) = 8500$$
$$5x - 16{,}000 + 2x = 8500$$
$$7x = 24{,}500$$
$$x = 3500, \ 8000 - x = 4500$$

Ans. $3500 at 5 dollars, $4500 at 2 percent

Method 2, using two letters

	Principal ($)	Rate of Interest	Annual Interest ($)
First	x	0.05	$0.05x$
Second	y	0.02	$0.02y$

$$M_2 \qquad x + y = 8000$$
$$M_{100} \quad 0.05x - 0.02y = 85$$
$$2x + 2y = 16{,}000$$
$$\text{Add:} \quad \underline{5x - 2y = 8{,}500}$$
$$7x = 24{,}500$$
$$x = 3500, \ y = 4500$$

10.5.   ALTERNATE METHODS IN A PROFIT AND LOSS INVESTMENT PROBLEM

Mrs. Mueller invested a total of $4000. On part of this she earned 4 percent. On the remainder she lost 3 percent. Combining her earnings and losses, she found her annual income to be $55. Find the amounts at each rate.

Solution: Method 1, using one letter

	Principal ($)	Rate of Interest	Annual Interest ($)
First	x	0.04	$0.04x$
Second	$4000 - x$	-0.03	$-0.03(4000 - x)$

$$M_{100} \quad 0.04x - 0.03(4000 - x) = 55$$
$$4x - 3(4000 - x) = 5500$$
$$7x - 12{,}000 = 5500$$
$$x = 2500, \ 4000 - x = 1500$$

Method 2, using two letters

	Principal ($)	Rate of Interest	Annual Interest ($)
First	x	0.04	$0.04x$
Second	y	-0.03	$-0.03y$

$$M_3 \qquad x + y = 4000 \rightarrow 3x + 3y = 12{,}000$$
$$M_{100} \quad 0.04x - 0.03y = 55 \rightarrow \underline{4x - 3y = 5{,}500}$$
$$7x = 17{,}500$$
$$x = 2500, \ y = 1500$$

Ans. Earnings on $2500 at 4 percent, losses on $1500 at 3 percent

10.6.   COMPLETE SOLUTION: ADDING A THIRD INVESTMENT

Mrs. Black has $3000 invested at 3 percent and $1000 at 4 percent. How much must she invest at 6 percent so that her annual income will be 5 percent of the entire investment?

Solution:

Let x = principal to be added at 6 percent
$4000 + x$ = entire principal at 5 percent

$$0.03(3000) + 0.04(1000) + 0.06x = 0.05(4000 + x)$$
$$90 + 40 + 0.06x = 200 + 0.05x$$
$$x = 7000 \quad \textit{Ans.} \ \$7000 \text{ at 6 percent}$$

11. MOTION PROBLEMS

The distance D traveled equals the rate of speed R multiplied by the time spent traveling T.

$$D = RT \qquad \text{Other forms:} \quad R = \frac{D}{T} \quad T = \frac{D}{R}$$

Thus, the distance traveled in 5 h at a rate of 30 mi/h is 150 mi, and the distance traveled in 5 s at a rate of 30 ft/s is 150 ft.

NOTE: In using $D = RT$, units for rate, time, and distance must be in agreement. Thus, if the rate is in miles per hour, use miles for distance and hours for time.

Uniform and Average Rates of Speed

Unless otherwise stated, *rate of speed* or simply *rate* may be taken to mean (1) uniform rate of speed or (2) average rate of speed.

(1) A uniform rate of speed is an unchanging or fixed rate of speed for each unit of time. Thus, a uniform rate of 40 mi/h for 3 h means that 40 mi was covered during each of the 3 h.

(2) An average rate of speed is a rate that is the average per unit of time for changing rates of speed. Thus, an average rate of 40 mi/h for 3 h may mean the average rate in a situation where 30 mi was covered during the first hour, 40 mi during the second hour, and 50 mi during the third hour.

11.1. REPRESENTATION OF DISTANCE

Write the expression that represents the distance in miles traveled; then simplify it:

		Representation	In Simplified Form
a) In 5 h at 20 mi/h and in 6 h more at R mi/h	*Ans.*	$5(20) + 6R$	$6R + 100$
b) In 6 h at 45 mi/h and in T h more at 50 mi/h		$6(45) + 50T$	$50T + 270$
c) In T h at 40 mi/h and in $(10 - T)$ h more at 30 mi/h		$40T + 30(10 - T)$	$10T + 300$
d) In 2 h at R mi/h and in 30 min more at 20 mi/h		$2R + \frac{1}{2}(20)$	$2R + 10$
e) In 4 h at R mi/h and in 8 h more at $(R + 10)$ mi/h		$4R + 8(R + 10)$	$12R + 80$
f) In 5 min at R mi/min and in 1 h more at 3 mi/min		$5R + 60(3)$	$5R + 180$

11.2. REPRESENTATION OF TIME: $T = \dfrac{D}{R}$

Write the expression that represents the time in hours needed to travel; then simplify it:

		Representation	In Simplified Form
a) 100 mi at 20 mi/h and 80 mi farther at R mi/h	*Ans.*	*a*) $\dfrac{100}{20} + \dfrac{80}{R}$	$5 + \dfrac{80}{R}$
b) 120 mi at 30 mi/h and D mi farther at 20 mi/h		*b*) $\dfrac{120}{30} + \dfrac{D}{20}$	$4 + \dfrac{D}{20}$
c) 60 mi at R mi/h and 80 mi farther at 2 mi/min		*c*) $\dfrac{60}{R} + \dfrac{80}{120}$	$\dfrac{60}{R} + \dfrac{2}{3}$

11.3. SEPARATION SITUATION

Two travelers start from the same place at the same time and travel **in opposite directions**.

Two planes leave the same airport at the same time and fly in opposite directions. The speed of the faster plane is 100 mi/h faster than the slower one. At the end of 5 h they are 2000 mi apart. Find the rate of each plane. See Fig. 9-21.

Solutions:

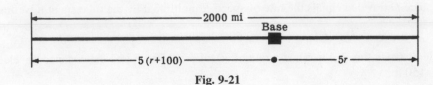

Fig. 9-21

ALTERNATE METHODS OF SOLUTION

Method 1

	Rate (mi/h)	Time (h)	Distance (mi)
Faster plane	$r + 100$	5	$5(r + 100)$
Slower plane	r	5	$5r$

The sum of the distances equals 2000 mi.
$$5r + 5(r + 100) = 2000$$
$$5r + 5r + 500 = 2000$$
$$10r = 1500$$

$r = 150, r + 100 = 250$

Ans. The rates are 150 and 250 mi/h.

Method 2

Rate of Separation (mi/h)	Time of Separation (h)	Distance of Separation (mi)
R	5	2000

Here R is the rate at which the two planes are separating from each other!

$$5R = 2000 \quad R = 400$$

Hence, the rate of separation is 400 mi/h. Since 400 mi/h is the sum of both rates,

$$r + (r + 100) = 400$$

$r = 150, r + 100 = 250$

11.4. **SEPARATION SITUATIONS**

In each of the following, two travelers start from the same place at the same time and travel in **opposite directions.** See Fig. 9-22. Using the letter indicated, express each statement as an equation; then solve and find each quantity represented.

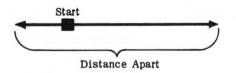

Fig. 9-22

a) After 5 h, both are 300 mi apart, one going 40 mi/h faster. Find the rate of the slower, using R in miles per hour.

b) At speeds of 40 and 20 mi/h, both travel until they are 420 mi apart. Find time of each, using T in hours.

c) At speeds in the ratio of 7 : 3, both are 360 mi apart at the end of 3 h. Find their respective rates in miles per hour, using $7x$ and $3x$ for these.

Ans. a) $5R + 5(R + 40) = 300, R = 10$
Slower rate is 10 mi/h.

b) $40T + 20T = 420, T = 7$
Time for each is 7 h.

c) $3(7x) + 3(3x) = 360, x = 12$
Faster rate is 84 mi/h.
Slower rate is 36 mi/h.

11.5.   CLOSURE SITUATION

Two travelers start from distant points at the same time and travel toward each other until they meet. See Fig. 9-23.

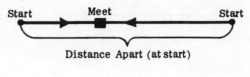

Start   Meet   Start

Distance Apart (at start)

Fig. 9-23

A car leaves Albany en route to New York at the same time as another car leaves New York for Albany. The car from Albany travels at an average rate of 40 mi/h, while the other averages 20 mi/h. If Albany and New York are 150 mi apart, how soon will the cars meet and how far will each have traveled? See Fig. 9-24.

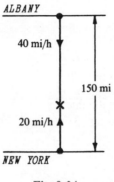

ALBANY

40 mi/h ▼

150 mi

20 mi/h ▲

NEW YORK

Fig. 9-24

Solutions:

ALTERNATE METHODS OF SOLUTION

Method 1

	Rate (mi/h)	Time (h)	Distance (mi)
From Albany	40	T	$40T$
From New York	20	T	$20T$

The sum of the distances is 150 mi

$$40T + 20T = 150$$
$$60T = 150$$
$$T = 2\tfrac{1}{2}$$

Method 2

Rate of Closure (mi/h)	Time of Closure (h)	Distance of Closure (mi)
40 + 20 or 60	T	150

Here, the cars are coming together at a rate of closure of 60 mi/h!

Hence, $60T = 150$
$$T = 2\tfrac{1}{2}$$

Ans.   The cars will meet in $2\tfrac{1}{2}$ h. The Albany car will have traveled 100 mi and the New York car 50 mi.

11.6. CLOSURE SITUATIONS

In each of the following, two travelers start from distant places at the same time and travel toward each other until they meet. See Fig. 9-25.

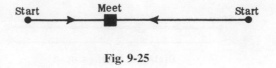

Fig. 9-25

Using the letter indicated, express each statement in an equation; then solve and find each quantity represented.

a) Starting 600 mi apart, they meet in 10 h, one traveling 20 mi/h faster. Find the rate of the slower in miles per hour, using R for this.

b) Starting 420 mi apart, they travel at rates of 27 and 33 mi/h. Find the time of each in hours, using T for this.

c) Starting 560 mi apart, they travel for 4 h, one rate being 4 times the other. Find the rate of the slower in miles per hour, using R for this.

d) Starting 480 mi apart, they travel at rates in the ratio of 4:3. They meet in 5 h. The slower is delayed for 1 h along the way. Find the rates of speed in miles per hour, using $4x$ and $3x$ for these.

Ans. *a)* $10R + 10(R + 20) = 600, R = 20$
Slower rate is 20 mi/h.

b) $27T + 33T = 420, T = 7$
Time for each is 7 h.

c) $4R + 4(4R) = 560, R = 28$
Slower rate is 28 mi/h.

d) $4(3x) + 5(4x) = 480, x = 15$
Slower rate is 45 mi/h.
Faster rate is 60 mi/h.

11.7   ROUND TRIP SITUATION

A traveler travels out and back to the starting place along the same road.

Henrietta drove from her home to Boston and back again along the same road in a total of 10 h. Her average rate going was 20 mi/h while her average rate on the return trip was 30 mi/h. How long did she take in each direction, and what distance did she cover each way? See Fig. 9-26.

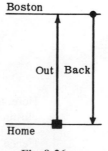

Fig. 9-26

Solutions:

<u>ALTERNATE METHODS OF SOLUTION</u>

Method 1, using $D = RT$

	Rate (mi/h)	Time (h)	Distance (mi)
Going	20	T	$20T$
Return	30	$10 - T$	$30(10 - T)$

The distance out equals the distance back.

$$20T = 30(10 - T)$$
$$20T = 300 - 30T$$
$$50T = 300$$
$$T = 6, \; 10 - T = 4$$

Method 2, using $T = \dfrac{D}{R}$

	Rate (mi/h)	Time (h)	Distance (mi)
Going	20	$\dfrac{D}{20}$	D
Return	30	$\dfrac{D}{30}$	D

The total time is 10 h.

$$\mathbf{M}_{60} \; \frac{D}{20} + \frac{D}{30} = 10 \qquad \text{LCD} = 60$$
$$3D + 2D = 600, \quad 5D = 600$$
$$D = 120$$

Ans.   Henrietta took 6 h going and 4 h back in going 120 mi each way.

11.8. ROUND TRIP SITUATIONS

In each situation, a traveler travels out and back to the starting place along the same road. See Fig. 9-27.

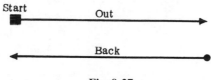

Fig. 9-27

Using the letter indicated, express each statement as an equation; then solve and find each quantity represented.

a) A traveler required a total of 12 h on a round trip, averaging 20 mi/h out and 30 mi/h back. Find the time in hours going, using T for this.

b) A traveler took 2 h more to travel back than to go out, averaging 50 mi/h out and 45 mi/h back. Find the time in hours going, using T for this.

c) A traveler traveled to his destination at an average rate of 25 mi/h. By traveling at 5 mi/h faster, he took 30 min less to return. Find his time in hours going out, using T for this.

d) After taking 5 h going out, a traveler came back in 3 h at a rate that was 20 mi/h faster. Find the average rate going in miles per hour, using R.

e) After taking 5 h going out, a traveler came back in 3 h at a rate that was 20 mi/h faster. Find the distance either way in miles, using D for this.

Ans.

a) $20T = 30(12 - T), \; T = 7.2$
Time out was 7.2 h.

b) $50T = 45(T + 2), \; T = 18$
Time out was 18 h.

c) $25T = 30(T - \frac{1}{2}), \; T = 3$
Time out was 3 h.

d) $5R = 3(R + 20), \; R = 30$
Rate out was 30 mi/h.

e) $\dfrac{D}{3} = \dfrac{D}{5} + 20, \; D = 150$
Each distance was 150 mi.

11.9. GAIN OR OVERTAKE SITUATION

After a traveler has begun her trip, a second one starts from the same place and, going in the same direction, overtakes the first. See Fig. 9-28.

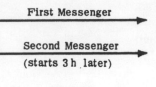

Fig. 9-28

A messenger going at 30 mi/h has been gone for 3 h. Another messenger, sent to overtake her, travels at 50 mi/h. How long will it take the second messenger to overtake the first, and what distance will he cover?

Solutions:

ALTERNATE METHODS OF SOLUTION

Method 1

	Rate (mi/h)	Time (h)	Distance (mi)
First	30	$T + 3$	$30(T + 3)$
Second	50	T	$50T$

The distances are equal.

$$30(T + 3) = 50T$$
$$30T + 90 = 50T, \quad 90 = 20T$$
$$4\tfrac{1}{2} = T$$

Method 2

Rate of Gain (mi/h)	Time of Gain (h)	Distance to Gain (mi)
$50 - 30$ or 20	T	$3(30)$ or 90

Here the second messenger is gaining at the rate of 20 mi/h. The first has a head start of 90 mi since she started 3 h earlier and her rate was 30 mi/h.

$$20T = 90$$
$$T = 4\tfrac{1}{2}$$

Ans.   To overtake the first, the second messenger must take $4\tfrac{1}{2}$ h and cover 225 mi.

11.10. GAIN OR OVERTAKE SITUATIONS

In each of the following, after a traveler has begun his trip, a second traveler starts from the same place and, going in the same direction, overtakes the first. See Fig. 9-29.

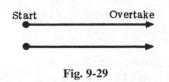

Fig. 9-29

Using the letter indicated, express each statement as an equation; then solve and find each quantity represented.

a) Their speeds are 20 and 40 mi/h. The second starts 3 h later. Find the time of the second in hours, using *T*.

b) The second starts 3 h later and travels 6 mi/h faster to overtake the first in 8 h. Find the rate of the first in miles per hour, using *R*.

c) The second travels 9 mi/h slower than twice the speed of the first and starts out $2\frac{1}{2}$ h later, overtaking in 4 h. Find the rate of the first in miles per hour, using *R*.

Ans.   *a*) $20(T + 3) = 40T$, $T = 3$
Time of second is 3 h.

        b) $11R = 8(R + 6)$, $R = 16$
Rate of first is 16 mi/h.

        c) $6\frac{1}{2}R = 4(2R - 9)$, $R = 24$
Rate of first is 24 mi/h.

11.11. MOTION PROBLEM: A TRIP IN TWO STAGES

In going 1200 mi, William used a train first and later a plane. The train going at 30 mi/h took 2 h longer than the plane traveling at 150 mi/h. How long did the trip take?

Solution: **See Fig. 9-30**

Fig. 9-30

	Rate (mi/h)	Time (h)	Distance (mi)
Plane trip	150	T	$150T$
Train trip	30	$T + 2$	$30T + 60$

The total distance is 1200 mi.

$30T + 60 + 150T = 1200$

$180T = 1140$, $T = 6\frac{1}{3}$ h

Ans.   The trip took $14\frac{2}{3}$ h: $6\frac{1}{3}$ by plane and $8\frac{1}{3}$ by train.

12. MATHEMATICAL MODELING

One of the most important changes in the mathematics curriculum over the last 10 years is the introduction into that curriculum of mathematical modeling. Certainly it is the case that problems and problem solving have been a significant part of that curriculum. How is modeling different?

According to the National Council of Teachers of Mathematics' publication, *Mathematical Modeling in the Secondary School Curriculum*, a mathematical model is "… a mathematical structure that approximates the features of phenomenon. The process of devising a mathematical model is called mathematical modeling" (Swetz and Hartzler, NCTM, Reston, Virginia, 1991). Thus, one can see that mathematical modeling does not in any way replace problem solving in the curriculum. Instead, it is a kind of problem solving.

Example:

One of the most significant applications of modeling in mathematics is in the area of population growth. Table 9-1 gives the population for a culture of bacteria from time $t = 0$ until $t = 5$ secs.

Table 9-1

Time (t)	Population (p) in millions
0	0
1	1
2	2
3	4
4	5
5	7

See Fig. 9-31 for the graph of these data with x axis representing t and y axis representing p.

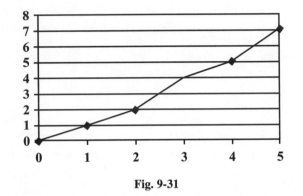

Fig. 9-31

Notice that for the times from 0 to 2, the graph in Fig. 9-31 is linear. In fact, it is approximately linear through $t=4$. Let us find the equation of the line that approximates these data. Since the data contains the points (1, 1) and (2, 2), a reasonable model for these data is the equation $y=x$. This equation, $y=x$ or, in function form,

$$p(x)=x$$

is a linear model. We can use this linear model to predict the population of this community.

For example, $p(x) = x$ predicts that,

$$p(3) = 3.$$

Since $p(3) = 4$, our model is off by 25%. $p(x) = x$ predicts that,

$$p(4) = 4$$

but $p(4)=5$, so the model is off by 20%.

Ask yourself whether the quadratic model,

$$p(x) = x^2/2$$

is a better model for these data. For example, what is the percent error for $t = 2, 3, 4$, and so on. What do these two models predict in terms of growth for various larger values of t? If the population at time 10 is 35, which is the better model?

Using the example above, we see that the critical steps in modeling are as follows:

a) Conjecture what model best fits the data given (or observed).
b) Analyze the model mathematically.
c) Draw reasonable conclusions based on the analysis in *b*).

In the preceding example, we:

a) Conjecture that a linear model provided a reasonable fit for given data.
b) Conjecture a model and analyze it mathematically.
c) Draw conclusions which include testing an additional model.

Supplementary Problems

NUMBER PROBLEMS HAVING ONE UNKNOWN

9.1. Express each statement as an equation; then find the unknown.   **(1.1)**

	Equation	Unknown
a) The sum of n and 12 is 21.	a) $n + 12 = 21$	$n = 9$
b) The result of adding m and 15 is $4m$.	b) $m + 15 = 4m$	$m = 5$
c) When p is subtracted from 40, 27 is the difference obtained.	c) $27 = 40 - p$	$p = 13$
d) Twenty increased by twice a is 32.	d) $20 + 2a = 32$	$a = 6$
e) Seven less than 3 times b is 23.	e) $3b - 7 = 23$	$b = 10$
f) Four times d exceeds 15 by d.	f) $4d - 15 = d$	$d = 5$
g) The product of 3 and $n + 6$ equals 20.	g) $3(n + 6) = 20$	$n = \dfrac{2}{3}$
h) Five more than half of x equals 29.	h) $\dfrac{x}{2} + 5 = 29$	$x = 48$
i) The difference between 15 and half of y is 6.	i) $15 - \dfrac{y}{2} = 6$	$y = 18$

(Ans. for column shown to left of Equation)*

9.2. Check each statement as indicated.   **(1.2)**

a) Is 5 the correct number for "the result of adding a number and 15 is 4 times the number"?

Ans. a) $5 + 15 \stackrel{?}{=} 4(5)$
$20 = 20$ Yes.

b) Is 11 the correct number for "7 less than 3 times the number is 23"?

b) $3(11) - 7 \stackrel{?}{=} 23$
$26 \neq 23$ No.

c) Is 2 the correct number for "the product of 3 and 6 more than a number equals 20"?

c) $3(6 + 2) \stackrel{?}{=} 20$
$24 \neq 20$ No.

d) Is 48 the correct number for "5 more than half a number equals 29"?

d) $\dfrac{48}{2} + 5 \stackrel{?}{=} 29$
$29 = 29$ Yes.

9.3. Find the number in each of the following: **(1.3)**

a) Four times the number increased by 2 equals 30.   *Ans.* a) 7
b) Seven times the number less 6 equals 29.   b) 5
c) The result of combining 5 times the number and the number is 24.   c) 4
d) If the number is added to 42, the sum is 8 times the number.   d) 6
e) If twice the number is subtracted from 13, the remainder is 8.   e) $2\frac{1}{2}$
f) If twice the number is increased by 6, the result is the same as decreasing 4 times the number by 14.   f) 10
g) Five times the number minus 12 equals 3 times the number increased by 10.   g) 11
h) The sum of one-half the number and one-third the number is 15.   h) 18

NUMBER PROBLEMS HAVING TWO UNKNOWNS

9.4. If n represents a number, represent another number that is: **(2.1)**

a) 4 more than 3 times n
b) 5 less than twice n
c) One-half of n increased by 3
d) Two-thirds of n decreased by 6

e) The sum of 3 times n and 20
f) 40 reduced by twice n
g) The product of $n + 4$ and 7
h) The quotient of $n - 2$ and 4

Ans. a) $3n + 4$ b) $2n - 5$ c) $\dfrac{n}{2} + 3$ d) $\dfrac{2}{3}n - 6$ e) $3n + 20$ f) $40 - 2n$

g) $7(n + 4)$ h) $\dfrac{n - 2}{4}$

9.5. Express each statement as an equation; then find the numbers. **(2.2)**

a) Two numbers are represented by n and $n - 8$. The first is 16 less than 3 times the second.

b) Two numbers are represented by n and $30 - n$. The first is 6 more than twice the second.

c) Two numbers are represented by n and $8n$. Half the second exceeds 3 times the first by 15.

Ans. a) $n = 3(n - 8) - 16$, 20, and 12

b) $n = 2(30 - n) + 6$, 22, and 8

c) $4n - 3n = 15$, 15, and 120

9.6. Using s for the smaller number and l for the larger, obtain two equations for each problem; then find l and s. **(2.3)**

a) The sum of two numbers is 20. Their difference is 11.
b) The sum of two numbers is 12. Twice the larger plus the smaller equals 21.
c) The difference of two numbers is 7. The larger is 1 less than twice the smaller.
d) Separate 50 into two parts such that the larger is 9 times the smaller.
e) Separate 42 into two parts such that the smaller is 3 less than one-half the larger.

Ans. a) $l + s = 20$ b) $l + s = 12$ c) $l - s = 7$ d) $l + s = 50$ e) $l + s = 42$

$l - s = 11$ $2l + s = 21$ $l = 2s - 1$ $l = 9s$ $s = \dfrac{l}{2} - 3$

$l = 15\frac{1}{2}, s = 4\frac{1}{2}$ $l = 9, s = 3$ $l = 15, s = 8$ $l = 45, s = 5$ $l = 30, s = 12$

9.7. Can you find two integers: **(2.4)**

a) Whose sum is 15 and whose difference is 5?
b) Whose sum is 20 and whose difference is 9?
c) Whose sum is 40 and the larger is 10 more than the smaller?

Ans. a) Yes, 10 and 5 b) No, the numbers $14\frac{1}{2}$ and $5\frac{1}{2}$ are not integers
c) Yes, 25 and 15

CONSECUTIVE-INTEGER PROBLEMS

9.8. Using n for the first integer, represent: **(3.1)**

a) Two consecutive odd integers and their sum
b) Four consecutive even integers and their sum
c) Five consecutive integers and their sum
d) Six consecutive odd integers and their sum

Ans. $n, n + 2$; sum $= 2n + 2$
$n, n + 2, n + 4, n + 6$; sum $= 4n + 12$
$n, n + 1, n + 2, n + 3, n + 4$; sum $= 5n + 10$
$n, n + 2, n + 4, n + 6, n + 8, n + 10$;
sum $= 6n + 30$

9.9. Using n for the middle integer, represent: **(3.2)**

a) Three consecutive odd integers and their sum
b) Five consecutive integers and their sum

Ans. $n - 2, n, n + 2$; sum $= 3n$
$n - 2, n - 1, n, n + 1, n + 2$; sum $= 5n$

9.10. Using n, $n + 1$, and $n + 2$ for three consecutive integers, express each statement as an equation; then find the integers.   **(3.3)**

  a) Their sum is 3 less than 5 times the first.
  b) The sum of the first and second is 6 more than the third.
  c) The sum of the first and third is 11 less than 3 times the second.
  d) Twice the sum of the first and second is 16 more than twice the third.
  e) Twice the third is 21 more than 3 times the sum of the first and second.
  f) The product of the second and third is 8 more than the square of the first.

  Ans. *a)* $3n + 3 = 5n - 3$; 3, 4, 5   *d)* $2(2n + 1) = 2(n + 2) + 16$; 9, 10, 11
        b) $2n + 1 = (n + 2) + 6$; 7, 8, 9   *e)* $2(n + 2) = 3(2n + 1) + 21$; -5, -4, -3
        c) $2n + 2 = 3(n + 1) - 11$; 10, 11, 12   *f)* $(n + 1)(n + 2) = n^2 + 8$; 2, 3, 4

9.11. Using n, $n + 2$, and $n + 4$ for three consecutive odd integers, express each statement as an equation; then find the integers.   **(3.4)**

  a) Their sum is 45.
  b) The sum of the first and second is 19 less than 3 times the third.
  c) Five times the sum of the first and third equals 18 more than 8 times the second.
  d) Twenty times the sum of the second and third equals 3 more than the first.
  e) Twice the sum of the first and second equals 4 times the sum of the second and third.

  Ans. *a)* $3n + 6 = 45$; 13, 15, 17   *d)* $20(2n + 6) = n + 3$; -3, -1, 1
        b) $2n + 2 = 3(n + 4) - 19$; 9, 11, 13   *e)* $2(2n + 2) = 4(2n + 6)$; -5, -3, -1
        c) $5(2n + 4) = 8(n + 2) + 18$; 7, 9, 11

9.12. Can you find: **(3.5)**

  a) Three consecutive integers whose sum is 36?
  b) Three consecutive odd integers whose sum is 33?
  c) Five consecutive even integers whose sum is 60?
  d) Five consecutive odd integers whose sum is 50?

  Ans. *a)* Yes; 11, 12, 13   *b)* Yes; 9, 11, 13   *c)* Yes; 8, 10, 12, 14, 16
        d) No. If n is used for the middle odd integer, $5n = 50$, $n = 10$, which is not odd.

AGE PROBLEMS

9.13. Represent the age of a person: **(4.1, 4.2)**

  a) In y years if present age is 50 years.   *Ans.* *a)* $50 + y$
  b) In 5 years if present age is F years.   *b)* $F + 5$
  c) In 5 years if age 10 years ago was S years.   *c)* $S + 10 + 5$ or $S + 15$
  d) 10 years ago if present age is J years.   *d)* $J - 10$
  e) y years ago if present age is 40 years.   *e)* $40 - y$

9.14. Using $5S$ and S to represent the present ages of a parent and child, express each statement as an equation; then find their present ages.   **(4.3)**

  a) In 7 years, the parent will be 3 times as old as the child will be then.
  b) Two years ago, the parent was 7 times as old as the child was then.
  c) In 3 years, the parent will be 1 year less than 4 times as old as the child will be then.
  d) Nine years ago, the parent was 3 times as old as the child will be 3 years hence.

  Ans. *a)* $5S + 7 = 3(S + 7)$; 35, 7   *c)* $5S + 3 = 4(S + 3) - 1$; 40, 8
        b) $5S - 2 = 7(S - 2)$; 30, 6   *d)* $5S - 9 = 3(S + 3)$; 45, 9

9.15. Obtain two equations for each problem, using J and C to represent the present ages of Juanita and Charles, respectively; then find their present ages. **(4.4)**

a) The sum of their present ages is 45. Juanita is 5 years older than Charles.

b) Juanita is 10 years older than Charles. Twelve years ago, Juanita was 3 times as old as Charles was then.

c) In 5 years, Juanita will be twice as old as Charles will be then. Four years ago, Juanita was 3 times as old as Charles was then.

d) At present Juanita is 6 times as old as Charles. Two years hence, Juanita will be 10 times as old as Charles was 3 years ago.

e) In 5 years, Juanita will be 25 years older than Charles will be then. Two years ago, Juanita was 7 times as old as Charles was last year.

Ans. a) $J + C = 45$ b) $J = C + 10$ c) $J + 5 = 2(C + 5)$ d) $J = 6C$

         $J = C + 5$ $J - 12 = 3(C - 12)$ $J - 4 = 3(C - 4)$ $J + 2 = 10(C - 3)$

         25, 20   27, 17   31, 13   48, 8

     e) $J + 5 = (C + 5) + 25$

        $J - 2 = 7(C - 1)$

        30, 5

9.16. George is now 8 years older than Harry. Find their present ages if: **(4.5)**

a) 14 years ago, George was twice as old as Harry was then.   *Ans.*   a) 30, 22

b) 2 years hence, George will be 3 times as old as Harry will be then.   b) 10, 2

c) 5 years hence, George will be twice as old as Harry was 2 years ago.   c) 25, 17

9.17. The sum of the present ages of Kathy and Samantha is 60 years. In 8 years, Kathy will be 5 years older than 3 times as old as Samantha was 3 years ago. **(4.5)**

Ans.   42 and 18 $[W + S = 60$ and $W + 8 = 3(S - 3) + 5]$

RATIO PROBLEMS

9.18. Express each ratio in lowest terms: **(5.1)**

a) 20¢ to 5¢	f) 50 percent to 25 percent	k) $\frac{1}{2}$ lb to $\frac{1}{4}$ lb
b) 5 dimes to 15 dimes	g) 15 percent to 75 percent	l) $2\frac{1}{2}$ days to $3\frac{1}{2}$ days
c) 30 lb to 25 lb	h) 33 percent to 77 percent	m) 5 ft to $\frac{1}{4}$ ft
d) 2 g to 14 g	i) $2.20 to $3.30	n) $\frac{1}{2}$ m to $1\frac{1}{2}$ m
e) 27 min to 21 min	j) $0.84 to $0.96	o) $16\frac{1}{2}$ cm to $5\frac{1}{2}$ cm

Ans.   a) 4   b) $\frac{1}{3}$ c) $\frac{6}{5}$ d) $\frac{1}{7}$ e) $\frac{9}{7}$ f) 2   g) $\frac{1}{5}$ h) $\frac{3}{7}$ i) $\frac{2}{3}$ j) $\frac{7}{8}$

       k) 2   l) $\frac{5}{7}$ m) 20   n) $\frac{1}{3}$ o) 3

9.19. Express each ratio in lowest terms: **(5.2)**

a) 1 year to 2 months	e) 2 yd to 2 ft	i) 100 lb to 1 ton
b) 2 weeks to 5 days	f) $2\frac{1}{3}$ yd to 2 ft	j) $2 to 25¢
c) 3 days to 3 weeks	g) $1\frac{1}{2}$ ft to 9 in	k) 2 quarters to 3 dimes
d) $\frac{1}{2}$ h to 20 min	h) 2 lb to 8 oz	l) 1 m^2 to 2 cm^2

Ans.   a) 6   b) $\frac{14}{5}$ c) $\frac{1}{7}$ d) $\frac{3}{2}$ e) 3   f) $\frac{7}{2}$ g) 2   h) 4   i) $\frac{1}{20}$ j) 8

       k) $\frac{5}{3}$ l) 5000

9.20. Express each ratio in lowest terms: **(5.3)**

a) 20¢ to 30¢ to $1   f) 2 h to $\frac{1}{2}$ h to 15 min

b) $3 to $1.50 to 25¢   g) 1 ton to 200 lb to 40 lb

c) 1 quarter to 1 dime to 1 nickel   h) 3 lb to 1 lb to 8 oz

d) 1 day to 4 days to 1 week   i) 1 gal to 1 qt to 1 pt

e) $\frac{1}{2}$ day to 9 h to 3 h

Ans. a) 2:3:10   b) 12:6:1   c) 5:2:1   d) 1:4:7   e) 4:3:1   f) 8:2:1

g) 50:5:1   h) 6:2:1   i) 8.2:1

9.21. Express each ratio in lowest terms: **(5.4)**

a) 60 to 70   Ans. $\frac{6}{7}$ g) 0.7 to 2.1   Ans. $\frac{1}{2}$ m) $7\frac{1}{2}$ to $2\frac{1}{2}$ Ans. 3

b) 84 to 7   12   h) 0.36 to 0.24 $\frac{3}{2}$ n) $1\frac{1}{2}$ to 12 $\frac{1}{8}$

c) 65 to 15 $\frac{13}{3}$ i) 0.002 to 0.007 $\frac{1}{7}$ o) 5 to $\frac{1}{3}$ 15

d) 125 to 500 $\frac{1}{4}$ j) 0.055 to 0.005   11   p) $\frac{1}{3}$ to $3\frac{1}{3}$ $\frac{1}{10}$

e) 630 to 105   6   k) 6.4 to 8   0.8 or $\frac{4}{5}$ q) $\frac{5}{6}$ to $1\frac{2}{3}$ $\frac{1}{2}$

f) 1760 to 990 $\frac{16}{9}$ l) 144 to 2.4   60   r) $\frac{7}{4}$ to $\frac{1}{8}$ 14

9.22. Express each ratio in lowest terms: **(5.5)**

a) x to $8x$ e) πab to πa^2 i) x to $4x$ to $10x$

b) $15c$ to 5   f) $4S$ to S^2 j) $15y$ to $10y$ to $5y$

c) $11d$ to 22   g) S^3 to $6S^2$ k) x^3 to x^2 to x

d) $2\pi r$ to πD h) $9r^2t$ to $6rt^2$ l) $12w$ to $10w$ to $8w$ to $2w$

Ans. a) $\dfrac{1}{8}$ b) $3c$ c) $\dfrac{d}{2}$ d) $\dfrac{2r}{D}$ e) $\dfrac{b}{a}$ f) $\dfrac{4}{S}$ g) $\dfrac{S}{6}$ h) $\dfrac{3r}{2t}$ i) 1:4:10,

j) 3:2:1   k) x^2:x:1   l) 6:5:4:1

9.23. Using x as their common factor, represent the numbers and their sum if: **(5.6)**

a) Two numbers have a ratio of 5:4   Ans. a) $5x$ and $4x$; sum = $9x$

b) Two numbers have a ratio of 9 to 1   b) $9x$ and x; sum = $10x$

c) Three numbers have a ratio of 2:5:11   c) $2x$, $5x$, and $11x$; sum = $18x$

d) Five numbers have a ratio of 1:2:2:3:7   d) x, $2x$, $2x$, $3x$, and $7x$; sum = $15x$

9.24. If two numbers in the ratio of 7 to 4 are represented by $7x$ and $4x$, express each statement as an
equation; then find x and the numbers.   **(5.7)**

a) The sum of the numbers is 99.   Ans. a) $11x = 99$, $x = 9$, 63, and 36

b) The difference of the numbers is 39.   b) $3x = 39$, $x = 13$, 91, and 52

c) Twice the smaller is 2 more than the larger.   c) $8x = 7x + 2$, $x = 2$, 14, and 8

d) The sum of the larger and one-half the smaller is 36.   d) $7x + 2x = 36$, $x = 4$, 28, and 16

9.25. The perimeter of a triangle is 60 in. Find each side.   **(5.8)**

a) The sides are in the ratio of 5:4:3.

b) The sides are in the ratio of 2:6:7.

c) Two sides are in the ratio of 3 to 2, and the third side is 25 in.

Ans. a) Using $5x$, $4x$, and $3x$ for sides: $12x = 60$, $x = 5$. Sides are 25, 20, 15 in.

b) Using $2x$, $6x$, and $7x$ for sides: $15x = 60$, $x = 4$. Sides are 8, 24, 28 in.

c) Using $3x$ and $2x$ for sides: $5x = 35$, $x = 7$. Sides are 21, 14, 25 in.

9.26. The ratio of Patti's money to Lee's money is 5:2. How much does each have according to the following?   **(5.9)**

a) They will have equal amounts if Patti gives Lee 24¢.

b) Patti will have twice as much as Lee if Patti gives Lee 30¢.

Ans. a) Using $5x$ and $2x$: $5x - 24 = 2x + 24$, $x = 16$. Amounts are 80¢ and 32¢.

b) Using $5x$ and $2x$: $5x - 30 = 2(2x + 30)$, $x = 90$. Amounts are \$4.50 and \$1.80.

9.27. An estate of \$8800 is to be divided among three heirs according to the conditions of a will. Find the amounts to be received: **(5.10)**

    a) The estate is to be divided in the ratio of 5:2:1.

    b) One heir is to get \$4000, and the others are to get the rest in the ratio of 7 to 5.

    Ans.   *a*) Using $5x$, $2x$, and x: $8x = 8800$, $x = 1100$. Amounts are \$5500, \$2200, \$1100.

           b) Using $7x$ and $5x$: $12x = 8800 - 4000$, $x = 400$. Amounts are \$2800, \$2000, \$4000.

9.28. The hourly wages of James and Stanley are in the ratio of 8:7. Find their hourly wages if: **(5.11)**

    a) In 3 h James earns 60¢ more than Stanley.

    b) In 5 h their combined earnings are \$22.50.

    Ans.   *a*) Using $8x$ and $7x$: $3(8x) = 3(7x) + 60$, $x = 20$. Wages are \$1.60 and \$1.40 per hour.

           b) Using $8x$ and $7x$: $5(8x + 7x) = 2250$, $x = 30$. Wages are \$2.40 and \$2.10 per hour.

ANGLE PROBLEMS

9.29. If two angles in the ratio of 5:4 are represented by $5x$ and $4x$, express each statement as an equation; then find x and the angles. **(6.1)**

    a) The angles are adjacent and form an angle of 45°.

    b) The angles are complementary.

    c) The angles are supplementary.

    d) The larger angle is 30° more than one-half the smaller.

    e) The smaller angle is 25° more than three-fifths the larger.

    f) The angles are the acute angles of a right triangle.

    g) The angles are two angles of a triangle whose third angle is their difference.

    h) The first angle is one of two congruent angles of an isosceles triangle. The remaining angle of the triangle is half of the other angle.

    Ans.   *a*) $5x + 4x = 45$, $x = 5$, 25° and 20°   *e*) $4x = 3x + 25$, $x = 25$, 125° and 100°

           b) $5x + 4x = 90$, $x = 10$, 50° and 40°   *f*) $5x + 4x = 90$, $x = 10$, 50° and 40°

           c) $5x + 4x = 180$, $x = 20$, 100° and 80°   *g*) $5x + 4x + x = 180$, $x = 18$, 90° and 72°

           d) $5x = 2x + 30$, $x = 10$, 50° and 40°   *h*) $5x + 5x + 2x = 180$, $x = 15$, 75° and 60°

9.30. If two angles are represented by a and b, obtain two equations for each problem; then find the angles. **(6.2)**

    a) The angles are adjacent, forming an angle of 75°. Their difference is 21°.

    b) The angles are complementary. One is 10° less than 3 times the other.

    c) The angles are supplementary. One is 20° more than 4 times the other.

    d) The angles are two angles of a triangle whose third angle is 50°. Twice the first added to 3 times the second equals 300°.

    Ans.   *a*) $a + b = 75$ *b*) $a + b = 90$ *c*) $a + b = 180$ *d*) $a + b = 130$

                 $a - b = 21$ $a = 3b - 10$ $a = 4b + 20$ $2a + 3b = 300$

                 48°, 27°   65°, 25°   148°, 32°   90°, 40°

9.31. If three angles in the ratio of 7:6:5 are represented by $7x$, $6x$, and $5x$, express each statement as an equation; then find x and the angles. **(6.3)**

    a) The first and second are adjacent and form an angle of 91°.

    b) The first and third are supplementary.

    c) The first and one-half the second are complementary.

    d) The angles are the three angles of a triangle.

    e) The sum of the second and third is 20° more than the first.

    f) The second is 12° more than one-third the sum of the first and third.

    Ans.   *a*) $7x + 6x = 91$, $x = 7$; 49°, 42°, and 35°   *d*) $7x + 6x + 5x = 180$, $x = 10$; 70°, 60°, and 50°

           b) $7x + 5x = 180$, $x = 15$; 105°, 90°, and 75°   *e*) $6x + 5x = 7x + 20$, $x = 5$; 35°, 30°, and 25°

           c) $7x + 3x = 90$, $x = 9$; 63°, 54°, and 45°   *f*) $6x = 4x + 12$, $x = 6$; 42°, 36°, and 30°

9.32. *a*)   One of two complementary angles is 5° less than 4 times the other. Find the angles.   **(6.4)**

       Ans.   19°, 71° [If x is the other angle, $x + (4x - 5) = 90$.]

     b)   One of two supplementary angles is 10° more than two-thirds of the other. Find the angles.

       Ans.   78°, 102° [If x is the other angle, $x + (\frac{2}{3}x + 10) = 180$.]

9.33. In a triangle, one angle is 9° less than twice another. The third angle is 18° more than twice their sum. Find the angles.   **(6.5)**

   Ans.   33°, 21°, 126° [If x is the second angle, $x + (2x - 9) + 6x = 180$.]

PERIMETER PROBLEMS

9.34. The length and width of a rectangle such as the one in Fig. 9-32 are represented by l and $2l - 20$, respectively. Express each statement as an equation; then find l and the dimensions of the rectangle.

                                                                               (7.1)

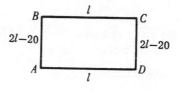

Fig. 9-32

a)   The perimeter is 50 km.
b)   The semiperimeter is 22 cm.
c)   A fence around the rectangle but not including BC is 25 ft.
d)   If each dimension is increased 10 rd, the perimeter of the rectangle will be 72 rd.
e)   If the width is unchanged and the length is doubled, the perimeter will be 128 in.
f)   The rectangle is a square.
g)   The perimeter of a square having the width as a side exceeds the perimeter of an equilateral triangle having the length as a side by 10 ft.

   Ans.   *a*) $6l - 40 = 50$, $l = 15$; 15 and 10 km   *e*) $2(2l) + 2(2l - 20) = 128$, $l = 21$; 21 and 22 in
              b) $3l - 20 = 22$, $l = 14$; 14 and 8 cm   *f*) $l = 2l - 20$, $l = 20$; 20 and 20
              c) $5l - 40 = 25$, $l = 13$; 13 and 6 ft   *g*) $4(2l - 20) = 3l + 10$, $l = 18$; 18 and 16 ft
              d) $2(l + 10) + 2(2l - 10) = 72$, $l = 12$; 12 and 4 rd

9.35. Using b for the base and a for each of the congruent sides of an isosceles triangle, obtain two equations for each problem; then find AB and AC. See Fig. 9-33.   **(7.2)**

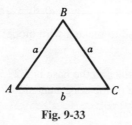

Fig. 9-33

a)   The perimeter is 32, and AB is 4 more than AC.
b)   The perimeter is 42, and AB is 9 less than twice AC.
c)   The perimeter is 28. The sum of twice AB and one-half of AC is 24.
d)   If each of the congruent sides is doubled, the new perimeter will be 43. If each of the congruent sides is increased by 10, the new perimeter will be 45.

Ans.   *a)* $2a + b = 32$ *b)* $2a + b = 42$ *c)* $2a + b = 28$

   $a = b + 4$ $a = 2b - 9$ $2a + \dfrac{b}{2} = 24$

   $AB = 12, AC = 8$ $AB = 15, AC = 12$ $AB = 10, AC = 8$

   d) $4a + b = 43$

   $2(a + 10) + b = 45$

   $AB = 9, AC = 7$

9.36. In a trapezoid having equal sides, the bases and sides are represented as shown in Fig. 9-34. Express each statement as an equation; then find x, AB, and AD. **(7.3)**

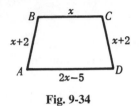

Fig. 9-34

a) The perimeter is 49 in.

b) The combined length of the upper and lower bases is 28 ft.

c) A fence around the trapezoid not including CD is 19 yd.

Ans.   *a)* $5x - 1 = 49$, $x = 10$, $AB = 12$ in, $AD = 15$ in

   b) $3x - 5 = 28$, $x = 11$, $AB = 13$ ft, $AD = 17$ ft

   c) $4x - 3 = 19$, $x = 5\frac{1}{2}$, $AB = 7\frac{1}{2}$ yd, $AD = 6$ yd

9.37. Three sides of a quadrilateral in the ratio of 2:3:5 are represented by $2x$, $3x$, and $5x$, respectively. If the base is 20, express each statement as an equation; then find x and AB. See Fig. 9-35. **(7.4)**

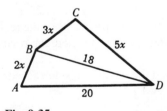

Fig. 9-35

a) The perimeter is 95.

b) AD is 40 less than the combined length of the other three sides.

c) BC is 12 less than the combined length of AD and AB.

d) The perimeter of triangle BCD exceeds that of triangle ABD by 22.

Ans.   *a)* $10x + 20 = 95$, $x = 7\frac{1}{2}$, $AB = 15$ *c)* $3x = (2x + 20) - 12$, $x = 8$, $AB = 16$

   b) $20 = 10x - 40$, $x = 6$, $AB = 12$ *d)* $8x + 18 = (2x + 38) + 22$, $x = 7$, $AB = 14$

9.38. A piece of wire 72 in long is to be used to make a wire model of an isosceles triangle. How long should the sides be if: **(7.5)**

a) The ratio of one of the congruent sides to the base is 3:2?

b) The base is to be 6 in less than one of the congruent sides?

Ans.   *a)* 27, 27, and 18 in   *b)* 26, 26, and 20 in

9.39. The side of one square is 7 km longer than the side of another. Find the sides of the squares. **(7.6)**

a) The sum of the two perimeters is 60 km.

b) The perimeter of the larger square is twice that of the smaller.

Ans.   *a)* 11 and 4 km   *b)* 14 and 7 km.

COIN OR STAMP PROBLEMS

9.40. Write the expression that represents the total value in cents of each; then simplify it: **(8.1, 8.2)**

  a) n nickels and $n - 5$ dimes

  b) n dimes and $3n - 10$ quarters

  c) n 5-cent and $(2n + 5)$ 10-cent stamps

  Ans.   a) $5n + 10(n - 5)$; $15n - 50$

  b) $10n + 25(3n - 10)$; $85n - 250$

  c) $5n + 10(2n + 5)$; $25n + 50$

9.41. Express each statement as an equation; then find n and the number of each coin or stamp.   **(8.3)**

  a) The total value of n nickels and $2n$ dimes is $1.25.

  b) The total value of n two-cent and $(n - 3)$ three-cent stamps is 91¢.

  c) The value of n quarters is $5.75 less than the value of $(n + 10)$ half-dollars.

  d) The value of n six-cent stamps equals that of $(2n - 7)$ ten-cent stamps.

  Ans.   a) $5n + 10(2n) = 125$, $n = 5$; 5 nickels and 10 dimes

  b) $2n + 3(n - 3) = 91$, $n = 20$; 20 two-cent and 17 three-cent stamps

  c) $25n = 50(n + 10) - 575$, $n = 3$; 3 quarters and 13 half-dollars

  d) $6n = 10(2n - 7)$, $n = 5$; 5 six-cent and 3 ten-cent stamps

9.42. Using x for the number of the first coin and y for the number of the second coin, obtain two equations for each problem; then find the number of each.   **(8.4)**

		Equations	Coins
a) Hilary has 25 coins, nickels and half-dollars. Their value is $6.65.	*Ans.*	$x + y = 25$ $5x + 50y = 665$	13 nickels 12 half-dollars
b) The value of a number of dimes and quarters is $2.05. If the quarters were replaced by nickels, the value would be $1.05.		$10x + 25y = 205$ $10x + 5y = 105$	8 dimes 5 quarters
c) The value of a number of nickels and dimes is $1.40. If the nickels were replaced by dimes and the dimes by nickels, the value would be $1.75.		$5x + 10y = 140$ $10x + 5y = 175$	14 nickels 7 dimes
d) The value of a number of nickels and quarters is $3.25. If the number of nickels were increased by 3 and the number of quarters were doubled, the value would be $5.90.		$5x + 25y = 325$ $5(x + 3) + 25(2y) = 590$	15 nickels 10 quarters

9.43. Jack has a number of coins worth $4.05. The collection consisted of 5 more nickels than dimes, and the number of quarters was 3 less than twice the number of dimes. How many of each had he?   **(8.5)**

  Ans.   Using d for the number of dimes: $5(d + 5) + 10d + 25(2d - 3) = 405$, $d = 7$
  12 nickels, 7 dimes, 11 quarters

9.44. Is it possible to change a $20 bill into an equal number of nickels, dimes, and quarters? If so, how many of each would be needed?   **(8.6)**

  Ans.   Yes. Using n for the number of nickels: $5n + 10n + 25n = 2000$, $n = 50$.
  50 nickels, 50 dimes, 50 quarters

9.45. At a show, the price of admission for a child was $1 and for an adult $2.50. Find the number of children and the number of adults if: **(8.7)**

  a) 40 were admitted and the total paid was $64.00

  b) Twice as many children as adults were admitted and the total paid was $40.50

  c) 10 more children than adults were admitted and the total paid was $62.50

  d) 15 more children than adults were admitted and the totals paid for each were equal

Ans.   *a)*   24 children and 16 adults $[c + 2.5(40 - c) = 64]$
         b)   18 children and 9 adults $[1(2a) + 2.5a = 40.5]$
         c)   25 children and 15 adults $(a + 10) + 2.5a = 62.50]$
         d)   25 children and 10 adults $[a + 15 = 2.5a]$

COST AND MIXTURE PROBLEMS

9.46.   Write the expression that represents the total value in cents; then simplify it: **(9.1)**

    a) n lb of tea valued at \$1.25 per pound and 5 lb of tea valued at \$1.40 per pound
    b) n stamps valued at 50¢ each and $n - 8$ stamps valued at 75¢ each
    c) $2n$ notebooks at 10¢ each and $3n - 5$ notebooks at 15¢ each
    d) d dozen pens at \$3.25 a dozen and $2d + 4$ doz. pens at \$2.80 a dozen

    Ans.   *a)* $125n + 5(140)$; $125n + 700$ *c)* $10(2n) + 15(3n - 5)$; $65n - 75$
            b) $50n + 75(n - 8)$; $125n - 600$ *d)* $325d + 280(2d + 4)$; $885d + 1120$

9.47.   Write the expression that represents the total cost in dollars; then simplify it: **(9.2)**

    a) n souvenirs priced at \$3 a piece and $n - 2$ souvenirs priced at \$1.50 each
    b) $2n$ tickets at \$3.30 each, $n/2$ tickets at \$4.40 each, and n tickets at \$1.60 each
    c)   12 chairs at \$7 each, n chairs at \$8.50 each, and $2n + 1$ chairs at \$10 each
    d) n boxes of cards at \$2.50 a box and $5n - 4$ boxes at \$3.75 a box

    Ans.   *a)* $3n + 1.50(n - 2)$; $4.5n - 3$ *c)* $7(12) + 8.50n + 10(2n + 1)$; $28.5n + 94$

            b) $3.30(2n) + 4.40\left(\dfrac{n}{2}\right) + 1.60n$; $10.4n$ *d)* $2.50n + 3.75(5n - 4)$; $21.25n - 15$

9.48.   Express each statement as an equation; then find n and the number of each kind: **(9.3)**

    a)   The cost of n lb of tea at 85¢ a pound and $(10 - n)$ lb of tea at 90¢ a pound is \$8.80.
    b)   The value of n tickets at 90¢ each and $3n + 2$ tickets at \$1.20 each is \$29.40.
    c)   John's earnings for n h at \$1.70 per hour and for $(n - 4)$ h at \$2.40 per hour are \$31.40.
    d)   The cost of $n + 2$ gifts at \$8 each. $2n$ gifts at \$10 each, and n gifts at \$15 each is \$231.

    Ans.   *a)* $85n + 90(10 - n) = 880$, $n = 4$, 4 lb at 85¢ and 6 lb at 90¢
            b) $90n + 120(3n + 2) = 2940$, $n = 6$, 6 tickets at 90¢ and 20 tickets at \$1.20
            c) $170n + 240(n - 4) = 3140$, $n = 10$, 10 h \$1.70 and 6 h at \$2.40
            d) $8(n + 2) + 10(2n) + 15n = 231$, $n = 5$, 7 gifts at \$8, 10 gifts at \$10, and 5 gifts at \$15

9.49.   A coffee merchant blended coffee worth 75¢ a pound with coffee worth 95¢ a pound. How many
       pounds of each grade did he use to make: **(9.4)**

		75¢ Grade	95¢ Grade
a) A mixture of 30 lb valued at 85¢ a pound?	*Ans.* *a)*	15 lb	15 lb
b) A mixture of 12 lb valued at 80¢ a pound?	*b)*	9 lb	3 lb
c) A mixture of 24 lb valued at 90¢ a pound?	*c)*	6 lb	18 lb

9.50.   At a game, tickets were sold at \$2.50, \$5.00, and \$8.00 each. The number sold at \$5.00 was 10 more
       than the number sold at \$8.00. The total receipts were \$165.00.   **(9.5)**
       Find the number sold at each price if:

    a)   There were 20 sold at \$2.50.
    b)   The number sold at \$2.50 equaled the sum of the other tickets.
    c)   The number sold at \$2.50 was 5 more than that at \$5.00.

    Ans.   *a)*   15 at \$5.00, 5 at \$8.00, 20 at \$2.50
            b)   15 at \$5.00, 5 at \$8.00, 20 at \$2.50
            c)   15 at \$5.00, 5 at \$8.00, 20 at \$2.50

INVESTMENT AND INTEREST PROBLEMS

9.51. Write the expression that represents the annual interest earned by each principal; then simplify it: **(10.1)**

    a) $1000 at 5 percent, $2000 at 3 percent, and P at 4 percent
    b) P at 2 percent, $2P$ at 5 percent, and $7500 at 6 percent
    c) P at 3 percent and $(2P - 600)$ at 7 percent
    d) P at 8 percent, $3P$ at 3 percent, and $(P + 2500)$ at 4 percent

    Ans. *a*) $0.05(1000) + 0.03(2000) + 0.04P;\ 0.04P + 110$
         b) $0.02P + 0.05(2P) + 0.06(7500);\ 0.12P + 450$
         c) $0.03P + 0.07(2P - 600);\ 0.17P - 42$
         d) $0.08P + 0.03(3P) + 0.04(P + 2500);\ 0.21P + 100$

9.52. Express each statement as an equation; then solve and state each principal.   **(10.2)**

    a) The total annual income from $800 at 3 percent and P at 4 percent is $56.
    b) The total annual interest from P at 2 percent and $(P + 2000)$ at 5 percent is $380.
    c) The annual interest from P at 6 percent equals that from $(2P - 3000)$ at 5 percent.
    d) The annual interest from $2P$ at $3\frac{1}{2}$ percent exceeds that from P at 5 percent by $40.

    Ans. *a*) $0.03(800) + 0.04P = 56,\ P = 800$; $800 at 4 percent
         b) $0.02P + 0.05\ (P + 2000) = 380,\ P = 4000$; $4000 at 2 percent, $6000 at 5 percent
         c) $0.06P = 0.05\ (2P - 3000),\ P = 3750$; $3750 at 6 percent, $4500 at 5 percent
         d) $0.03\frac{1}{2}(2P) = 0.05P + 40,\ P = 2000$; $4000 at $3\frac{1}{2}$ percent, $2000 at 5 percent

9.53. Ms. Horowitz invested two sums of money, the first at 5 percent and the second at 6 percent. State the amount invested at each rate if the ratio of the investments was: **(10.3)**

    a) 3:1 and the interest from the 5 percent investment exceeded that from the 6 percent investment by $180
    b) 4:3 and the total interest was $342

    Ans. *a*) Using $3x$ and x: $0.05(3x) - 0.06x = 180,\ x = 2000$; $6000 at 5 percent and $2000 at 6 percent
         b) Using $4x$ and $3x$: $0.05(4x) + 0.06(3x) = 342,\ x = 900$; $3600 at 5 percent and $2700 at 6 percent

9.54. A total of $1200 is invested, partly at 3 percent and the rest at 5 percent. Find the amounts invested at each rate if: **(10.4)**

    a) The total interest is $54.   *Ans.* *a*) $300 at 3 percent, $900 at 5 percent
    b) The annual amounts of interest are both equal.   *b*) $750 at 3 percent, $450 at 5 percent
    c) The annual interest from the 3 percent investment   *c*) $800 at 3 percent, $400 at 5 percent
        exceeds that from the 5 percent investment by $4.

9.55. On a total investment of $5400, Mr. Adams lost 4 percent on one part and earned 5 percent on the other. How large was each investment if: **(10.5)**

    a) His losses equaled his earnings?   *Ans.* $3000 at −4 percent and $2400 at 5 percent
    b) His net income was $144? $1400 at −4 percent and $4000 at 5 percent

9.56. Mrs. Howard has $400 invested at 2 percent and $600 at 3 percent. State the amount she must invest at 6 percent so that her annual income will be *a*) 4 percent of the entire investment, *b*) 5 percent of the entire investment.   **(10.6)**

    Ans. Using x for the added investment:
         a) $0.02(400) + 0.03(600) + 0.06x = 0.04(1000 + x)$; $700 to be added at 6 percent
         b) $0.02(400) + 0.03(600) + 0.06x = 0.05(1000 + x)$; $2400 to be added at 6 percent

MOTION PROBLEMS

9.57. Write the expression, in simplified form, that represents the distance in miles traveled in: **(11.1)**

a) 7 h at 20 mi/h and in *T* h more at 42 mi/h

b) 10 h at *R* mi/h and in 12 h more at 10 mi/h

c) *T* h at 25 mi/h and in (*T* + 8) h more at 15 mi/h

d) 5 h at *R* mi/h and in $3\frac{1}{2}$ h more at (2*R* + 8) mi/h

Ans. *a*) 7(20) + 42*T*; 42*T* + 140

b) 10*R* + 12(10); 10*R* + 120

c) 25*T* + 15(*T* + 8); 40*T* + 120

d) 5*R* + $3\frac{1}{2}$(2*R* + 8); 12*R* + 28

9.58. Write the expression, in simplified form, that represents the time in hours needed to travel. **(11.2)**

a) 240 mi at 40 mi/h and *D* mi farther at 27 mi/h

b) 100 mi at *R* mi/h and 75 mi farther at 15 mi/h

c) *D* mi at 15 mi/h and 3*D* mi farther at 45 mi/h

d) 40 mi at 4 mi min and 50 mi farther at *R* mi/h

Ans. *a*) $\dfrac{240}{40} + \dfrac{D}{27}; \dfrac{D}{27} + 6$

b) $\dfrac{100}{R} + \dfrac{75}{15}; \dfrac{100}{R} + 5$

c) $\dfrac{D}{15} + \dfrac{3D}{45}; \dfrac{2D}{15}$

d) $\dfrac{40}{240} + \dfrac{50}{R}; \dfrac{50}{R} + \dfrac{1}{6}$

9.59. Two trains leave the same terminal at the same time and travel in opposite directions. After 8 h they are 360 mi apart. The speed of the faster train is 3 mi/h less than twice that of the slower train. Find the rate of each train. **(11.3)**

Ans. Using *R* for speed of slower train in miles per hour: 8*R* + 8(2*R* − 3) = 360, rates are 16 and 29 mi/h.

9.60. In each situation, the travelers start from the same place at the same time and travel in *opposite directions*. Using the letter indicated, express each statement in an equation; then solve and find each quantity represented. **(11.4)**

a) At speeds of 25 and 15 mi/h, they travel until they are 210 mi apart. Find the time of each, using *T* in hours.

b) After 7 h both are 707 mi apart, one going 39 mi/h slower. Find the faster rate, using *R* in miles per hour for this.

c) At speeds in the ratio of 6:5, both are 308 mi apart in 14 h. Find their respective rates, using 6*x* and 5*x* in mi/h for these.

Ans. *a*) 25*T* + 15*T* = 210, time is $5\frac{1}{4}$ h. *b*) 7*R* + 7(*R* − 39) = 707, faster rate is 70 mi/h.

c) 14(6*x*) + 14(5*x*) = 308, *x* = 2, rates are 12 and 10 mi/h.

9.61. Two planes leave from points 1925 mi apart at the same time and fly toward each other. Their average speeds are 225 and 325 mi/h. How soon will the planes meet and how far will each have traveled? **(11.5)**

Ans. Using *T* for the time of travel: 225*T* + 325*T* = 1925. Planes will meet in $3\frac{1}{2}$ h.

The distances traveled will be $787\frac{1}{2}$ and $1137\frac{1}{2}$ mi.

9.62. In each situation, the travelers start from distant places and travel toward each other until they meet. Using the letter indicated, express each statement in an equation; then solve and find each quantity represented. **(11.6)**

a) Starting 297 mi apart, they travel at 38 and 28 mi/h. Find the time of each in hours, using *T*.

b) Starting 630 mi apart, they meet in 5 h, one traveling 46 mi/h faster than the other. Find the rate of the faster in miles per hour, using *R* for this.

c) Starting 480 mi apart, they meet in 6 h, one traveling 5 times as fast as the other. Find the rate of the slower, in miles per hour, using *R* for this.

Ans. *a*) 38*T* + 28*T* = 297, time is $4\frac{1}{2}$ h.

b) 5*R* + 5(*R* − 46) = 630, rate of faster is 86 mi/h.

c) 6*R* + 6(5*R*) = 480, rate of slower is $13\frac{1}{3}$ mi/h.

9.63. A plane traveled from its base to a distant point and back again along the same route in a total of 8 h. Its average rate going was 180 mi/h, and its average rate returning was 300 mi/h. How long did it take in each direction and what was the distance covered each way?   **(11.7)**

 Ans. Using T for the time in hours going, $180T = 300(8 - T)$, time going was 5 h.
 The distance covered each way was 900 mi.

9.64. In each situation, a traveler traveled out and back to the starting place along the same road. Using the letter indicated, express each statement as an equation, then solve and find each quantity represented.
                                                                                                  (11.8)

 a) A traveler traveled for 3 h at an average rate of 44 mi/h and returned in $5\frac{1}{2}$ h. Find the average rate returning in miles per hour, using R.
 b) A traveler took 3 h less to return than to go. If he averaged 45 mi/h out and 54 mi/h back, find his time returning, using T for this.
 c) A traveler required a total of 8 h for a round trip. If she averaged 24 mi/h out and 36 mi/h back, find her time returning, using T.
 d) After taking 3 h going out, a traveler came back in 5 h, averaging 28 mi/h slower on the way back. Find the average rate going in miles per hour, using R for this.

 Ans. *a)* $3(44) = 5\frac{1}{2}R$, rate returning 24 mi/h   *c)* $24(8 - T) = 36T$, time back 3.2 h
      b) $45(T + 3) = 54T$, time back 15 h   *d)* $3R = 5(R - 28)$, rate going 70 mi/h

9.65. A boat travels at 24 mi/h. A patrol boat starts 3 h later from the same place and travels at 32 mi/h in the same direction. How long will the patrol boat need to overtake the first boat, and what distance will it cover?   **(11.9)**

 Ans. Using T for the time of the patrol boat in hours, $32T = 24(T + 3)$.
 The time needed to overtake is 9 h and the distance to cover is 288 mi.

9.66. In each situation, after a traveler has begun her or his trip, a second traveler starts from the same place along the same road and, going in the same direction, overtakes the first.   **(11.10)**

 a) The second starts $2\frac{1}{2}$ h later and travels for 6 h at 34 mi/h. Find the rate of the first in miles per hour, using R for this.
 b) Their speeds are 15 and 24 mi/h. The second starts 2 h after the first. Find the time of the first, using T.
 c) The first travels 18 mi/h slower than twice the speed of the second. If the second starts out 2 h later and overtakes the first in 5 h, find the rate of the second in miles per hour, using R for this.
 d) The ratio of their rates is 4:5. The first was delayed 2 h along the way and also made a detour of 10 mi extra. The second overtook the first in 6 h, after starting 4 h later. Find their rates in miles per hour, using $4x$ and $5x$ for these.

 Ans. *a)* $8\frac{1}{2}R = 6(34)$; rate of first 24 mi/h
      b) $15T - 24(T - 2)$; time of first $5\frac{1}{3}$ h
      c) $7(2R - 18) = 5R$; rate of second 14 mi/h
      d) $8(4x) - 10 = 6(5x)$, $x = 5$; rates 20 and 25 mi/h

9.67. On a trip, Phillip used a boat first and later a train. The boat trip took 3 h longer than the train ride. How long did the trip take if the speeds of the boat and train were: **(11.11)**

 a) 20 and 30 mi/h, respectively, and the trip was 210 mi?
 b) 10 and 40 mi/h, respectively, and the trip was 255 mi?

 Ans. Using T for the time of the train:
      a) $20(T + 3) + 30T = 210$, $T = 3$, total time 9 h
      b) $10(T + 3) + 40T = 255$, $T = 4\frac{1}{2}$, total time 12 h

9.68.   For the following data from the U.S. Census: **(12.0)**

Year	Population of U.S.
1950	150,697,000
1940	131,669,000
1930	122,775,000
1920	105,711,000
1910	91,972,000
1900	75,995,000
1890	62,948,000
1880	50,156,000
1870	38,558,000
1860	31,443,000
1850	23,192,000
1840	17,069,000
1830	12,866,000
1820	9,638,000
1810	7,240,000
1800	5,308,000
1790	3,929,000

(a) Graph these data, using the vertical axis for population (in millions).

(b) For which years is the graph almost linear? *Ans.* From 1880 to 1900.

(c) Find the equation of an approximating linear model for these data.

(d) Use your model in (c) to predict the population in 1980.
       (**NOTE.**   The actual population in 1980 was 227 million.)

(e) What is the percent error in your model for 1980?

(f) Construct a quadratic model for these data.
       (***Hint***: Review Chapter 13 before attempting this.)

(g) Which is the better model: a linear function or a quadratic function for the above data? Why?

CHAPTER 10

Products
and Factoring

1. UNDERSTANDING FACTORS AND PRODUCTS

A Product and Its Factors

A product is the result obtained by multiplying two or more numbers.

The **factors of the product** are the numbers being multiplied.

Thus, 2, x, and y are the factors of the product $2xy$.

To find the product of a monomial and a polynomial, multiply the monomial by every term of the polynomial.

Thus, $3(4x + 2y) = 12x + 6y$.

Two factors of any number are 1 and the number itself.

Thus, 1 and 31 are factors of 31, while 1 and x are factors of x.

A **prime number** is a whole number which has no whole number factors except 1 and itself.

Thus, 2, 5, 17, and 31 are prime numbers.

Note how products may be divided into factors or into prime factors:

Product		Factors		Prime Factors
36	=	$4 \cdot 9$	=	$2 \cdot 2 \cdot 3 \cdot 3$
36	=	$6 \cdot 6$	=	$2 \cdot 3 \cdot 2 \cdot 3$
$10a^2$	=	$10 \cdot a^2$	=	$2 \cdot 5 \cdot a \cdot a$
$2ax + 6a$	=	$2(ax + 3a)$	=	$2a(x + 3)$

To factor a number or expression is to find its factors, not including 1 and itself.

Thus, $2ax + 6a$ may be factored into $2(ax + 3a)$ or further into $2a(x + 3)$.

To factor a polynomial (two or more terms) completely, continue factoring until the polynomial factors cannot be factored further.

Thus, $2ax + 6a$ may be factored completely into $2a(x + 3)$.

213

NOTE: Monomial factors need not be factored further.

Thus, $6ax^2 + 12x^2$ may be factored completely into $6x^2 (a + 2)$; 6 and x^2 need not be further factored.

1.1.   FINDING THE PRODUCT OF MONOMIAL FACTORS

Find each product:

a) $5 \cdot 7 \cdot x$	Ans. a) $35x$	d) $4(3a)(10b)$	Ans. d) $120ab$
b) $-3xxx$	b) $-3x^3$	e) $(5x^2)(-11y)$	e) $-55x^2y$
c) $8x^2 \cdot x^3 \cdot x^4$	c) $8x^9$	f) $(2ab)(3ac)(4ad)$	f) $24a^3bcd$

1.2.   FINDING THE PRODUCT OF TWO FACTORS, A MONOMIAL AND A POLYNOMIAL

Find each product

a) $4(a + b)$	Ans. $4a + 4b$	f) $P(1 + rn)$	Ans. $P + Prn$
b) $y(w + z)$	$wy + yz$	g) $\pi r(r + h)$	$\pi r^2 + \pi rh$
c) $3(4y - 1)$	$12y - 3$	h) $a(x - y + 1)$	$ax - ay + a$
d) $-7x(x - 2)$	$-7x^2 + 14x$	i) $-3(a + b - 2)$	$-3a - 3b + 6$
e) $3a^2(3a - 5)$	$9a^3 - 15a^2$		

1.3.   FINDING PRODUCTS INVOLVING FRACTIONS AND DECIMALS

Find each product:

a) $6(5 + \frac{1}{2})$	Ans. a) $30 + 3 = 33$	f) $24\left(\dfrac{x}{2} + \dfrac{x}{3}\right)$	Ans. f) $12x + 8x = 20x$
b) $12(\frac{1}{3} + \frac{3}{4})$	b) $4 + 9 = 13$	g) $20\left(\dfrac{x}{5} - \dfrac{x}{4}\right)$	g) $4x - 5x = -x$
c) $14(10 - \frac{1}{2})$	c) $140 - 2 = 138$	h) $9\left(\dfrac{x}{3} - \dfrac{2x}{9}\right)$	h) $3x - 2x = x$
d) $0.02\frac{1}{2}(1000 - 40)$	d) $25 - 1 = 24$	i) $0.05(2000 - x)$	i) $100 - 0.05x$
e) $0.03(2000 + 250)$	e) $60 + 7.50 = 67.50$	j) $0.03\frac{1}{2}(4000 - 2x)$	j) $140 - 0.07x$

---

2.   FACTORING A POLYNOMIAL HAVING A COMMON MONOMIAL FACTOR

A common monomial factor of a polynomial is a monomial that is a factor of each term of the polynomial.

Thus, 7 and a^2 are common monomial factors of $7a^2x + 7a^2y - 7a^2z$.

The **highest common (monomial) factor (HCF) of a polynomial** is the product of all its common monomial factors.

Thus, $7ab$ is the highest common monomial factor of $7abc + 7abd$.

To Factor a Polynomial Having a Common Monomial Factor

Factor: $7ax^2 + 14bx^2 - 21cx^2$

Procedure:	**Solution:**

1. Use the highest common (monomial) factor (HCF) as one factor:

1. HCF is $7x^2$.

2. Find the other factor by dividing each term of the polynomial by the HCF:

2. Divide each term by HCF:

$$\frac{7ax^2 + 14bx^2 - 21cx^2}{7x^2} = a + 2b - 3c$$

Hence, by factoring

$$7ax^2 + 14bx^2 - 21cx^2 = 7x^2\,(a + 2b - 3c)$$

NOTE: A fraction may be used as a common factor if it is the numerical coefficient of each term of the original polynomial.

Thus, by factoring, $\frac{1}{2}x + \frac{1}{2}y = \frac{1}{2}(x + y)$.

2.1.   FACTORING POLYNOMIALS HAVING A COMMON MONOMIAL FACTOR

Factor, removing the highest common factor:

a) $5a - 5b$	*Ans.*	$5(a - b)$	f) $S - Snd$	*Ans.*	$S(1 - nd)$
b) $\frac{1}{2}h + \frac{1}{2}k$		$\frac{1}{2}(h + k)$	g) $\pi R^2 - \pi r^2$		$\pi(R^2 - r^2)$
c) $2ay - 2by$		$2y(a - b)$	h) $5bx + 10by - 15b$		$5b(x + 2y - 3)$
d) $9x^2 - 9x$		$9x(x - 1)$	i) $4x^3 + 8x^2 - 24x$		$4x(x^2 + 2x - 6)$
e) $x^5 + 3x^2$		$x^2(x^3 + 3)$			

2.2.   FACTORING NUMERICAL POLYNOMIALS

Evaluate each, using factoring:

a) $7(14) - 6\frac{1}{2}(14)$	*Ans.* a)	$14(7 - 6\frac{1}{2}) = 14(\frac{1}{2}) = 7$
b) $6(8\frac{1}{2}) + 4(8\frac{1}{2})$	b)	$8\frac{1}{2}(6 + 4) = 8\frac{1}{2}(10) = 85$
c) $17\frac{1}{4}(10) - 15\frac{1}{4}(10)$	c)	$10(17\frac{1}{4} - 15\frac{1}{4}) = 10(2) = 20$
d) $0.03(800) + 0.03(750) - 0.03(550)$	d)	$0.03(800 + 750 - 550) = 0.03(1000) = 30$
e) $0.02\frac{1}{2}(8500) - 0.02\frac{1}{2}(7500) - 0.02\frac{1}{2}(1000)$	e)	$0.02\frac{1}{2}(8500 - 7500 - 1000) = 0.02\frac{1}{2}(0) = 0$
f) $8.4(3^3) + 5.3(3^3) - 3.7(3^3)$	f)	$3^3(8.4 + 5.3 - 3.7) = 3^3(10) = 270$

3.   SQUARING A MONOMIAL

The **square of a number** is the product of the number multiplied by itself. The number is used twice as a factor.

Thus, the square of 7 or $7^2 = (7)(7) = 49$.

Also, the square of -7 or $(-7)^2 = (-7)(-7) = 49$.

Opposites have the same square.

Thus, both $+7$ and -7 have the same square, 49; that is, $(+7)^2 = (-7)^2$.

To square a fraction, square both its numerator and its denominator.

Thus, the square of $\frac{2}{3}$ or $\left(\frac{2}{3}\right)^2 = \left(\frac{2}{3}\right)\left(\frac{2}{3}\right) = \frac{2^2}{3^2} = \frac{4}{9}$.

In general, $\left(\frac{a}{b}\right)^2 = \frac{a^2}{b^2}$.

To **square a monomial**, square its numerical coefficient, keep each base, and double the exponent of each base.

Thus, $(5ab^3)^2 = (5ab^3)(5ab^3) = 25a^2b^6$.

3.1. SQUARING NUMBERS

Find each square:

a) $(+8)^2$	Ans.	64	f) $(\frac{1}{5})^2$	Ans.	$\frac{1}{25}$	k) $(0.002)^2$	Ans.	0.000004
b) $(-8)^2$		64	g) $(-\frac{1}{5})^2$		$\frac{1}{25}$	l) $(-1.2)^2$		1.44
c) $(0.3)^2$		0.09	h) $(\frac{10}{11})^2$		$\frac{100}{121}$	m) $(1.25)^2$		1.5625
d) $(-0.3)^2$		0.09	i) $(\frac{5}{3})^2$ or $(1\frac{2}{3})^2$		$\frac{25}{9}$ or $2\frac{7}{9}$	n) $(-0.125)^2$		0.015625
e) $(-0.07)^2$		0.0049	j) $(-1\frac{2}{3})^2$		$\frac{25}{9}$ or $2\frac{7}{9}$	o) $(-0.101)^2$		0.010201

3.2. SQUARING MONOMIALS

Find each square:

a) $(r^3)^2$	Ans.	r^6	g) $\left(-\dfrac{ab^2}{c^7}\right)^2$	Ans.	$\dfrac{a^2b^4}{c^{14}}$
b) $(x^4)^2$		x^8	h) $(-\frac{10}{3}w^{20})^2$		$\frac{100}{9}w^{40}$
c) $\left(\dfrac{a^5}{b^3}\right)^2$		$\dfrac{a^{10}}{b^6}$	i) $(-8a^2b^3c)^2$		$64a^4b^6\,c^2$
d) $\left(\dfrac{x^{11}}{y}\right)^2$		$\dfrac{x^{22}}{y^2}$	j) $(0.1x^{100})^2$		$0.01x^{200}$
e) $(2ab)^2$		$4a^2b^2$	k) $\left(\dfrac{rst}{uv}\right)^2$		$\dfrac{r^2s^2t^2}{u^2v^2}$
f) $(-4x^4)^2$		$16x^8$	l) $(-\frac{7}{10}x^7y^{10})^2$		$\frac{49}{100}x^{14}y^{20}$

3.3. FINDING AREAS OF SQUARES: $A = s^2$

Find the area of a square whose side is:

a) 3 ft	Ans.	9 ft^2	d) $8x$	Ans.	$64x^2$
b) $\frac{3}{4}$ yd		$\frac{9}{16}$ yd^2	e) $1.5y^2$		$2.25y^4$
c) 1.3 mi		1.69 mi^2	f) $\frac{2}{3}m^3$		$\frac{4}{9}m^6$

4. FINDING THE SQUARE ROOT OF A MONOMIAL

The **square root of a number** is one of its two equal factors.

Thus, the square root of 49 is either $+7$ or -7, since $49 = (+7)(+7) = (-7)(-7)$.

A positive number has two square roots which are opposites of each other; that is, they have the same absolute value but differ in sign.

Thus, $\frac{25}{36}$ has two square roots, $+\frac{5}{6}$ and $-\frac{5}{6}$.

The **principal square root of a number** is its positive square root.

Thus, the principal square root of 81 is $+9$.

The symbol $\sqrt{}$ is used to indicate the principal or positive square root of a number.

Thus, $\sqrt{81} = 9$, $\sqrt{\frac{49}{64}} = \frac{7}{8}$, $\sqrt{0.09} = 0.3$.

To find the principal square root of a monomial, find the principal square root of its numerical coefficient, keep each base, and use half the exponent of each base.

Thus, $\sqrt{16y^{16}} = 4y^8$.

The **principal square root of a fraction** is the principal square root of its numerator divided by the principal square root of its denominator.

Thus, $\sqrt{\dfrac{100}{121}} = \dfrac{10}{11}$ $\sqrt{\dfrac{x^8}{y^6}} = \dfrac{x^4}{y^3}$.

4.1.   FINDING PRINCIPAL SQUARE ROOTS

Find each principal square root:

a) $\sqrt{100}$	Ans. 10	e) $\sqrt{a^6}$	Ans. a^3	i) $\sqrt{900a^2b^2}$	Ans. $30ab$			
b) $\sqrt{2500}$	50	f) $\sqrt{a^8b^{12}}$	a^4b^6	j) $\sqrt{0.09c^{20}}$	$0.3c^{10}$			
c) $\sqrt{0.0001}$	0.01	g) $\sqrt{100s^{100}}$	$10s^{50}$	k) $\sqrt{36r^{30}}$	$6r^{18}$			
d) $\sqrt{\dfrac{81}{400}}$	$\dfrac{9}{20}$	h) $\sqrt{\dfrac{x^{18}}{y^2}}$	$\dfrac{x^9}{y}$	l) $\sqrt{\dfrac{169}{x^{10}}}$	$\dfrac{13}{x^5}$			

4.2.   FINDING SIDES OF SQUARES: $s = \sqrt{A}$

Find the side of a square whose area is:

a) 49 in^2	Ans. 7 in	c) $\frac{25}{81}$ yd^2	Ans. $\frac{5}{9}$ yd	e) $1.21b^4$	Ans. $1.1b^2$
b) 0.0004 ft^2	0.02 ft	d) $169a^2$	$13a$	f) $\dfrac{100}{169x^2y^2}$	$\dfrac{10}{13xy}$

5. FINDING THE PRODUCT OF THE SUM AND DIFFERENCE OF TWO NUMBERS

$$(x + y)(x - y) = x^2 - y^2$$

If the sum of two numbers is multiplied by their difference, the product is the square of the first minus the square of the second.

Thus, $(x + 5)(x - 5) = x^2 - 25$.

Note below how the middle term drops out in each case.

(1) Multiply $x + y$ by $x - y$:

$$
\begin{array}{r}
x + y \\
x - y \\
\hline
x^2 + xy \\
- xy - y^2 \\
\hline
\text{Ans. } x^2 \qquad - y^2
\end{array}
$$

(2) Multiply $a^3 + 8$ by $a^3 - 8$:

$$
\begin{array}{r}
a^3 + 8 \\
a^3 - 8 \\
\hline
a^6 + 8a^3 \\
- 8a^3 - 64 \\
\hline
\text{Ans. } a^6 \qquad - 64
\end{array}
$$

(3) Multiply 103 by 97:

$$
\begin{array}{r}
103 = 100 + 3 \\
97 = 100 - 3 \\
\hline
10{,}000 + 300 \\
- 300 - 9 \\
\hline
\text{Ans.} \quad 10{,}000 \qquad - 9 = 9991
\end{array}
$$

Note, in (3) above, a new method for multiplying 103 by 97. This process of multiplying the sum of two numbers by their difference is a valuable shortcut in arithmetic in cases such as the following:

(a) $33 \times 27 = (30 + 3)(30 - 3) = 900 - 9$ or 891
(b) $2\frac{1}{2} \times 1\frac{1}{2} = (2 + \frac{1}{2})(2 - \frac{1}{2}) = 4 - \frac{1}{4}$ or $3\frac{3}{4}$
(c) $9.8 \times 10.2 = (10 - 0.2)(10 + 0.2) = 100 - 0.04$ or 99.96

5.1. MULTIPLYING THE SUM OF TWO NUMBERS BY THEIR DIFFERENCE

Find each product:

(First + Second)(First − Second)	1. (First)^2	2. (Second)^2	3. Product (Ans.)
a) $(m + 7)(m - 7)$	m^2	49	$m^2 - 49$
b) $(8 + 3x)(8 - 3x)$	64	$9x^2$	$64 - 9x^2$
c) $(11a + 5b)(11a - 5b)$	$121a^2$	$25b^2$	$121a^2 - 25b^2$
d) $(x^2 + y^3)(x^2 - y^3)$	x^4	y^6	$x^4 - y^6$
e) $(1 + y^4z^5)(1 - y^4z^5)$	1	y^8z^{10}	$1 - y^8z^{10}$
f) $(ab + \frac{2}{3})(ab - \frac{2}{3})$	a^2b^2	$\frac{4}{9}$	$a^2b^2 - \frac{4}{9}$
g) $(0.2 + \frac{3}{5}p)(0.2 - \frac{3}{5}p)$	0.04	$\frac{9}{2.5}p^2$	$0.04 - \frac{a}{25}p^2$

5.2. MULTIPLYING TWO NUMBERS BY THE SUM-PRODUCT METHOD

Multiply, using the sum and difference of two numbers:

a) 18×22 *Ans.* $18 \times 22 = (20 - 2)(20 + 2) = 400 - 4 = 396$
b) 25×35 $25 \times 35 = (30 - 5)(30 + 5) = 900 - 25 = 875$
c) 0.96×1.04 $0.96 \times 1.04 = (1 - 0.04)(1 + 0.04) = 1 - 0.0016 = 0.9984$
d) 7.1×6.9 $7.1 \times 6.9 = (7 + 0.1)(7 - 0.1) = 49 - 0.01 = 48.99$
e) $2\frac{1}{4} \times 1\frac{3}{4}$ $2\frac{1}{4} \times 1\frac{3}{4} = (2 + \frac{1}{4})(2 - \frac{1}{4}) = 4 - \frac{1}{16} = 3\frac{15}{16}$

5.3. MULTIPLYING A MONOMIAL BY SUM AND DIFFERENCE FACTORS

Find each product. (First, multiply the sum and difference factors.)

a) $2(x + 5)(x - 5)$ b) $x^2(1 - 3y)(1 + 3y)$ c) $t^2 (t + u)(t - u)$
d) $24(m^2 + \frac{1}{2})(m^2 - \frac{1}{2})$

Solutions:

a) $2(x^2 - 25)$ b) $x^2(1 - 9y^2)$ c) $t^2(t^2 - u^2)$ d) $24(m^4 - \frac{1}{4})$

Ans. $2x^2 - 50$ *Ans.* $x^2 - 9x^2y^2$ *Ans.* $t^4 - t^2u^2$ *Ans.* $24m^4 - 6$

6. FACTORING THE DIFFERENCE OF TWO SQUARES

$$x^2 - y^2 = (x - y)(x + y)$$

The expression $x^2 - y^2$ is the difference of two squares x^2 and y^2.
The factor $x + y$ is the sum of the principal square roots of x^2 and y^2, while the other factor, $x - y$, is their difference.

To Factor the Difference of Two Squares

Factor: $c^2 - 49$

Procedure:

1. **Obtain principal square root** of each square:
2. **One factor is the sum of the principal square roots. The other factor is their difference:**

Solutions:

1. Principal square roots: $\sqrt{c^2} = c$, $\sqrt{49} = 7$
2. Factors:

Ans. $(c + 7)(c - 7)$

6.1. **PROCEDURE FOR FACTORING THE DIFFERENCE OF TWO SQUARES**

Factor:

	First Square $-$ Second Square	1. $\sqrt{\text{First Square}}$	2. $\sqrt{\text{Second Square}}$	3. Factors (Ans.)
a)	$36 - b^2$	6	b	$(6 + b)(6 - b)$
b)	$1 - 25y^2$	1	$5y$	$(1 + 5y)(1 - 5y)$
c)	$a^2b^2 - 100$	ab	10	$(ab + 10)(ab - 10)$
d)	$x^6y^8 - z^{10}$	x^3y^4	z^5	$(x^3y^4 + z^5)(x^3y^4 - z^5)$
e)	$16 - x^{16}$	4	x^8	$(4 + x^8)(4 - x^8)$
f)	$u^4 - \frac{9}{25}$	u^2	$\frac{3}{5}$	$(u^2 + \frac{3}{5})(u^2 - \frac{3}{5})$
g)	$(a + b)^2 - c^2$	$a + b$	c	$(a + b + c)(a + b - c)$

6.2. **COMPLETE FACTORING INVOLVING THE DIFFERENCE OF TWO SQUARES**

Factor completely. (*Hint*: Remove the highest common monomial factor first.)

a) $10 - 40x^2$　　　　b) $75y^3 - 27y$　　　　c) $\pi R^2 - \pi r^2$　　　　d) $5abc^4 - 80ab$

Solutions:

a)　　10 is HCF
　　　$10(1 - 4x^2)$
Ans.　$10(1 + 2x)(1 - 2x)$

b)　　$3y$ is HCF
　　　$3y(25y^2 - 9)$
Ans.　$3y(5y + 3)(5y - 3)$

c)　　π is HCF
　　　$\pi(R^2 - r^2)$
Ans.　$\pi(R + r)(R - r)$

d)　　$5ab$ is HCF
　　　$5ab(c^4 - 16)$
Ans.　$5ab(c^2 + 4)(c + 2)(c - 2)$

7. FINDING THE PRODUCT OF TWO BINOMIALS WITH LIKE TERMS

Two Methods of Multiplying $3x + 5$ by $2x + 4$

Method 1 (Usual Method)　Using method 1, each separate option is shown. The middle term of the answer, $22x$, is obtained by adding the cross products $12x$ and $10x$. The arrows indicate the cross products.

$$3x + 5$$
$$2x + 4$$
$$6x^2 + 10x$$
$$\underline{+12x + 20}$$
Ans. $6x^2 + 22x + 20$

Method 2 (Method by Inspection) Using method 2, the answer is written **by inspection**. The middle term of the answer, $22x$, is obtained **mentally** by adding the product of the outer terms to the product of the inner terms. The arrows indicate the outer and inner products.

$$\overset{\overset{\displaystyle 12x}{\longmapsto}}{(3x + 5)} \; (2x + 4)$$
$$\underset{\underset{\displaystyle 10x}{\longmapsto}}{}$$

Ans. $6x^2 + 22x + 20$

To Multiply Two Binomials by Inspection

Multiply: $(3x + 5)(2x + 4)$

Procedure:

1. To obtain the first term of the product, **multiply the first terms:**

    Key: First

2. To obtain the middle term of the product, **add the product of the outer terms to the product of the inner terms:**

    Key: Outer + inner

3. To obtain the last term of the product, **multiply the last terms:**

    Key: Last

4. **Combine** the results to obtain answer:

Solution:

1. Multiply **first** terms:

    $$(3x)(2x) = 6x^2$$

2. Add **outer** and **inner** products:

    $$\overset{\overset{\displaystyle 12x}{\longmapsto}}{(3x + 5)(2x + 4)}$$
    $$\overset{}{10x}$$

    $$12x + 10x = 22x$$

3. Multiply **last** terms:

    $$(+5)(+4) = +20$$

4. Combine: $6x^2 + 22x + 20$ **Ans.**

Since a **mixed number** is the sum of an integer and a fraction, it may be expressed as a binomial. Thus, $3\frac{1}{2}$ and $4\frac{1}{3}$ are mixed numbers.

Two mixed numbers may be multiplied in the same way as two binomials. Thus,

$$3\tfrac{1}{2} \times 4\tfrac{1}{3} = (3 + \tfrac{1}{2})(4 + \tfrac{1}{3}) = 12 + 3 + \tfrac{1}{6} = 15\tfrac{1}{6} \quad \textit{Ans.}$$

7.1. PRODUCTS OF BINOMIALS BY STEPS

Multiply, showing each separate product:

Binomial × Binomial	First	Outer + Inner	Last	Product (Ans.)
a) $(x + 7)(x + 3)$	x^2	$(+3x) + (+7x)$	$+21$	$x^2 + 10x + 21$
b) $(x - 7)(x - 3)$	x^2	$(-3x) + (-7x)$	$+21$	$x^2 - 10x + 21$
c) $(x + 7)(x - 3)$	x^2	$(-3x) + (+7x)$	-21	$x^2 + 4x - 21$
d) $(3w - 5)(4w + 7)$	$12w^2$	$(+21w) + (-20w)$	-35	$12w^2 + w - 35$
e) $(a + 3b)(a - 8b)$	a^2	$(-8ab) + (+3ab)$	$-24b^2$	$a^2 - 5ab - 24b^2$

7.2. PRODUCTS OF TWO BINOMIALS MENTALLY

Multiply mentally:

a) $(x + 7)(x + 7)$ *Ans.* $x^2 + 14x + 49$

b) $(x - 3)(x - 3)$ $x^2 - 6x + 9$

c) $(3y + 1)(2y + 1)$ *Ans.* $6y^2 + 5y + 1$

d) $(2c - b)(5c - b)$ $10c^2 - 7bc + b^2$

e) $(3 - 7d)(1 - d)$ $3 - 10d + 7d^2$

f) $(2r + 3s)(3r + 2s)$ $6r^2 + 13rs + 6s^2$

g) $(ab + 4)(ab - 3)$ $a^2b^2 + ab - 12$

h) $(p - qr)(p - 3qr)$ $p^2 - 4pqr + 3q^2r^2$

i) $(pqr - 7)(pqr - 10)$ $p^2q^2r^2 - 17pqr + 70$

j) $(t^2 + 5)(t^2 + 9)$ $t^4 + 14t^2 + 45$

k) $(10 - p^2)(6 - v^2)$ $60 - 16v^2 + v^4$

l) $(c^2d - 5g)(c^2d + 3g)$ $c^4d^2 - 2c^2dg - 15g^2$

7.3.   REPRESENTING AREAS OF RECTANGLES: $A = LW$
Represent the area of a rectangle whose dimensions are:

a) $l + 8$ and $l - 2$ *Ans.* $A = (l + 8)(l - 2) = l^2 + 6l - 16$

b) $2l - 1$ and $l + 5$ $A = (2l - 1)(l + 5) = 2l^2 + 9l - 5$

c) $w + 3$ and $w + 7$ $A = (w + 3)(w + 7) = w^2 + 10w + 21$

d) $2w - 1$ and $3w + 1$ $A = (2w - 1)(3w + 1) = 6w^2 - w - 1$

e) $3s - 5$ and $s + 8$ $A = (3s - 5)(s + 8) = 3s^2 + 19s - 40$

8.   FACTORING TRINOMIALS IN FORM OF $x^2 + bx + c$

A trinomial in the form of $x^2 + bx + c$ may or may not be factorable into binomial factors. If factoring is possible, use the following procedure.

To Factor a Trinomial in Form of $x^2 + bx + c$

Factor: a) $x^2 + 6x + 5$ b) $x^4 - 6x^2 + 8$

Procedure:

Solutions:

1. Obtain the factors x and x of x^2. Use each as the first term of each binomial:

a) **1.** Factor x^2:
   (x, x)

b) **1.** Factor x^4:
   (x^2, x^2)

2. Select from the factors of the last term c those factors whose sum $= b$, the coefficient of x. Use each as the second term of each binomial:

2. Factor $+5$:
Select $(+5, +1)$
since sum $= +6$.
Discard $(-5, -1)$.

2. Factor $+8$:
Select $(-4, -2)$
since sum $= -6$.
Discard $(+4, +2)$,
$(-8, -1)$ and $(+8, +1)$

3. Form binomial factors from factors obtained in Steps 1 and 2:

3. $(x + 5)(x + 1)$ *Ans.*

3. $(x^2 - 4)(x^2 - 2)$ *Ans.*

8.1.   **FACTORING TRINOMIALS IN FORM OF** $x^2 + bx + c$

Factor each trinomial:

Trinomial Form: $x^2 + bx + c$	1. Factors of x^2, Each $= \sqrt{x^2}$	2. Factors of c Whose Sum $= b$	3. Binomial Factors (Ans.) (Combine 1 and 2)
a) $x^2 + 4x + 3$	x, x	$+3, +1$	$(x + 3)(x + 1)$
b) $x^2 - 4x + 3$	x, x	$-3, -1$	$(x - 3)(x - 1)$
c) $y^2 + 4y - 12$	y, y	$+6, -2$	$(y + 6)(y - 2)$
d) $w^2 - w - 12$	w, w	$-4, +3$	$(w - 4)(w + 3)$
e) $r^2 + 6rs + 5s^2$	r, r	$+5s, + s$	$(r + 5s)(r + s)$
f) $a^2b^2 - 12ab + 20$	ab, ab	$-10, -2$	$(ab - 10)(ab - 2)$
g) $x^4 - 5x^2 - 14$	x^2, x^2	$-7, +2$	$(x^2 - 7)(x^2 + 2)$

9. FACTORING A TRINOMIAL IN FORM OF $ax^2 + bx + c$

A trinomial in the form of $ax^2 + bx + c$ may or may not be factorable into binomial factors. If this is possible, use the following procedure.

To Factor a Trinomial in Form of $ax^2 + bx + c$

Factor: a) $2x^2 - 11x + 5$ b) $3a^2 + 10ab + 7b^2$

Procedure: **Solutions:**

1. Factor ax^2, **the first term.***
   Use each as the first term of
   each binomial factor:
2. **Select from the factors of c,
   the last term,** those factors to be
   used as the second term of each
   binomial such that the middle
   term, bx, results:

3. **Form binomial factors** from
   factors obtained in Steps 1 and 2
   and test for middle term, bx:

a)   **1.** Factor $2x^2$:

       $(2x, x)$

   2. Factor $+5$:
     Select $(-1, -5)$ to
     obtain middle term,
     $-11x$. Discard
     $(+1, +5)$.

   3. $(2x - 1)(x - 5)$
      Middle term, $-11x$,
      results.

   Ans. $(2x - 1)(x - 5)$

b)   **1.** Factor $3a^2$:

       $(3a, a)$

   2. Factor $+7b^2$:
     Select $(+7b, +b)$ to
     obtain middle term, b
     $10ab$. Discard
     $(-7b, -b)$.

   3. $(3a + 7b)(a + b)$
      Middle term, $10ab$,
      results.

   Ans. $(3a + 7b)(a + b)$

* When the first term ax^2 is positive, use positive factors.   Thus, in $2x^2 - 11x + 5$, do not use $- 2x$ and $- x$ as factors of $2x^2$.

9.1.   FACTORING TRINOMIALS IN FORM OF $ax^2 + bx + c$

Factor each trinomial:

Trinomial Form: $ax^2 + bx + c$	Factors of ax^2	Factors of c to Obtain Middle Term, bx	Binomial Factors (Ans.) (Test Middle Term)
a) $5x^2 + 11x + 2$	$5x, x$	$+1, +2$	$(5x + 1)(x + 2)$ $10x$
b) $4w^2 + 7w + 3$	$4w, w$ Discard $(2w, 2w)$	$+3, +1$	$(4w + 3)(w + 1)$ $+4w$
c) $4y^2 - 8w + 3$	$2y, 2y$ Discard $(4y, y)$	$-3, -1$	$(2y - 3)(2y - 1)$ $2y$
d) $4w^2 + 13wx + 3x^2$	$4w, w$ Discard $(2w, 2w)$	$+x, +3x$	$(4w + x)(w + 3x)$ $+12wx$
e) $8 + 15h - 2h^2$	$8, 1$ Discard $(4, 2)$	$-h, +2h$	$(8 - h)(1 + 2h)$ $+16h$

10.   SQUARING A BINOMIAL

$$(x + y)^2 - x^2 + 2xy + y^2$$

The square of a binomial is a perfect-square trinomial.

Thus, the square of $x + y$ or $(x + y)^2$ is the perfect-square trinomial $x^2 + 2xy + y^2$.

To Square a Binomial

Square: $3x + 5$

Procedure:

1. **Square the first term** to obtain the first term of the trinomial:
2. **Double the product of both terms** to obtain the middle term of the trinomial:
3. **Square the last term** to obtain the last term of the trinomial:
4. **Combine results:**

Solution:

1. Square $3x$: $(3x)^2 = 9x^2$

2. Double $(3x)(+5)$:
$$2(3x)(+5) = 30x$$

3. Square $+5$: $5^2 = 25$

4. Combine: $9x^2 + 30x + 25$ **Ans.**

11.   FACTORING A PERFECT-SQUARE TRINOMIAL

$$x^2 + 2xy + y^2 = (x + y)(x + y) = (x + y)^2$$
$$x^2 - 2xy + y^2 = (x - y)(x - y) = (x - y)^2$$

The factors of a perfect-square trinomial are two equal binomials

Thus, the factors of the perfect-square trinomial $x^2 + 2xy + y^2$ are $x + y$ and $x + y$.

A **perfect-square trinomial** has:

(1) Two terms which are positive perfect squares.

(2) A remaining term which is double the product of the square roots of the other two terms. This term may be positive or negative.

Thus, $x^2 + 14x + 49$ and $x^2 - 14x + 49$ are perfect-square trinomials.

The last term of each binomial factor has the same sign as the middle term of the perfect-square trinomial.

Thus, $x^2 - 14x + 49 = (x - 7)^2$ and $x^2 + 14x + 49 = (x + 7)^2$.

To Factor a Perfect-Square Trinomial

Factor: $4x^2 - 20x + 25$

Procedure:

1. Find the principal square root of the first term. This becomes the first term of each binomial:

2. Find the principal square root of the last term, and prefix the sign of the middle term. This becomes the last term of each binomial:

3. Form binomial from results in Steps 1 and 2. The answer is the binomial squared:

Solution:

1. Find $\sqrt{4x^2}$:

$$\sqrt{4x^2} = 2x$$

2. Find $\sqrt{25}$:

$\sqrt{25} = 5$. Prefix minus before 5 since middle term is negative.

3. Form $2x - 5$.

Ans. $4x^2 - 20x + 25 = (2x - 5)^2$

11.1. FACTORING A PERFECT-SQUARE TRINOMIAL

Factor each perfect-square trinomial:

Perfect-Square Trinomial	$\sqrt{}$ First Term	Sign of Middle Term	$\sqrt{}$ Last Term	(Binomial)² Ans.
a) $25y^2 - 10y + 1$	$5y$	$-$	1	$(5y - 1)^2$
b) $25y^2 + 10y + 1$	$5y$	$+$	1	$(5y + 1)^2$
c) $9a^2 + 42a + 49$	$3a$	$+$	7	$(3a + 7)^2$
d) $9a^2 - 42a + 49$	$3a$	$-$	7	$(3a - 7)^2$
e) $49 + 14ab + a^2b^2$	7	$+$	ab	$(7 + ab)^2$
f) $16x^6 - 24x^3 + 9$	$4x^3$	$-$	3	$(4x^3 - 3)^2$
g) $16t^2 + 4t + \frac{1}{4}$	$4t$	$+$	$\frac{1}{2}$	$(4t + \frac{1}{2})^2$
h) $a^4b^8 - 2a^2b^4c^5 + c^{10}$	a^2b^4	$-$	c^5	$(a^2b^4 - c^5)^2$

11.2. REPRESENTING THE SIDE OF A SQUARE: $s = \sqrt{A}$

Represent the side of a square whose area is:

a) $(l^2 - 14w + 49)\,\text{in}^2$ b) $(w^2 + 20w + 100)\,\text{ft}^2$ c) $4s^2 + 4s + 1$ d) $9s^2 - 30s + 25$

Ans. a) $(l - 7)\,\text{in}$ b) $(w + 10)\,\text{ft}$ c) $2s + 1$ d) $3s - 5$

12. COMPLETELY FACTORING POLYNOMIALS

To factor an expression completely, continue factoring until the polynomial factors cannot be factored further. Thus, to factor $5x^2 - 5$ completely, first factor it into $5(x^2 - 1)$. Then factor further into $5(x - 1)(x + 1)$.

If an expression has a common monomial factor:

1. **Remove its highest common factor (HCF).**
2. **Continue factoring its polynomial factors** until no further factors remain.

Procedure to Completely Factor Expressions Having Common Monomial Factor

Factor completely: *a)* $8a^2 - 50$ *b)* $3y^3 - 60y^2 + 300y$ *c)* $10y^2 - 15xy^2 + 5x^2y^2$

Procedure: **Solutions:**

1. Remove highest common factor (HCF):
   1. HCF = 2
   $2(4a^2 - 25)$
   1. HCF = 3y
   $3y(y^2 - 20y + 100)$
   1. HCF = $5y^2$
   $5y^2(2 - 3x + x^2)$

2. Continue factoring polynomial factors:
   2. Factor $4a^2 - 25$:
   $2(2a + 5)(2a - 5)$
   2. Factor $y^2 - 20y + 100$:
   $3y(y - 10)(y - 10)$
   2. Factor $2 - 3x + x^2$:
   $5y^2(2 - x)(1 - x)$

   Ans. $2(2a + 5)(2a - 5)$ **Ans.** $3y(y - 10)^2$ **Ans.** $5y^2(2 - x)(1 - x)$

Procedure to Completely Factor Expressions Having No Common Monomial Factor

Factor completely: *a)* $x^4 - 1$ *b)* $16a^4 - 81$

Procedure: **Solutions:**

1. Factor into polynomials:
   1. $(x^2 + 1)(x^2 - 1)$
   1. $(4a^2 + 9)(4a^2 - 9)$
2. Continue factoring:
   2. $(x^2 + 1)(x + 1)(x - 1)$ **Ans.**
   2. $(4a^2 + 9)(2a + 3)(2a - 3)$ **Ans.**

12.1. Factoring Completely Expressions Having Common Monomial Factor
Factor completely:

Polynomial	Remove HCF	Continue Factoring
a) $3b^2 - 27$	$3(b^2 - 9)$	$3(b + 3)(b - 3)$ *Ans.*
b) $a^5 - 16a^3$	$a^3(a^2 - 16)$	$a^3(a + 4)(a - 4)$ *Ans.*
c) $a^5 - 16a$	$a(a^4 - 16)$	$a(a^2 + 4)(a^2 - 4) = a(a^2 + 4)(a + 2)(a - 2)$ *Ans.*
d) $5x^2 + 10x + 5$	$5(x^2 + 2x + 1)$	$5(x + 1)^2$ *Ans.*
e) $x^5 - 6x^4y + 9x^3y^2$	$x^3(x^2 - 6xy + 9y^2)$	$x^3(x - 3y)^2$ *Ans.*
f) $y^5 - 4y^3 + 3y$	$y(y^4 - 4y^2 + 3)$	$y(y^2 - 3)(y^2 - 1) = y(y^2 - 3)(y + 1)(y - 1)$ *Ans.*

12.2. FACTORING COMPLETELY EXPRESSIONS HAVING NO COMMON MONOMIAL FACTOR

Factor completely:

Polynomial	Factor into Polynomials	Continue Factoring
a) $1 - a^4$	$(1 + a^2)(1 - a^2)$	$(1 + a^2)(1 + a)(1 - a)$ *Ans.*
b) $x^4 - 16y^4$	$(x^2 + 4y^2)(x^2 - 4y^2)$	$(x^2 + 4y^2)(x + 2y)(x - 2y)$ *Ans.*
c) $p^8 - q^{12}$	$(p^4 + q^6)(p^4 - q^6)$	$(p^4 + q^6)(p^2 + q^3)(p^2 - q^3)$ *Ans.*

13. THE VARIABLE: DIRECT AND INVERSE VARIATION

A **variable** in algebra is a letter which may represent any number of a set of numbers under discussion when the set contains more than one number.

Thus, in $y = 2x$, if x represents 1, then y represents 2; if x represents 2, then y represents 4; and if x represents 6, then y represents 12. These three pairs of corresponding values may be tabulated as shown. The set of numbers being discussed is the set of all numbers.

y	2	4	12
x	1	2	6

A **constant** is any letter or number which has a fixed value that does not change under discussion.

Thus, 5 and π are constants.

Throughout this chapter, as is customary in mathematics, the letter k is used to represent a constant while x, y, and z represent variables.

Measuring the Change in a Variable

As a variable x changes in value from one number x_1 to a second number x_2, the change may be measured by finding the difference $x_2 - x_1$, or by finding the ratio $\dfrac{x_2}{x_1}$. A change in value may be measured by subtracting or dividing.

Thus, the change in the speed of an automobile from 20 to 60 mi/h may be expressed as follows:

(1) The second speed is $60 - 20$ or **40 mi/h faster** than the first speed.
(2) The second speed is $\frac{60}{20}$, or **three times as fast** as the first speed.

Any formula may be considered as a relationship of its variables. Mathematics and science abound in formulas which have exactly the same general structure: $z = xy$. To illustrate, study the following:

Formula	Rule
$D = RT$	Distance = rate $\times$ time
$A = LW$	Area = length $\times$ width
$T = PR$	Interest = price $\times$ rate of interest
$C = NP$	Cost = number $\times$ price
$F = PA$	Force = pressure $\times$ area
$M = DV$	Mass = density $\times$ volume

Note, in each formula which has the form $z = xy$, that there are three variables. Furthermore, one of the variables is the product of the other two. In this chapter, we study the relation of these formulas to the two basic types of variation. These types of variation are **direct variation** and **inverse variation**.

13.1. USING DIVISION TO MEASURE THE CHANGE IN A VARIABLE

Using division, find the ratio which indicates the change in each variable and express this ratio in a sentence.

a) The price of a suit changes from \$100 to \$120.
b) The speed of an automobile changes from 10 to 40 mi/h.
c) A salary changes from \$300 per week to \$400 per week.
d) Length of a line changes from 12 to 4 ft.

Solutions:

a) $\dfrac{\$120}{\$100} = \dfrac{6}{5}$. Hence, second price is **six-fifths** of first price.

b) $\dfrac{40 \text{ mi/h}}{10 \text{ mi/h}} = 4$. Hence, second rate is **4 times (quadruple)** the first rate.

c) $\dfrac{\$400 \text{ per week}}{\$300 \text{ per week}} = \dfrac{4}{3}$. Hence, second salary is **four-thirds** of first salary.

d) $\dfrac{4 \text{ ft}}{12 \text{ ft}} = \dfrac{1}{3}$. Hence, second length is **one-third** of first length.

13.2. MULTIPLYING OR DIVIDING VARIABLES

Complete each:

a) If x is **doubled**, it will change from 13 to ( ) or from ( ) to 90.
b) If y is **tripled**, it will change from 17 to ( ) or from ( ) to 87.
c) If z is **halved**, it will change from 7 to ( ) or from ( ) to 110.
d) If s is **multiplied** by $\frac{5}{4}$, it will change from 16 to ( ) or from ( ) to 55.

Ans. *a)* 26, 45   *b)* 51, 29   *c)* $3\frac{1}{2}$, 220   *d)* 20, 44

14. UNDERSTANDING DIRECT VARIATION: $y = kx$ or $\dfrac{y}{x} = k$

Direct Variation Formula with Constant Ratio k

$$y = kx \quad \text{or} \quad \frac{y}{x} = k$$

Direct Variation Formula without Constant

If x varies from x_1 to x_2 while y varies from y_1 to y_2, then:

$$\frac{x_2}{x_1} = \frac{y_2}{y_1}$$

Rule 1.   **If $y = kx$ or $\dfrac{y}{x} = k$, then:**

(1) x and y vary directly as each other; that is, x varies directly as y, and y varies directly as x.

(2) x and y are directly proportional to each other; that is, $\dfrac{x_2}{x_1} = \dfrac{y_2}{y_1}$ as x varies from x_1 to x_2, and y varies from y_1 to y_2.

Thus:

1. If $y = 4x$, then y and x vary directly as each other.
2. If $I = 0.06P$, I and P vary directly as each other.
3. The perimeter of a square equals 4 times its side; that is, $P = 4S$. For a square, P and S vary directly as each other.

Multiplication and Division in Direct Variation

Rule 2. If x and y vary directly as each other, and a value of either x or y is multiplied by a number, then the corresponding value of the other is multiplied by the same number.

Thus, if $y = 5x$ and x is tripled, then y is tripled, as shown.

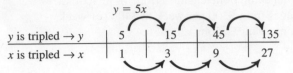

$$y = 5x$$

| y is tripled → y | 5 | 15 | 45 | 135 |
| x is tripled → x | 1 | 3 | 9 | 27 |

Rule 3. If x and y vary directly as each other, and a value of either x or y is divided by a number, then the corresponding value of the other is divided by the same number.

Thus, if $y = 10x$ and x is halved (divided by 2), then y is halved, as shown.

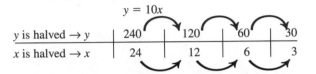

$$y = 10x$$

| y is halved → y | 240 | 120 | 60 | 30 |
| x is halved → x | 24 | 12 | 6 | 3 |

Direct Variation Applied to a Rectangle

To understand direct variation more fully, note how it applies to a rectangle:

If the length of a rectangle is fixed and the width is tripled, then the area is tripled. As a result, the new area $ABCD$ is 3 times the area of the old (shaded) rectangle, as shown in Fig. 10-1.

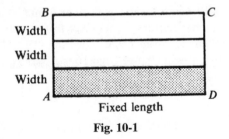

Fig. 10-1

14.1. RULE 1. DIRECT VARIATION IN FORMULAS

(1) Express each equation in the form $y = kx$.

(2) State the variables that vary directly as each other.

a) $120R = P$ c) $0.05 = \dfrac{1}{P}$ e) $A = \frac{1}{2}bh$ when b is constant

b) $\dfrac{C}{D} = \pi$ d) $f = \dfrac{n}{d}$ when $f = \dfrac{2}{3}$ f) $V = \pi R^2 H$ when $R = 8$

Ans.   a) (1)$P = 120R$, (2) P and R d) (1) $n = \frac{2}{3}d$, (2) n and d

b) (1) $C = \pi D$, (2) C and D e) (1) $A = kh$, (2) A and h

c) (1) $I = 0.05P$, (2) I and P f) (1) $V = 64\pi H$, (2) V and H

14.2. RULES 2 AND 3. MULTIPLICATION AND DIVISION IN DIRECT VARIATION

Complete each:

a) If $y = 8x$ and x is tripled, then   *Ans.* y is tripled.

b) If $C = \pi D$ and D is quadrupled, then C is quadrupled.

c) If $C = \pi D$ and C is halved, then D is halved.

d) If $D = RT$, $T = 12$ and R is divided by 4, then D is divided by 4.

e) If $A = LW$, L is constant and W is doubled, then A is doubled.

14.3. **APPLYING DIRECT VARIATION TO STATEMENTS**

(1) Complete each statement, and (2) state the formula to which direct variation applies.

   a) At a uniform speed, doubling time will ...
   b) If the time of travel is constant, tripling rate will ...
   c) If a rectangle has a fixed width, to multiply its area by 5 ...
   d) If the value of a fraction is constant, and the numerator is halved,...

Ans. a) (1) Double distance, (2) $D = kT$ since $D = RT$ and R is constant.
   b) (1) Triple distance, (2) $D = kR$ since $D = RT$ and T is constant.
   c) (1) Multiply its length by 5, (2) $A = kL$ since $A = LW$ and W is constant.

   d) (1) The denominator is halved, (2) $n = kd$ since $f = \dfrac{n}{d}$ and f is constant.

14.4. **FINDING VALUES FOR DIRECTLY VARYING VARIABLES**

a) If y and x vary directly as each other, and $y = 36$ when $x = 9$, find y when $x = 27$.

b) If y and x vary directly as each other, and $y = 25$ when $x = 10$, find x when $y = 5$.

Solution: Proportion Method

If y and x vary directly,

	x	y
Second values	27	y
First values	9	36

$$\frac{y_2}{y_1} = \frac{x_2}{x_1}$$

$$\frac{y}{36} = \frac{27}{9}$$

Ans. 108

$$y = \frac{27}{9}(36) - 108$$

Solution: Proportion Method

If y and x vary directly,

	x	y
Second values	x	5
First values	10	25

$$\frac{x_2}{x_1} = \frac{y_2}{y_1}$$

$$\frac{x}{10} = \frac{5}{25}$$

Ans. 2

$$x = \frac{5}{25}(10) = 2$$

a)   **Ratio Method**

(Note the equal ratios in the table.)

	x	y
Second values	$\dfrac{27}{9}\left[\tfrac{3}{1}\right]$	$\dfrac{y}{36}\left[\tfrac{3}{1}\right]$
First values		

If 2nd x value is **three times** 1st x value,
then 2nd y value is **three times** 1st y value.
Hence, $y = 3(36) = 108$. *Ans.* 108

b)   **Ratio Method**

(Note the equal ratios in the table.)

	x	y
Second values	$\dfrac{x}{10}\left[\tfrac{1}{5}\right]$	$\dfrac{5}{25}\left[\tfrac{1}{5}\right]$
First values		

If second y value is **one-fifth** of first y value,
then second x value is **one-fifth** of first x value.
Hence, $x = \tfrac{1}{5}(10) = 2$. *Ans.* 2

Equation Method

If y and x vary directly, $kx = y$.
Since $y = 36$ when $x = 9$, $9k = 36$, $k = 4$.
Since $k = 4$, $y = 4x$.
When $x = 27$, $y = 4(27) = 108$. *Ans.* 108

Equation Method

If y and x vary directly, $kx = y$.
Since $y = 25$ when $x = 10$, $10k = 25$,
$k = 2\tfrac{1}{2}$.
Since $k = 2\tfrac{1}{2}$, $y = 2\tfrac{1}{2}x$.
When $y = 5$, $5 = 2\tfrac{1}{2}x$, $x = 2$. *Ans.* 2

14.5.   APPLYING DIRECT VARIATION TO A MOTION PROBLEM

Henry traveled a distance of 124 mi at 40 mi/h. If he had taken the same amount of time, how far would he have traveled at 50 mi/h?

Solution:

	Rate (mi/h)	Distance (mi)
Second trip	50	D
First trip	40	124

($\frac{5}{4}$ for Rate, $\frac{5}{4}$ for Distance)

Proportion Method

Since $D = RT$ and the time T is constant, D and R vary directly as each other.

Hence, $\dfrac{D_2}{D_1} = \dfrac{R_2}{R_1}$, $\dfrac{D}{124} = \dfrac{50}{40}$, $\dfrac{D}{124} = \dfrac{5}{4}$,

$$D = \frac{5}{4}(124) = 155 \qquad Ans. \quad 155 \text{ mi}$$

Ratio Method

(Note use of the equal ratios in the table.) If the second rate is **five-fourths** of the first rate, then the second distance is **five-fourths** of the first distance. Hence, $D = \dfrac{5}{4}(124) = 155$.

Ans.   155 mi

14.6.   APPLYING DIRECT VARIATION TO AN INTEREST PROBLEM

In a bank, the annual interest on $4500 is $180. At the same rate, what is the annual interest on $7500?

Solution:

	Principal ($)	Annual Interest ($)
Second principal	7500	I
First principal	4500	180

($\frac{5}{3}$ for Principal, $\frac{5}{3}$ for Annual Interest)

Proportion Method

Since $I = PR$ and the rate R is constant, I and P vary directly as each other.

Hence, $\dfrac{I_2}{I_1} = \dfrac{P_2}{P_1}$, $\dfrac{I}{180} = \dfrac{7500}{4500}$, $\dfrac{I}{180} = \dfrac{5}{3}$,

$$I = \frac{5}{3}(180) = 300 \qquad Ans. \quad \$300$$

Ratio Method

If the second principal is **five-thirds** of the first principal, then the second interest is **five-thirds** of the first interest. Hence, $I = \frac{5}{3}(180) = 300$.

Ans. $300

15.   UNDERSTANDING INVERSE VARIATION: $xy = k$

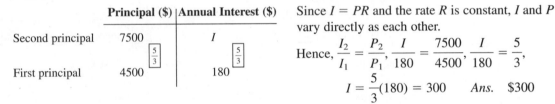

Inverse Variation Formula with Constant Product k

$$xy = k$$

Inverse Variation Formula without Constant

If x varies from x_1 to x_2, while y varies from y_1 to y_2, then:

$$\frac{x_2}{x_1} = \frac{y_1}{y_2} \qquad \text{or} \qquad \frac{y_2}{y_1} = \frac{x_1}{x_2}$$

Rule 1.   If $xy = k$, then:

(1) x and y **vary inversely** as each other: that is, x varies inversely as y, and y varies inversely as x.

(2) x and y are **inversely proportional** to each other; that is $\dfrac{x_2}{x_1} = \dfrac{y_1}{y_2}$ or $\dfrac{y_2}{y_1} = \dfrac{x_1}{x_2}$, as x varies from x_1 to x_2, and y varies from y_1 to y_2.

Thus,

1. If $xy = 12$, then y and x vary inversely as each other.
2. If $PR = 150$, then P and R vary inversely as each other.
3. For a fixed distance of 120 mi to be traveled, the motion formula $D = RT$ becomes $120 = RT$. In such a case, R and T vary inversely as each other.

Multiplication and Division in Inverse Variation

Rule 2. If x and y vary inversely as each other, and a value of either x or y is **multiplied** by a number, then the corresponding value of the other is **divided** by the same number.

Thus, if $xy = 24$ and y is doubled, then x is halved, as shown.

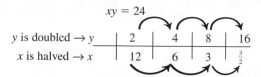

$$xy = 24$$

y is doubled $\rightarrow y$	2	4	8	16
x is halved $\rightarrow x$	12	6	3	$\frac{3}{2}$

Rule 3. If x and y vary inversely as each other and either x or y is divided by a number, then the other is **multiplied** by the same number.

Thus, if $xy = 250$, and y is divided by 5, then x is multiplied by 5, as shown.

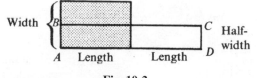

y is divided by 5 $\rightarrow y$	250	50	10	2
x is multiplied by 5 $\rightarrow x$	1	5	25	125

Inverse Variation Applied to a Rectangle

To understand inverse variation more fully, note how it applies to a rectangle:

If the length of a rectangle is doubled and the width is halved, its area remains constant. As a result, the new area $ABCD$, equals the old (shaded) area. See Fig. 10-2.

Fig. 10-2

15.1. INVERSE VARIATION IN FORMULAS

(1) Express each equation in the form $xy = k$.
(2) State the variables that vary inversely as each other.

a) $5RT = 500$	*Ans.*	(1) $RT = 100$, (2) R and T
b) $\frac{1}{2}PV = 7$		(1) $PV = 14$, (2) P and V
c) $f = \dfrac{n}{d}$ when $n = 3$		(1) $fd = 3$, (2) f and d
d) $A = \frac{1}{2}bh$ when A is constant		(1) $bh = k$, (2) b and h

15.2. RULES 2 AND 3: MULTIPLICATION AND DIVISION IN INVERSE VARIATION
Complete each:

a) If $xy = 25$ and x is tripled, then	*Ans.*	y is divided by 3.
b) If $50 = RT$ and R is divided by 10, then		T is multiplied by 10.
c) If $BH = 5$ and B is multiplied by $\frac{3}{2}$, then		H is divided by $\frac{3}{2}$.
d) If $A = LW$, A is constant and L is doubled, then		W is halved.
e) If $PV = k$ and V is quadrupled, then		P is divided by 4.

15.3. APPLYING INVERSE VARIATION TO STATEMENTS

(1) Complete each statement.

(2) State the formula to which inverse variation applies.

    a) Over the same distance, doubling the speed …

    b) For a fixed area, multiplying the length of a rectangle by 5 …

    c) For an enclosed gas at a constant temperature, dividing the pressure by 7 …

    d) If the total cost is the same, tripling the price of an article …

Ans.   *a*) (1) Halves the time, (2) $RT = k$ since $RT = D$ and D is constant

       b) (1) Divides the width by 5, (2) $LW = k$ since $LW = A$ and A is constant

       c) (1) Multiplies the volume by 7, (2) $PV = k$ (Boyle's law)

       d) (1) Divides the number of these articles purchased by 3, (2) $NP = k$ since $NP = C$ and C is constant

15.4. FINDING VALUES FOR INVERSELY VARYING VARIABLES

a) If y and x vary inversely as each other and $y = 10$ when $x = 6$, find y when $x = 15$.

b) If y and x vary inversely as each other and $y = 12$ when $x = 4$, find x when $y = 8$.

Solution:

Proportion Method   Proportion Method

a) If y and x vary inversely,

$$\frac{y_2}{y_1} = \frac{x_1}{x_2}$$

$$\frac{y}{10} = \frac{6}{15}$$

$$y = \frac{6}{15}(10) = 4 \quad \textit{Ans.} \quad 4$$

	x	y
Second values	15	y
First values	6	10

b) If y and x vary inversely,

$$\frac{x_2}{x_1} = \frac{y_1}{y_2}$$

$$\frac{x}{4} = \frac{12}{8}$$

$$x = \frac{12}{8}(4) = 6 \quad \textit{Ans.} \quad 6$$

	x	y
Second values	x	8
First values	4	12

Ratio Method   Ratio Method

(Note the inverse ratios in the table.)

	x	y
Second values	$\dfrac{15}{6}\ \boxed{\tfrac{5}{2}}$	$\dfrac{y}{10}\ \boxed{\tfrac{2}{5}}$
First values		

If second x value is **five-halves** of first x value, then second y value is **two-fifths** of first y value.

    Hence, $y = \frac{2}{5}(10) = 4.$ *Ans.* 4

(Note the inverse ratios in the table.)

	x	y
Second values	$\dfrac{x}{4}\ \boxed{\tfrac{3}{2}}$	$\dfrac{8}{12}\ \boxed{\tfrac{2}{3}}$
First values		

If second y value is **two-thirds** of first y value, then second x value is **three-halves** of first x value.

    Hence, $x = \frac{3}{2}(4) = 6.$ *Ans.* 6

Equation Method   Equation Method

If x and y vary inversely, $k = xy$.

Since $y = 10$ when $x = 6$,

$k = (6)(10) = 60$.

Since $k = 60$, $xy = 60$.

When $x = 15$, $15y = 60$, $y = 4$.   *Ans.* 4

If x and y vary inversely, $k = xy$.

Since $y = 12$ when $x = 4$,

$k = (4)(12) = 48$.

Since $k = 48$, $xy = 48$.

When $y = 8$, $8x = 48$, $x = 6$.   *Ans.* 6

Supplementary Problems

10.1. Find each product: (1.1)

$a)$ $3 \cdot x \cdot 4 \cdot x$　　　　$c)$ $(-2)5xy^3x^4$　　　　$e)$ $(-5ab^2)(2a^2b)$　　　　$g)$ $-3 \cdot \dfrac{1}{x} \cdot \dfrac{1}{y}$

$b)$ $(-3)4xyy$　　　　$d)$ $(-3x)(-7x^2)y$　　　　$f)$ $(3a^4b)(10ac^4)$　　　　$h)$ $5 \cdot \dfrac{1}{a^2} \cdot \dfrac{1}{b^2} \cdot c$

Ans. $a)$ $12x^2$　　　　$c)$ $-10x^5y^3$　　　　$e)$ $-10a^3b^3$　　　　$g)$ $-\dfrac{3}{xy}$

$b)$ $-12xy^2$　　　　$d)$ $21x^3y$　　　　$f)$ $30a^5bc^4$　　　　$h)$ $\dfrac{5c}{a^2b^2}$

10.2. Find each product: (1.2)

$a)$ $3(x - 3)$　　　$d)$ $a(b + c)$　　　$g)$ $x^2(x - 3)$　　　$j)$ $a(b - c + d)$
$b)$ $-5(a + 4)$　　$e)$ $-3c(d - g)$　　$h)$ $-3x(x^2 - 5x)$　　$k)$ $-3x(x^2 - 2x + 5)$
$c)$ $-8(r - 7)$　　$f)$ $2gh(3k - 5)$　　$i)$ $7a^3(5a^2 - 8a)$　　$l)$ $2b^2c(b - 3c - 10c^2)$

Ans. $a)$ $3x - 9$　　　$d)$ $ab + ac$　　　$g)$ $x^3 - 3x^2$　　　$j)$ $ab - ac + ad$
$b)$ $-5a - 20$　　$e)$ $-3cd + 3cg$　　$h)$ $-3x^3 + 15x^2$　　$k)$ $-3x^3 + 6x^2 - 15x$
$c)$ $-8r + 56$　　$f)$ $6ghk - 10gh$　　$i)$ $35a^5 - 56a^4$　　$l)$ $2b^3c - 6b^2c^2 - 20b^2c^3$

10.3. Find each product: (1.3)

$a)$ $4(8 - \frac{1}{2})$　　　　　$d)$ $50\left(\dfrac{2x}{5} - \dfrac{3x}{10}\right)$　　　　$g)$ $0.06(2000 + 500 + 25)$

$b)$ $20(5 - 0.7)$　　　　$e)$ $12\left(\dfrac{x}{4} + \dfrac{x}{3} - \dfrac{x}{2}\right)$　　　$h)$ $0.03(7000 - x)$

$c)$ $36(1\frac{1}{4} + \frac{2}{9})$　　　　$f)$ $70\left(2 - \dfrac{5r}{14}\right)$　　　　$i)$ $0.05\frac{1}{2}(3000 + 2x)$

Ans. $a)$ $32 - 2 = 30$　　　　$d)$ $20x - 15x = 5x$　　　　$g)$ $120 + 30 + 1.50 = 151.50$
$b)$ $100 - 14 = 86$　　$e)$ $3x + 4x - 6x = x$　　　$h)$ $210 - 0.03x$
$c)$ $45 + 8 = 53$　　　$f)$ $140 - 25r$　　　　$i)$ $165 + 0.11x$

10.4. Factor, removing highest common factor: (2.1)

$a)$ $3a - 21b$　　　$e)$ $p + prt$　　　$i)$ $\pi rh + \pi r^2$　　　$m)$ $8x^2 - 16x + 32$
$b)$ $-5c - 15d$　　$f)$ $7hm - 7h$　　$j)$ $\pi R^2 + \pi r^2$　　$n)$ $10x^3 + 20x^2 - 55x$
$c)$ $xy + 2xz$　　　$g)$ $6V^2 + 3V$　　$k)$ $\frac{1}{2}bh + \frac{1}{2}b'h$　　$o)$ $a^2bc + ab^2c - abc^2$
$d)$ $5RS - 10RT$　　$h)$ $y^5 + y^4$　　　$l)$ $\frac{1}{4}mnr - \frac{1}{4}mn$　　$p)$ $x^2y^3 - x^2y^2 + x^2y$

Ans. $a)$ $3(a - 7b)$　　　$e)$ $p(1 + rt)$　　　$i)$ $\pi r(h + r)$　　　$m)$ $8(x^2 - 2x + 4)$
$b)$ $-5(c + 3d)$　　$f)$ $7h(m - 1)$　　$j)$ $\pi(R^2 + r^2)$　　$n)$ $5x(2x^2 + 4x - 11)$
$c)$ $x(y + 2z)$　　　$g)$ $3V(2V + 1)$　　$k)$ $\frac{1}{2}b(b + b')$　　$o)$ $abc(a + b - c)$
$d)$ $5R(S - 2T)$　　$h)$ $y^4(y + 1)$　　　$l)$ $\frac{1}{4}mn(r - 1)$　　$p)$ $x^2y(y^2 - y + 1)$

10.5. Evaluate each, using factoring: (2.2)

$a)$ $6(11) + 4(11)$　　　　　$d)$ $2.7(0.3) - 0.7(0.3)$　　　$g)$ $10(8^2) - 7(8^2)$
$b)$ $8(11\frac{1}{2}) + 12(11\frac{1}{2})$　　　$e)$ $18(7^2) + 2(7^2)$　　　$h)$ $\frac{1}{2}(13)(6) + \frac{1}{2}(7)(6)$
$c)$ $21(2\frac{1}{3}) - 12(2\frac{1}{3})$　　　$f)$ $\frac{3}{4}(11) - \frac{3}{4}(3)$　　　$i)$ $5^3(28) - 5^3(25)$

Ans. $a)$ $11(6 + 4) = 11(10) = 110$　　　　$d)$ $0.3(2.7 - 0.7) = 0.3(2) = 0.6$　　$g)$ $8^2(10 - 7) = 8^2(3) = 192$
$b)$ $11\frac{1}{2}(8 + 12) = 11\frac{1}{2}(20) = 230$　　$e)$ $7^2(18 + 2) = 7^2(20) = 980$　　$h)$ $\frac{1}{2}(6)(13 + 7) = \frac{1}{2}(6)(20) = 60$
$c)$ $2\frac{1}{3}(21 - 12) = 2\frac{1}{3}(9) = 21$　　$f)$ $\frac{3}{4}(11 - 3) = \frac{3}{4}(8) = 6$　　$i)$ $5^3(28 - 25) = 5^3(3) = 375$

10.6. Find each square: **(3.1)**

a) 6^2 d) 600^2 g) $\left(-\frac{3}{7}\right)^2$ j) 35^2

b) $(-6)^2$ e) $\left(\frac{1}{6}\right)^2$ h) $\left(-\frac{9}{2}\right)^2$ k) 3.5^2

c) $(0.6)^2$ f) $\left(\frac{2}{5}\right)^2$ i) $\left(1\frac{1}{3}\right)^2$ l) $(-0.35)^2$

Ans.   a) 36   d) 360,000   g) $\frac{9}{49}$ j) 1225

b) 36   e) $\frac{1}{36}$ h) $8\frac{1}{4}$ k) 1225

c) 0.36   f) $\frac{4}{25}$ i) $\frac{16}{9}$ or $1\frac{7}{9}$ l) 0.1225

10.7. Find each square: **(3.2)**

a) $(b^2)^2$ d) $(5a^5)^2$ g) $(rs^2t^3)^2$

b) $(w^5)^2$ e) $(-10b^{10})^2$ h) $(-0.1r^4s^9)^2$

c) $\left(\dfrac{1}{V^7}\right)^2$ f) $\left(\dfrac{0.3}{C^3}\right)^2$ i) $\left(-\dfrac{5r^5}{3t^3}\right)^2$

Ans.   a) b^4 d) $25a^{10}$ g) $r^2s^4t^6$

b) w^{10} e) $100b^{20}$ h) $0.01r^8s^{18}$

c) $\dfrac{1}{V^{14}}$ f) $\dfrac{0.09}{C^6}$ i) $\dfrac{25r^{10}}{9t^6}$

10.8. Find the area of a square whose side is: **(3.3)**

a) 5 yd   c) 2.5 mi   e) $0.7\,y^2$

b) $\dfrac{7}{8}$ rd   d) $\dfrac{x}{7}$ in   f) $\dfrac{5m^2}{3p^3}$

Ans.   a) $25\,\text{yd}^2$ c) $6.25\,\text{mi}^2$ e) $0.49y^4$

b) $\dfrac{49}{64}\,\text{rd}^2$ d) $\dfrac{x^2}{49}\,\text{in}^2$ f) $\dfrac{25m^4}{9p^6}$

10.9. Find each principal square root: **(4.1)**

a) $\sqrt{144}$ d) $\sqrt{c^6}$ g) $\sqrt{36x^{36}}$ j) $\sqrt{0.0001h^{50}}$

b) $\sqrt{1.44}$ e) $\sqrt{p^8q^{10}}$ h) $\sqrt{16a^4b^{16}}$ k) $\sqrt{0.0144m^{14}}$

c) $\sqrt{\dfrac{900}{49}}$ f) $\sqrt{\dfrac{r^{14}}{s^{20}}}$ i) $\sqrt{\dfrac{0.01}{d^{100}}}$ l) $\sqrt{\dfrac{25h^{10}}{64k^{16}}}$

Ans.   a) 12   d) c^3 g) $6x^{18}$ j) $0.01h^{25}$

b) 1.2   e) p^4q^5 h) $4a^2b^8$ k) $0.12m^7$

c) $\dfrac{30}{7}$ f) $\dfrac{r^7}{s^{10}}$ i) $\dfrac{0.1}{d^{50}}$ l) $\dfrac{5h^5}{8k^8}$

10.10. Find the side of a square whose area is: **(4.2)**

a) $100\,\text{ft}^2$ *Ans.* 10 ft   d) $9x^2\text{in}^2$ *Ans.* $3x$ in   g) $100x^{20}$ *Ans.* $10x^{10}$

b) $0.25\,\text{mi}^2$ 0.5 mi   e) $2.25b^8\,\text{yd}^2$ $1.5b^4$ yd   h) $169(a+b)^2$ $13(a+b)$

c) $\dfrac{16}{81}\,\text{rd}^2$ $\dfrac{4}{9}$ rd   f) $2500c^{10}\text{cm}^2$ $50c^5$ cm   i) $\dfrac{4x^2y^2}{9}$ $\dfrac{2xy}{3}$

10.11. Find each product: **(5.1)**

a) $(s+4)(s-4)$ *Ans.* s^2-16 g) $(a^2+b^2)(a^2-b^2)$ *Ans.* a^4-b^4

b) $(10-t)(10+t)$ $100-t^2$ h) $(ab+c^2)(ab-c^2)$ $a^2b^2-c^4$

c) $(2x+1)(2x-1)$ $4x^2-1$ i) $(v+\frac{1}{5})(v-\frac{1}{5})$ $v^2-\frac{1}{25}$

d) $(3y-7z)(3y+7z)$ $9y^2-49z^2$ j) $(d-1.2)(d+1.2)$ $d^2-1.44$

$e)$ $(rs + 1)(rs - 1)$ Ans. $r^2s^2 - 1$ $k)$ $\left(3x^2 - \dfrac{2}{y}\right)\left(3x^2 + \dfrac{2}{y}\right)$ Ans. $9x^4 - \dfrac{4}{y^2}$

$f)$ $(1 - 8x^2)(1 + 8x^2)$ $1 - 64x^4$ $l)$ $(3c^3 + 0.1)(3c^3 - 0.1)$ $9c^6 - 0.01$

10.12. Multiply, using the sum difference of two numbers: **(5.2)**

$a)$ 21×19 $c)$ $10\frac{1}{3} \times 9\frac{2}{3}$ $e)$ 89×91 $g)$ $3\frac{1}{2} \times 2\frac{1}{2}$
$b)$ 3.4×2.6 $d)$ 17×23 $f)$ 10.3×9.7 $h)$ 7.5×8.5

Ans. $a)$ $20^2 - 1^2 = 399$ $c)$ $10^2 - (\frac{1}{3})^2 = 99\frac{8}{9}$ $e)$ $90^2 - 1^2 = 8099$ $g)$ $3^2 - (\frac{1}{2})^2 = 8\frac{3}{4} = 8.75$
$b)$ $3^2 - 0.4^2 = 8.84$ $d)$ $20^2 - 3^2 = 391$ $f)$ $10^2 - 0.3^2 = 99.91$ $h)$ $8^2 - (\frac{1}{2})^2 = 63\frac{3}{4} = 63.75$

10.13. Find each product: **(5.3)**

$a)$ $3(x + 2)(x - 2)$ $d)$ $x(x - 3)(x + 3)$ $g)$ $36(w^2 - \frac{1}{3})(w^2 + \frac{1}{3})$
$b)$ $a^2(b + c)(b - c)$ $e)$ $q^2(10 - q)(10 + q)$ $h)$ $ab(c^3 + 1)(c^3 - 1)$

$c)$ $5(1 - d)(1 + d)$ $f)$ $160\left(1 - \dfrac{q}{4}\right)\left(10 + \dfrac{q}{4}\right)$

Ans. $a)$ $3x^2 - 12$ $d)$ $x^3 - 9x$ $g)$ $36w^4 - 4$
$b)$ $a^2b^2 - a^2c^2$ $e)$ $100q^2 - q^4$ $h)$ $abc^6 - ab$
$c)$ $5 - 5d^2$ $f)$ $160 - 10q^2$

10.14. Factor: **(6.1)**

$a)$ $r^2 - 25$ $g)$ $x^4 - 9$ $m)$ $\dfrac{9}{v^2} - 0.25$

$b)$ $64 - u^2$ $h)$ $25 - 16y^{16}$ $n)$ $\dfrac{x^4}{81} - \dfrac{25}{49}$

$c)$ $81 - c^2d^2$ $i)$ $d^2 - 0.01e^2$ $o)$ $\dfrac{p^2}{q^2} - \dfrac{r^2}{16}$

$d)$ $9x^2 - 1600$ $j)$ $0.09A^2 - 49$ $p)$ $k^6 - 25m^{10}$
$e)$ $t^2 - \frac{4}{9}$ $k)$ $B^2 - 0.0001$
$f)$ $100t^2 - 225s^2$ $l)$ $R^2S^2 - 1.21$

Ans. $a)$ $(r + 5)(r - 5)$ $g)$ $(x^2 + 3)(x^2 - 3)$ $m)$ $\left(\dfrac{3}{v} + 0.5\right)\left(\dfrac{3}{v} - 0.5\right)$

$b)$ $(8 + u)(8 - u)$ $h)$ $(5 - 4y^8)(5 + 4y^8)$ $n)$ $\left(\dfrac{x^2}{9} + \dfrac{5}{7}\right)\left(\dfrac{x^2}{9} - \dfrac{5}{7}\right)$

$c)$ $(9 + cd)(9 - cd)$ $i)$ $(d + 0.1\,e)(d - 0.1e)$ $o)$ $\left(\dfrac{p}{q} + \dfrac{r}{4}\right)\left(\dfrac{p}{q} - \dfrac{r}{4}\right)$

$d)$ $(3x + 40)(3x - 40)$ $j)$ $(0.3A + 7)(0.3A - 7)$ $p)$ $(k^3 + 5m^5)(k^3 - 5m^5)$
$e)$ $(t + \frac{2}{3})(t - \frac{2}{3})$ $k)$ $(B + 0.01)(B - 0.01)$
$f)$ $(10t + 15s)(10t - 15s)$ $l)$ $(RS + 1.1)(RS - 1.1)$

10.15. Factor completely: **(6.2)**

$a)$ $3x^2 - 3$ $d)$ $y^5 - y^3$ $g)$ $12x^4 - 12$
$b)$ $5x^3 - 45x$ $e)$ $\pi R^3 - 25\pi R$ $h)$ $15x^4 - 240$
$c)$ $175 - 7y^2$ $f)$ $\frac{1}{3}\pi R^2h - \frac{1}{3}\pi r^2h$ $i)$ $x^7 - 81x^3$

Ans. $a)$ $3(x + 1)(x - 1)$ $d)$ $y^3(y + 1)(y - 1)$ $g)$ $12(x^2 + 1)(x + 1)(x - 1)$
$b)$ $5x(x + 3)(x - 3)$ $e)$ $\pi R(R + 5)(R - 5)$ $h)$ $15(x^2 + 4)(x + 2)(x - 2)$

$c)$ $7(5 + y)(5 - y)$ $f)$ $\dfrac{\pi h}{3}(R + r)(R - r)$ $i)$ $x^3(x^2 + 9)(x + 3)(x - 3)$

10.16. Multiply, showing each separate product: **(7.1)**

	First	Outer	+	Inner	Last	Combine (Ans.)
a) $(x + 5)(x + 1)$	x^2	x	+	$5x$	$+5$	$x^2 + 6x + 5$
b) $(x + 8)(x - 2)$	x^2	$-2x$	+	$8x$	-16	$x^2 + 6x - 16$
c) $(x - 7)(x - 6)$	x^2	$-6x$	+	$(-7x)$	$+42$	$x^2 - 13x + 42$
d) $(x - 10)(x + 9)$	x^2	$+9x$	+	$(-10x)$	-90	$x^2 - x - 90$
e) $(3a - 1)(4a + 1)$	$12a^2$	$+3a$	+	$(-4a)$	-1	$12a^2 - a - 1$
f) $(5b + 2)(5b - 2)$	$25b^2$	$-10b$	+	$10b$	-4	$25b^2 - 4$
g) $(c - 3)(2c + 8)$	$2c^2$	$+8c$	+	$(-6c)$	-24	$2c^2 + 2c - 24$
h) $(r - 4s)(r - 11s)$	r^2	$-11rs$	+	$(-4rs)$	$+44s^2$	$r^2 - 15rs + 44s^2$
i) $(3s + 2t)(3s - 2t)$	$9s^2$	$-6st$	+	$6st$	$-4t^2$	$9s^2 - 4t^2$
j) $(x^2 + 5)(x^2 + 8)$	x^4	$+8x^2$	+	$5x^2$	$+40$	$x^4 + 13x^2 + 40$
k) $(3w^2 + 2x)(w^2 - x)$	$3w^4$	$-3w^2x$	+	$2w^2x$	$-2x^2$	$3w^4 - w^2x - 2x^2$
l) $(2w^3 - 3)(6w^3 + 9)$	$12w^6$	$+18w^3$	+	$(-18w^3)$	-27	$12w^6 - 27$

10.17. Multiply mentally: **(7.2)**

a) $(3c + 1)(4c + 5)$ d) $(pq - 8)(pq + 11)$ g) $(d^2 + 6)(d^2 + 1)$

b) $(2 + 7c)(1 - c)$ e) $(2AB + 7)(2AB - 7)$ h) $(8 - g^2)(3 - 2g^2)$

c) $(c + 3d)(c + 12d)$ f) $(3x - 2y)(4x - 9y)$ i) $(4c^3 + 1)(c^3 - 2)$

Ans. a) $12c^2 + 19c + 5$ d) $p^2q^2 + 3pq - 88$ g) $d^4 + 7d^2 + 6$

b) $2 + 5c - 7c^2$ e) $4A^2B^2 - 49$ h) $24 - 19g^2 + 2g^4$

c) $c^2 + 15cd + 36d^2$ f) $12x^2 - 35xy + 18y^2$ i) $4c^6 - 7c^3 - 2$

10.18. Represent the area of a rectangle whose dimensions are: **(7.3)**

a) $l - 3$ and $l - 8$ c) $w + 10$ and $w - 12$ e) $13 - 2s$ and $2 - s$

b) $2l + 5$ and $3l - 1$ d) $6 - w$ and $8 - w$ f) $7s + 1$ and $9s + 2$

Ans. a) $l^2 - 11l + 24$ c) $w^2 - 2w - 120$ e) $26 - 17s + 2s^2$

b) $6l^2 + 13l - 5$ d) $48 - 14w + w^2$ f) $63s^2 + 23s + 2$

10.19. Factor each trinomial: **(8.1)**

a) $a^2 + 7a + 10$ g) $x^2 - x - 2$ m) $a^2b^2 + 8ab + 15$

b) $b^2 + 8b + 15$ h) $y^2 - 3y - 4$ n) $d^2e^2 - 15def + 36f^2$

c) $r^2 - 12r + 27$ i) $w^2 + 2w - 8$ o) $x^4 + 5x^2 + 4$

d) $s^2 - 14s + 33$ j) $w^2 + 7w - 18$ p) $y^4 - 6y^2 - 7$

e) $h^2 - 27h + 50$ k) $x^2 + 14xy + 24y^2$ q) $x^4 + 8x^2y^2 + 16y^4$

f) $m^2 + 19m + 48$ l) $c^2 - 17cd + 30d^2$ r) $x^4y^4 - 10x^2y^2 + 25$

Ans. a) $(a + 5)(a + 2)$ g) $(x - 2)(x + 1)$ m) $(ab + 5)(ab + 3)$

b) $(b + 5)(b + 3)$ h) $(y - 4)(y + 1)$ n) $(de - 12f)(de - 3f)$

c) $(r - 9)(r - 3)$ i) $(w + 4)(w - 2)$ o) $(x^2 + 4)(x^2 + 1)$

d) $(s - 11)(s - 3)$ j) $(w + 9)(w - 2)$ p) $(y^2 - 7)(y^2 + 1)$

e) $(h - 25)(h - 2)$ k) $(x + 12y)(x + 2y)$ q) $(x^2 + 4y^2)(x^2 + 4y^2)$

f) $(m + 16)(m + 3)$ l) $(c - 15d)(c - 2d)$ r) $(x^2y^2 - 5)(x^2y^2 - 5)$

10.20. Factor each trinomial: **(9.1)**

a) $5x^2 + 6x + 1$ Ans. $(5x + 1)(x + 1)$ g) $7h^2 + 10h + 3$ Ans. $(7h + 3)(h + 1)$

b) $5x^2 - 6x + 1$ $(5x - 1)(x - 1)$ h) $7h^2 - 10h + 3$ $(7h - 3)(h - 1)$

c) $5x^2 + 4x - 1$ $(5x - 1)(x + 1)$ i) $7h^2 - 20h - 3$ $(7h + 1)(h - 3)$

d) $5x^2 - 4x - 1$ $(5x + 1)(x - 1)$ j) $7h^2 - 22h + 3$ $(7h - 1)(h - 3)$

e) $5 + 8x + 3x^2$ $(5 + 3x)(1 + x)$ k) $7h^4 - 15h^2 + 2$ $(7h^2 - 1)(h^2 - 2)$

f) $5 + 14x - 3x^2$ $(5 - x)(1 + 3x)$ l) $3 + 4h^2 - 7h^4$ $(3 + 7h^2)(1 - h^2)$

10.21. Factor each trinomial: **(9.1)**

a) $4a^2 + 5a + 1$ f) $7x^2 - 15x + 2$ k) $5c^2 + 11cd + 2d^2$

b) $2a^2 + 3a + 1$ g) $7x^2 + 13x - 2$ l) $5c^2d^2 + 7cd + 2$

c) $3a^2 + 4a + 1$ h) $3 - 10y + 7y^2$ m) $2x^4 - 9x^2 + 7$

d) $5 + 7b + 2b^2$ i) $3 + 4y - 7y^2$ n) $35 - 17r^2 + 2r^4$

e) $4 + 16b + 15b^2$ j) $3x^2 + 10xy + 7y^2$ o) $3t^6 + 8t^3 - 3$

Ans. a) $(4a + 1)(a + 1)$ f) $(7x - 1)(x - 2)$ k) $(5c + d)(c + 2d)$

b) $(2a + 1)(a + 1)$ g) $(7x - 1)(x + 2)$ l) $(5cd + 2)(cd + 1)$

c) $(3a + 1)(a + 1)$ h) $(3 - 7y)(1 - y)$ m) $(2x^2 - 7)(x^2 - 1)$

d) $(5 + 2b)(1 + b)$ i) $(3 + 7y)(1 - y)$ n) $(5 - r^2)(7 - 2r^2)$

e) $(2 + 3b)(2 + 5b)$ j) $(3x + 7y)(x + y)$ o) $(t^3 + 3)(3t^3 - 1)$

10.22. Find each square: **(10.1)**

a) $(b + 4)^2$ f) $(2d - 3)^2$ k) $(6ab - c)^2$

b) $(b - 4)^2$ g) $(x - 4y)^2$ l) $(8 - 3abc)^2$

c) $(5 - c)^2$ h) $(5x + 4y)^2$ m) $(x^2 + 2y^2)^2$

d) $(5 - 2c)^2$ i) $(7xy + 2)^2$ n) $(x^3 - 8)^2$

e) $(3d + 2)^2$ j) $(2xy - 7)^2$ o) $(1 - 5x^5)^2$

Ans. a) $b^2 + 8b + 16$ f) $4d^2 - 12d + 9$ k) $36a^2b^2 - 12abc + c^2$

b) $b^2 - 8b + 16$ g) $x^2 - 8xy + 16y^2$ l) $64 - 48abc + 9a^2b^2c^2$

c) $25 - 10c + c^2$ h) $25x^2 + 40xy + 16y^2$ m) $x^4 + 4x^2y^2 + 4y^4$

d) $25 - 20c + 4c^2$ i) $49x^2y^2 + 28xy + 4$ n) $x^6 - 16x^3 + 64$

e) $9d^2 + 12d + 4$ j) $4x^2y^2 - 28xy + 49$ o) $1 - 10x^5 + 25x^{10}$

10.23. Find each square by expressing each number as a binomial: **(10.2)**

a) 31^2 Ans. $(30 + 1)^2 = 961$ f) $(6.5)^2$ Ans. $(6 + 0.5)^2 = 42.25$

b) 42^2 $(40 + 2)^2 = 1764$ g) $(8\frac{1}{4})^2$ $(8 + \frac{1}{4})^2 = 68\frac{1}{16}$

c) 29^2 $(30 - 1)^2 = 841$ h) $(19\frac{3}{4})^2$ $(20 - \frac{1}{4})^2 = 390\frac{1}{16}$

d) 55^2 $(50 + 5)^2 = 3025$ i) 9.8^2 $(10 - 0.2)^2 = 96.04$

e) 5.5^2 $(5 + 0.5)^2 = 30.25$

10.24. Represent the area of a square whose side is: **(10.3)**

a) $(l + 7)$ ft   d) $(2w - 3)$ cm   g) $a + b$

b) $(l - 2)$ in   e) $(10 - w)$ mi   h) $c + 2d$

c) $(2l + 9)$ rd   f) $7 - 5w$ i) $15 - S^2$

Ans. a) $(l^2 + 14l + 49)$ ft^2 d) $(4w^2 - 12w + 9)$ cm^2 g) $a^2 + 2ab + b^2$

b) $(l^2 - 4l + 4)$ in^2 e) $(100 - 20w + w^2)$ mi^2 h) $c^2 - 4cd + 4d^2$

c) $(4l^2 + 36l + 81)$ rd^2 f) $49 - 70w + 25w^2$ i) $225 - 30S^2 + S^4$

10.25. Factor. [In (g) to (i), rearrange before factoring.]   **(11.1)**

a) $d^2 + 10d + 25$ d) $a^2b^2 - 40ab + 400$ g) $x^2 + 100y^2 + 20xy$

b) $9 - 6h + h^2$ e) $25x^2 - 30x + 9$ h) $16ab + a^2b^2 + 64$

c) $144 - 24r + r^2$ f) $49x^2 + 28xy + 4y^2$ i) $x^4 + 4y^4 - 4x^2y^2$

Ans. a) $(d + 5)^2$ d) $(ab - 20)^2$ g) $x^2 + 20xy + 100y^2 = (x + 10y)^2$

b) $(3 - h)^2$ e) $(5x - 3)^2$ h) $a^2b^2 + 16ab + 64 = (ab + 8)^2$

c) $(12 - r)^2$ f) $(7x + 2y)^2$ i) $x^4 - 4x^2y^2 + 4y^4 = (x^2 - 2y^2)^2$

10.26. Represent the side of a square whose area is: **(11.2)**

a) $(81l^2 + 18l + 1)$ in^2 c) $(4w^2 - 36w + 81)$ mi^2 e) $s^2 - 12s + 36$

b) $(9 - 30l + 25l^2)$ ft^2 d) $(9 + 42s + 49s^2)$ cm^2 f) $s^4 - 24s^2 + 144$

Ans. a) $(9l + 1)$ in   c) $(2w - 9)$ mi   e) $s - 6$

b) $(3 - 5l)$ ft   d) $(3 + 7s)$ cm   f) $s^2 - 12$

10.27. Factor completely: **(12.1)**

$\quad$ a) $b^3 - 16b$ $\qquad\qquad$ d) $2x^3 + 8x^2 + 8x$ $\qquad$ g) $r^3 - 2r^2 - 15r$

$\quad$ b) $3b^2 - 75$ $\qquad\qquad$ e) $x^4 - 12x^3 + 36x^2$ $\qquad$ h) $24 - 2r - 2r^2$

$\quad$ c) $3abc^2 - 3abd^2$ $\qquad$ f) $100 - 40x + 4x^2$ $\qquad$ i) $x^2y^2z - 16xyz + 64z$

Ans. $\quad$ a) $b(b + 4)(b - 4)$ $\qquad$ d) $2x(x + 2)^2$ $\qquad\qquad$ g) $r(r - 5)(r + 3)$

$\qquad$ b) $3(b + 5)(b - 5)$ $\qquad$ e) $x^2(x - 6)^2$ $\qquad\qquad$ h) $2(3 - r)(4 + r)$

$\qquad$ c) $3ab(c + d)(c - d)$ $\qquad$ f) $4(5 - x)^2$ $\qquad\qquad$ i) $z(xy - 8)^2$

10.28. Factor completely: **(12.2)**

$\quad$ a) $a^4 - b^4$ $\qquad\qquad$ c) $81 - a^4$ $\qquad\qquad$ e) $a^8 - 1$

$\quad$ b) $1 - c^{12}$ $\qquad\qquad$ d) $16 - 81b^4$ $\qquad\qquad$ f) $1 - a^{18}$

Ans. $\quad$ a) $(a^2 + b^2)(a + b)(a - b)$ $\qquad\qquad$ d) $(4 + 9b^2)(2 + 3b)(2 - 3b)$

$\qquad$ b) $(1 + c^6)(1 + c^3)(1 - c^3)$ $\qquad\qquad$ e) $(a^4 + 1)(a^2 + 1)(a + 1)(a - 1)$

$\qquad$ c) $(9 + a^2)(3 + a)(3 - a)$ $\qquad\qquad$ f) $(1 + a^8)(1 + a^4)(1 + a^2)(1 + a)(1 - a)$

10.29. Using division, find the ratio which indicates the change in each variable and express this ratio in a sentence. $\qquad$ **(13.1)**

$\quad$ a) The speed of a plane changes from 200 to 100 mi/h.

$\quad$ b) A salary changes from \$50 per week to \$60 per week.

$\quad$ c) The length of a line changes from 2 ft to 1 ft 6 in.

$\quad$ d) The price of a radio set changes from \$100 to \$175.

$\quad$ e) John's monthly income remains at \$275 per month.

$\quad$ f) The bus fare changes from 25¢ to 40¢.

Ans. $\quad$ a) $\frac{1}{2}$; second speed is **one-half** of first. $\qquad$ d) $\frac{7}{4}$; second price is **seven-fourths** of first.

$\qquad$ b) $\frac{6}{5}$; second salary is **six-fifths** of first. $\qquad$ e) 1; second income **is equal** to first.

$\qquad$ c) $\frac{3}{4}$; second length is **three-quarters** of first. $\qquad$ f) $\frac{8}{5}$; second fare is **eight-fifths** of first.

10.30. Complete each: **(13.2)**

$\quad$ a) If x is tripled, it will change from 7 to ( ) or from ( ) to 72. $\qquad$ *Ans.* $\quad$ 21, 24

$\quad$ b) If y is multiplied by $\frac{3}{2}$, it will change from 12 to ( ) or from ( ) to 48. $\qquad$ 18, 32

$\quad$ c) If z is divided by 5, it will change from 35 to ( ) or from ( ) to 75. $\qquad$ 7, 375

$\quad$ d) If k which equals 5.7 remains constant, the new value is ( ). $\qquad$ 5, 7

$\quad$ e) If r is divided by $\frac{2}{5}$, it will change from 40 to ( ) or from ( ) to 30. $\qquad$ 100, 12

10.31. (1) Express each equation in the form $y = kx$. (2) State the variables that vary directly as each other.   **(14.1)**

$\quad$ a) $\dfrac{C}{R} = 2\pi$ $\qquad$ c) $24 = \dfrac{P}{T}$ $\qquad$ e) $V = \frac{1}{3}Bh$ when $h = 60$

$\quad$ b) $8s = p$ $\qquad\qquad$ d) $f = \dfrac{n}{d}$ when $d = 5$ $\qquad$ f) $I = PRT$ when R and T are constant

Ans. $\quad$ a) (1) $C = 2\pi R$, (2) C and R $\qquad$ d) (1) $f = \dfrac{1}{5}n$, (2) f and n

$\qquad$ b) (1) $p = 8s$, (2) p and s $\qquad\qquad$ e) (1) $V = 20B$, (2) V and B

$\qquad$ c) (1) $P = 24T$, (2) P and T $\qquad\qquad$ f) (1) $l = kP$, (2) l and P

10.32. Complete each: **(14.2)**

$\quad$ a) If $b = 5c$ and c is doubled, then ... *Ans.* $\quad$ b is doubled.

$\quad$ b) If $\dfrac{c}{d} = 1.5$ and d is halved, then ... *Ans.* $\quad$ c is halved.

$\quad$ c) If $h = \frac{7}{5}p$ and p is divided by 5, then ... *Ans.* $\quad$ h is divided by 5.

$\quad$ d) If $LW = A$, L is constant and W is multiplied by 7, then ... *Ans.* $\quad$ A is multiplied by 7.

$\quad$ e) If $NP = C$, $P = 150$ and C is multiplied by $3\frac{1}{2}$, then ... *Ans.* $\quad$ N is multiplied by $3\frac{1}{2}$.

10.33. (1) Complete each statement. (2) State the formula to which direct variation applies.   **(14.3)**

        a) If the time of travel remains fixed, halving rate will …
        b) Doubling the length of a rectangle which has a constant width will …
        c) If one-third as many articles are purchased at the same price per article, then the cost is …
        d) At the same rate of interest, to obtain 3 times as much annual interest …
        e) If the circumference of a circle is quadrupled, then its radius …

Ans. *a*) (1) halve distance, (2) $D = kR$ since $D = RT$ and T is constant
      b) (1) double area, (2) $A = kL$ since $A = LW$ and W is constant
      c) (1) one-third as much, (2) $C = kN$ since $C = NP$ and P is constant
      d) (1) triple principal, (2) $l = kP$ since $l = PR$ and R is constant
      e) (1) is quadrupled, (2) $C = 2\pi R$

10.34. Find each missing value: **(14.4)**

        a) If y varies directly as x and $y = 10$ when $x = 5$, find y when $x = 15$.   *Ans.* 30
        b) If r varies directly as s and $r = 80$ when $s = 8$, find r when $s = 6$.   60
        c) If L varies directly as A and $L = 6$ when $A = 21$, find L when $A = 28$.   8
        d) If D varies directly as T and $D = 100$ when $T = 2$, find T when $D = 300$.   6
        e) If N varies directly as C and $C = 25$ when $N = 10$, find C when $N = 16$.   40

10.35. A pilot flew 800 mi at 120 mph. In the same time,   **(14.5)**

        a) how many miles would she have traveled at 150 mi/h?   *Ans.* 1000 mi
        b) how fast must she go to fly 1200 mi?   180 mi/h

10.36. A motorist finds in traveling 100 mi that he is consuming gas at the rate of 15 mi/gal. If he uses the same number of gallons, find: **(14.5)**

        a) how far he could travel if gas were consumed at the rate of 12 mi/gal?   *Ans.* 80 mi
        b) the rate of consumption if he covers 120 mi?   18 mi/gal

10.37. A saleswoman earned \$25 in commission when she sold \$500 worth of tools. If her rate of commission remains fixed,   **(14.6)**

        a) how much would her commission be if she sold \$700 worth of tools?   *Ans.* \$35
        b) how much must she sell to earn \$45 in commission?   \$900

10.38. The annual dividend on \$6000 worth of stock is \$360. At the same dividend rate, what is the annual dividend on \$4500?   *Ans.* \$270   **(14.6)**

10.39. (1) Express each equation in the form $xy = k$. (2) State the variables that vary inversely as each other.   **(15.1)**

        a) $3LW = 27$ *d*) $f = \dfrac{n}{d}$ when $n = 25$

        b) $\dfrac{PV}{10} = 7$ *e*) $A = \frac{1}{2}Dd$ when A is constant

        c) $\dfrac{1}{y} = x$ *f*) $A = \pi ab$ when $A = 30$

Ans. *a*) (1) $LW = 9$, (2) L and W *d*) (1) $fd = 25$, (2) f and d
      b) (1) $PV = 70$, (2) P and V *e*) (1) $Dd = k$, (2) D and d

      c) (1) $xy = 1$, (2) x and y *f*) (1) $ab = \dfrac{30}{\pi}$, (2) a and b

10.40. Complete each: **(15.2)**

        a) If $pv = 20$ and p is doubled, then … *Ans.* v is halved.
        b) If $RT = 176$ and R is tripled, then … *Ans.* T is divided by 3 or multiplied by $\frac{1}{3}$.
        c) If $240 = RP$ and P is multiplied by $\frac{4}{3}$, then … *Ans.* R is divided by $\frac{4}{3}$ or multiplied by $\frac{3}{4}$.
        d) If $IR = E$, E is constant, and I is divided by 15, then … *Ans.* R is multiplied by 15.

10.41. (1) Complete each statement. (2) State the formula to which inverse variation applies.　　　**(15.3)**

　　a) For a fixed area, tripling the width of a rectangle …
　　b) Taking twice as long to cover the same distance requires …
　　c) If the pressure of an enclosed gas at constant temperature is reduced to one-half, then its volume …
　　d) Quadrupling the length of a triangle with a fixed area …
　　e) If the numerator remains the same, halving the denominator will …

　Ans.　*a*) (1) divides the length by 3, (2) $LW = k$ since $A = LW$ and A is constant
　　　　b) (1) half the rate, (2) $RT = k$ since $D = RT$ and D is constant
　　　　c) (1) is doubled, (2) $PV = k$ (Boyle's law)
　　　　d) (1) divides the height by 4 (2) $bh = k$ since $A = \frac{1}{2}bh$ and A is constant
　　　　e) (1) double the fraction, (2) $fd = k$ since $f = \dfrac{n}{d}$ and n is constant

10.42. Find each missing value:　　　　　　　　　　　　　　　　　　　　　　　　　　　**(15.4)**

　　a) If y varies inversely as x, and $y = 15$ when $x = 2$, find y when $x = 6$.　　*Ans.*　5
　　b) If t varies inversely as r, and $t = 15$ when $r = 8$, find t when $r = 24$.　　　　　5
　　c) If b varies inversely as h, and $b = 21$ when $h = 3$, find b when $h = 9$.　　　　　7
　　d) If R varies inversely as T, and $R = 36$ when $T = 2\frac{1}{2}$, find R when $T = 2$.　　　45
　　e) If P varies inversely as V, and $P = 12$ when $V = 40$, find V when $P = 10$.　　　48

CHAPTER 11

Fractions

1. UNDERSTANDING FRACTIONS

The **terms of a fraction** are its numerator and its denominator.

The Various Meanings of "Fraction"

Meaning 1. A fraction may mean **division.**

Thus, $\frac{3}{4}$ may mean 3 divided by 4, or $3 \div 4$.

When a fraction means division, its numerator is the **dividend** and its denominator is the **divisor.**

Thus, if $\frac{14}{5}$ means $14 \div 5$, then 14 is the dividend and 5 is the divisor.

Meaning 2. A fraction may mean **ratio.**

Thus, $\frac{3}{4}$ may mean the ratio of 3 to 4, or 3:4.

When a fraction means the ratio of two quantities, the quantities must have a common unit.

Thus, the ratio of 3 days to 2 weeks equals 3:14, or $\frac{3}{14}$. This is found by changing 2 weeks to 14 days and eliminating the common unit.

Meaning 3. A fraction may mean **a part of a whole thing** or **a part of a group of things.**

Thus, $\frac{3}{4}$ may mean three-fourths of a dollar, or 3 out of 4 dollars.

Zero Numerators or Zero Denominators

(1) **When the numerator of a fraction is zero**, the value of the fraction is zero provided the denominator is not zero also.

Thus, $\dfrac{0}{3} = 0$. Also, $\dfrac{x}{3} = 0$ if $x = 0$.

In $\dfrac{x-5}{3}$, if $x = 5$, then the fraction equals zero. However, $\dfrac{0}{0}$ is meaningless. To see this, suppose that $\dfrac{0}{0} = x$. Then $0 \cdot x = 0$, which is true for all x.

(2) Since division by zero is impossible, a fraction with a zero denominator has no meaning.

Thus, $3 \div 0$ is impossible. Hence, $\dfrac{3}{0}$ is meaningless. To see this, suppose that $\dfrac{3}{0} = x$.

Then $0 \cdot x = 3$ or $0 = 3$, which is impossible.

Also, if $x = 0$, $5 \div x$ is impossible and $\dfrac{5}{x}$ is meaningless.

1.1.   MEANING 1: FRACTIONS MEANING DIVISION

Express each in fractional form:

a) 10 divided by 17   *Ans.* $\dfrac{10}{17}$ *c*) $(x + 5) \div 3$ *Ans.* $\dfrac{x + 5}{3}$

b) 5 divided by *a* $\dfrac{5}{a}$ *d*) The quotient of *x* and $x - 2$ $\dfrac{x}{x - 2}$

1.2.   MEANING 2: FRACTIONS MEANING RATIO

Express each in fractional form. [In (*d*) to (*f*), a common unit must be used.]

a) Ratio of 5 to 8   *Ans.* $\dfrac{5}{8}$ *d*) Ratio of *a* ft to 5 yd   *Ans.* $\dfrac{a}{15}$

b) Ratio of 15 to *t* $\dfrac{15}{t}$ *e*) Ratio of 7 min to 1 h $\dfrac{7}{60}$

c) Ratio of *a* ft to *b* ft $\dfrac{a}{b}$ *f*) Ratio of 3¢ to one quarter $\dfrac{3}{25}$

1.3.   MEANING 3: FRACTIONS MEANING PARTS OF A WHOLE OR OF A GROUP

Express each in fractional form:

a) 3 of 32 equal parts of an inch   *Ans.* $\dfrac{3}{32}$ *d*) 5 members out of a team of 21   *Ans.* $\dfrac{5}{21}$

b) 3 of *n* equal parts of a circle $\dfrac{3}{n}$ *e*) *n* students out of a group of 30 $\dfrac{n}{30}$

c) *a* out of *b* dollars $\dfrac{a}{b}$ *f*) *s* squads out of *t* squads $\dfrac{s}{t}$

1.4.   FRACTIONS HAVING ZERO NUMERATORS

State the value of:

a) $\dfrac{0}{10}$ *Ans.* 0   *d*) *x* when $\dfrac{x}{5} = 0$ *Ans.* 0

b) $\dfrac{x}{10}$ when $x = 0$ 0   *e*) *x* when $\dfrac{x - 7}{10} = 0$ 7

c) $\dfrac{x - 2}{10}$ when $x = 2$ 0   *f*) *x* when $\dfrac{3x + 6}{12} = 0$ -2

1.5.   FRACTIONS HAVING ZERO DENOMINATORS

For what value of *x* is the fraction meaningless?

a) $\dfrac{3}{x}$ *Ans.* When $x = 0$ *d*) $\dfrac{x + 4}{4x - 16}$ *Ans.* When $4x - 16 = 0$ or $x = 4$

b) $\dfrac{10}{3x}$ *Ans.* When $3x = 0$ or $x = 0$ e) $\dfrac{8x}{5x + 5}$ *Ans.* When $5x + 5 = 0$ or $x = -1$

c) $\dfrac{10}{x - 3}$ When $x - 3 = 0$ or $x = 3$ f) $\dfrac{2x - 12}{2x + 12}$ When $2x + 12 - 0$ or $x = -6$

2. CHANGING FRACTIONS TO EQUIVALENT FRACTIONS

Equivalent fractions are fractions having the same value although they have different numerators and denominators.

Thus, $\dfrac{2}{3}, \dfrac{20}{30}$, and $\dfrac{20x}{30x}$ are equivalent fractions, since $\dfrac{2}{3} = \dfrac{20}{30} - \dfrac{20x}{30x}$ (so long as $x \neq 0$).

To obtain equivalent fractions, use one of the following fraction rules:

Rule 1. The value of a fraction is not changed if its numerator and denominator are **multiplied** by the same number except zero.

Thus, $\dfrac{3}{4} = \dfrac{30}{40}$. Here both 3 and 4 are multiplied by 10.

Also, $\dfrac{5}{7} = \dfrac{5x}{7x}$. Here both 5 and 7 are multiplied by x, where $x \neq 0$.

Rule 2. The value of a fraction is not changed if its numerator and denominator are **divided** by the same number except zero.

Thus, $\dfrac{400}{500} = \dfrac{4}{5}$. Here both 40 and 50 are divided by 100.

Also, $\dfrac{7a^2}{9a^2} = \dfrac{7}{9}$. Here both $7a^2$ and $9a^2$ are divided by a^2, where $a \neq 0$.

2.1. USING MULTIPLICATION TO OBTAIN EQUIVALENT FRACTIONS

Change each to equivalent fractions by multiplying its numerator and denominator (terms) by 2, 5, x, $4x$, $x - 3$, x^2, and $x^2 + x - 3$ (assume that the multipliers are non-zero):

	Multiply Both Numerator and Denominator by						
	2	5	x	$4x$	$x - 3$	x^2	$x^2 + x - 3$
$a)$ $\dfrac{2}{3}$ *Ans.*	$\dfrac{2}{3} -$	$\dfrac{4}{6} -$	$\dfrac{10}{15} =$	$\dfrac{2x}{3x} =$	$\dfrac{8x}{12x} =$	$\dfrac{2x - 6}{3x - 9} =$	$\dfrac{2x^2}{3x^2} =$ $\dfrac{2x^2 + 2x - 6}{3x^2 + 3x - 9}$
$b)$ $\dfrac{a}{7}$	$\dfrac{a}{7} =$	$\dfrac{2a}{14} =$	$\dfrac{5a}{35} =$	$\dfrac{ax}{7x} =$	$\dfrac{4ax}{28x} =$	$\dfrac{ax - 3a}{7x - 21} =$	$\dfrac{ax^2}{7x^2} =$ $\dfrac{ax^2 + ax - 3a}{7x^2 + 7x - 21}$
$c)$ $\dfrac{3}{x}$	$\dfrac{3}{x} =$	$\dfrac{6}{2x} =$	$\dfrac{15}{5x} =$	$\dfrac{3x}{x^2} =$	$\dfrac{12x}{4x^2} =$	$\dfrac{3x - 9}{x^2 - 3x} =$	$\dfrac{3x^2}{x^3} =$ $\dfrac{3x^2 + 3x - 9}{x^3 + x^2 - 3x}$

2.2. USING DIVISION TO OBTAIN EQUIVALENT FRACTIONS

Change each to equivalent fractions by dividing its numerator and denominator (terms) by 2, 5, x, $4x$, x^2, and $20x^2$, where $x \neq 0$.

Divide Both Numerator and Denominator by					
2	5	x	$4x$	x^2	$20x^2$

a) $\dfrac{20x^2}{80x^2}$ *Ans.* $\dfrac{20x^2}{80x^2} = \dfrac{10x^2}{40x^2} = \dfrac{4x^2}{16x^2} = \dfrac{20x}{80x} = \dfrac{5x}{20x} = \dfrac{20}{80} = \dfrac{1}{4}$

b) $\dfrac{40x^3}{60x^2}$ $\dfrac{40x^3}{60x^2} = \dfrac{20x^3}{30x^2} = \dfrac{8x^3}{12x^2} = \dfrac{40x^2}{60x} = \dfrac{10x^2}{15x} = \dfrac{40x}{60} = \dfrac{2x}{3}$

2.3. OBTAINING MISSING TERMS

Show how to obtain each missing term:

a) $\dfrac{3}{7} = \dfrac{30}{?}$ To get 30, multiply 3 by 10.   *a)* $\dfrac{3}{7} = \dfrac{30}{70}$ *Ans.*   70
Hence, multiply 7 by 10 to get 70.

b) $\dfrac{27}{33} = \dfrac{9}{?}$ To get 9, divide 27 by 3.   *b)* $\dfrac{27}{33} = \dfrac{9}{11}$ 11
Hence, divide 33 by 3 to get 11.

c) $\dfrac{3a}{b} = \dfrac{3ab}{?}$ To get $3ab$, multiply $3a$ by b.   *c)* $\dfrac{3a}{b} = \dfrac{3ab}{b^2}$ b^2
Hence, multiply b by b to get b^2.

d) $\dfrac{6ac}{3c^2} = \dfrac{?}{c}$ To get c, divide $3c^2$ by $3c$ *d)* $\dfrac{6ac}{3c^2} = \dfrac{2a}{c}$ $2a$
Hence, divide $6ac$ by $3c$ to get $2a$.

e) $\dfrac{5}{x-5} = \dfrac{25}{?}$ To get 25, multiply 5 by 5.   *e)* $\dfrac{5}{x-5} = \dfrac{25}{5x-25}$ $5x-25$
Hence, multiply $x - 5$ by 5 to get $5x - 25$.

3.   RECIPROCALS AND THEIR USES

Reciprocal of a Number

The reciprocal of a number is 1 divided by the number. Thus, the reciprocal of 5 is $\dfrac{1}{5}$, and the reciprocal of a is $\dfrac{1}{a}$, as long as $a \neq 0$.

Also, the reciprocal of $\dfrac{2}{3}$ is $\dfrac{3}{2}$ since $\dfrac{3}{2} = 1 \div \dfrac{2}{3}$.

Rules of Reciprocals

Rule 1.   The fractions $\dfrac{a}{b}$ and $\dfrac{b}{a}$ are reciprocals of each other; that is, the reciprocal of a fraction is the fraction inverted.

Rule 2.   The product of two reciprocals is 1.

Thus, $\dfrac{2}{3} \cdot \dfrac{3}{2} = 1$, $\dfrac{a}{b} \cdot \dfrac{b}{a} = 1$.

Rule 3.   To divide by a number or a fraction, multiply by its reciprocal.

Thus, $8 \div \dfrac{2}{3} = 8 \cdot \dfrac{3}{2}$ or 12, $7 \div 5 = 7 \cdot \dfrac{1}{5}$ or $\dfrac{7}{5}$.

Rule 4.   To solve an equation for an unknown having a fractional coefficient, multiply both members by the reciprocal fraction.

Thus, if $\dfrac{2}{3}x = 10$, multiply both sides by $\dfrac{3}{2}$. Then $\dfrac{3}{2} \cdot \dfrac{2}{3}x = \dfrac{3}{2} \cdot 10$ and $x = 15$.

3.1.   RULE 1: $\dfrac{a}{b}$ AND $\dfrac{b}{a}$ ARE RECIPROCALS OF EACH OTHER (as long as $a, b \neq 0$).

State the reciprocal of each:

a) $\dfrac{1}{15}$ *Ans.* $\dfrac{15}{1}$ or 15   e) $\dfrac{27}{31}$ *Ans.* $\dfrac{31}{27}$ i) 0.5   *Ans.* $\dfrac{10}{5}$ or 2

b) $\dfrac{1}{x}$ $\dfrac{x}{1}$ or x f) $\dfrac{3}{5}$ $\dfrac{5}{3}$ or $1\frac{2}{3}$ j) 0.04 $\dfrac{100}{4}$ or 25

c) 20 $\dfrac{1}{20}$ g) $\dfrac{3x}{10y}$ $\dfrac{10y}{3x}$ k) $4\frac{2}{3}$ $\dfrac{3}{14}$

d) y $\dfrac{1}{y}$ h) $\dfrac{x+3}{y-10}$ $\dfrac{y-10}{x+3}$ l) $0.001x$ $\dfrac{1000}{x}$

3.2.   RULE 2:   THE PRODUCT OF TWO RECIPROCALS IS 1

Supply the missing entry:

a) $\left(\dfrac{2}{7}\right)\left(\dfrac{7}{2}\right) = ?$ *Ans.* 1   d) $\left(\dfrac{x}{17}\right)\left(\dfrac{17}{x}\right)\left(\dfrac{2}{5}\right) = (?)$ *Ans.* $\dfrac{2}{5}$

b) $\dfrac{3}{8}(?) = 1$ $\dfrac{8}{3}$ e) $\left(\dfrac{8}{7}\right)\left(\dfrac{7}{8}\right) + ? = 5$ 4

c) $10\left(\dfrac{5}{9}\right)\left(\dfrac{9}{5}\right) = ?$ 10   f) $3\left(\dfrac{3}{5}\right)\left(\dfrac{5}{3}\right) + \left(\dfrac{8}{15}\right)(?) - 4$ $\dfrac{15}{8}$

3.3.   RULE 3:   USING RECIPROCALS TO CHANGE DIVISION TO MULTIPLICATION

Show how to change each division to multiplication by using the reciprocal of a number:

a) $8 \div \dfrac{4}{5}$ b) $\dfrac{2}{x} \div y$ c) $\dfrac{1}{r} \div \dfrac{s}{t}$ d) $\dfrac{3}{x} \div \dfrac{y}{7}$

Ans.   a) $8 \cdot \dfrac{5}{4} = 10$ b) $\dfrac{2}{x} \cdot \dfrac{1}{y} = \dfrac{2}{xy}$ c) $\dfrac{1}{r} \cdot \dfrac{t}{s} = \dfrac{t}{rs}$ d) $\dfrac{3}{x} \cdot \dfrac{7}{y} = \dfrac{21}{xy}$

3.4.   RULE 4:   USING RECIPROCALS TO SOLVE EQUATIONS

Solve:

a) $\dfrac{3}{4}x = 21$ b) $2\frac{1}{3}y = 70$ c) 30% of $P = 24$ d) $\dfrac{ax}{3} = 5a$

Solutions:

$\dfrac{3}{4}x = 21$ $2\frac{1}{3}y = 70$ 30% of $P = 24$ $\dfrac{ax}{3} = 5a$

$\mathbf{M}_{4/3}$ $\dfrac{4}{3} \cdot \dfrac{3}{4}x = \dfrac{4}{3} \cdot 21$ $\mathbf{M}_{3/7}$ $\dfrac{3}{7} \cdot \dfrac{7}{3}y = \dfrac{3}{7} \cdot 70$ $\mathbf{M}_{10/3}$ $\dfrac{10}{3} \cdot \dfrac{3}{10}P = \dfrac{10}{3} \cdot 24$ $\mathbf{M}_{3/a}$ $\dfrac{3}{a} \cdot \dfrac{ax}{3} = \dfrac{3}{a} \cdot 5a$

Ans. $x = 28$ $y = 30$ $P = 80$ $x = 15$

4. REDUCING FRACTIONS TO LOWEST TERMS

A fraction is reduced to lowest terms when its numerator and denominator have no common factor except 1.

Thus, $\dfrac{3x}{7x}$ is not in lowest terms because x is a common factor of the numerator and denominator. After x has been eliminated by division, the resulting fraction $\frac{3}{7}$ will be in lowest terms.

To reduce a fraction to lowest terms, use the following rule:

RULE: The value of a fraction is not changed if its numerator and denominator are divided by the same number except zero.

To Reduce a Fraction to Lowest Terms

Reduce: $\quad a)\ \dfrac{3ab^2c}{3ab^2d} \qquad b)\ \dfrac{8a+8b}{12a+12b} \qquad c)\ \dfrac{2a^2-2b^2}{5a+5b}$

Procedure:

Solutions:

1. Factor its terms:
(Numerator and denominator)

$a)\ \dfrac{3ab^2c}{3ab^2d} \qquad b)\ \dfrac{8(a+b)}{12(a+b)} \qquad c)\ \dfrac{2(a+b)(a-b)}{5(a+b)}$

2. Divide both terms by every common factor: (This is commonly called canceling.)

$\dfrac{\cancel{3ab^2}c}{\cancel{3ab^2}d} \qquad \dfrac{\overset{2}{\cancel{8}}(a+b)^{(1)}}{\underset{3}{\cancel{12}}(a+b)_{(1)}} \qquad \dfrac{2(a+b)^{1}(a-b)}{5(a+b)_{1}}$

$Ans.\ \dfrac{c}{d} \qquad\qquad Ans.\ \dfrac{2}{3} \qquad\qquad Ans.^*\ \dfrac{2(a-b)}{5}$

Rule 1. If two expressions are exactly alike or have the same value, their quotient is 1.

Thus, $\dfrac{5abc}{5abc}=1, \qquad \dfrac{a+b}{b+a}=1, \qquad \dfrac{\overset{4}{\cancel{8}}(x^2+\overset{1}{\cancel{x-5}})}{\cancel{2}(x^2+\cancel{x-5})}=4.$

Binomials Which Are Negatives of Each Other

Binomials such as $x-y$ and $y-x$ are negatives of each other if x and y have different values.
Thus, if $x=5$ and $y=2$, then $x-y=+3$ and $y-x=-3$.
Hence, either $x-y=-(y-x)$ or $x-y=-(x-y)$.

Rule 2. If two binomials are negatives of each other, their quotient is -1.

Thus, $\dfrac{x-y}{y-x}=-1, \qquad \dfrac{(5+x)^{1}(5-x)^{-1}}{(x+5)(x-5)}=-1, \qquad \dfrac{(a-b)^{-1}(7-c)^{-1}}{(b-a)(c-7)}=1.$

Warning! How Not to Reduce a Fraction to Lowest Terms

1. Do not subtract the same number from the numerator and denominator.

Thus, $\dfrac{5}{6}$ does **not** equal $\dfrac{5-4}{6-4}$ or $\dfrac{1}{2}$. Also, $\dfrac{n+1}{n+2}$ does **not** equal $\dfrac{\cancel{n}+1}{\cancel{n}+2}$ or $\dfrac{1}{2}$.

2. Do not add the same number to both numerator and denominator.

Thus, $\dfrac{1}{2}$ does **not** equal $\dfrac{1+3}{2+3}$ or $\dfrac{4}{5}$. Also, $\dfrac{x-3}{y-3}$ does **not** equal $\dfrac{x\cancel{-3}}{y\cancel{-3}}$ or $\dfrac{x}{y}$.

* *Note:* In answers, algebraic factors which remain need not be multiplied out.

4.1. REDUCING FRACTIONS WHOSE TERMS HAVE COMMON MONOMIAL FACTORS

Reduce to lowest terms:

a) $\dfrac{39rs}{52rs}$ b) $\dfrac{32a^3b^3}{64a^2b}$ c) $\dfrac{5x-35}{15x}$ d) $\dfrac{21a^2}{14a^2-7ab}$

Solutions:

a) $\dfrac{\overset{3\,(1)}{\cancel{39rs}}}{\underset{4\,(1)}{\cancel{52rs}}}$ b) $\dfrac{\overset{1\quad ab^2}{\cancel{32a^3b^3}}}{\underset{2\quad(1)}{\cancel{64a^2b}}}$ c) $\dfrac{\overset{1}{\cancel{5}}(x-7)}{\underset{3}{\cancel{15x}}}$ d) $\dfrac{\overset{3a}{\cancel{21a^2}}}{\underset{1}{\cancel{7a}}(2a-b)}$

Ans. $\dfrac{3}{4}$ $\dfrac{ab^2}{2}$ $\dfrac{x-7}{3x}$ $\dfrac{3a}{2a-b}$

4.2. RULE 1: REDUCING FRACTIONS WHOSE TERMS HAVE A COMMON BINOMIAL FACTOR

Reduce to lowest terms:

a) $\dfrac{2x+6}{3ax+9a}$ b) $\dfrac{x^2+x}{2+2x}$ c) $\dfrac{3x+3y}{3y^2-3x^2}$ d) $\dfrac{(b+c)^2}{-acx-abx}$

Solutions:

a) $\dfrac{2(x+3)}{3a(x+3)}$ b) $\dfrac{x(x+1)}{2(1+x)}$ c) $\dfrac{3(x+y)}{3(y+x)(y-x)}$ d) $\dfrac{(b+c)(b+c)}{-ax(c+b)}$

a) $\dfrac{2(\overset{1}{\cancel{x+3}})}{3a(\cancel{x+3})}$ b) $\dfrac{x(\overset{1}{\cancel{x+1}})}{2(\cancel{1+x})}$ c) $\dfrac{\overset{1}{\cancel{3}}(\overset{(1)}{\cancel{x+y}})}{\cancel{3}(\cancel{y+x})(y-x)}$ d) $\dfrac{(b+c)(\overset{1}{\cancel{b+c}})}{-ax(\cancel{c+b})}$

Ans. $\dfrac{2}{3a}$ $\dfrac{x}{2}$ $\dfrac{1}{y-x}$ $-\dfrac{b+c}{ax}$

4.3. RULE 2: REDUCING FRACTIONS HAVING BINOMIAL FACTORS WHICH ARE NEGATIVES OF EACH OTHER

Reduce to lowest terms:

a) $\dfrac{4-y}{3y-12}$ b) $\dfrac{d^2-49}{14-2d}$ c) $\dfrac{5-5r}{10rt-10t}$ d) $\dfrac{(w-x)^2}{x^2-w^2}$

Solutions:

a) $\dfrac{(\overset{-1}{\cancel{4-y}})}{3(\cancel{y-4})}$ b) $\dfrac{(\overset{-1}{\cancel{d-7}})(d+7)}{2(\cancel{7-d})}$ c) $\dfrac{\overset{(1)}{\cancel{5}}(\overset{-1}{\cancel{1-r}})}{\underset{(2)}{\cancel{10}}t(\cancel{r-1})}$ d) $\dfrac{(\overset{-1}{\cancel{w-x}})(w-x)}{(\cancel{x-w})(x+w)}$

Ans. $-\dfrac{1}{3}$ $-\dfrac{d+7}{2}$ $-\dfrac{1}{2t}$ $-\dfrac{w-x}{x+w}$ or $\dfrac{x-w}{x+w}$

4.4. FRACTIONS HAVING AT LEAST ONE TRINOMIAL TERM

Reduce to lowest terms:

a) $\dfrac{b^2+3b}{b^2+10b+21}$ b) $\dfrac{x^2-9x+20}{4x-x^2}$ c) $\dfrac{y^2+2y-15}{2y^2-12y+18}$

Solutions:

$$a) \quad \frac{b(\overset{1}{\cancel{b+3}})}{(\cancel{b+3})(b+7)} \qquad\qquad b) \quad \frac{(x-5)(\overset{-1}{\cancel{x-4}})}{x(\cancel{4-x})} \qquad\qquad c) \quad \frac{(y+5)(\overset{1}{\cancel{y-3}})}{2(y-3)(\cancel{y-3})}$$

$$Ans. \qquad \frac{b}{b+7} \qquad\qquad\qquad -\frac{x-5}{x} \text{ or } \frac{5-x}{x} \qquad\qquad \frac{y+5}{2(y-3)}$$

5. MULTIPLYING FRACTIONS

To Multiply Fractions Having No Cancelable Common Factor

$$\text{Multiply:} \quad a) \quad \frac{3}{5} \cdot \frac{7}{11} \qquad b) \quad \frac{x}{3} \cdot \frac{5}{r} \qquad c) \quad \frac{a}{4} \cdot \frac{9}{2} \cdot \frac{c}{d} \cdot \frac{7}{a+c}$$

Procedure:

Solutions:

1. **Multiply numerators** to obtain numerator of answer:

$$a) \quad \frac{3(7)}{5(11)} \qquad b) \quad \frac{x(5)}{3(r)} \qquad c) \quad \frac{a(9)(c)(7)}{4(2)(d)(a+c)}$$

2. **Multiply denominators** to obtain denominator of answer:

$$\textbf{Ans.} \quad \frac{21}{55} \qquad\qquad \textbf{Ans.} \quad \frac{5x}{3r} \qquad \textbf{Ans.} \quad \frac{63ac}{8d(a+c)}$$

To Multiply Fractions Having a Cancelable Common Factor

$$\text{Multiply:} \quad a) \quad \frac{3}{5} \cdot \frac{10}{7} \cdot \frac{77}{6} \qquad b) \quad \frac{7x}{a} \cdot \frac{2a}{7+7x} \qquad c) \quad \frac{5x}{a^2-b^2} \cdot \frac{3a+3b}{x}$$

Procedure:

Solutions:

1. **Factor those numerators and denominators which are polynomials:**

$$a) \quad \frac{3}{5} \cdot \frac{10}{7} \cdot \frac{77}{6} \qquad b) \quad \frac{7x}{a} \cdot \frac{2a}{7(1+x)} \qquad c) \quad \frac{5x}{(a+b)(a-b)} \cdot \frac{3(a+b)}{x}$$

2. **Divide out factors common** to any numerator and any denominator:

$$\frac{\overset{1}{\cancel{3}}}{\underset{1}{\cancel{5}}} \cdot \frac{\overset{2}{\cancel{10}}}{\underset{1}{\cancel{7}}} \cdot \frac{\overset{11}{\cancel{77}}}{\underset{2}{\cancel{6}}} \qquad \frac{\overset{1}{\cancel{7}}x}{\underset{1}{\cancel{a}}} \cdot \frac{\overset{1}{\cancel{2}}a}{\underset{1}{\cancel{7}}(1+x)} \qquad \frac{5\cancel{x}}{(\cancel{a+b})(a-b)} \cdot \frac{3(\overset{1}{\cancel{a+b}})}{\underset{1}{\cancel{x}}}$$

3. **Multiply** remaining factors:

$$\textbf{Ans.} \quad 11 \qquad\qquad \textbf{Ans.} \quad \frac{2x}{1+x} \qquad\qquad \textbf{Ans.} \quad \frac{15}{a-b}$$

5.1. MULTIPLYING FRACTIONS HAVING NO CANCELABLE COMMON FACTOR

Multiply:

$$a) \quad \frac{1}{5} \cdot \frac{2}{5} \cdot \frac{4}{7} \qquad b) \quad \frac{2}{d} \cdot \frac{c}{5} \cdot \frac{x}{y} \qquad c) \quad 5xy \cdot \frac{3c}{ab} \qquad d) \quad \frac{a}{4} \cdot \frac{3}{r} \cdot \frac{r+5}{a-2}$$

Solutions:

$$a) \quad \frac{1(2)(4)}{5(5)(7)} \qquad b) \quad \frac{2(c)(x)}{d(5)(y)} \qquad c) \quad \frac{5xy \cdot 3c}{ab} \qquad d) \quad \frac{a(3)(r+5)}{4(r)(a-2)}$$

$$Ans. \qquad \frac{8}{175} \qquad\qquad \frac{2cx}{5dy} \qquad\qquad \frac{15cxy}{ab} \qquad\qquad \frac{3a(r+5)}{4r(a-2)}$$

5.2. **MULTIPLYING FRACTIONS HAVING A CANCELABLE COMMON FACTOR**

Multiply:

$a)$ $\dfrac{9}{5} \cdot \dfrac{7}{3} \cdot \dfrac{15}{14}$ $b)$ $\dfrac{12c}{d} \cdot \dfrac{h}{3} \cdot \dfrac{d^2}{h^2}$ $c)$ $\dfrac{c}{a+b} \cdot \dfrac{5a+5b}{2c^2+2c}$ $d)$ $\dfrac{9x}{3x-15} \cdot \dfrac{(x-5)^2}{2(5-x)}$

<u>Solutions:</u>

$a)$ $\dfrac{\overset{3}{\cancel{9}}}{\underset{(1)}{\cancel{5}}} \cdot \dfrac{\overset{1}{\cancel{7}}}{\underset{(1)}{\cancel{3}}} \cdot \dfrac{\overset{3}{\cancel{15}}}{\underset{(2)}{\cancel{14}}}$ $b)$ $\dfrac{\overset{4}{\cancel{12}}c}{\underset{(1)}{\cancel{d}}} \cdot \dfrac{\overset{1}{\cancel{h}}}{\underset{(1)}{\cancel{3}}} \cdot \dfrac{\overset{d}{\cancel{d^2}}}{\cancel{h^2}_h}$ $c)$ $\dfrac{\overset{1}{\cancel{c}}}{\cancel{(a+b)}} \cdot \dfrac{5\cancel{(a+b)}^{1}}{2\cancel{c}(c+1)}$ $d)$ $\dfrac{\overset{3}{\cancel{9}x}}{\cancel{3}\cancel{(x-5)}} \cdot \dfrac{\cancel{(x-5)}^{-1}\,\cancel{(x-5)}}{2\cancel{(5-x)}}$

Ans. $\dfrac{9}{2}$ $\dfrac{4cd}{h}$ $\dfrac{5}{2(c+1)}$ $-\dfrac{3x}{2}$

5.3. **MORE DIFFICULT MULTIPLICATION OF FRACTIONS**

Multiply:

$a)$ $\dfrac{y^2+6y+5}{7y^2-63} \cdot \dfrac{7y+21}{(5+y)^2}$ $b)$ $\dfrac{a+4}{4a} \cdot \dfrac{2a-8}{4+a} \cdot \dfrac{a^2-4}{24-12a} \cdot \dfrac{4a^2}{4-a}$

<u>Solutions:</u>

$a)$ $\dfrac{\cancel{(y+5)}^{1}\,(y+1)}{\cancel{7}(y-3)\cancel{(y+3)}} \cdot \dfrac{\cancel{7}\cancel{(y+3)}^{(1)\ (1)}}{\cancel{(5+y)}(5+y)}$ $b)$ $\dfrac{a+4}{\cancel{4a}} \cdot \dfrac{\overset{(1)}{\cancel{2}}\cancel{(a-4)}^{1}}{\cancel{(4+a)}} \cdot \dfrac{(a+2)\cancel{(a-2)}^{-1}}{\cancel{12}\underset{6}{}(2-a)} \cdot \dfrac{\overset{a}{\cancel{4a^2}}}{\cancel{4-a}}$

Ans. $\dfrac{y+1}{(y-3)(5+y)}$ or $\dfrac{y+1}{(y-3)(y+5)}$ $\dfrac{a(a+2)}{6}$

6. DIVIDING FRACTIONS

Rule.   To divide by a fraction, invert the fraction and multiply.

To Divide Fractions

Divide: $a)$ $\dfrac{2}{3} \div 5$ $b)$ $\dfrac{9}{4} \div \dfrac{a}{3}$ $c)$ $\dfrac{14}{x} \div 2\tfrac{1}{3}$ $d)$ $\dfrac{a^2}{b^2} \div \dfrac{a}{b}$

<u>Procedure:</u>	<u>Solutions:</u>
1. Invert fraction **which is divisor:**	$a)$ Invert $\dfrac{5}{1}$ $b)$ Invert $\dfrac{a}{3}$ $c)$ $2\tfrac{1}{3}=\dfrac{7}{3}\cdot$ Invert $\dfrac{7}{3}$ $d)$ Invert $\dfrac{a}{b}$
2. Multiply resulting **fractions:**	$\dfrac{2}{3}\cdot\dfrac{1}{5}$ $\dfrac{9}{4}\cdot\dfrac{3}{a}$ $\dfrac{\overset{2}{\cancel{14}}}{x}\cdot\dfrac{3}{\underset{1}{\cancel{7}}}$ $\dfrac{\overset{a}{\cancel{a^2}}}{\cancel{b^2}_b}\cdot\dfrac{\overset{1}{\cancel{b}}}{\underset{1}{\cancel{a}}}$ *Ans.* $\dfrac{2}{15}$ *Ans.* $\dfrac{27}{4a}$ *Ans.* $\dfrac{6}{x}$ *Ans.* $\dfrac{a}{b}$

6.1. **DIVIDING FRACTIONS HAVING MONOMIAL TERMS**

Divide:

$a)$ $2\tfrac{3}{4} \div 22$ $b)$ $\dfrac{8}{x^3} \div \dfrac{12}{x^2}$ $c)$ $b^2 \div \dfrac{7}{b^3}$ $d)$ $\dfrac{5y}{7} \cdot \dfrac{2x}{y} \div \dfrac{x^5}{42}$

Solutions:

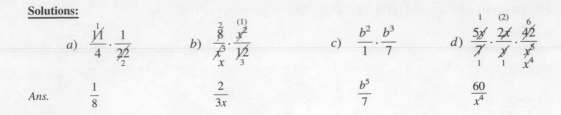

Ans. $\dfrac{1}{8}$ $\dfrac{2}{3x}$ $\dfrac{b^5}{7}$ $\dfrac{60}{x^4}$

6.2. DIVIDING FRACTIONS HAVING POLYNOMIAL TERMS

Divide:

$$a)\ \frac{a^2 - 100}{8} \div \frac{2a + 20}{20} \qquad b)\ \frac{5a^2}{b^2 - 36} \div \frac{25ab - 25a}{b^2 - 7b + 6} \qquad c)\ \frac{4x^2 - 1}{9x - 3x^2} \div \frac{2x^2 - 7x - 4}{x^2 - 7x + 12}$$

Solutions:

$$a)\ \frac{a^2 - 100}{8} \cdot \frac{20}{2a + 20} \qquad\qquad\qquad b)\ \frac{5a^2}{b^2 - 36} \cdot \frac{b^2 - 7b + 6}{25ab - 25a}$$

$$\frac{\overset{1}{(a + 10)}\ (a - 10)}{\underset{(2)}{8}} \cdot \frac{\overset{5}{20}}{2(a + 10)} \qquad\qquad \frac{\overset{(1)a}{5a^2}}{(b + 6)(b - 6)} \cdot \frac{\overset{1}{(b - 6)}\ \overset{1}{(b - 1)}}{\underset{5\ (1)}{25a(b - 1)}}$$

Ans. $\dfrac{5(a - 10)}{4}$ $\dfrac{a}{5(b + 6)}$

$$c)\ \frac{4x^2 - 1}{9x - 3x^2} \cdot \frac{x^2 - 7x + 12}{2x^2 - 7x - 4}$$

$$\frac{\overset{1}{(2x + 1)}\ (2x - 1)}{3x(3 - x)} \cdot \frac{\overset{-1}{(x - 3)}\ \overset{1}{(x - 4)}}{(2x + 1)(x - 4)}$$

Ans. $-\dfrac{2x - 1}{3x}$ or $\dfrac{1 - 2x}{3x}$

7. ADDING OR SUBTRACTING FRACTIONS HAVING THE SAME DENOMINATOR

To Combine (Add or Subtract) Fractions Having the Same Denominator

Combine: $a)\ \dfrac{2a}{15} + \dfrac{7a}{15} - \dfrac{4a}{15} \qquad b)\ \dfrac{5a}{3} - \dfrac{2a - 9}{3} \qquad c)\ \dfrac{7}{x - 2} - \dfrac{5 + x}{x - 2}$

Procedure: **Solutions:**

1. **Keep denominator
   and combine numerators:*** $a)\ \dfrac{2a + 7a - 4a}{15} \qquad b)\ \dfrac{5a - (2a - 9)}{3} \qquad c)\ \dfrac{7 - (5 + x)}{x - 2}$

2. **Reduce resulting
   fraction:** $\dfrac{\overset{1}{5a}}{\underset{3}{15}} \qquad\qquad \dfrac{3a + 9}{3} = \dfrac{\overset{(1)}{3}(a + 3)}{\underset{(1)}{3}} \qquad \dfrac{\overset{-1}{2 - x}}{x - 2}$

Ans. $\dfrac{a}{3}$ *Ans.* $a + 3$ *Ans.* -1

* *Note:* In combining numerators, enclose each polynomial numerator in parentheses preceded by the sign of its fraction.

7.1. COMBINING FRACTIONS HAVING SAME MONOMIAL DENOMINATOR

Combine:

a) $2\frac{4}{5} + \frac{4}{5} - 1\frac{3}{5}$ b) $\frac{8}{3c} - \frac{1}{3c} + \frac{11}{3c}$ c) $\frac{5x}{8} - \frac{x}{8} - \frac{4}{8}$ d) $\frac{x+5}{3x} - \frac{1-x}{3x} - \frac{7x+4}{3x}$

Solutions:

a) $\dfrac{14 + 4 - 8}{5}$ b) $\dfrac{8 - 1 + 11}{3c}$ c) $\dfrac{5x - (x - 4)}{8}$ d) $\dfrac{(x + 5) - (1 - x) - (7x + 4)}{3x}$

$\dfrac{\overset{2}{\cancel{10}}}{\underset{1}{\cancel{5}}}$ $\dfrac{\overset{6}{\cancel{18}}}{\underset{1}{\cancel{3c}}}$ $\dfrac{4x + 4}{8} = \dfrac{\overset{1}{\cancel{4}}(x + 1)}{\underset{2}{\cancel{8}}}$ $\dfrac{x + 5 - 1 + x - 7x - 4}{3x} = -\dfrac{\overset{1}{\cancel{5x}}}{\underset{1}{\cancel{3x}}}$

Ans. 2 $\dfrac{6}{c}$ $\dfrac{x + 1}{2}$ $-\dfrac{5}{3}$

7.2. COMBINING FRACTIONS HAVING SAME POLYNOMIAL DENOMINATOR

Combine:

a) $\dfrac{10x}{2x - 6} - \dfrac{9x + 3}{2x - 6}$ b) $\dfrac{2b}{a - b} - \dfrac{2a}{a - b}$ c) $\dfrac{5}{x^2 + 3x - 4} + \dfrac{7x - 8}{x^2 + 3x - 4} - \dfrac{3x + 1}{x^2 + 3x - 4}$

Solutions:

a) $\dfrac{10x - (9x + 3)}{2x - 6}$ b) $\dfrac{2b - 2a}{a - b}$ c) $\dfrac{5 + (7x - 8) - (3x + 1)}{x^2 + 3x - 4}$

$\dfrac{\overset{1}{\cancel{x - 3}}}{2\cancel{(x - 3)}}$ $\dfrac{2\overset{-1}{\cancel{(b - a)}}}{\cancel{a - b}}$ $\dfrac{4x - 4}{x^2 + 3x - 4}$ or $\dfrac{4\overset{1}{\cancel{(x - 1)}}}{\cancel{(x - 1)}(x + 4)}$

Ans. $\dfrac{1}{2}$ -2 $\dfrac{4}{x + 4}$

8. ADDING OR SUBTRACTING FRACTIONS HAVING DIFFERENT DENOMINATORS

The Lowest Common Denominator

The lowest common denominator (LCD) of two or more fractions is the smallest number divisible without remainder by their denominators.

Thus, 12 is the **LCD** of $\frac{1}{3}$ and $\frac{1}{4}$. Of the common denominators 12, 24, 36, and so on, the lowest or smallest is 12.

To Find the LCD

Rule 1. If no two denominators have a common factor, find the **LCD** by multiplying all the denominators.

Thus, $3ax$ is the LCD of $\dfrac{1}{3}, \dfrac{1}{a}$, and $\dfrac{1}{x}$.

Rule 2. If two of the denominators have a common factor, find the LCD by multiplying the common factor by the remaining factors.

Thus, for $\dfrac{1}{3xy}$ and $\dfrac{1}{5xy}$, the LCD, $15xy$, is obtained by multiplying the common factor xy by the remaining factors, 3 and 5.

Rule 3. If there is a common literal factor with more than one exponent, use its highest exponent in the LCD.

Thus, $3y^5$ is the LCD of $\dfrac{1}{3y}$, $\dfrac{1}{y^2}$, and $\dfrac{1}{y^5}$.

To Combine (Add or Subtract) Fractions Having Different Denominators

Combine: a) $\dfrac{1}{3} + \dfrac{3}{4} - \dfrac{1}{12}$ b) $\dfrac{5}{x} + \dfrac{3}{y}$

Procedure: **Solutions:**

1. Find the LCD: a)   LCD = 12   b) LCD = xy

2. Change each fraction to an equivalent fraction whose denominator is LCD:
(Note how each fraction is multiplied by 1.)

$\dfrac{1}{3} \cdot \dfrac{4}{4} + \dfrac{3}{4} \cdot \dfrac{3}{3} - \dfrac{1}{12}$ $\dfrac{5}{x} \cdot \dfrac{y}{y} + \dfrac{3}{y} \cdot \dfrac{x}{x}$

$\dfrac{4}{12} + \dfrac{9}{12} - \dfrac{1}{12}$ $\dfrac{5y}{xy} + \dfrac{3x}{xy}$

3. Combine fractions having the same denominator and reduce, if necessary:

$\dfrac{12}{12}$

1   *Ans.* $\dfrac{5y + 3x}{xy}$ *Ans.*

8.1.   RULE 1.   COMBINING FRACTIONS WHOSE DENOMINATORS HAVE NO COMMON FACTOR

Combine: a) $\dfrac{2}{5} + \dfrac{3}{4}$ b) $\dfrac{3}{x} - \dfrac{3}{x+1}$ c) $r + 2 - \dfrac{4r - 1}{2r}$

Procedure: **Solutions:**

1. Find LCD (rule 1):   LCD = $(5)(4) = 20$ LCD = $x(x+1)$ LCD = $2r$

2. Change to equivalent fractions having same LCD:

$\dfrac{2}{5} \cdot \dfrac{4}{4} + \dfrac{3}{4} \cdot \dfrac{5}{5}$ $\dfrac{3}{x} \cdot \dfrac{(x+1)}{(x+1)} - \dfrac{3}{(x+1)} \cdot \dfrac{x}{x}$ $\dfrac{(r+2)}{1} \cdot \dfrac{2r}{2r} - \dfrac{4r - 1}{2r}$

$\dfrac{8}{20} + \dfrac{15}{20}$ $\dfrac{3x + 3}{x(x+1)} - \dfrac{3x}{x(x+1)}$ $\dfrac{2r^2 + 4r}{2r} - \dfrac{4r - 1}{2r}$

3. Combine fractions: $\dfrac{23}{20}$ or $1\frac{3}{20}$ *Ans.* $\dfrac{3}{x(x+1)}$ *Ans.* $\dfrac{2r^2 + 1}{2r}$ *Ans.*

8.2.   RULE 2.   COMBINING FRACTIONS WHOSE DENOMINATORS HAVE A COMMON FACTOR

Combine: a) $\dfrac{5}{6} - \dfrac{1}{12} + \dfrac{3}{2}$ b) $\dfrac{2}{5a} + \dfrac{7}{5a + 5}$ c) $\dfrac{a + 4}{3a} - \dfrac{2 - 4a}{6a}$

Procedure: **Solutions:**

1. Find LCD (rule 2):   LCD = 12   LCD = $5a(a + 1)$ LCD = $6a$

2. Change to equivalent fractions having same LCD:

$\dfrac{5}{6} \cdot \dfrac{2}{2} - \dfrac{1}{12} + \dfrac{3}{2} \cdot \dfrac{6}{6}$ $\dfrac{2}{5a} \cdot \dfrac{(a+1)}{(a+1)} + \dfrac{7}{5(a+1)} \cdot \dfrac{a}{a}$ $\dfrac{(a+4)}{3a} \cdot \dfrac{2}{2} - \dfrac{2 - 4a}{6a}$

$\dfrac{10}{12} - \dfrac{1}{12} + \dfrac{18}{12}$ $\dfrac{2a + 2}{5a(a+1)} + \dfrac{7a}{5a(a+1)}$ $\dfrac{2a + 8}{6a} - \dfrac{2 - 4a}{6a}$

3. Combine fractions: $\dfrac{27}{12} = \dfrac{9}{4}$ *Ans.* $\dfrac{9a + 2}{5a(a+1)}$ *Ans.* $\dfrac{6a + 6}{6a} = \dfrac{a + 1}{a}$ *Ans.*

8.3.   RULE 3.   COMBINING FRACTIONS WHOSE LCD INCLUDES BASE WITH HIGHEST EXPONENT

Combine:

a) $\dfrac{5}{x} + \dfrac{7}{x^2}$ b) $\dfrac{3}{x^3} - \dfrac{2}{x^7} + \dfrac{1}{x^5}$ c) $\dfrac{s^2}{9r^2} - \dfrac{s^3}{12r^3}$ d) $\dfrac{2}{a^2b} + \dfrac{3}{ab^2}$

Solutions:

a)   LCD $= x^2$ b)   LCD $= x^7$ c)   LCD $= 36r^3$ d) LCD $= a^2b^2$

$\dfrac{5}{x} \cdot \dfrac{x}{x} + \dfrac{7}{x^2}$ $\dfrac{3}{x^3} \cdot \dfrac{x^4}{x^4} - \dfrac{2}{x^7} + \dfrac{1}{x^5} \cdot \dfrac{x^2}{x^2}$ $\dfrac{s^2}{9r^2} \cdot \dfrac{4r}{4r} - \dfrac{s^3}{12r^3} \cdot \dfrac{3}{3}$ $\dfrac{2}{a^2b} \cdot \dfrac{b}{b} + \dfrac{3}{ab^2} \cdot \dfrac{a}{a}$

$\dfrac{5x}{x^2} + \dfrac{7}{x^2}$ $\dfrac{3x^4}{x^7} - \dfrac{2}{x^7} + \dfrac{x^2}{x^7}$ $\dfrac{4rs^2}{36r^3} - \dfrac{3s^3}{36r^3}$ $\dfrac{2b}{a^2b^2} + \dfrac{3a}{a^2b^2}$

Ans. $\dfrac{5x+7}{x^2}$ $\dfrac{3x^4+x^2-2}{x^7}$ $\dfrac{4rs^2-3s^3}{36r^3}$ $\dfrac{2b+3a}{a^2b^2}$

8.4.   COMBINING FRACTIONS HAVING BINOMIAL DENOMINATORS

Combine:

a) $\dfrac{3x}{x-2} + \dfrac{5x}{x+2}$ b) $\dfrac{3}{2y+4} - \dfrac{5}{3y+6}$ c) $\dfrac{2a-3}{a^2-25} - \dfrac{7}{5a-25}$

Solutions: (Factor denominators first)

a)

    LCD $= (x-2)(x+2)$

    $\dfrac{3x}{x-2} \cdot \dfrac{x+2}{x+2} + \dfrac{5x}{x+2} \cdot \dfrac{x-2}{x-2}$

    $\dfrac{3x^2+6x}{(x-2)(x+2)} + \dfrac{5x^2-10x}{(x+2)(x-2)}$

Ans $\dfrac{8x^2-4x}{(x-2)(x+2)}$

b) $\dfrac{3}{2(y+2)} - \dfrac{5}{3(y+2)}$

    LCD $= 6(y+2)$

    $\dfrac{3}{2(y+2)} \cdot \dfrac{3}{3} - \dfrac{5}{3(y+2)} \cdot \dfrac{2}{2}$

    $\dfrac{9}{6(y+2)} - \dfrac{10}{6(y+2)}$

    $-\dfrac{1}{6(y+2)}$

c) $\dfrac{2a-3}{(a+5)(a-5)} - \dfrac{7}{5(a-5)}$

    LCD $= 5(a+5)(a-5)$

    $\dfrac{2a-3}{(a+5)(a-5)} \cdot \dfrac{5}{5} - \dfrac{7}{5(a-5)} \cdot \dfrac{a+5}{a+5}$

    $\dfrac{10a-15}{5(a+5)(a-5)} - \dfrac{(7a+35)}{5(a+5)(a-5)}$ Ans. $\dfrac{3a-50}{5(a+5)(a-5)}$

9.   SIMPLIFYING COMPLEX FRACTIONS

A complex fraction is a fraction containing at least one other fraction within it.

Thus, $\dfrac{\frac{3}{4}}{2}$, $\dfrac{5}{\frac{2}{3}}$, and $\dfrac{x+\frac{1}{2}}{x-\frac{1}{4}}$ are complex fractions.

To Simplify a Complex Fraction: LCD — Multiplication Method

Simplify: $\quad a)\ \dfrac{\frac{2}{3}}{\frac{3}{4}} \qquad\qquad b)\ \dfrac{\frac{1}{2}-\frac{1}{3}}{5} \qquad\qquad c)\ \dfrac{x-\frac{1}{3}}{\frac{3}{5}+\frac{7}{10}}$

Procedure:

Solutions:

$$\text{LCD} = 12 \qquad\qquad \text{LCD} = 6 \qquad\qquad \text{LCD} = 30$$

1. Find LCD of fractions in complex fraction:

$$\dfrac{\frac{2}{3}}{\frac{3}{4}}\cdot\dfrac{12}{12} \qquad\qquad \dfrac{\frac{1}{2}-\frac{1}{3}}{5}\cdot\dfrac{6}{6} \qquad\qquad \dfrac{x-\frac{1}{3}}{\frac{3}{5}+\frac{7}{10}}\cdot\dfrac{30}{30}$$

2. Multiply both numerator and denominator by LCD and reduce, if necessary:

$$Ans.\ \dfrac{8}{9} \qquad\qquad \dfrac{3-2}{30}=\dfrac{1}{30} \qquad\qquad \dfrac{30x-10}{18+21}=\dfrac{30x-10}{39}$$

To Simplify a Complex Fraction: Combining — Division Method

Simplify: $\quad a)\ \dfrac{\frac{1}{2}-\frac{1}{3}}{\frac{1}{2}+\frac{1}{3}} \qquad b)\ \dfrac{x-\frac{1}{3}}{x+\frac{1}{3}} \qquad c)\ \dfrac{\frac{x}{2}+\frac{x}{5}}{2x-\frac{3x}{10}} \qquad d)\ \dfrac{1+\frac{2}{y}}{1-\frac{4}{y^2}}$

Procedure:

Solutions:

1. Combine terms of numerator:

$$\dfrac{1}{2}-\dfrac{1}{3}=\dfrac{1}{6} \qquad x-\tfrac{1}{3}=\dfrac{3x-1}{3} \qquad \dfrac{x}{2}+\dfrac{x}{5}=\dfrac{7x}{10} \qquad 1+\dfrac{2}{y}=\dfrac{y+2}{y}$$

2. Combine terms of denominator:

$$\dfrac{1}{2}+\dfrac{1}{3}=\dfrac{5}{6} \qquad x+\tfrac{1}{3}=\dfrac{3x+1}{3} \qquad 2x-\dfrac{3x}{10}=\dfrac{17}{10} \qquad 1-\dfrac{4}{y^2}=\dfrac{y^2-4}{y^2}$$

3. Divide new numerator by new denominator:

$$\dfrac{1}{6}\div\dfrac{5}{6} \qquad \dfrac{3x-1}{3}\div\dfrac{3x+1}{3} \qquad \dfrac{7x}{10}\div\dfrac{17x}{10} \qquad \dfrac{y+2}{y}\div\dfrac{y^2-4}{y^2}$$

$$\dfrac{1}{6}\cdot\dfrac{6}{5} \qquad \dfrac{3x-1}{3}\cdot\dfrac{3}{3x+1} \qquad \dfrac{7x}{10}\cdot\dfrac{10}{17x} \qquad \dfrac{y+2}{y}\cdot\dfrac{y^2}{y^2-4}$$

$$Ans.\ \dfrac{1}{5} \qquad\qquad \dfrac{3x-1}{3x+1} \qquad\qquad \dfrac{7}{17} \qquad\qquad \dfrac{y}{y-2}$$

9.1. **SIMPLIFYING NUMERICAL COMPLEX FRACTIONS**

Simplify:

$$a)\ \dfrac{4-\frac{1}{3}}{5} \qquad b)\ \dfrac{5+\frac{2}{5}}{7-\frac{1}{10}} \qquad c)\ \dfrac{\frac{1}{2}-\frac{1}{4}}{\frac{3}{8}+\frac{1}{16}} \qquad d)\ \dfrac{1-\frac{1}{6}+\frac{2}{3}}{\frac{2}{9}+3-\frac{1}{2}}$$

Solutions:

$a)\quad \text{LCD} = 3 \qquad\quad b)\ \textbf{Division method} \quad c)\ \text{LCD} = 16 \qquad\quad d)\ \text{LCD} = 18$

$$\dfrac{(4-\frac{1}{3})}{5}\cdot\dfrac{3}{3} \qquad \dfrac{27}{5}\div\dfrac{69}{10} \qquad \dfrac{(\frac{1}{2}-\frac{1}{4})}{(\frac{3}{8}+\frac{1}{16})}\cdot\dfrac{16}{16} \qquad \dfrac{(1-\frac{1}{6}+\frac{2}{3})}{(\frac{2}{9}+3-\frac{1}{2})}\cdot\dfrac{18}{18}$$

$$\dfrac{12-1}{15} \qquad \dfrac{\overset{9}{\cancel{27}}}{\underset{(1)}{\cancel{5}}}\cdot\dfrac{\overset{2}{\cancel{10}}}{\underset{23}{\cancel{69}}} \qquad \dfrac{8-4}{6+1} \qquad \dfrac{18-3+12}{4+54-9}$$

$$Ans.\ \dfrac{11}{15} \qquad\qquad \dfrac{18}{23} \qquad\qquad \dfrac{4}{7} \qquad\qquad \dfrac{27}{49}$$

9.2. SIMPLIFYING COMPLEX FRACTIONS

Simplify:

$a)$ $\dfrac{x - \frac{1}{2}}{4}$

$b)$ $\dfrac{\frac{x}{3} - \frac{x}{5}}{\frac{1}{2}}$

$c)$ $\dfrac{\frac{1}{x} + \frac{1}{x^2}}{\frac{4}{x^3}}$

$d)$ $\dfrac{\frac{1}{2x} - \frac{4}{y}}{\frac{1}{x} + \frac{2}{3y}}$

Solutions:

$a)$ LCD $= 2$

$$\dfrac{x - \frac{1}{2}}{4} \cdot \dfrac{2}{2}$$

$b)$ **Division method**

$$\dfrac{2x}{15} \div \dfrac{1}{2}$$

$c)$ LCD $= x^3$

$$\dfrac{\frac{1}{x} + \frac{1}{x^2}}{\frac{4}{x^3}} \cdot \dfrac{x^3}{x^3}$$

$d)$ LCD $= 6xy$

$$\dfrac{\frac{1}{2x} - \frac{4}{y}}{\frac{1}{x} + \frac{2}{3y}} \cdot \dfrac{6xy}{6xy}$$

Ans. $\dfrac{2x - 1}{8}$

$\dfrac{2x}{15} \cdot \dfrac{2}{1} = \dfrac{4x}{15}$

$\dfrac{x^2 + x}{4}$

$\dfrac{3y - 24x}{6y + 4x}$

Supplementary Problems

11.1. Express each in fractional form:　　　　　　　　　　　　　　　**(1.1 to 1.3)**

$a)$ $10 \div 19$

$b)$ $n \div (p + 3)$

$c)$ 34 divided by x

$d)$ Ratio of 25 to 11

$e)$ Ratio of 7¢ to a dime

$f)$ Quotient of $r - 7$ and 30

$g)$ 3 of 8 equal portions of a pie

$h)$ 3 out of 8 equal pies

$i)$ Ratio of 1 s to 1 h

Ans. $a)$ $\dfrac{10}{19}$　$b)$ $\dfrac{n}{p + 3}$　$c)$ $\dfrac{34}{x}$　$d)$ $\dfrac{25}{11}$　$e)$ $\dfrac{7}{10}$　$f)$ $\dfrac{r - 7}{30}$　$g)$ $\dfrac{3}{8}$　$h)$ $\dfrac{3}{8}$　$i)$ $\dfrac{1}{3600}$

11.2. State the value of x when:　　　　　　　　　　　　　　　**(1.4 to 1.5)**

$a)$ $\dfrac{x}{10} = 0$　　　　　　*Ans.*　0

$b)$ $\dfrac{x - 3}{5} = 0$　　　　　　3

$c)$ $\dfrac{2x + 5}{7x} = 0$　　　　　$-2\frac{1}{2}$

$d)$ $\dfrac{3x - 12}{3x + 12} = 0$　　　　　4

$e)$ $\dfrac{3x + 12}{3x - 12} = 0$　　　　　-4

$f)$ $\dfrac{1}{3x}$ is meaningless　　　*Ans.*　0

$g)$ $\dfrac{5}{x - 10}$ is meaningless　　　10

$h)$ $\dfrac{x}{5x + 20}$ is meaningless　　　-4

$i)$ $\dfrac{2x + 1}{2x - 1}$ is meaningless　　　$\dfrac{1}{2}$

11.3. Change each to equivalent fractions by performing the operations indicated on its numerator and denominator (terms):　　　　　　　　　　　　　　　**(2.1, 2.2)**

$a)$ Multiply terms of $\frac{1}{2}$ by 5.

$b)$ Multiply terms of $\frac{2}{3}$ by $5x$.

$c)$ Multiply terms of $\dfrac{a}{7}$ by $a + 2$.

$d)$ Divide terms of $\frac{10}{15}$ by 5.

$e)$ Divide terms of $\dfrac{3x}{6x^2}$ by $3x$.

$f)$ Divide terms of $\dfrac{5(a - 2)^2}{7(a - 2)}$ by $a - 2$.

Ans. $a)$ $\dfrac{5}{10}$　$b)$ $\dfrac{10x}{15x}$　$c)$ $\dfrac{a(a + 2)}{7(a + 2)} = \dfrac{a^2 + 2a}{7a + 14}$　$d)$ $\dfrac{2}{3}$　$e)$ $\dfrac{1}{2x}$　$f)$ $\dfrac{5(a - 2)}{7}$

11.4. Change $\dfrac{x}{5}$ to equivalent fractions by multiplying its terms by: **(2.1)**

   a) 6   *b)* y *c)* x^2 *d)* $3x - 2$ *e)* $a + b$

Ans. *a)* $\dfrac{6x}{30}$ *b)* $\dfrac{xy}{5y}$ *c)* $\dfrac{x^3}{5x^2}$ *d)* $\dfrac{3x^2 - 2x}{15x - 10}$ *e)* $\dfrac{ax + bx}{5a + 5b}$

11.5. Change $\dfrac{12a^2}{36a^3}$ to equivalent fractions by dividing its terms by   **(2.2)**

   a) 2   *b)* 12   *c)* $4a$ *d)* a^2 *e)* $12a^2$

Ans. *a)* $\dfrac{6a^2}{18a^3}$ *b)* $\dfrac{a^2}{3a^3}$ *c)* $\dfrac{3a}{9a^2}$ *d)* $\dfrac{12}{36a}$ *e)* $\dfrac{1}{3a}$

11.6. Obtain each missing term: **(2.3)**

a) $\dfrac{3}{7} = \dfrac{?}{28}$ Ans. 12   *d)* $\dfrac{10}{x} = \dfrac{10ab}{?}$ Ans. abx *g)* $\dfrac{7ac}{14cd} = \dfrac{a}{?}$ Ans. $2d$

b) $\dfrac{5}{9} = \dfrac{20}{?}$ 36   *e)* $\dfrac{2a}{3} = \dfrac{?}{30bc}$ $20abc$ *h)* $\dfrac{a^2 - b^2}{2a + 2b} = \dfrac{?}{2}$ $a - b$

c) $\dfrac{7a}{8} = \dfrac{?}{8a}$ $7a^2$ *f)* $\dfrac{x + 2}{3} = \dfrac{?}{3(x - 2)}$ $x^2 - 4$ *i)* $\dfrac{24(a + 2)}{36(a + 2)^2} = \dfrac{?}{3(a + 2)}$ 2

11.7. State the reciprocal of each: **(3.1)**

a) $\dfrac{1}{100}$ Ans. 100   *d)* $\dfrac{3}{x}$ Ans. $\dfrac{x}{3}$ *g)* $\dfrac{3x}{2}$ Ans. $\dfrac{2}{3x}$ *j)* $10\%x$ Ans. $\dfrac{10}{x}$

b) 100 $\dfrac{1}{100}$ *e)* $\dfrac{2}{7}$ $\dfrac{7}{2}$ or $3\frac{1}{2}$ *h)* $\dfrac{2}{3}x$ $\dfrac{3}{2x}$ *k)* 25%   4 or 400%

c) $\dfrac{1}{3x}$ $3x$ *f)* $1\frac{1}{3}$ $\dfrac{3}{4}$ *i)* 0.3 $\dfrac{10}{3}$ *l)* 100%   1 or 100%

11.8. Supply the missing entry: **(3.2)**

a) $\left(\dfrac{3}{4}\right)\left(\dfrac{4}{3}\right) = ?$ *c)* $\left(\dfrac{3}{7}\right)\left(\dfrac{7}{3}\right)(?)\left(\dfrac{2}{5}\right) = 1$ *e)* $\left(\dfrac{7}{4}\right)\left(\dfrac{4}{7}y\right) = (?)$

b) $\left(\dfrac{8}{5}\right)(?) = 1$ *d)* $\left(\dfrac{2}{x}\right)(?) = 1$ *f)* $(?)\left(\dfrac{6}{5}x\right) = x$

Ans.   *a)* 1   *b)* $\dfrac{5}{8}$ *c)* $\dfrac{5}{2}$ *d)* $\dfrac{x}{2}$ *e)* y *f)* $\dfrac{5}{6}$

11.9. Show how to change each division to multiplication by using the reciprocal of a number: **(3.3)**

a) $12 \div \dfrac{6}{7}$ *c)* $20y \div \dfrac{y}{x}$ *e)* $(3a - 6b) \div \dfrac{3}{2}$

b) $\dfrac{6}{7} \div 12$ *d)* $3x^2 \div \dfrac{x}{y}$ *f)* $(x^3 + 8x^2 - 14x) \div \dfrac{x}{10}$

Ans.   *a)* $12 \cdot \dfrac{7}{6} = 14$ *b)* $\dfrac{6}{7} \cdot \dfrac{1}{12} = \dfrac{1}{14}$ *c)* $20y\left(\dfrac{x}{y}\right) = 20x$ *d)* $3x^2\left(\dfrac{y}{x}\right) = 3xy$

   e) $\dfrac{2}{3}(3a - 6b) = 2a - 4b$ *f)* $\dfrac{10}{x}(x^3 + 8x^2 - 14x) = 10x^2 + 80x - 140$

11.10.   Solve for x, indicating each multiplier of both sides: **(3.4)**

a) $\dfrac{5}{9}x = 35$ c) $1\frac{1}{3}x = 20$ e) $\dfrac{bx}{2} = 7b$

b) $\dfrac{8}{3}x = 20$ d) $0.3x = \dfrac{9}{5}$ f) $\dfrac{3x}{c} = 6c$

Ans.   a) $\mathbf{M}_{9/5}$, $x = 63$ c) $\mathbf{M}_{3/4}$, $x = 15$ e) $\mathbf{M}_{2/b}$, $x = 14$

        b) $\mathbf{M}_{3/8}$, $x = 7\frac{1}{2}$ d) $\mathbf{M}_{10/3}$, $x = 6$ f) $\mathbf{M}_{c/3}$, $x = 2c^2$

11.11.   Reduce to lowest terms: **(4.1)**

a) $\dfrac{21}{35}$ b) $\dfrac{42}{24}$ c) $\dfrac{3ab}{9ac}$ d) $\dfrac{15d^2}{5d^3}$ e) $\dfrac{8g - 4h}{12g}$ f) $\dfrac{10rs}{5s^2 + 20s}$

Ans.   a) $\dfrac{3}{5}$ b) $\dfrac{7}{4}$ c) $\dfrac{b}{3c}$ d) $\dfrac{3}{d}$ e) $\dfrac{2g - h}{3g}$ f) $\dfrac{2r}{s + 4}$

11.12.   Reduce to lowest terms: **(4.2)**

a) $\dfrac{5(x + 5)}{11(x + 5)}$ c) $\dfrac{3a - 3b}{6a - 6b}$ e) $\dfrac{(l - 5)^2}{8l - 40}$ g) $\dfrac{km + kn}{n^2 + nm}$

b) $\dfrac{8ab(2c - 3)}{4ac(2c - 3)}$ d) $\dfrac{a^2 - 9}{7a + 21}$ f) $\dfrac{c^2 - d^2}{(c - d)^2}$ h) $\dfrac{cd + 5d}{5x + cx}$

Ans.   a) $\dfrac{5}{11}$ c) $\dfrac{1}{2}$ e) $\dfrac{l - 5}{8}$ g) $\dfrac{k}{n}$

        b) $\dfrac{2b}{c}$ d) $\dfrac{a - 3}{7}$ f) $\dfrac{c + d}{c - d}$ h) $\dfrac{d}{x}$

11.13.   Reduce to lowest terms: **(4.3)**

a) $\dfrac{6 - x}{5x - 30}$ c) $\dfrac{y^2 - 4}{2a - ay}$ e) $\dfrac{1 - 2p}{2p^2 - p}$ g) $\dfrac{ac^2 - ad^2}{d^2 - cd}$

b) $\dfrac{x^2 - 3x}{21 - 7x}$ d) $\dfrac{2b - 3ab}{9a^2 - 4}$ f) $\dfrac{3w - 3x}{2x^2 - 2w^2}$ h) $\dfrac{(r - s)^2}{s^2 - r^2}$

Ans.   a) $-\dfrac{1}{5}$ c) $-\dfrac{y + 2}{a}$ e) $-\dfrac{1}{p}$ g) $-\dfrac{a(c + d)}{d}$

        b) $-\dfrac{x}{7}$ d) $-\dfrac{b}{3a + 2}$ f) $-\dfrac{3}{2(x + w)}$ h) $-\dfrac{r - s}{r + s}$ or $\dfrac{s - r}{s + r}$

11.14.   Reduce to lowest terms: **(4.4)**

a) $\dfrac{c^2 + 5c}{c^2 + 12c + 35}$ c) $\dfrac{4a^2 - b^2}{4a^2 - 4ab + b^2}$ e) $\dfrac{2x^2 - 2}{x^2 - 4x - 5}$ g) $\dfrac{2w^2 - 14w + 24}{12w^2 - 32w - 12}$

b) $\dfrac{b^2 + 6b - 7}{b^2 - 49}$ d) $\dfrac{9 - 3s}{s^2 - 5s + 6}$ f) $\dfrac{y^2 + 8y + 12}{y^2 - 3y - 10}$ h) $\dfrac{2a^2 - 9a - 5}{6a^2 + 7a + 2}$

Ans.   a) $\dfrac{c}{c + 7}$ c) $\dfrac{2a + b}{2a - b}$ e) $\dfrac{2(x - 1)}{x - 5}$ g) $\dfrac{w - 4}{2(3w + 1)}$

        b) $\dfrac{b - 1}{b - 7}$ d) $-\dfrac{3}{s - 2}$ f) $\dfrac{y + 6}{y - 5}$ h) $\dfrac{a - 5}{3a + 2}$

11.15. Multiply: **(5.1, 5.2)**

$a)$ $\frac{3}{4} \times 5$ $c)$ $\frac{3}{4} \times \frac{28}{5}$ $e)$ $\frac{3}{5} \times \frac{5}{8} \times \frac{8}{3}$

$b)$ $\frac{2}{5} \times \frac{3}{7} \times 11$ $d)$ $\frac{20}{6} \times \frac{9}{4} \times \frac{12}{13}$ $f)$ $\frac{2}{9} \times \frac{3}{4} \times \frac{3}{8} \times \frac{16}{17}$

Ans. $a)$ $\frac{15}{4}$ $b)$ $\frac{66}{35}$ $c)$ $\frac{21}{5}$ $d)$ $\frac{90}{13}$ $e)$ 1 $f)$ $\frac{1}{17}$

11.16. Multiply: **(5.1, 5.2)**

$a)$ $\frac{3b}{c} \cdot \frac{b^2}{5}$ $e)$ $\frac{a+b}{7} \cdot \frac{a-b}{2}$ $i)$ $\frac{15b+15c}{16x^2-9} \cdot \frac{8x+6}{5b+5c}$

$b)$ $7ab \cdot \frac{a}{2} \cdot \frac{3}{c}$ $f)$ $\frac{a+b}{5} \cdot \frac{15}{11a+11b}$ $j)$ $\frac{(v+w)^2}{bv+bw} \cdot \frac{b^2}{3v+3w}$

$c)$ $\frac{x^2}{y} \cdot \frac{3}{y^2} \cdot x^3$ $g)$ $\frac{x^2}{y} \cdot \frac{y^2}{3} \cdot \frac{1}{x^3}$ $k)$ $\frac{r^4-s^4}{5r+5s} \cdot \frac{25}{2r^2+2s^2}$

$d)$ $\frac{m+n}{3} \cdot \frac{m+n}{mn}$ $h)$ $\frac{6r+12s}{2rs-4r} \cdot \frac{7s-14}{3r+6s}$ $l)$ $\frac{5ax+10a}{ab-10b} \cdot \frac{200-2a^2}{ax+2a}$

Ans. $a)$ $\frac{3b^3}{5c}$ $e)$ $\frac{a^2-b^2}{14}$ $i)$ $\frac{6}{4x-3}$

$b)$ $\frac{21a^2b}{2c}$ $f)$ $\frac{3}{11}$ $j)$ $\frac{b}{3}$

$c)$ $\frac{3x^5}{y^3}$ $g)$ $\frac{y}{3x}$ $k)$ $\frac{5(r-s)}{2}$

$d)$ $\frac{(m+n)^2}{3mn}$ $h)$ $\frac{7}{r}$ $l)$ $-\frac{10(10+a)}{b}$

11.17. Multiply: **(5.3)**

$a)$ $\frac{a^2-4}{(a-2)^2} \cdot \frac{a^2-9a+14}{a^3+2a^2}$ $c)$ $\frac{2c+b}{b^2} \cdot \frac{3b^2-3bc}{bc+2c^2} \cdot \frac{5bc-10c^2}{b^2-3bc+2c^2}$

$b)$ $\frac{(x+3)^2}{x^2-7x-30} \cdot \frac{7x^2-700}{x^3+3x^2}$ $d)$ $\frac{r^2-s^2}{3-r} \cdot \frac{r^2+3rs}{4s-4r} \cdot \frac{rs-3s}{r^2+4rs+3s^2}$

Ans. $a)$ $\frac{a-7}{a^2}$ $b)$ $\frac{7(x+10)}{x^2}$ $c)$ $\frac{15}{b}$ $d)$ $\frac{rs}{4}$

11.18. Divide: **(6.1)**

$a)$ $10 \div \frac{2}{3}$ $c)$ $3x \div \frac{12x}{7}$ $e)$ $\frac{ab}{4} \div \frac{b^2}{6}$ $g)$ $\frac{r^2s}{16t} \div \frac{rs^2}{8t^2}$

$b)$ $\frac{3}{4} \div 36$ $d)$ $\frac{5}{y} \div \frac{7}{y}$ $f)$ $\frac{x}{2} \div \frac{ax}{6}$ $h)$ $\frac{2x}{y} \div \frac{10y}{3x}$

Ans. $a)$ 15 $b)$ $\frac{1}{48}$ $c)$ $\frac{7}{4}$ $d)$ $\frac{5}{7}$ $e)$ $\frac{3a}{2b}$ $f)$ $\frac{3}{a}$ $g)$ $\frac{rt}{2s}$ $h)$ $\frac{3x^2}{5y^2}$

11.19. Divide: **(6.2)**

$a)$ $\frac{3x}{x+2} \div \frac{5x}{x+2}$ $g)$ $\frac{2x^2+2y^2}{x^2-4y^2} \div \frac{x^2+y^2}{x+2y}$

$b)$ $\frac{2(y+3)}{7} \div \frac{8(y+3)}{63}$ $h)$ $\frac{s^2+st}{rs-st} \div \frac{s^2-t^2}{r^2-t^2}$

c) $\dfrac{2a - 10}{3b - 21} \div \dfrac{a - 5}{4b - 28}$

i) $\dfrac{a^3 - 64a}{2a^2 + 16a} \div \dfrac{a^2 - 9a + 8}{a^2 + 4a - 5}$

d) $\dfrac{3(a^2 + 5)}{b(a^2 - 3)} \div \dfrac{18(a^2 + 5)}{7(a^2 - 3)}$

j) $\dfrac{a^3 - ab^2}{6a + 6b} \div \dfrac{a^2 - 4ab + 3b^2}{12a - 36b}$

e) $\dfrac{d^2 - 1}{3d + 9} \div \dfrac{d^2h - h}{dh + 3h}$

k) $\dfrac{3p + p^2}{12 + 3p} \div \dfrac{9p - p^3}{12 - p - p^2}$

f) $\dfrac{c^2 - cd}{3c^2 + 3cd} \div \dfrac{7c - 7d}{c^2 - d^2}$

l) $\dfrac{c^2 + 16c + 64}{2c^2 - 128} \div \dfrac{3c^2 + 30c + 48}{c^2 - 6c - 16}$

Ans. a) $\dfrac{3}{5}$ c) $\dfrac{8}{3}$ e) $\dfrac{1}{3}$ g) $\dfrac{2}{x - 2y}$ i) $\dfrac{a + 5}{2}$ k) $\dfrac{1}{3}$

      b) $\dfrac{9}{4}$ d) $\dfrac{7}{6b}$ f) $\dfrac{c - d}{21}$ h) $\dfrac{r + t}{s - t}$ j) $2a$ l) $\dfrac{1}{6}$

11.20. Combine: **(7.1)**

a) $\frac{4}{9} + \frac{2}{9}$ d) $\frac{3}{5} + \frac{4}{5} - \frac{1}{5}$ g) $3\frac{5}{12} + 2\frac{3}{12}$

b) $\frac{10}{13} - \frac{7}{13}$ e) $\frac{6}{15} - \frac{2}{15} - \frac{2}{15}$ h) $10\frac{2}{7} - 8\frac{1}{7}$

c) $\frac{3}{11} - \frac{8}{11}$ f) $\frac{3}{8} - \frac{1}{8} + \frac{7}{8}$ i) $\frac{5}{16} + \frac{15}{16} - \frac{7}{16} + \frac{3}{16}$

Ans. a) $\frac{6}{9} = \frac{2}{3}$ b) $\frac{3}{13}$ c) $-\frac{5}{11}$ d) $\frac{6}{5}$ e) $\frac{2}{15}$ f) $\frac{9}{8}$ g) $5\frac{2}{3}$ h) $2\frac{1}{7}$ i) 1

11.21. Combine: **(7.1)**

a) $\dfrac{x}{3} + \dfrac{5x}{3}$ c) $\dfrac{x^2}{6} + \dfrac{x^2}{6}$ e) $\dfrac{17}{3s} - \dfrac{14}{3s}$ g) $\dfrac{5x}{y} + \dfrac{x}{y} - \dfrac{3}{y}$

b) $\dfrac{7y}{5} - \dfrac{3y}{5}$ d) $\dfrac{10}{r} + \dfrac{3}{r}$ f) $\dfrac{5a}{b^2} - \dfrac{3a}{b^2}$ h) $\dfrac{4p}{3q} - \dfrac{p}{3q} + \dfrac{8}{3q} - \dfrac{2}{3q}$

Ans. a) $\dfrac{6x}{3} = 2x$ b) $\dfrac{4y}{5}$ c) $\dfrac{x^2}{3}$ d) $\dfrac{13}{r}$ e) $\dfrac{1}{s}$ f) $\dfrac{2a}{b^2}$ g) $\dfrac{6x - 3}{y}$ h) $\dfrac{p + 2}{q}$

11.22. Combine: **(7.1)**

a) $\dfrac{a + 2}{8} + \dfrac{7a - 2}{8}$ c) $\dfrac{2y - 7}{5} - \dfrac{2y - 10}{5}$ e) $\dfrac{b + 8}{5a} + \dfrac{4b + 7}{5a}$

b) $\dfrac{5x}{7} - \dfrac{5x - 35}{7}$ d) $\dfrac{8}{x} + \dfrac{2}{x} - \dfrac{3x + 10}{x}$ f) $\dfrac{x + 3}{2x^2} + \dfrac{3x - 3}{2x^2}$

Ans. a) $\dfrac{8a}{8} = a$ b) $\dfrac{35}{7} = 5$ c) $\dfrac{3}{5}$ d) $-\dfrac{3x}{x} = -3$ e) $\dfrac{5b + 15}{5a} = \dfrac{b + 3}{a}$ f) $\dfrac{4x}{2x^2} = \dfrac{2}{x}$

11.23. Combine: **(7.2)**

a) $\dfrac{5}{x + 3} + \dfrac{x - 2}{x + 3}$ d) $\dfrac{2}{c + 3} - \dfrac{c + 5}{c + 3}$ g) $\dfrac{3a}{2(a + b)} + \dfrac{a + 4b}{2(a + b)}$

b) $\dfrac{2x}{x + y} + \dfrac{2y}{x + y}$ e) $\dfrac{7}{d^2 + 7} + \dfrac{d^2}{d^2 + 7}$ h) $\dfrac{7a - 2}{a(b + c)} - \dfrac{a - 2}{a(b + c)}$

c) $\dfrac{5p}{p - q} - \dfrac{5q}{p - q}$ f) $\dfrac{6}{2x - 3} - \dfrac{4x}{2x - 3}$ i) $\dfrac{3x}{x^2 - 4} - \dfrac{x + 4}{x^2 - 4}$

Ans. *a)* $\dfrac{x+3}{x+3} = 1$ *d)* $\dfrac{-c-3}{c+3} = -1$ *g)* $\dfrac{4a+4b}{2(a+b)} = 2$

b) $\dfrac{2x+2y}{x+y} = 2$ *e)* 1 *h)* $\dfrac{6a}{a(b+c)} = \dfrac{6}{b+c}$

c) $\dfrac{5p-5q}{p-q} = 5$ *f)* $\dfrac{6-4x}{2x-3} = -2$ *i)* $\dfrac{2x-4}{x^2-4} = \dfrac{2}{x+2}$

11.24. Combine: **(8.1)**

a) $\dfrac{2}{3} + \dfrac{1}{2}$ *b)* $\dfrac{3}{4} - \dfrac{1}{5}$ *c)* $\dfrac{a}{7} - \dfrac{b}{2}$ *d)* $\dfrac{4}{3x} + \dfrac{5}{2y}$ *e)* $\dfrac{a}{b} + \dfrac{b}{a}$

Ans. *a)* $\dfrac{7}{6}$ *b)* $\dfrac{11}{20}$ *c)* $\dfrac{2a-7b}{14}$ *d)* $\dfrac{8y+15x}{6xy}$ *e)* $\dfrac{a^2+b^2}{ab}$

11.25. Combine: **(8.1)**

a) $\dfrac{4}{5} + \dfrac{2}{x+3}$ *c)* $\dfrac{8c}{3c+1} - \dfrac{1}{2}$ *e)* $\dfrac{x}{2-x} - \dfrac{x}{2+x}$

b) $\dfrac{3}{a} + \dfrac{2}{5-a}$ *d)* $\dfrac{2}{y+1} + \dfrac{3}{y-1}$ *f)* $\dfrac{4}{a+5} + \dfrac{2}{a+1}$

Ans. *a)* $\dfrac{4x+22}{5(x+3)}$ *b)* $\dfrac{15-a}{a(5-a)}$ *c)* $\dfrac{13c-1}{2(3c+1)}$ *d)* $\dfrac{5y+1}{y^2-1}$ *e)* $\dfrac{2x^2}{4-x^2}$ *f)* $\dfrac{6a+14}{(a+5)(a+1)}$

11.26. Combine: **(8.2)**

a) $\dfrac{1}{3} + \dfrac{3}{4} + \dfrac{1}{6}$ *c)* $\dfrac{11}{10} - \dfrac{2}{5} + \dfrac{1}{2}$ *e)* $\dfrac{5x}{8} - \dfrac{x}{12}$ *g)* $\dfrac{p}{8} + \dfrac{q}{3} - \dfrac{r}{6}$

b) $\dfrac{7}{8} - \dfrac{1}{2} - \dfrac{1}{4}$ *d)* $\dfrac{a}{3} - \dfrac{a}{12}$ *f)* $\dfrac{c}{6} + \dfrac{2c}{9}$ *h)* $\dfrac{4}{3x} + \dfrac{3}{2x} + \dfrac{1}{6x}$

Ans. *a)* $\dfrac{5}{4}$ *b)* $\dfrac{1}{8}$ *c)* $\dfrac{6}{5}$ *d)* $\dfrac{a}{4}$ *e)* $\dfrac{13x}{24}$ *f)* $\dfrac{7c}{18}$ *g)* $\dfrac{3p+8q-4r}{24}$ *h)* $\dfrac{3}{x}$

11.27. Combine: **(8.3)**

a) $\dfrac{3}{a} + \dfrac{4}{a^2}$ *d)* $\dfrac{1}{3} + \dfrac{2}{x} + \dfrac{1}{x^2}$ *g)* $\dfrac{8}{r^2s} + \dfrac{4}{rs^2}$

b) $\dfrac{5}{b^2} - \dfrac{2}{b}$ *e)* $\dfrac{2}{x} - \dfrac{5}{x^2} + \dfrac{3}{x^3}$ *h)* $\dfrac{10}{a^2b^2} - \dfrac{4}{ab^2} + \dfrac{5}{a^2b}$

c) $\dfrac{7}{2b} + \dfrac{2}{3b^2}$ *f)* $\dfrac{1}{2x^3} + \dfrac{3}{x^2} - \dfrac{7}{5x}$ *i)* $\dfrac{1}{a^2b^2} + \dfrac{3}{a^2} + \dfrac{2}{b^2}$

Ans. *a)* $\dfrac{3a+4}{a^2}$ *b)* $\dfrac{5-2b}{b^2}$ *c)* $\dfrac{21b+4}{6b^2}$ *d)* $\dfrac{x^2+6x+3}{3x^2}$ *e)* $\dfrac{2x^2-5x+3}{x^3}$

f) $\dfrac{5+30x-14x^2}{10x^3}$ *g)* $\dfrac{8s+4r}{r^2s^2}$ *h)* $\dfrac{10-4a+5b}{a^2b^2}$ *i)* $\dfrac{1+3b^2+2a^2}{a^2b^2}$

11.28. Combine: **(8.4)**

a) $\dfrac{7}{2} - \dfrac{8}{h+3}$ *d)* $\dfrac{2}{p+2} - \dfrac{2}{p+3}$ *g)* $\dfrac{4x}{x^2-36} - \dfrac{4}{x+6}$

b) $\dfrac{4a}{3b+6} - \dfrac{a}{b+2}$

e) $\dfrac{3a}{a+2} + \dfrac{a}{a-2}$

h) $\dfrac{8}{7-y} - \dfrac{8y}{49-y^2}$

c) $\dfrac{10}{x^2+x} + \dfrac{2}{3x^2+3x}$

f) $\dfrac{2}{3x+3y} - \dfrac{3}{5x+5y}$

i) $\dfrac{3t^2}{9t^2-16s^2} - \dfrac{t}{3t+4s}$

Ans. a) $\dfrac{7h+5}{2(h+3)}$ b) $\dfrac{a}{3b+6}$ c) $\dfrac{32}{3(x^2+x)}$ d) $\dfrac{2}{(p+2)(p+3)}$ e) $\dfrac{4a^2-4a}{a^2-4}$

f) $\dfrac{1}{15(x+y)}$ g) $\dfrac{24}{x^2-36}$ h) $\dfrac{56}{49-y^2}$ i) $\dfrac{4st}{9t^2-16s^2}$

11.29. Simplify: (9.1)

a) $\dfrac{5+\frac{2}{3}}{2}$

c) $\dfrac{\frac{1}{4}+\frac{1}{3}}{2}$

e) $\dfrac{\frac{3}{14}-\frac{2}{7}}{5}$

g) $\dfrac{2+\frac{1}{5}+\frac{3}{4}}{10}$

b) $\dfrac{4+\frac{1}{3}}{4-\frac{1}{3}}$

d) $\dfrac{5}{\frac{1}{2}-\frac{1}{4}}$

f) $\dfrac{\frac{1}{8}+\frac{5}{16}}{\frac{7}{12}}$

h) $\dfrac{\frac{4}{5}-\frac{1}{6}}{\frac{4}{5}+\frac{1}{3}}$

Ans. a) $\dfrac{17}{6}$ b) $\dfrac{13}{11}$ c) $\dfrac{7}{24}$ d) 20   e) $-\dfrac{1}{70}$ f) $\dfrac{3}{4}$ g) $\dfrac{59}{200}$ h) $\dfrac{19}{34}$

11.30. Simplify: (9.2)

a) $\dfrac{y+\frac{1}{3}}{2}$

c) $\dfrac{5p+2}{\frac{2}{3}}$

e) $\dfrac{5h}{\frac{1}{2}-\frac{1}{5}}$

g) $\dfrac{\frac{3}{x}+\frac{1}{6x}}{\frac{7}{3x}}$

b) $\dfrac{2a-1}{\frac{1}{2}}$

d) $\dfrac{\frac{x}{4}-\frac{2x}{9}}{3}$

f) $\dfrac{\frac{r}{4}}{\frac{7}{8}-\frac{r}{2}}$

h) $\dfrac{\frac{a}{x}+\frac{x}{a}}{\frac{a}{x}-\frac{x}{a}}$

Ans. a) $\dfrac{3y+1}{6}$ b) $4a-2$ c) $\dfrac{15p+6}{2}$ d) $\dfrac{x}{108}$ e) $\dfrac{50h}{3}$ f) $\dfrac{2r}{7-4r}$

g) $\dfrac{19}{14}$ h) $\dfrac{a^2+x^2}{a^2-x^2}$

CHAPTER 12

Roots and Radicals

1. UNDERSTANDING ROOTS AND RADICALS

The square root of a number is one of its two equal factors.

Thus, $+5$ is a square root of 25 since $(+5)(+5) = 25$.

Also -5 is another square root of 25 since $(-5)(-5) = 25$.

RULE. A positive number has two square roots which are **opposites** of each other (same absolute value but unlike signs).

Thus, $+10$ and -10 are the two square roots of 100.

To indicate both square roots, the symbol $\pm$, which combines $+$ and $-$, may be used.

Thus, the square root of 49, $+7$ and -7, may be written together as ± 7.

Read ± 7 as "plus or minus 7."

The principal square root of a number is its positive square root.

Thus, the principal square root of 36 is $+6$.

The symbol $\sqrt{}$ indicates the principal square root of a number.

Thus, $\sqrt{9} = $ principal square root of $9 = +3$.

To indicate the negative square root of a number, place the minus sign before $\sqrt{}$.

Thus, $-\sqrt{16} = -4$.

NOTE: Unless otherwise stated, whenever a square root of a number is to be found, it is understood that the principal or positive square root is required.

Radical, Radical Sign, Radicand, Index

(1) **A radical** is an indicated root of a number or an expression.

Thus, $\sqrt{5}$, $\sqrt[3]{8x}$, and $\sqrt[4]{7x^3}$ are radicals.

(2) The symbols $\sqrt{}$, $\sqrt[3]{}$, and $\sqrt[4]{}$ are **radical signs**.

(3) The radicand is the number or expression under the radical sign.

Thus, 80 is the radicand of $\sqrt[3]{80}$ or $\sqrt[4]{80}$.

(4) **The index of a root** is the small number written above and to the left of the radical sign $\sqrt{}$. The index indicates which root is to be taken. In square roots, the index 2 is not indicated but is understood.

Thus, $\sqrt[3]{8}$ indicates the third root or cube root of 8.

The cube root of a number is one of its three equal factors.

Thus, 3 is a cube root of 27 since $(3)(3)(3) = 27$.

This is written $3 = \sqrt[3]{27}$ and read, "3 is a cube root of 27."

Also,   3 is a cube root of -27 since $(-3)(-3)(-3) = -27$.

Square and Cube Roots of Negative Numbers: Imaginary Numbers

The cube root of a negative number is a negative number. For example, $\sqrt[3]{-8} = -2$, since $(-2)(-2)(-2) = -8$. However, the square root of a negative number is not a negative number. For example, the square root of -16 is **not** -4, since $(-4)(-4) = 16$, not -16. Numbers such as $\sqrt{-16}$ or $\sqrt{-100}$ are called **imaginary numbers**, an unfortunate name given their great importance in science, engineering, and higher mathematics.

In general, the nth root of a number is one of its n equal factors.

Thus, 2 is the fifth root of 32 since it is one of the five equal factors of 32.

1.1.   OPPOSITE SQUARE ROOTS OF A POSITIVE NUMBER

State the square roots of:

a) 25   b) $25x^2$ c) x^4 d) $81a^2b^2$ e) $\dfrac{49}{64}$ f) $\dfrac{9a^2}{16b^2}$

Ans.   a) ± 5 b) $\pm 5x$ c) $\pm x^2$ d) $\pm 9ab$ e) $\pm\dfrac{7}{8}$ f) $\pm\dfrac{3a}{4b}$

1.2.   PRINCIPAL SQUARE ROOT OF A NUMBER

Find each principal (positive) square root:

a) $\sqrt{36}$ b) $\sqrt{0.09}$ c) $\sqrt{1.44x^2}$ d) $\sqrt{\dfrac{25}{4}}$ e) $\sqrt{\dfrac{x^2}{400}}$ f) $\sqrt{\dfrac{b^6}{c^8}}$

Solutions:

Ans.   a) 6   b) 0.3   c) 1.2x   d) $\dfrac{5}{2}$ e) $\dfrac{x}{20}$ f) $\dfrac{b^3}{c^4}$

1.3.   NEGATIVE SQUARE ROOT OF A NUMBER

Find each negative square root:

a) $-\sqrt{100}$ b) $-\sqrt{0.16}$ c) $-\sqrt{400c^2}$ d) $\sqrt{\dfrac{9}{4}}$ e) $-\sqrt{\dfrac{81}{r^2}}$ f) $\sqrt{\dfrac{p^8}{q^{12}}}$

Solutions:

Ans.   a) -10 b) -0.4 c) $-20c$ d) $-\dfrac{3}{2}$ e) $-\dfrac{9}{r}$ f) $-\dfrac{p^4}{q^6}$

1.4.   EVALUATING NUMERICAL SQUARE ROOT EXPRESSIONS

Find the value of:

a) $3\sqrt{4} + \sqrt{25} - 2\sqrt{9}$ b) $\sqrt{\tfrac{25}{9}} + \sqrt{\tfrac{64}{9}}$ c) $4\sqrt{\tfrac{81}{4}}$ d) $\tfrac{1}{3}\sqrt{36} + \tfrac{2}{5}\sqrt{100}$

Solutions:

a) $3(2) + 5 - 2(3)$ b) $\tfrac{5}{3} + \tfrac{8}{3}$ c) $4(\tfrac{9}{2})$ d) $\tfrac{1}{3}(6) + \tfrac{2}{5}(10)$

Ans. $= 5$ $= \tfrac{13}{3}$ $= 18$ $= 6$

2. UNDERSTANDING RATIONAL AND IRRATIONAL NUMBERS

Rational Numbers

A **rational number** is one that can be expressed as the quotient or **ratio** of two integers. (Note the word **ratio** in **rational**.)

Thus, $\frac{3}{5}$ and $-\frac{3}{5}$ are rational numbers.

Kinds of Rational Numbers

1. **All integers,** that is, all positive and negative whole numbers and zero.

   Thus, 5 is rational since 5 can be expressed as $\frac{5}{1}$.

2. **Fractions** whose numerator and denominator are integers, after simplification.

   Thus, $\frac{1.5}{2}$ is rational because it equals $\frac{3}{4}$ when simplified. However, $\frac{\sqrt{2}}{3}$ is not rational.

3. **Decimals** which have a limited number of decimal places.

   Thus, 3.14 is rational since it can be expressed as $\frac{314}{100}$.

4. **Decimals** which have an unlimited number of decimal places and the digits continue to repeat themselves. Thus, 0.6666... is rational since it can be expressed as $\frac{2}{3}$.

   NOTE: ... is a symbol meaning "continued without end."

5. **Square root expressions** whose radicand is a perfect square, such as $\sqrt{25}$; **cube root expressions** whose radicand is a perfect cube, such as $\sqrt[3]{27}$; etc.

Irrational Numbers

An **irrational number** is one that cannot be expressed as the ratio of two integers. Thus,

(1) $\sqrt{2}$ is an irrational number since it cannot equal a fraction whose terms are integers. However, we can approximate $\sqrt{2}$ to any number of decimal places. For example, $\sqrt{2} = 1.4142$ to the nearest ten-thousandth. Such approximations are rational numbers.

(2) π is also irrational. An approximation of π has been carried out to thousands of places.

(3) Other examples of irrational numbers are $\frac{\sqrt{5}}{2}$, $\frac{2}{\sqrt{5}}$, and $2 + \sqrt{7}$.

	Number												
	-3	0	$\sqrt{100}$	20%	$0.333...$	0.333	$\sqrt{0.09}$	$\frac{\sqrt{25}}{12}$	$\sqrt{7}$	$\frac{2}{3+\sqrt{4}}$	$-\sqrt{\frac{32}{2}}$	$-\frac{\sqrt{32}}{2}$	$\sqrt[3]{25}$
Positive Integer			√										
Negative Integer	√										√		
Rational Number	√	√	√	√	√	√	√	√		√	√		
Expressed as Ratio of Two Integers	$\frac{-3}{1}$	$\frac{0}{1}$	$\frac{10}{1}$	$\frac{1}{5}$	$\frac{1}{3}$	$\frac{333}{1000}$	$\frac{3}{10}$	$\frac{5}{12}$		$\frac{2}{5}$	$\frac{-4}{1}$		
Irrational Number									√			√	√

Fig. 12-1

Surds

A **surd** is either an indicated root of a number which can only be approximated or a polynomial involving one or more such roots. A surd is irrational but not imaginary.

Thus, $\sqrt{5}$, $2 + \sqrt{5}$, and $10 + \sqrt[3]{7} - 2\sqrt{3}$ are surds. $\sqrt{5}$ is a monomial surd; $2 + \sqrt{5}$ is a binomial surd. However, $\sqrt{-4}$ is not a surd since it is imaginary.

2.1. Classifying Rational and Irrational Numbers

Classify each number in the table in Fig. 12-1. If rational, express it as the ratio of two integers.

2.2. Expressing Rational Numbers as the Ratio of Two Integers

Express each of the following rational numbers as the ratio of two integers:

a) 10   *Ans.* $\dfrac{10}{1}$ *e)* $\dfrac{\sqrt{100}}{7}$ *Ans.* $\dfrac{10}{7}$ *i)* $\dfrac{\sqrt{16}}{\sqrt{121}}$ *Ans.* $\dfrac{4}{11}$

b) -7 $\dfrac{-7}{1}$ *f)* $\dfrac{8}{\sqrt{49}}$ $\dfrac{8}{7}$ *j)* $\sqrt{\dfrac{18}{8}}$ $\dfrac{3}{2}$

c) 6.3 $\dfrac{63}{10}$ *g)* $3\sqrt{25}$ $\dfrac{15}{1}$ *k)* $\sqrt{6\tfrac{1}{4}}$ $\dfrac{5}{2}$

d) $-5\tfrac{1}{2}$ $\dfrac{-11}{2}$ *h)* $\tfrac{1}{3}\sqrt{0.81}$ $\dfrac{3}{10}$ *l)* $\sqrt{0.0001}$ $\dfrac{1}{100}$

3. FINDING THE SQUARE ROOT OF A NUMBER BY USING A GRAPH

Approximate square roots of numbers can be obtained by using a graph of $x = \sqrt{y}$.

The table of values used to graph $x = \sqrt{y}$ from $x = 0$ to $x = 8$ is in Fig. 12-2.

Table of values used to graph $x = \sqrt{y}$ from $x = 0$ to $x = 8$:

If $x =$	0	1	$1\tfrac{1}{2}$	2	$2\tfrac{1}{2}$	3	$3\tfrac{1}{2}$	4	$4\tfrac{1}{2}$	5	$5\tfrac{1}{2}$	6	$6\tfrac{1}{2}$	7	$7\tfrac{1}{2}$	8
then $y =$	0	1	$2\tfrac{1}{4}$	4	$6\tfrac{1}{4}$	9	$12\tfrac{1}{4}$	16	$20\tfrac{1}{4}$	25	$30\tfrac{1}{4}$	36	$42\tfrac{1}{4}$	49	$56\tfrac{1}{4}$	64

Fig. 12-2

In the table, each x value is the principal square root of the corresponding y value; that is, $x = \sqrt{y}$. On the graph, the x value of each point is the principal square root of its y value. See Fig. 12-3.

To Find the Square Root of a Number Graphically

Find $\sqrt{27}$ graphically.

Procedure:

1. Find the number on the y axis:
2. From the number proceed horizontally to the curve:
3. From the curve proceed vertically to the x axis:
4. Read the approximate square root value on the x axis:

Solution:

1. Find 27 on the y axis.
2. From 27 follow the horizontal line to point A on the curve.
3. From A follow the vertical line to the x axis.
4. Read 5.2 on the x axis.

Ans. $\sqrt{27} = 5.2$ approximately.

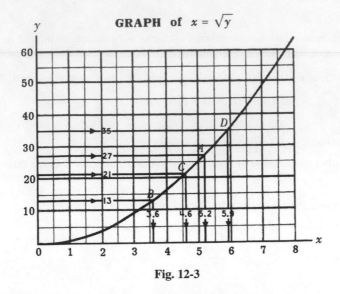

Fig. 12-3

To find the square of a number graphically, proceed in the reverse direction; from the x axis to the curve, then to the y axis.

Thus, to square 7.2, follow this path: $x = 7.2$ to curve, then to $y = 52$.

Ans. $7.2 = \sqrt{52}$ approximately.

Alternately, you may use a graphing calculator to sketch $x = \sqrt{y}$.

3.1. FINDING SQUARE ROOTS GRAPHICALLY

From the graph of $x = \sqrt{y}$, find the square root of each number to the nearest tenth, indicating the path followed:

	Number	Square Root	Path
a)	13	3.6	$y = 13$ to B to $x = 3.6$
b)	21	4.6	$y = 21$ to C to $x = 4.6$
c)	35	5.9	$y = 35$ to D to $x = 5.9$

4.   FINDING THE SQUARE ROOT OF A NUMBER BY USING A TABLE OR CALCULATOR

Approximate square roots of numbers can be obtained by using a calculator. The square root values obtained from a calculator are more precise than those from the graph of $x = \sqrt{y}$.

Multiplying or Dividing the Radicand in a Square Root by 100

Rule 1.   If the radicand of a square root is multiplied by 100, the square root is multiplied by 10.

Rule 2.   If the radicand of a square root is divided by 100, the square root is divided by 10.

Thus, apply rule 1.   Thus, apply rule 2:

$$\sqrt{9} = 3 \qquad\qquad\qquad\qquad \sqrt{9} = 3$$

$$\sqrt{9\,00} = 30 \qquad\qquad\qquad \sqrt{0.09} = 0.3$$

$$\sqrt{9\,00\,00} = 300 \qquad\qquad \sqrt{0.00\,09} = 0.03$$

$$\sqrt{9\,00\,00\,00} = 3000 \qquad \sqrt{0.00\,00\,09} = 0.003$$

NOTE: Each underlining above indicates a multiplication or division of the radicand by 100.

4.1.   FINDING SQUARE ROOTS BY USING A CALCULATOR
Find the approximate value of each square root, using a calculator.

a) $\sqrt{5}$ *Ans.* 2.24   c) $\sqrt{500}$ *Ans.* 22.36   e) $\sqrt{562}$ *Ans.* 23.71

b) $\sqrt{50}$ 7.07   d) $\sqrt{262}$ 16.19   f) $\sqrt{862}$ 29.36

4.2.   EVALUATING BY USING A SQUARE ROOT TABLE
Find the value of each to the nearest hundredth, using a calculator.

a) $3\sqrt{2}$ b) $2\sqrt{300} - 25$ c) $\dfrac{\sqrt{45}}{3} + \dfrac{1}{\sqrt{2}}$

Solutions:

a)   Since $\sqrt{2} = 1.414$,   b)   Since $\sqrt{300} = 17.32$,   c)   Since $\sqrt{45} = 6.708$ and

$3\sqrt{2} = 3(1.414)$ $2\sqrt{300} - 25$ $\sqrt{2} = 1.414,\ \dfrac{\sqrt{45}}{3} + \dfrac{1}{\sqrt{2}}$

$= 4.242$ $= 2(17.32) - 25$ $= \dfrac{6.708}{3} + \dfrac{1}{1.414}$

Ans.   4.24 $= 9.64$ $= 2.236 + 0.707$

                             Ans.   9.64 $= 2.943$

                                                       Ans.   2.94

4.3.   RULE 1. MULTIPLYING THE RADICAND OF A SQUARE ROOT BY A MULTIPLE OF 100
Using a calculator and applying **rule 1**, find the approximate value of each square root. Round calculator values to the nearest thousandth.

a) $\sqrt{50,000}$ and $\sqrt{5,000,000}$ b) $\sqrt{4000}$ and $\sqrt{400,000}$ c) $\sqrt{53,400}$ and $\sqrt{5,340,000}$

Solutions:

a)   Since $\sqrt{5} = 2.236$,   b) Since $\sqrt{40} = 6.325$,   c)   Since $\sqrt{534} = 23.108$

$\sqrt{5\ 00\ 00} = 223.6$ *Ans.* $\sqrt{40\ 00} = 63.25$ *Ans.* $\sqrt{534\ 00} = 231.08$ *Ans.*

$\sqrt{5\ 00\ 00\ 00} = 2236.$ *Ans.* $\sqrt{40\ 00\ 00} = 632.5$ *Ans.* $\sqrt{534\ 00\ 00} = 2310.8$ *Ans.*

4.4.   RULE 2. DIVIDING THE RADICAND OF A SQUARE ROOT BY A MULTIPLE OF 100
Using a calculator and applying rule 2, find the approximate value of each square root. Round calculator values to the nearest thousandth.

a) $\sqrt{2.55}$ and $\sqrt{0.0255}$ b) $\sqrt{0.37}$ and $\sqrt{0.0037}$ c) $\sqrt{0.05}$ and $\sqrt{0.0005}$

Solutions:

a) Since $\sqrt{255} = 15.969$,   b) Since $\sqrt{37} = 6.083$,   c) Since $\sqrt{5} = 2.236$,

$\sqrt{2.55} = 1.5969$ *Ans.* $\sqrt{0.37} = 0.6083$ *Ans.* $\sqrt{0.05} = 0.2236$ *Ans.*

$\sqrt{0.02\ 55} = 0.15969$ *Ans.* $\sqrt{0.00\ 37} = 0.06083$ *Ans.* $\sqrt{0.00\ 05} = 0.02236$ *Ans.*

5.   SIMPLIFYING THE SQUARE ROOT OF A PRODUCT

$$\boxed{\ \sqrt{ab} = \sqrt{a}\sqrt{b} \qquad \sqrt{abc} = \sqrt{a}\sqrt{b}\sqrt{c}\ }$$

Rule 1. The square root of a product of two or more numbers equals the product of the separate square roots of these numbers.

$$\text{Thus, } \sqrt{3600} = \sqrt{(36)(100)} = \sqrt{36}\sqrt{100} = (6)(10) = 60$$

To Simplify the Square Root of a Product

Simplify $\sqrt{72}$.

Procedure:

1. Factor the radicand, choosing perfect-square factors:

2. Form a separate square root for each factor:

3. Extract the square roots of each perfect square:

4. Multiply the factors outside the radical:

Solution:

1. $\sqrt{(4)(9)(2)}$

2. $\sqrt{4}\sqrt{9}\sqrt{2}$

3. $(2)(3)\sqrt{2}$

4. $6\sqrt{2}$

Simplifying Square Roots of Powers

Rule 2. To find the square root of a power, keep the base and take one-half of the exponent.

$$\text{Thus, } \sqrt{x^6} = x^3 \text{ since } x^3 \cdot x^3 = x^6.$$

Rule 3. To find the square root of the product of powers, keep each base and take one-half of the exponent.

$$\text{Thus, } \sqrt{x^2 y^4} = xy^2 \text{ since } \sqrt{x^2 y^4} = \sqrt{x^2}\,\sqrt{y^4} = xy^2.$$

5.1. **RULE 1.   FINDING EXACT SQUARE ROOT THROUGH SIMPLIFICATION**

Find each square root through simplification:

$a)$ $\sqrt{1600}$ $b)$ $\sqrt{256}$ $c)$ $\sqrt{225}$ $d)$ $\sqrt{324}$ $e)$ $\frac{1}{2}\sqrt{576}$

Solutions:

$a)$ $\sqrt{(16)(100)}$ $b)$ $\sqrt{(4)(64)}$ $c)$ $\sqrt{(25)(9)}$ $d)$ $\sqrt{(9)(36)}$ $e)$ $\frac{1}{2}\sqrt{(4)(144)}$

    $\sqrt{16}\sqrt{100}$ $\sqrt{4}\sqrt{64}$ $\sqrt{25}\sqrt{9}$ $\sqrt{9}\sqrt{36}$ $\frac{1}{2}\sqrt{4}\sqrt{144}$

      $(4)(10)$ $(2)(8)$ $(5)(3)$ $(3)(6)$ $\frac{1}{2}(2)(12)$

Ans.   40   16   15   18   12

5.2. **RULE 1.   SIMPLIFYING SQUARE ROOTS OF NUMBERS**

Simplify:

$a)$ $\sqrt{75}$ $b)$ $\sqrt{90}$ $c)$ $\sqrt{112}$ $d)$ $5\sqrt{288}$ $e)$ $\frac{1}{3}\sqrt{486}$ $f)$ $\frac{1}{2}\sqrt{1200}$

Solutions:

$a)$ $\sqrt{(25)(3)}$ $c)$ $\sqrt{(16)(7)}$ $e)$ $\frac{1}{3}\sqrt{(81)(6)}$

      $\sqrt{25}\sqrt{3}$ $\sqrt{16}\sqrt{7}$ $\frac{1}{3}\sqrt{81}\sqrt{6}$

Ans. $5\sqrt{3}$ $4\sqrt{7}$ $3\sqrt{6}$

$b)$ $\sqrt{(9)(10)}$ $d)$ $5\sqrt{(144)(2)}$ $f)$ $\frac{1}{2}\sqrt{(400)(3)}$
      $\sqrt{9}\sqrt{10}$ $5\sqrt{144}\sqrt{2}$ $\frac{1}{2}\sqrt{400}\sqrt{3}$

Ans. $3\sqrt{10}$ $60\sqrt{2}$ $10\sqrt{3}$

5.3. EVALUATING SIMPLIFIED SQUARE ROOTS

Using $\sqrt{2} = 1.414$ or $\sqrt{3} = 1.732$, evaluate to the nearest tenth:

$a)$ $\sqrt{72}$ $b)$ $\frac{1}{3}\sqrt{243}$ $c)$ $3\sqrt{450}$ $d)$ $\frac{3}{4}\sqrt{4800}$ $e)$ $1.2\sqrt{363}$

Solutions:

$a)$ $\sqrt{36\cdot2}$ $b)$ $\frac{1}{3}\sqrt{81\cdot3}$ $c)$ $3\sqrt{9\cdot25\cdot2}$ $d)$ $\frac{3}{4}\sqrt{16\cdot100\cdot3}$ $e)$ $1.2\sqrt{121\cdot3}$

       $\sqrt{36}\sqrt{2}$ $\frac{1}{3}\sqrt{81}\sqrt{3}$ $3\sqrt{9}\sqrt{25}\sqrt{2}$ $\frac{3}{4}\sqrt{16}\sqrt{100}\sqrt{3}$ $1.2\sqrt{121}\sqrt{3}$

       $6\sqrt{2}$ $\frac{1}{3}(9)\sqrt{3}$ $3(3)(5)\sqrt{2}$ $\frac{3}{4}(4)(10)\sqrt{3}$ $1.2(11)\sqrt{3}$

       $6(1.414)$ $3(1.732)$ $45(1.414)$ $30(1.732)$ $(13.2)(1.732)$

Ans. 8.5 5.2 63.6 52.0 22.9

5.4. RULES 2 AND 3. SIMPLIFYING SQUARE ROOTS OF POWERS
Simplify:

$a)$ $\sqrt{49b^8}$ $b)$ $\sqrt{900c^{10}d^{16}}$ $c)$ $\sqrt{98b^6}$ $d)$ $2\sqrt{25t^3}$

Solutions:

$a)$ $\sqrt{49}\sqrt{b^8}$ $b)$ $\sqrt{900}\sqrt{c^{10}}\sqrt{d^{16}}$ $c)$ $\sqrt{49}\sqrt{2}\sqrt{b^6}$ $d)$ $2\sqrt{25}\sqrt{t^2}\sqrt{t}$

Ans. $7b^4$ $30c^5d^8$ $7b^3\sqrt{2}$ $10t\sqrt{t}$

6. SIMPLIFYING THE SQUARE ROOT OF A FRACTION

$$\sqrt{\frac{a}{b}} = \frac{\sqrt{a}}{\sqrt{b}}$$

RULE.   The square root of a fraction equals the square root of the numerator divided by the square root of the denominator.

$$\text{Thus,}\quad \sqrt{\frac{25}{64}} = \frac{\sqrt{25}}{\sqrt{64}} = \frac{5}{8}$$

To simplify the square root of a fraction whose denominator is not a perfect square, **change the fraction to an equivalent fraction** which has a denominator that is the smallest perfect square. Thus,

$$\sqrt{\frac{1}{8}} = \sqrt{\frac{2}{16}} = \sqrt{\frac{2}{16}} = \frac{\sqrt{2}}{4} \text{ or } \frac{1}{4}\sqrt{2}$$

6.1. FINDING EXACT SQUARE ROOTS INVOLVING FRACTIONS
Simplify:

$a)$ $\sqrt{2\frac{1}{4}}$ $b)$ $6\sqrt{\frac{1600}{9}}$ $c)$ $5\sqrt{\frac{144}{625}}$ $d)$ $\frac{8}{5}\sqrt{\frac{225}{256}}$ $e)$ $\sqrt{\frac{9}{49}}\sqrt{\frac{49}{400}}$

Solutions:

a) $\sqrt{\dfrac{9}{4}}$ b) $\dfrac{6\sqrt{1600}}{\sqrt{9}}$ c) $\dfrac{5\sqrt{144}}{\sqrt{625}}$ d) $\dfrac{8}{5}\cdot\dfrac{\sqrt{225}}{\sqrt{256}}$ e) $\dfrac{\sqrt{9}}{\sqrt{49}}\cdot\dfrac{\sqrt{49}}{\sqrt{400}}$

$\dfrac{\sqrt{9}}{\sqrt{4}}$ $\dfrac{6(40)}{3}$ $\dfrac{5(12)}{25}$ $\dfrac{8}{5}\cdot\dfrac{(15)}{(16)}$ $\dfrac{3}{7}\cdot\dfrac{7}{20}$

Ans. $\dfrac{3}{2}$ 80 $\dfrac{12}{5}$ $\dfrac{3}{2}$ $\dfrac{3}{20}$

6.2.   **SIMPLIFYING SQUARE ROOTS OF FRACTIONS WHOSE DENOMINATORS ARE PERFECT SQUARES**

Simplify:

a) $6\sqrt{\dfrac{7}{9}}$ b) $\dfrac{2}{3}\sqrt{\dfrac{11}{100}}$ c) $\dfrac{1}{5}\sqrt{\dfrac{75}{16}}$ d) $x\sqrt{\dfrac{200}{x^2}}$ e) $10\sqrt{\dfrac{y}{25}}$

Solutions:

a) $\dfrac{6\sqrt{7}}{\sqrt{9}}$ b) $\dfrac{2\sqrt{11}}{3\sqrt{100}}$ c) $\dfrac{\sqrt{25}\sqrt{3}}{5\sqrt{16}}$ d) $\dfrac{x\sqrt{100}\sqrt{2}}{\sqrt{x^2}}$ e) $\dfrac{10\sqrt{y}}{\sqrt{25}}$

$\dfrac{6\sqrt{7}}{3}$ $\dfrac{2\sqrt{11}}{3(10)}$ $\dfrac{5\sqrt{3}}{5(4)}$ $\dfrac{10x\sqrt{2}}{x}$ $\dfrac{10\sqrt{y}}{5}$

Ans. $2\sqrt{7}$ $\dfrac{\sqrt{11}}{15}$ $\dfrac{\sqrt{3}}{4}$ $10\sqrt{2}$ $2\sqrt{y}$

6.3.   **SIMPLIFYING SQUARE ROOTS OF FRACTIONS WHOSE DENOMINATORS ARE NOT PERFECT SQUARES**

Simplify:

a) $\sqrt{\dfrac{1}{5}}$ b) $\sqrt{\dfrac{3}{8}}$ c) $\sqrt{\dfrac{a}{3}}$ d) $\sqrt{\dfrac{x^2}{27}}$ e) $\sqrt{\dfrac{3}{y}}$

Solutions:

a) $\sqrt{\dfrac{1}{5}\cdot\dfrac{5}{5}}$ b) $\sqrt{\dfrac{3}{8}\cdot\dfrac{2}{2}}$ c) $\sqrt{\dfrac{a}{3}\cdot\dfrac{3}{3}}$ d) $\sqrt{\dfrac{x^2}{27}\cdot\dfrac{3}{3}}$ e) $\sqrt{\dfrac{3}{y}\cdot\dfrac{y}{y}}$

$\sqrt{\dfrac{5}{25}}$ $\sqrt{\dfrac{6}{16}}$ $\sqrt{\dfrac{3a}{9}}$ $\sqrt{\dfrac{3x^2}{81}}$ $\sqrt{\dfrac{3y}{y^2}}$

$\dfrac{\sqrt{5}}{\sqrt{25}}$ $\dfrac{\sqrt{6}}{\sqrt{16}}$ $\dfrac{\sqrt{3a}}{\sqrt{9}}$ $\dfrac{\sqrt{3x^2}}{\sqrt{81}}$ $\dfrac{\sqrt{3y}}{y^2}$

Ans. $\dfrac{\sqrt{5}}{5}$ $\dfrac{\sqrt{6}}{4}$ $\dfrac{\sqrt{3a}}{3}$ $\dfrac{x\sqrt{3}}{9}$ $\dfrac{\sqrt{3y}}{y}$

6.4.   **EVALUATING SQUARE ROOTS AFTER SIMPLIFICATION**

Using $\sqrt{3} = 1.732$ or $\sqrt{5} = 2.236$, evaluate each to the nearest hundredth:

a) $\sqrt{\dfrac{3}{4}}$ b) $\sqrt{\dfrac{4}{3}}$ c) $30\sqrt{\dfrac{20}{9}}$ d) $10\sqrt{\dfrac{4}{5}}$ e) $\sqrt{\dfrac{49}{20}}$

Solutions:

$$a)\quad \frac{\sqrt{3}}{\sqrt{4}} \qquad b)\quad \sqrt{\frac{4}{3}\cdot\frac{3}{3}} \qquad c)\quad \frac{30\sqrt{20}}{\sqrt{9}} \qquad d)\quad 10\sqrt{\frac{4}{5}\cdot\frac{5}{5}} \qquad e)\quad \sqrt{\frac{49}{20}\cdot\frac{5}{5}}$$

$$\frac{\sqrt{3}}{2} \qquad \frac{\sqrt{4}\sqrt{3}}{\sqrt{9}} \qquad 10\sqrt{4}\sqrt{5} \qquad \frac{10\sqrt{4}\sqrt{5}}{5} \qquad \frac{\sqrt{49}\sqrt{5}}{\sqrt{100}}$$

$$\frac{1.732}{2} \qquad \frac{2}{3}(1.732) \qquad 20(2.236) \qquad 4(2.236) \qquad \frac{7}{10}(2.236)$$

Ans. 0.87   1.15   44.7   8.94   1.6

7. ADDING AND SUBTRACTING SQUARE ROOTS OF NUMBERS

Like radicals are radicals having the same index and the same radicand.

Thus:

Like Radicals	Unlike Radicals
$5\sqrt{3}$ and $2\sqrt{3}$	$5\sqrt{3}$ and $2\sqrt{5}$
$8\sqrt{x}$ and $-3\sqrt{x}$	$8\sqrt{x}$ and $-3\sqrt{y}$
$7\sqrt[3]{6}$ and $3\sqrt[3]{6}$	$7\sqrt[3]{6}$ and $3\sqrt{6}$

RULE: **To combine (add or subtract) like radicals,** keep the common radical and combine their coefficients.

Thus, $5\sqrt{3} + 2\sqrt{3} - 4\sqrt{3} = (5 + 2 - 4)\sqrt{3} = 3\sqrt{3}$.

NOTE: Combining like radicals involves the same process as that for combining like terms. Hence, to combine $5\sqrt{3}$ and $2\sqrt{3}$, think of combining $5x$ and $2x$ when $x = \sqrt{3}$.

Unlike radicals may be combined into one radical if like radicals can be obtained by simplifying. Thus, $\sqrt{50} + \sqrt{32} = 5\sqrt{2} + 4\sqrt{2} = 9\sqrt{2}$.

7.1. COMBINING LIKE RADICALS
Combine. (Keep radical and combine coefficients.)

$$a)\ 3\sqrt{2} + 2\sqrt{2} \qquad d)\ \tfrac{1}{2}\sqrt{5} + 1\tfrac{1}{2}\sqrt{5} \qquad g)\ 2x\sqrt{7} + 3x\sqrt{7}$$

$$b)\ 5\sqrt{3} - 10\sqrt{3} \qquad e)\ 7\tfrac{3}{4}\sqrt{10} - 5\tfrac{3}{4}\sqrt{10} \qquad h)\ 20\sqrt{a} - 13\sqrt{a}$$

$$c)\ \sqrt{6} + 9\sqrt{6} \qquad f)\ \sqrt{2x} + 3\sqrt{2x} \qquad i)\ 5a\sqrt{y} - 2a\sqrt{y}$$

Solutions:

Ans.
$$a)\ (3+2)\sqrt{2} = 5\sqrt{2} \qquad d)\ (\tfrac{1}{2}+1\tfrac{1}{2})\sqrt{5} = 2\sqrt{5} \qquad g)\ (2x+3x)\sqrt{7} = 5x\sqrt{7}$$

$$b)\ (5-10)\sqrt{3} = -5\sqrt{3} \qquad e)\ (7\tfrac{3}{4} - 5\tfrac{3}{4})\sqrt{10} = 2\sqrt{10} \qquad h)\ (20-13)\sqrt{a} = 7\sqrt{a}$$

$$c)\ (1+9)\sqrt{6} = 10\sqrt{6} \qquad f)\ (1+3)\sqrt{2x} = 4\sqrt{2x} \qquad i)\ (5a-2a)\sqrt{y} = 3a\sqrt{y}$$

7.2. COMBINING UNLIKE NUMERICAL RADICALS AFTER SIMPLIFICATION
Combine after simplifying:

$$a)\ \sqrt{5} + \sqrt{20} + \sqrt{45} \qquad b)\ \sqrt{27} + 3\sqrt{12} \qquad c)\ \sqrt{8} - 6\sqrt{\tfrac{1}{2}} \qquad d)\ \sqrt{4a} + \sqrt{25a} + \sqrt{\tfrac{a}{4}}$$

Solutions:

$a)$ $\sqrt{5} + 2\sqrt{5} + 3\sqrt{5}$ $b)$ $3\sqrt{3} + 3(2\sqrt{3})$ $c)$ $2\sqrt{2} - 6(\frac{1}{2}\sqrt{2})$ $d)$ $2\sqrt{a} + 5\sqrt{a} + \frac{1}{2}\sqrt{a}$

 $\phantom{\sqrt{5}+}$ $3\sqrt{3} + 6\sqrt{3}$ $2\sqrt{2} - 3\sqrt{2}$

Ans. $6\sqrt{5}$ $9\sqrt{3}$ $-\sqrt{2}$ $7\frac{1}{2}\sqrt{a}$

8. MULTIPLYING SQUARE ROOTS OF NUMBERS

$$\boxed{\sqrt{a}\sqrt{b} = \sqrt{ab} \qquad \sqrt{a}\sqrt{b}\sqrt{c} = \sqrt{abc}}$$

Rule 1.   The product of the square roots of two or more nonnegative numbers equals the square root of their product.

Thus, $\sqrt{2}\sqrt{3}\sqrt{6} = \sqrt{36} = 6$. However, $\sqrt{-5}\sqrt{-5} = -5$. $\sqrt{-5}\sqrt{-5} \neq \sqrt{25}$ or 5.

Rule 2.   The square of the square root of a number equals the number.

Thus, $(\sqrt{7})^2 = (\sqrt{7})(\sqrt{7}) = 7$. In general, $(\sqrt{x})^2 = x$. Hence, when you are squaring the square root of a number, merely eliminate the radical sign.

To Multiply Square Root Monomials

Multiply: $a)$ $3\sqrt{2} \cdot 2\sqrt{6}$ $b)$ $\frac{2}{3}\sqrt{2x} \cdot 6\sqrt{2y}$

Procedure:

1. **Multiply** coefficients and radicals separately:
2. **Multiply** the resulting products:
3. **Simplify**, if possible:

Solutions:

$a)$ $3 \cdot 2\sqrt{2}\sqrt{6}$ $b)$ $\frac{2}{3} \cdot 6\sqrt{2x}\sqrt{2y}$

 $6\sqrt{12}$ $4\sqrt{4xy}$

 $6(2\sqrt{3})$ $4 \cdot 2\sqrt{xy}$

Ans. $12\sqrt{3}$ **Ans.** $8\sqrt{xy}$

8.1. MULTIPLYING SQUARE ROOT MONOMIALS

Multiply:

$a)$ $\sqrt{2}\sqrt{5}\sqrt{6}\sqrt{15}$ $b)$ $(2\sqrt{3})(3\sqrt{5})(5\sqrt{10})$ $c)$ $\sqrt{\dfrac{x}{7}}\sqrt{14x}$ $d)$ $(4\sqrt{m})(2\sqrt{25m})$

Solutions:

$a)$ $\sqrt{2 \cdot 5 \cdot 6 \cdot 15}$ $b)$ $2 \cdot 3 \cdot 5\sqrt{3 \cdot 5 \cdot 10}$ $c)$ $\sqrt{\left(\dfrac{x}{7}\right)(14x)}$ $d)$ $4 \cdot 2\sqrt{(m)(25m)}$

 $\sqrt{900}$ $30\sqrt{150} = 30(5\sqrt{6})$ $\sqrt{2x^2}$ $8\sqrt{25m^2} = 8(5m)$

Ans. 30 $150\sqrt{6}$ $x\sqrt{2}$ $40m$

8.2. RULE 2. SQUARING LIKE RADICALS

Multiply:

$a)$ $(2\sqrt{17})(3\sqrt{17})$ $b)$ $\left(3\sqrt{\dfrac{2}{3}}\right)^2$ $c)$ $\left(\dfrac{5}{2}\sqrt{8}\right)^2$ $d)$ $(2a\sqrt{2a})^2$ $e)$ $\left(\dfrac{x}{3}\sqrt{\dfrac{3}{x}}\right)^2$

Solutions:

a) $2 \cdot 3(\sqrt{17})^2$ b) $3^2\left(\sqrt{\dfrac{2}{3}}\right)^2$ c) $\left(\dfrac{5}{2}\right)^2(\sqrt{8})^2$ d) $(2a)^2(\sqrt{2a})^2$ e) $\left(\dfrac{x}{3}\right)^2\left(\sqrt{\dfrac{3}{x}}\right)^2$

     $6(17)$ $9\left(\dfrac{2}{3}\right)$ $\dfrac{25}{4}(8)$ $4a^2(2a)$ $\dfrac{x^2}{9}\left(\dfrac{3}{x}\right)$

Ans.   105   6   50 $8a^3$ $\dfrac{x}{3}$

8.3.   **MULTIPLYING A MONOMIAL BY A POLYNOMIAL**

Multiply:

a) $\sqrt{3}(\sqrt{3} + \sqrt{27})$ b) $3\sqrt{2}(\sqrt{2} + \sqrt{5})$ c) $\dfrac{\sqrt{5}}{2}(4\sqrt{3} - 6\sqrt{20})$

Solutions:

a) $\sqrt{3}\sqrt{3} + \sqrt{3}\sqrt{27}$ b) $3\sqrt{2}\sqrt{2} + 3\sqrt{2}\sqrt{5}$ c) $\left(\dfrac{\sqrt{5}}{2}\right)(4\sqrt{3}) - \left(\dfrac{\sqrt{5}}{2}\right)(6\sqrt{20})$

    $3 + \sqrt{81}$ $3(2) + 3\sqrt{10}$ $2\sqrt{15} - 3\sqrt{100}$

Ans.   12 $6 + 3\sqrt{10}$ $\overset{2}{\sqrt{}}15 - 30$

8.4.   **MULTIPLYING BINOMIALS CONTAINING RADICALS**

Multiply:

a) $(5 + \sqrt{2})(5 - \sqrt{2})$ b) $(\sqrt{10} + \sqrt{3})(\sqrt{10} - \sqrt{3})$ c) $(\sqrt{7} + \sqrt{2})(\sqrt{7} + \sqrt{2})$

Procedure:	**Solutions:**		
1. Multiply first terms:	a) $5 \cdot 5 = 25$	b) $\sqrt{10}\sqrt{10} = 10$	c) $\sqrt{7}\sqrt{7} = 7$
2. Add products of outer and inner terms:	$5\sqrt{2} - 5\sqrt{2} = 0$	$\sqrt{30} - \sqrt{30} = 0$	$\sqrt{14} + \sqrt{14} = 2\sqrt{14}$
3. Multiply last terms:	$(\sqrt{2})(-\sqrt{2}) = -2$	$(\sqrt{3})(-\sqrt{3}) = -3$	$\sqrt{2}\sqrt{2} = 2$
4. Combine results:	*Ans.* 23	*Ans.* 7	*Ans.* $9 + 2\sqrt{14}$

9.   DIVIDING SQUARE ROOTS OF NUMBERS

$$\boxed{\dfrac{\sqrt{a}}{\sqrt{b}} = \sqrt{\dfrac{a}{b}} \text{ assuming } b \neq 0}$$

RULE: The square root of a number **divided** by the square root of another number equals the square root of their quotient.

$$\text{Thus, } \dfrac{\sqrt{6}}{\sqrt{2}} = \sqrt{\dfrac{6}{2}} = \sqrt{3}.$$

To Divide Square Root Monomials

Divide: a) $\dfrac{14\sqrt{40}}{2\sqrt{5}}$ b) $\dfrac{6\sqrt{x^4}}{3\sqrt{x}}$

Procedure: **Solutions:**

1. **Divide** coefficients and radicals separately:

$$a) \quad \frac{14}{2}\sqrt{\frac{40}{5}} \qquad\qquad b) \quad \frac{6}{3}\sqrt{\frac{x^4}{x}}$$

2. **Multiply** the resulting quotients:

$$7\sqrt{8} \qquad\qquad 2\sqrt{x^3}$$

3. **Simplify**, if possible:

$$7(2\sqrt{2}) \qquad\qquad 2(\sqrt{x^2}\sqrt{x})$$

$$\textit{Ans.} \quad 14\sqrt{2} \qquad\qquad \textit{Ans.} \quad 2x\sqrt{x}$$

9.1.   **DIVIDING SQUARE ROOT MONOMIALS**
Divide:

$$a) \quad \frac{8\sqrt{8}}{2\sqrt{2}} \qquad b) \quad \frac{6\sqrt{2}}{12\sqrt{8}} \qquad c) \quad \frac{10\sqrt{56}}{4\sqrt{7}} \qquad d) \quad \frac{9\sqrt{x^3}}{3\sqrt{x}} \qquad e) \quad \frac{5a\sqrt{5a}}{10a\sqrt{10a}}$$

Solutions:

$$a) \quad \frac{8}{2}\sqrt{\frac{8}{2}} \qquad b) \quad \frac{6}{12}\sqrt{\frac{2}{8}} \qquad c) \quad \frac{10}{4}\sqrt{\frac{56}{7}} \qquad d) \quad \frac{9}{3}\sqrt{\frac{x^3}{x}} \qquad e) \quad \frac{5a}{10a}\sqrt{\frac{5a}{10a}}$$

$$4\sqrt{4} = 4(2) \qquad \frac{1}{2}\sqrt{\frac{1}{4}} = \frac{1}{2}\left(\frac{1}{2}\right) \qquad \frac{5}{2}\sqrt{8} = \frac{5}{2}(2\sqrt{2}) \quad 3\sqrt{x^2} \qquad \frac{1}{2}\sqrt{\frac{1}{2}} = \frac{1}{2}\left(\frac{1}{2}\sqrt{2}\right)$$

$$\textit{Ans.} \quad 8 \qquad\qquad \frac{1}{4} \qquad\qquad\qquad 5\sqrt{2} \qquad\qquad 3x \qquad\qquad \frac{1}{4}\sqrt{2}$$

9.2.   **DIVIDING A POLYNOMIAL BY A MONOMIAL**
Divide:

$$a) \quad \frac{\sqrt{50} + \sqrt{98}}{\sqrt{2}} \qquad b) \quad \frac{2\sqrt{24} + \sqrt{96}}{\sqrt{3}} \qquad c) \quad \frac{18\sqrt{40} - 10\sqrt{80}}{2\sqrt{10}} \qquad d) \quad \frac{15\sqrt{y^5} - 6\sqrt{y^3}}{3\sqrt{y}}$$

Solutions:

$$a) \quad \frac{\sqrt{50}}{\sqrt{2}} + \frac{\sqrt{98}}{\sqrt{2}} \qquad b) \quad \frac{2\sqrt{24}}{\sqrt{3}} + \frac{\sqrt{96}}{\sqrt{3}} \qquad c) \quad \frac{18\sqrt{40}}{2\sqrt{10}} - \frac{10\sqrt{80}}{2\sqrt{10}} \qquad d) \quad \frac{15\sqrt{y^5}}{3\sqrt{y}} - \frac{6\sqrt{y^3}}{3\sqrt{y}}$$

$$\sqrt{25} + \sqrt{49} \qquad 2\sqrt{8} + \sqrt{32} \qquad 9\sqrt{4} - 5\sqrt{8} \qquad 5\sqrt{\frac{y^5}{y}} - 2\sqrt{\frac{y^3}{y}}$$

$$5 + 7 \qquad\qquad 2(2\sqrt{2}) + 4\sqrt{2} \qquad 18 - 5(2\sqrt{2}) \qquad 5\sqrt{y^4} - 2\sqrt{y^2}$$

$$\textit{Ans.} \quad 12 \qquad\qquad 8\sqrt{2} \qquad\qquad 18 - 10\sqrt{2} \qquad\qquad 5y^2 - 2y$$

10.   RATIONALIZING THE DENOMINATOR OF A FRACTION

To rationalize the denominator of a fraction is to change the denominator from an irrational number to a rational number. To do this when the denominator is a monomial surd, multiply both terms of the fraction by the smallest surd which will make the denominator rational.

Thus, to rationalize the denominator of $\dfrac{4}{\sqrt{8}}$, multiply by $\dfrac{\sqrt{2}}{\sqrt{2}}$:

$$\frac{4}{\sqrt{8}} \cdot \frac{\sqrt{2}}{\sqrt{2}} = \frac{4\sqrt{2}}{\sqrt{16}} = \frac{4\sqrt{2}}{4} = \sqrt{2}$$

Rationalizing the denominator simplifies evaluation.

Thus, $\dfrac{4}{\sqrt{8}} = \sqrt{2} = 1.414$. Otherwise, $\dfrac{4}{\sqrt{8}} = \dfrac{4}{2.828}$, thus leading to a lengthy division.

10.1. RATIONALIZING THE DENOMINATOR

Rationalize each denominator:

a) $\dfrac{12}{\sqrt{6}}$ c) $\dfrac{24}{\sqrt{8}}$ e) $\dfrac{1}{\sqrt{c^3}}$

b) $\dfrac{\sqrt{3}}{\sqrt{10}}$ d) $\dfrac{3}{\sqrt{x}}$ f) $\dfrac{3+\sqrt{2}}{\sqrt{2}}$

Solutions:

a) $\dfrac{12}{\sqrt{6}} \cdot \dfrac{\sqrt{6}}{\sqrt{6}}$ c) $\dfrac{24}{\sqrt{8}} \cdot \dfrac{\sqrt{2}}{\sqrt{2}}$ e) $\dfrac{1}{\sqrt{c^3}} \cdot \dfrac{\sqrt{c}}{\sqrt{c}}$

$\dfrac{12\sqrt{6}}{6}$ $\dfrac{24\sqrt{2}}{\sqrt{16}}$ $\dfrac{\sqrt{c}}{\sqrt{c^4}}$

Ans. $2\sqrt{6}$ Ans. $6\sqrt{2}$ Ans. $\dfrac{\sqrt{c}}{c^2}$

b) $\dfrac{\sqrt{3}}{\sqrt{10}} \cdot \dfrac{\sqrt{10}}{\sqrt{10}}$ d) $\dfrac{3}{\sqrt{x}} \cdot \dfrac{\sqrt{x}}{\sqrt{x}}$ f) $\dfrac{\sqrt{2}}{\sqrt{2}} \cdot \dfrac{3+\sqrt{2}}{\sqrt{2}}$

Ans. $\dfrac{\sqrt{30}}{10}$ Ans. $\dfrac{3\sqrt{x}}{x}$ $\dfrac{\sqrt{2}(3+\sqrt{2})}{2}$

Ans. $\dfrac{3\sqrt{2}+2}{2}$

10.2. EVALUATING FRACTIONS WITH IRRATIONAL DENOMINATORS

Evaluate to the nearest tenth, after rationalizing the denominator:

a) $\dfrac{7}{\sqrt{7}}$ b) $\dfrac{20}{\sqrt{50}}$ c) $\dfrac{3}{\sqrt{6}}$ d) $\dfrac{1}{\sqrt{12}}$ e) $\dfrac{3+\sqrt{5}}{\sqrt{5}}$

Solutions:

a) $\dfrac{7}{\sqrt{7}} \cdot \dfrac{\sqrt{7}}{\sqrt{7}}$ b) $\dfrac{20}{\sqrt{50}} \cdot \dfrac{\sqrt{2}}{\sqrt{2}}$ c) $\dfrac{3}{\sqrt{6}} \cdot \dfrac{\sqrt{6}}{\sqrt{6}}$ d) $\dfrac{1}{\sqrt{12}} \cdot \dfrac{\sqrt{3}}{\sqrt{3}}$ e) $\dfrac{3+\sqrt{5}}{\sqrt{5}} \cdot \dfrac{\sqrt{5}}{\sqrt{5}}$

$\dfrac{7\sqrt{7}}{7}$ $\dfrac{20\sqrt{2}}{10}$ $\dfrac{3\sqrt{6}}{6}$ $\dfrac{\sqrt{3}}{\sqrt{36}}$ $\dfrac{\sqrt{5}(3+\sqrt{5})}{5}$

$\sqrt{7}$ $2\sqrt{2}$ $\dfrac{\sqrt{6}}{2}$ $\dfrac{\sqrt{3}}{6}$ $\dfrac{3\sqrt{5}+5}{5}$

2.646 $2(1.414)$ $\dfrac{2.449}{2}$ $\dfrac{1.732}{6}$ $\dfrac{3(2.236)+5}{5}$

Ans. 2.6 2.8 1.2 0.3 2.3

11. SOLVING RADICAL EQUATIONS

Radical equations are equations in which the unknown is included in a radicand.

Thus, $2\sqrt{x} + 5 = 9$ is a radical equation.

but $2x + \sqrt{5} = 9$ is not a radical equation,

To Solve a Radical Equation

Solve: $\sqrt{2x} + 5 = 9$

Procedure:

1. **Isolate** the term containing the radical:

2. **Square** both sides:
3. **Solve** for the unknown:
4. **Check** the roots obtained in the original equation:

Solution:

1. $\quad\quad \sqrt{2x} + 5 = 9$
    $\quad\quad\quad\quad \sqrt{2x} = 4$

2. By squaring, $2x = 16$
3. $\quad\quad\quad\quad\quad x = 8$
4. $\quad$ Check for $x = 8$:

$$\sqrt{2x} + 5 \overset{?}{=} 9$$
$$\sqrt{16} + 5 \overset{?}{=} 9$$
$$9 = 9 \quad Ans. \ x = 8$$

NOTE: In the solution of equations, the symbol "**Sq**" means "square both sides."

An extraneous root of an equation is a value of the unknown which satisfies, **not the original equation**, but a later equation obtained from the original equation. It must be rejected.

Thus, if $\sqrt{2x} = -4$, when both sides are squared, $2x = 16$ or $x = 8$.

However, $x = 8$ does not satisfy $\sqrt{2x} = -4$, the original equation.

Hence, 8 is an *extraneous root* of $\sqrt{2x} = -4$ and must be rejected.

11.1. SOLVING SIMPLE RADICAL EQUATIONS

Solve and check:

a) $\sqrt{5x} = 10$ 　　　　*b)* $\sqrt{\dfrac{x}{3}} = 2$ 　　　*c)* $\sqrt{7x + 5} = 3$ 　　　*d)* $\sqrt{2x - 8} = -3$

Solutions:

a) By squaring,

$\quad\quad 5x = 100$

$\quad\quad\quad x = 20$

Check: $\quad \sqrt{5x} = 10$

$\quad\quad\quad\quad \sqrt{100} \overset{?}{=} 10$

$\quad\quad\quad\quad\quad 10 = 10$

$Ans.\quad\quad x = 20$

b) By squaring,

$\quad\quad \dfrac{x}{3} = 4$

$\quad\quad\quad x = 12$

Check: $\quad \sqrt{\dfrac{x}{3}} = 2$

$\quad\quad\quad\quad \sqrt{4} \overset{?}{=} 2$

$\quad\quad\quad\quad 2 = 2$

$\quad\quad Ans.\quad\quad x = 12$

c) By squaring,

$$7x + 5 = 9$$

$$x = \frac{4}{7}$$

Check: $\sqrt{7x + 5} = 3$

$$\sqrt{4 + 5} \stackrel{?}{=} 3$$

$$3 = 3$$

Ans. $x = \dfrac{4}{7}$

d) By squaring,

$$4(x - 8) = 9$$

$$x = \frac{41}{4}$$

Check: $2\sqrt{x - 8} = -3$

$$2\sqrt{\frac{9}{4}} \stackrel{?}{=} -3$$

$$3 \neq -3$$

Ans.   Reject $x = \dfrac{41}{4}$. (No root.)

11.2. ISOLATING THE RADICAL IN SOLVING RADICAL EQUATIONS

Solve and check:

a) $3\sqrt{x} - 6 = 9$

b) $5\sqrt{x} = \sqrt{x} - 12$

c) $\sqrt{3y + 4} - 2 = 3$

Solutions: (Isolate the radical term first.)

$\mathbf{A_6}$ $3\sqrt{x} - 6 = 9$

$\mathbf{D_3}$ $3\sqrt{x} = 15$

$\mathbf{Sq}$ $\sqrt{x} = 5$

                $x = 25$

Check: $3\sqrt{x} - 6 = 9$

        $3\sqrt{25} - 6 \stackrel{?}{=} 9$

              $9 = 9$

Ans. $x = 25$

$\mathbf{S_{\sqrt{x}}}$ $5\sqrt{x} = \sqrt{x} - 12$

$\mathbf{D_4}$ $4\sqrt{x} = -12$

$\mathbf{Sq}$ $\sqrt{x} = -3$

            $x = 9$

Check: $5\sqrt{x} = \sqrt{x} - 12$

          $5\sqrt{9} \stackrel{?}{=} \sqrt{9} - 12$

            $15 \neq -9$

Ans.   Reject $x = 9$.   (No root.)

$\mathbf{A_2}$ $\sqrt{3y + 4} - 2 = 3$

$\mathbf{Sq}$ $\sqrt{3y + 4}$ $= 5$

            $3y + 4 = 25$

                $y = 7$

Check: $\sqrt{3y + 4} - 2 = 3$

          $\sqrt{25} - 2 \stackrel{?}{=} 3$

              $3 = 3$

Ans. $y = 7$

11.3. TRANSFORMING RADICAL EQUATIONS

Solve for the letter indicated:

a) Solve for x:

$$\sqrt{\frac{x}{2}} = y$$

b) Solve for y:

$$\sqrt{2y - 5} = x$$

c) Solve for A:

$$r = \sqrt{\frac{A}{\pi}}$$

d) Solve for x:

$$\sqrt{x + 4} = y + 2$$

Solutions:

a) By squaring,

$$\frac{x}{2} = y^2$$

Ans. $x = 2y^2$

b) By squaring,

$$2y - 5 = x^2$$

Ans. $y = \dfrac{x^2 + 5}{2}$

c) By squaring,

$$r^2 = \frac{A}{\pi}$$

Ans. $A = \pi r^2$

d) By squaring,

$$x + 4 = y^2 + 4y + 4$$

Ans. $x = y^2 + 4y$

Supplementary Problems

12.1. State the square roots of: **(1.1)**

        a) 100 b) $100y^2$ c) $4c^2d^2$ d) $\dfrac{4}{81}$ e) $\dfrac{25r^2}{36s^2}$ f) $\dfrac{x^4z^{16}}{y^6}$

    Ans. a) ±10 b) $\pm10y$ c) $\pm2cd$ d) $\pm\dfrac{2}{9}$ e) $\pm\dfrac{5r}{6s}$ f) $\pm\dfrac{x^2z^8}{y^3}$

12.2. Find each principal square root: **(1.2)**

    a) $\sqrt{400}$ b) $\sqrt{0.04}$ c) $\sqrt{1600h^2}$ d) $\sqrt{1\frac{9}{16}}$ e) $\sqrt{\dfrac{16r^2}{s^2t^2}}$ f) $\sqrt{\dfrac{64m^8}{p^{14}}}$

    Ans. a) 20 b) 0.2 c) $40h$ d) $\dfrac{5}{4}$ e) $\dfrac{4r}{st}$ f) $\dfrac{8m^4}{p^7}$

12.3. Find each negative square root: **(1.3)**

    a) $-\sqrt{\dfrac{1}{49}}$ b) $-\sqrt{a^2b^4}$ c) $-\sqrt{r^{10}t^{20}}$ d) $-\sqrt{\dfrac{121}{25w^{12}}}$ e) $-\sqrt{\dfrac{225x^{16}}{y^{36}}}$

    Ans. a) $-\dfrac{1}{7}$ b) $-ab^2$ c) $-r^5t^{10}$ d) $-\dfrac{11}{5w^6}$ e) $-\dfrac{15x^8}{y^{18}}$

12.4. Find the value of: **(1.4)**

    a) $\sqrt{81}-\sqrt{25}$ *Ans.* 4   d) $\sqrt{\dfrac{4}{9}}+\sqrt{\dfrac{1}{9}}$ *Ans.* 1   g) $4\sqrt{x^2}-x\sqrt{4}$ *Ans.* $2x$

    b) $2\sqrt{9}+3\sqrt{4}$ 12   e) $8\sqrt{\dfrac{1}{16}}+4\sqrt{\dfrac{9}{16}}$ 5   h) $\sqrt{16x^2}+16\sqrt{x^2}$ $20x$

    c) $\frac{1}{2}\sqrt{36}-\frac{1}{3}\sqrt{81}$ 0   f) $\left(-\sqrt{\dfrac{4}{9}}\right)\left(-\sqrt{\dfrac{81}{100}}\right)$ $\dfrac{3}{5}$ i) $\sqrt{\dfrac{16}{x^2}}+\dfrac{16}{\sqrt{x^2}}$ $\dfrac{20}{x}$

12.5. Express each of the following rational numbers as the ratio of two integers: **(2.2)**

    a) 5 b) -13 c) 0.07 d) $-11\frac{2}{3}$ e) $\dfrac{\sqrt{25}}{3}$ f) $\dfrac{17}{\sqrt{144}}$ g) $\dfrac{16-\sqrt{16}}{\sqrt{25}}$ h) $\sqrt{\dfrac{50}{32}}$

    Ans. a) $\dfrac{5}{1}$ b) $\dfrac{-13}{1}$ c) $\dfrac{7}{100}$ d) $\dfrac{-35}{3}$ e) $\dfrac{5}{3}$ f) $\dfrac{17}{12}$ g) $\dfrac{12}{5}$ h) $\dfrac{5}{4}$

12.6. From the graph of $x=\sqrt{y}$, find the square root of each number to the nearest tenth: **(3.1)**

    a) $\sqrt{17}$ b) $\sqrt{23}$ c) $\sqrt{37}$ d) $\sqrt{45}$ e) $\sqrt{48}$ f) $\sqrt{58}$ g) $\sqrt{61}$

    Ans. a) 4.1   b) 4.8   c) 6.1   d) 6.7   e) 6.9   f) 7.6   g) 7.8

12.7. Find the approximate value of each square root, to the nearest hundredth using a calculator: **(4.1)**

    a) $\sqrt{7}$ d) $\sqrt{46}$ g) $\sqrt{275}$ j) $\sqrt{528}$ m) $\sqrt{780}$

    b) $\sqrt{70}$ e) $\sqrt{94}$ h) $\sqrt{329}$ k) $\sqrt{576}$ n) $\sqrt{892}$

    c) $\sqrt{700}$ f) $\sqrt{138}$ i) $\sqrt{486}$ l) $\sqrt{747}$ o) $\sqrt{917}$

Ans. *a)* 2.65   *d)* 6.78   *g)* 16.58   *j)* 22.98   *m)* 27.93

       b) 8.37   *e)* 9.70   *h)* 18.14   *k)* 24.00   *n)* 29.87

       c) 26.46   *f)* 11.75   *i)* 22.05   *l)* 27.33   *o)* 30.28

12.8. Find the value of each to the nearest tenth, using a calculator: **(4.2)**

$a)$ $2\sqrt{3}$ $c)$ $\frac{1}{2}\sqrt{140}$ $e)$ $100 - 10\sqrt{83}$ $g)$ $\dfrac{\sqrt{3} + \sqrt{2}}{5}$

$b)$ $10\sqrt{50}$ $d)$ $\frac{2}{5}\sqrt{280}$ $f)$ $\dfrac{3 + \sqrt{2}}{5}$ $h)$ $\dfrac{10 - \sqrt{2}}{3}$

Ans. *a)* 3.5   *b)* 70.7   *c)* 5.9   *d)* 6.7   *e)* 8.9   *f)* 0.9   *g)* 0.6   *h)* 2.9

12.9. Find the approximate value of each, using a calculator and rule 1:   **(4.3)**

$a)$ $\sqrt{3000}$ $c)$ $\sqrt{19,000}$ $e)$ $\sqrt{5000} + \sqrt{500}$ to nearest tenth

$b)$ $\sqrt{30,000}$ $d)$ $\sqrt{190,000}$ $f)$ $\sqrt{8300} - \sqrt{830}$ to nearest tenth

Ans.   *a)* 54.77   *c)* 137.8   *e)* 70.71 + 22.36 = 93.07 or 93.1

        b) 173.2   *d)* 435.9   *f)* 91.10 − 28.81 = 62.29 or 62.3

12.10. Find the approximate value of each, using a calculator and rule 2:   **(4.4)**

$a)$ $\sqrt{0.03}$ $d)$ $\sqrt{3.85}$ $g)$ $\sqrt{55} + \sqrt{0.55}$ to nearest hundredth

$b)$ $\sqrt{0.0003}$ $e)$ $\sqrt{0.0385}$ $h)$ $\sqrt{427} - \sqrt{4.27}$ to nearest hundredth

$c)$ $\sqrt{0.30}$ $f)$ $\sqrt{0.3}$ $i)$ $\sqrt{0.02} + \sqrt{0.0002}$ to nearest hundredth

Ans.   *a)* 0.1732   *b)* 0.01732   *c)* 0.5477   *d)* 1.962   *e)* 0.1962

        f) 0.5477 (= $\sqrt{0.30}$)   *g)* 8.16   *h)* 18.59   *i)* 0.16

12.11. Simplify: **(5.2)**

$a)$ $\sqrt{63}$ $d)$ $\frac{1}{2}\sqrt{32}$ $g)$ $\sqrt{17,500}$ $j)$ $\frac{3}{7}\sqrt{392}$

$b)$ $\sqrt{96}$ $e)$ $\frac{1}{5}\sqrt{300}$ $h)$ $\sqrt{4500}$ $k)$ $\frac{2}{5}\sqrt{250}$

$c)$ $\sqrt{448}$ $f)$ $\frac{7}{8}\sqrt{320}$ $i)$ $\sqrt{21,600}$ $l)$ $\frac{5}{3}\sqrt{999}$

Ans.   *a)* $3\sqrt{7}$ *b)* $4\sqrt{6}$ *c)* $8\sqrt{7}$ *d)* $2\sqrt{2}$ *e)* $2\sqrt{3}$ *f)* $7\sqrt{5}$ *g)* $50\sqrt{7}$

        h) $30\sqrt{5}$ *i)* $60\sqrt{6}$ *j)* $6\sqrt{2}$ *k)* $2\sqrt{10}$ *l)* $5\sqrt{111}$

12.12. Using $\sqrt{2} = 1.414$ or $\sqrt{5} = 2.236$, evaluate to the nearest tenth: **(5.3)**

$a)$ $\frac{2}{3}\sqrt{18}$ *Ans.* $2\sqrt{2} = 2.8$ $c)$ $10\sqrt{20}$ *Ans.* $20\sqrt{5} = 44.7$ $e)$ $\frac{1}{4}\sqrt{8000}$ *Ans.* $10\sqrt{5} = 22.4$

$b)$ $\frac{1}{3}\sqrt{45}$ $\sqrt{5} = 2.2$ $d)$ $5\sqrt{98}$ $35\sqrt{2} = 49.5$ $f)$ $\frac{1}{2}\sqrt{16,200}$ $45\sqrt{2} = 63.6$

12.13. Simplify: **(5.4)**

$a)$ $\sqrt{49m^2}$ *Ans.* $7m$ $e)$ $\sqrt{7r^2s^2}$ *Ans.* $rs\sqrt{7}$ $i)$ $\sqrt{400a^8b^{10}}$ *Ans.* $20a^4b^5$

$b)$ $\sqrt{100a^2b^2}$ $10ab$ $f)$ $\sqrt{4s^2t^2u}$ $2st\sqrt{u}$ $j)$ $\sqrt{25x^3}$ $5x\sqrt{x}$

$c)$ $\sqrt{64r^2p^2q^2}$ $8rpq$ $g)$ $\sqrt{81x^4}$ $9x^2$ $k)$ $\sqrt{9u^5w^{12}}$ $3u^2w^6\sqrt{u}$

$d)$ $\sqrt{4p^2q}$ $2p\sqrt{q}$ $h)$ $\sqrt{144y^6}$ $12y^3$ $l)$ $\sqrt{27t^3x^5}$ $3tx^2\sqrt{3tx}$

12.14. Simplify: **(6.1)**

a) $\sqrt{6\frac{1}{4}}$ Ans. $\frac{5}{2}$ c) $\sqrt{\frac{2500}{81}}$ Ans. $\frac{50}{9}$ e) $\sqrt{\frac{441}{169}}$ Ans. $\frac{21}{13}$ g) $\sqrt{\frac{4}{9}}\sqrt{\frac{25}{16}}$ Ans. $\frac{5}{6}$

b) $\sqrt{7\frac{1}{9}}$ $\frac{8}{3}$ d) $\sqrt{\frac{9}{6400}}$ $\frac{3}{80}$ f) $\sqrt{\frac{289}{10,000}}$ $\frac{17}{100}$ h) $\sqrt{\frac{36}{225}}\sqrt{\frac{1}{4}}\sqrt{\frac{25}{16}}$ $\frac{1}{4}$

12.15. Simplify: **(6.2)**

a) $\sqrt{\frac{5}{16}}$ Ans. $\frac{\sqrt{5}}{4}$ c) $\sqrt{\frac{8}{25}}$ Ans. $\frac{2}{5}\sqrt{2}$ e) $10\sqrt{\frac{x}{25}}$ Ans. $2\sqrt{x}$ g) $\frac{1}{y}\sqrt{\frac{y^3}{900}}$ Ans. $\frac{\sqrt{y}}{30}$

b) $9\sqrt{\frac{13}{9}}$ $3\sqrt{13}$ d) $21\sqrt{\frac{32}{49}}$ $12\sqrt{2}$ f) $\sqrt{\frac{150}{x^4}}$ $\frac{5}{x^2}\sqrt{6}$ h) $y^4\sqrt{\frac{45}{y^6}}$ $3y\sqrt{5}$

12.16. Simplify: **(6.3)**

a) $\sqrt{\frac{1}{7}}$ Ans. $\sqrt{\frac{7}{7}}$ c) $\sqrt{\frac{5}{18}}$ Ans. $\sqrt{\frac{10}{6}}$ e) $\sqrt{\frac{1}{b}}$ Ans. $\frac{1}{b}\sqrt{b}$ g) $\sqrt{\frac{2s}{g}}$ Ans. $\frac{1}{g}\sqrt{2sg}$

b) $\sqrt{\frac{3}{11}}$ $\sqrt{\frac{33}{11}}$ d) $5\sqrt{\frac{7}{20}}$ $\sqrt{\frac{35}{2}}$ f) $\sqrt{\frac{1}{c^3}}$ $\frac{1}{c^2}\sqrt{c}$ h) $\sqrt{\frac{x^3}{50}}$ $\frac{x}{10}\sqrt{2x}$

12.17. Using $\sqrt{2} = 1.414$ or $\sqrt{3} = 1.732$, evaluate each to the nearest hundredth: **(6.4)**

a) $\sqrt{\frac{1}{2}}$ c) $\sqrt{\frac{9}{50}}$ e) $16\sqrt{\frac{9}{32}}$ g) $\sqrt{2\frac{1}{12}}$

b) $\sqrt{\frac{1}{3}}$ d) $\sqrt{\frac{1}{12}}$ f) $45\sqrt{\frac{4}{27}}$ h) $\sqrt{2\frac{13}{18}}$

Ans.   a) $\frac{1}{2}\sqrt{2} = 0.71$ c) $\frac{3}{10}\sqrt{2} = 0.42$ e) $6\sqrt{2} = 8.48$ g) $\frac{5}{6}\sqrt{3} = 1.44$

b) $\frac{1}{3}\sqrt{3} = 0.58$ d) $\sqrt{3} = 1.73$ f) $10\sqrt{3} = 17.32$ h) $\frac{7}{8}\sqrt{2} = 1.65$

12.18. Combine: **(7.1)**

a) $5\sqrt{3} + \sqrt{3}$ d) $5\sqrt{5} - 2\sqrt{5} + \sqrt{10}$ g) $12\sqrt{b} - 2\sqrt{a} + 6\sqrt{b}$

b) $7\sqrt{11} - 10\sqrt{11}$ e) $x\sqrt{15} + 2x\sqrt{15}$ h) $5b\sqrt{c} + 3b\sqrt{c} + c\sqrt{b}$

c) $3\sqrt{2} + 4\sqrt{3} - 2\sqrt{2}$ f) $11\sqrt{a} + \sqrt{a} - 3\sqrt{a}$

Ans.   a) $6\sqrt{3}$ d) $3\sqrt{5} + \sqrt{10}$ g) $18\sqrt{b} - 2\sqrt{a}$

b) $-3\sqrt{11}$ e) $3x\sqrt{15}$ h) $8b\sqrt{c} + c\sqrt{b}$

c) $\sqrt{2} + 4\sqrt{3}$ f) $9\sqrt{a}$

12.19. Simplify and combine: **(7.2)**

a) $\sqrt{2} + \sqrt{32}$ e) $2\sqrt{27} - 4\sqrt{12}$ i) $8\sqrt{3} - 6\sqrt{\frac{1}{3}}$

b) $2\sqrt{5} + 3\sqrt{20}$ f) $\sqrt{700} - 2\sqrt{63}$ j) $10\sqrt{18} + 20\sqrt{\frac{1}{2}}$

c) $2\sqrt{28} - 3\sqrt{7}$ g) $-3\sqrt{90} - 5\sqrt{40}$ k) $6\sqrt{\frac{5}{12}} - \sqrt{60}$

d) $x\sqrt{3} + x\sqrt{27}$ h) $\sqrt{9y} + 2\sqrt{25y}$ l) $\sqrt{\frac{a^2}{2}} - \sqrt{\frac{a^2}{8}}$

Ans.   a) $\sqrt{2} + 4\sqrt{2} = 5\sqrt{2}$ e) $6\sqrt{3} - 8\sqrt{3} = -2\sqrt{3}$ i) $8\sqrt{3} - 2\sqrt{3} = 6\sqrt{3}$

b) $2\sqrt{5} + 6\sqrt{5} = 8\sqrt{5}$ f) $10\sqrt{7} - 6\sqrt{7} = 4\sqrt{7}$ j) $30\sqrt{2} + 10\sqrt{2} = 40\sqrt{2}$

c) $4\sqrt{7} - 3\sqrt{7} = \sqrt{7}$ g) $-9\sqrt{10} - 10\sqrt{10} = -19\sqrt{10}$ k) $\sqrt{15} - 2\sqrt{15} = -\sqrt{15}$

d) $x\sqrt{3} + 3x\sqrt{3} = 4x\sqrt{3}$ h) $3\sqrt{y} + 10\sqrt{y} = 13\sqrt{y}$ l) $\frac{a}{2}\sqrt{2} - \frac{a}{4}\sqrt{2} = \frac{a}{4}\sqrt{2}$

12.20. Multiply: **(8.1)**

a) $4\sqrt{2}\cdot5\sqrt{3}$ d) $\sqrt{2}\sqrt{3}\sqrt{12}$ g) $\sqrt{21}\sqrt{3}$ j) $\frac{1}{2}\sqrt{3}\cdot6\sqrt{18}$

b) $7\sqrt{7}\cdot10\sqrt{10}$ e) $3\sqrt{6}\cdot5\sqrt{24}$ h) $\sqrt{10}\sqrt{20}$ k) $\frac{2}{3}\sqrt{7}\cdot15\sqrt{14}$

c) $4\sqrt{3}\cdot3\sqrt{2}\cdot6\sqrt{5}$ f) $10\sqrt{\frac{1}{3}}\cdot7\sqrt{75}$ i) $\sqrt{8}\sqrt{12}\sqrt{10}$ l) $\frac{3}{2}\sqrt{24}\cdot4\sqrt{3}$

Ans.   a) $20\sqrt{6}$ d) $6\sqrt{2}$ g) $\sqrt{63} = 3\sqrt{7}$ j) $3\sqrt{54} = 9\sqrt{6}$

     b) $70\sqrt{70}$ e) $15\sqrt{144} = 180$ h) $\sqrt{200} = 10\sqrt{2}$ k) $10\sqrt{98} = 70\sqrt{2}$

     c) $72\sqrt{30}$ f) $70\sqrt{25} = 350$ i) $\sqrt{960} = 8\sqrt{15}$ l) $6\sqrt{72} = 36\sqrt{2}$

12.21. Multiply: **(8.1)**

a) $2\sqrt{a}\cdot4\sqrt{b}$ d) $\sqrt{pq}\sqrt{p}\sqrt{q}$ g) $\sqrt{5a}\sqrt{3a}$ j) $\sqrt{\dfrac{y}{x}}\sqrt{\dfrac{x}{y}}$

b) $c\sqrt{5}\cdot d\sqrt{7}$ e) $\sqrt{4m}\sqrt{25m}$ h) $\sqrt{\frac{1}{2}b}\sqrt{40b}$ k) $3x\sqrt{x}\cdot5x\sqrt{x}$

c) $r\sqrt{x}\cdot s\sqrt{y}$ f) $\sqrt{3n}\sqrt{27n^3}$ i) $\sqrt{8h^3}\sqrt{2k^2}$ l) $y^2\sqrt{y}\cdot y\sqrt{y^2}$

Ans.   a) $8\sqrt{ab}$ d) $\sqrt{p^2q^2} = pq$ g) $\sqrt{15a^2} = a\sqrt{15}$ j) 1

     b) $cd\sqrt{35}$ e) $\sqrt{100m^2} = 10m$ h) $\sqrt{20b^2} = 2b\sqrt{5}$ k) $15x^2\sqrt{x^2} = 15x^3$

     c) $rs\sqrt{xy}$ f) $\sqrt{81n^4} = 9n^2$ i) $\sqrt{16h^3k^2} = 4hk\sqrt{h}$ l) $y^3\sqrt{y^3} = y^4\sqrt{y}$

12.22. Multiply: **(8.2)**

a) $5\sqrt{10}\cdot\sqrt{10}$ d) $(\sqrt{14})^2$ g) $\sqrt{7x}\cdot\sqrt{7x}$ j) $5\sqrt{2b}\cdot4\sqrt{2b}$

b) $\sqrt{12}\cdot\frac{1}{2}\sqrt{12}$ e) $(3\sqrt{6})^2$ h) $x\sqrt{5}\cdot y\sqrt{5}$ k) $(2x\sqrt{2x})^2$

c) $2\sqrt{\dfrac{2}{3}}\cdot3\sqrt{\dfrac{3}{2}}$ f) $(\frac{1}{2}\sqrt{8})^2$ i) $(7\sqrt{2a})^2$ l) $\left(c\sqrt{\dfrac{5}{c}}\right)^2$

Ans.   a) $5\cdot10 = 50$ d) 14 g) $7x$ j) $20(2b) = 40b$

     b) $\frac{1}{2}\cdot12 = 6$ e) $9\cdot6 = 54$ h) $5xy$ k) $4x^2(2x) = 8x^3$

     c) 6 f) $\frac{1}{4}\cdot8 = 2$ i) $49(2a) = 98a$ l) $c^2\left(\dfrac{5}{c}\right) = 5c$

12.23. Multiply: **(8.3)**

a) $\sqrt{2}(5 + \sqrt{2})$ c) $3\sqrt{3}(\sqrt{27} + 7\sqrt{3})$ e) $\sqrt{6}(\sqrt{6x} + \sqrt{24x})$

b) $\sqrt{5}(2\sqrt{5} - 3\sqrt{20})$ d) $\sqrt{x}(\sqrt{16x} - \sqrt{9x})$ f) $\sqrt{8}(\sqrt{2x} + \sqrt{18y})$

Ans.   a) $5\sqrt{2} + 2$ c) $27 + 63 = 90$ e) $6\sqrt{x} + 12\sqrt{x} = 18\sqrt{x}$

     b) $10 - 30 = -20$ d) $4x - 3x = x$ f) $4\sqrt{x} + 12\sqrt{y}$

12.24. Multiply: **(8.4)**

a) $(6 + \sqrt{3})(6 - \sqrt{3})$ d) $(3 + \sqrt{2})^2$ g) $(2 - \sqrt{6})(3 - \sqrt{6})$

b) $(\sqrt{5} - 2)(\sqrt{5} + 2)$ e) $(\sqrt{11} - 3)^2$ h) $(\sqrt{7} + 3)(\sqrt{7} + 2)$

c) $(\sqrt{7} + \sqrt{2})(\sqrt{7} - \sqrt{2})$ f) $(\sqrt{3} + \sqrt{5})^2$ i) $(4\sqrt{2} - \sqrt{5})(3\sqrt{2} + 2\sqrt{5})$

Ans. *a)* $36 - 3 = 33$ *d)* $9 + 6\sqrt{2} + 2 = 11 + 6\sqrt{2}$ *g)* $6 - 5\sqrt{6} + 6 = 12 - 5\sqrt{6}$

   b) $5 - 4 = 1$ *e)* $11 - 6\sqrt{11} + 9 = 20 - 6\sqrt{11}$ *h)* $7 + 5\sqrt{7} + 6 = 13 + 5\sqrt{7}$

   c) $7 - 2 = 5$ *f)* $3 + 2\sqrt{15} + 5 = 8 + 2\sqrt{15}$ *i)* $24 + 5\sqrt{10} - 10 = 14 - 5\sqrt{10}$

12.25. Divide: **(9.1)**

a) $\dfrac{15\sqrt{60}}{5\sqrt{15}}$ *Ans.* $3\sqrt{4} = 6$ *d)* $\dfrac{\sqrt{120}}{2\sqrt{5}}$ *Ans.* $\dfrac{\sqrt{24}}{2} = \sqrt{6}$ *g)* $\dfrac{5c\sqrt{5c}}{c\sqrt{c}}$ *Ans.* $5\sqrt{5}$

b) $\dfrac{24\sqrt{2}}{3\sqrt{32}}$ $8\sqrt{\dfrac{1}{16}} = 2$ *e)* $\dfrac{3\sqrt{24}}{\sqrt{2}}$ $3\sqrt{12} = 6\sqrt{3}$ *h)* $\dfrac{3x\sqrt{x^5}}{9x\sqrt{x}}$ $\tfrac{1}{3}\sqrt{x^4} = \dfrac{x^2}{3}$

c) $\dfrac{6\sqrt{3}}{3\sqrt{27}}$ $2\sqrt{\dfrac{1}{9}} = \dfrac{2}{3}$ *f)* $\dfrac{14\sqrt{a^3}}{2\sqrt{a}}$ $7\sqrt{a^2} = 7a$ *i)* $\dfrac{a\sqrt{ab}}{ab\sqrt{b}}$ $\dfrac{1}{b}\sqrt{a}$

12.26. Divide: **(9.2)**

a) $\dfrac{\sqrt{200} + \sqrt{50}}{\sqrt{5}}$ *Ans.* $\sqrt{40} + \sqrt{10} = 3\sqrt{10}$ *d)* $\dfrac{\sqrt{20} + \sqrt{80}}{2\sqrt{5}}$ *Ans.* $\dfrac{\sqrt{4}}{2} + \dfrac{\sqrt{16}}{2} = 3$

b) $\dfrac{9\sqrt{200} - 12\sqrt{32}}{3\sqrt{2}}$ $3\sqrt{100} - 4\sqrt{16} = 14$ *e)* $\dfrac{\sqrt{27a} - 2\sqrt{3a}}{\sqrt{a}}$ $\sqrt{27} - 2\sqrt{3} = \sqrt{3}$

c) $\dfrac{8\sqrt{2} - 4\sqrt{8}}{\sqrt{32}}$ $8\sqrt{\dfrac{1}{16}} - 4\sqrt{\dfrac{1}{4}} = 0$ *f)* $\dfrac{\sqrt{y^7} - \sqrt{y^5}}{\sqrt{y^3}}$ $\sqrt{y^4} - \sqrt{y^2} = y^2 - y$

12.27. Rationalize each denominator: **(10.1)**

a) $\dfrac{2}{\sqrt{5}}$ *Ans.* $\dfrac{2\sqrt{5}}{5}$ *e)* $\dfrac{12}{\sqrt{6}}$ *Ans.* $2\sqrt{6}$ *i)* $\dfrac{16 - 2\sqrt{2}}{\sqrt{8}}$ *Ans.* $4\sqrt{2} - 1$

b) $\dfrac{12}{\sqrt{2}}$ $6\sqrt{2}$ *f)* $\dfrac{\sqrt{2}}{\sqrt{3}}$ $\sqrt{\dfrac{6}{3}}$ *j)* $\dfrac{5}{\sqrt{y}}$ $\dfrac{5\sqrt{y}}{y}$

c) $\dfrac{10}{\sqrt{50}}$ $\sqrt{2}$ *g)* $\dfrac{4 + \sqrt{2}}{\sqrt{2}}$ $2\sqrt{2} + 1$ *k)* $\dfrac{2a}{\sqrt{10}}$ $\dfrac{a\sqrt{10}}{5}$

d) $\dfrac{\sqrt{5}}{\sqrt{15}}$ $\dfrac{\sqrt{3}}{3}$ *h)* $\dfrac{3\sqrt{3} - 6}{\sqrt{3}}$ $3 - 2\sqrt{3}$ *l)* $\dfrac{3x}{\sqrt{x}}$ $3\sqrt{x}$

12.28. Evaluate to the nearest tenth, after rationalizing the denominator: **(10.2)**

a) $\dfrac{2}{\sqrt{2}}$ *Ans.* $\sqrt{2} = 1.4$ *d)* $\dfrac{40}{\sqrt{8}}$ *Ans.* $10\sqrt{2} = 14.1$

b) $\dfrac{12}{\sqrt{6}}$ $2\sqrt{6} = 4.9$ *e)* $\dfrac{30}{\sqrt{45}}$ $2\sqrt{5} = 4.5$

c) $\dfrac{10\sqrt{2}}{\sqrt{5}}$ $2\sqrt{10} = 6.3$ *f)* $\dfrac{9\sqrt{5}}{\sqrt{15}}$ $3\sqrt{3} = 5.2$

g) $\dfrac{12 - \sqrt{3}}{\sqrt{3}}$ $Ans.$ $4\sqrt{3} - 1 = 5.9$ i) $\dfrac{\sqrt{20} + \sqrt{5}}{\sqrt{10}}$ $Ans.$ $\tfrac{3}{2}\sqrt{2} = 2.1$

h) $\dfrac{1 + \sqrt{5}}{\sqrt{5}}$ $\dfrac{\sqrt{5} + 5}{5} = 1.4$

12.29. Solve: **(11.1)**

a) $\sqrt{3x} = 6$ $Ans.$ $x = 12$ g) $\sqrt{c + 3} = 4$ $Ans.$ $c = 13$

b) $3\sqrt{x} = 15$ $x = 25$ h) $\sqrt{d - 5} = 3$ $d = 14$

c) $8 = 2\sqrt{y}$ $y = 16$ i) $\sqrt{3h} = 5$ $h = \frac{25}{3}$

d) $5\sqrt{a} = 2$ $a = \frac{4}{25}$ j) $\sqrt{10 - x} = 2$ $x = 6$

e) $\tfrac{1}{2}\sqrt{b} = 5$ $b = 100$ k) $\sqrt{2r - 5} = 3$ $r = 7$

f) $\tfrac{2}{3}\sqrt{c} = 8$ $c = 144$ l) $\sqrt{54 - 3s} = 6$ $s = 6$

12.30. Solve: **(11.2)**

a) $2\sqrt{a} - 3 = 7$ d) $8\sqrt{r} - 5 = \sqrt{r} + 9$ g) $\sqrt{2x - 7} + 8 = 11$

b) $8 + 3\sqrt{b} = 20$ e) $20 - 3\sqrt{t} = \sqrt{t} - 4$ h) $22 = 17 + \sqrt{40 - 3y}$

c) $7 - \sqrt{2b} = 3$ f) $2\sqrt{5x} - 3 = 7$ i) $30 - \sqrt{20 - 2y} = 26$

$Ans.$ a) $a = 25$ b) $b = 16$ c) $b = 8$ d) $r = 4$ e) $t = 36$ f) $x = 5$ g) $x = 8$

h) $y = 5$ i) $y = 2$

12.31. Solve and show, by checking, that each value for the unknown is extraneous: **(11.2)**

a) $\sqrt{a} = -2$ c) $5 + \sqrt{c} = 4$ e) $2\sqrt{t} + 5 = \sqrt{t}$ g) $\sqrt{y + 5} = -4$

b) $-\sqrt{b} = 3$ d) $6 - \sqrt{r} = 10$ f) $10 - \sqrt{3v} = 12$ h) $3 - \sqrt{12 - x} = 8$

12.32. Solve for the letter indicated: **(11.3)**

a) $a = \sqrt{2b}$ for b d) $s = \dfrac{\sqrt{t}}{3}$ for v g) $R = \dfrac{1}{2}\sqrt{\dfrac{A}{\pi}}$ for A

b) $g = 3\sqrt{h}$ for h e) $2u = \sqrt{\dfrac{v}{2}}$ for v h) $R = \sqrt{\dfrac{3V}{\pi H}}$ for V

c) $g - \sqrt{3h}$ for h f) $h = \sqrt{3g} \quad 4$ for g

$Ans.$ a) $b = \dfrac{a^2}{2}$ c) $h = \dfrac{g^2}{3}$ e) $v = 8u^2$ g) $A = 4\pi R^2 H$

b) $h = \dfrac{g^2}{9}$ d) $t = 9s^2$ f) $g = \dfrac{h^2 + 4}{3}$ h) $V = \tfrac{1}{3}\pi R^2 H$

12.33. Graph $y = 3\dfrac{\sqrt{x}}{1}$ using a graphing calculator.

CHAPTER 13

Quadratic Equations

1. UNDERSTANDING QUADRATIC EQUATIONS IN ONE UNKNOWN

A quadratic equation in one unknown is an equation in which the highest power of the unknown is the second.

Thus, $2x^2 + 3x - 5 = 0$ is a quadratic equation in x.

Standard Quadratic Equation Form: $ax^2 + bx + c = 0$

The **standard form of a quadratic equation** in one unknown is $ax^2 + bx + c = 0$, where a, b, and c represent known numbers and x represents the unknown number. The number a cannot equal zero. Thus, $3x^2 - 5x + 6 = 0$ is in standard form. Here, $a = 3$, $b = -5$, and $c = 6$.

To Transform a Quadratic Equation into Standard Form $ax^2 + bx + c = 0$

(1) Remove parentheses. Thus, $x(x + 1) - 5 = 0$ becomes $x^2 + x - 5 = 0$.

(2) Clear of fractions. Thus, $x - 4 + \dfrac{3}{x} = 0$ becomes $x^2 - 4x + 3 = 0$.

(3) Remove radical signs. Thus, $\sqrt{x^2 - 3x} = 2$ becomes $x^2 - 3x - 4 = 0$.

(4) Collect like terms. Thus, $x^2 + 7x = 2x + 6$ becomes $x^2 + 5x - 6 = 0$.

1.1. EXPRESSING QUADRATIC EQUATIONS IN STANDARD FORM

Express each quadratic equation in standard form $ax^2 + bx + c = 0$, so that a has a positive value:

a) $x(x + 1) = 6$ b) $\dfrac{3}{x} + x = 4$ c) $x^2 = 25 - 5x - 15$ d) $2x = \sqrt{x + 3}$

Solutions:

a) Remove ( ):

$$x^2 + x = 6$$

Ans. $x^2 + x - 6 = 0$

b) Clear fractions:

$$3 + x^2 = 4x$$

Ans. $x^2 - 4x + 3 = 0$

c) Collect like terms:

$$x^2 = 10 - 5x$$

Ans. $x^2 + 5x - 10 = 0$

d) Remove radical signs:

Squaring: $4x^2 = x + 3$

Ans. $4x^2 - x - 3 = 0$

1.2.   **VALUES OF *a*, *b*, AND *c* IN STANDARD QUADRATIC EQUATION FORM**
Express each quadratic equation in standard form, so that *a* has a positive value. Then state the values of *a*, *b*, and *c*.

		Standard Form	*a*	*b*	*c*
a) $x^2 - 9x = 10$	*Ans.*	$x^2 - 9x - 10 = 0$	1	-9	-10
b) $5x^2 - 125$		$5x^2 - 125 = 0$	5	0	-125
c) $2x^2 = 8x$		$2x^2 - 8x = 0$	2	-8	0
d) $2x^2 + 9x = 2x - 3$		$2x^2 + 7x + 3 = 0$	2	$+7$	$+3$
e) $x(x + 3) = 10$		$x^2 + 3x - 10 = 0$	1	$+3$	-10
f) $5(x^2 + 2) = 7(x + 3)$		$5x^2 - 7x - 11 = 0$	5	-7	-11
g) $x - 5 = \dfrac{7}{x}$		$x^2 - 5x - 7 = 0$	1	-5	-7
h) $\sqrt{2x^2 - 1} = x + 2$		$x^2 - 4x - 5 = 0$	1	-4	-5

2. SOLVING QUADRATIC EQUATIONS BY FACTORING

Rule 1.   Every quadratic equation has two roots.

Thus, $x^2 = 9$ has two roots, 3 and -3; that is, $x = \pm 3$.

Rule 2.   If the product of two factors is zero, then one or the other of the factors or both must equal zero.

Thus: In $5(x - 3) = 0$, the factor $x - 3 = 0$.

In $(x - 2)(x - 3) = 0$, either $x - 2 = 0$ or $x - 3 = 0$, or both equal zero.

In $(x - 3)(x - 3) = 0$, both factors $x - 3 = 0$.

In $(x - 3)(y - 5) = 0$, either $x - 3 = 0$, $y - 5 = 0$ or both equal zero.

To Solve a Quadratic Equation by Factoring

Solve: $x(x - 4) = 5$

Procedure:

1. **Express in form** $ax^2 + bx + c = 0$:
2. **Factor** $ax^2 + bx + c$:
3. **Let each factor equal zero:**
4. **Solve** each resulting equation:
5. **Check** each root in original equation:

Solution:

1. S_5 $x^2 - 4x = 5$
                   $x^2 - 4x - 5 = 0$

2. $(x - 5)(x + 1) = 0$

3. $x - 5 = 0$ $x + 1 = 0$

4. $x = 5$ $x = -1$

5.   **Check** in $x(x - 4) = 5$:

If $x = 5$, $5(1) \overset{?}{=} 5$ If $x = -1$, $(-1)(-5) \overset{?}{=} 5$
                   $5 = 5$ $5 = 5$

Ans. $x = 5$ or -1

2.1.   SOLVING QUADRATIC EQUATIONS BY FACTORING

$$\text{Solve:} \quad a)\ x^2 - x = 6 \qquad\qquad b)\ x + \frac{16}{x} = 8$$

Procedure:	**Solutions:**	
1. Express as $ax^2 + bx + c = 0$:	**1.** $x^2 - x - 6 = 0$	**1.** $x^2 + 16 = 8x$
		$x^2 - 8x + 16 = 0$
2. Factor $ax^2 + bx + c$:	**2.** $(x - 3)(x + 2) = 0$	**2.** $(x - 4)(x - 4) = 0$
3. Let each factor equal zero:	**3.** $x - 3 = 0$ $x + 2 = 0$	**3.** $x - 4 = 0$ $x - 4 = 0$
4. Solve:	**4.** $x = 3$ $x = -2$	**4.** $x = 4$ $x = 4$
5. Check (in original equation):	**5.** (Check to be done by the student.)	
	Ans. $x = 3$ or -2	**Ans.** $x = 4$ (equal roots)

2.2.   SOLVING QUADRATIC EQUATIONS IN STANDARD FORM
Solve by factoring:

$$a)\ x^2 + 9x + 20 = 0 \qquad b)\ a^2 - 49 = 0 \qquad c)\ 3r^2 - 21r = 0$$

Solutions:

a) $(x + 4)(x + 5) = 0$	b) $(a + 7)(a - 7) = 0$	c) $3r(r - 7) = 0$
$x + 4 = 0$ $x + 5 = 0$	$a + 7 = 0$ $a - 7 = 0$	$3r = 0$ $r - 7 = 0$
$x = -4$ $x = -5$	$a = -7$ $a = +7$	$r = 0$ $r = +7$
Ans. $x = -4$ or -5	*Ans.* $a = -7$ or $+7$	*Ans.* $r = 0$ or $+7$

2.3.   MORE DIFFICULT SOLUTIONS BY FACTORING
Solve by factoring:

$$a)\ 2 + \frac{5}{x} = \frac{12}{x^2} \qquad b)\ (y - 8)(2y - 3) = 34 \qquad c)\ x - 6 = \sqrt{x}$$

Solutions:

a) Clear of fractions:	b) Remove ( ):	c) Remove radical sign:
$2x^2 + 5x = 12$	$2y^2 - 19y + 24 = 34$	**Sq** $x^2 - 12x + 36 = x$
$2x^2 + 5x - 12 = 0$	$2y^2 - 19y - 10 = 0$	$x^2 - 13x + 36 = 0$
$(2x - 3)(x + 4) = 0$	$(2y + 1)(y - 10) = 0$	$(x - 9)(x - 4) = 0$
$2x - 3 = 0$ $x + 4 = 0$	$2y + 1 = 0$ $y - 10 = 0$	$x - 9 = 0$ $x - 4 = 0$
$x = \frac{3}{2}$ $x = -4$	$y = -\frac{1}{2}$ $y = 10$	$x = 9$ $x = 4$
Ans. $x = \frac{3}{2}$ or -4	*Ans.* $y = -\frac{1}{2}$ or 10	*Ans.* $x = 9$ (Show that 4 is extraneous.)

2.4.   NUMBER PROBLEMS INVOLVING QUADRATIC EQUATIONS

a) Find a number whose square is 2 less than 3 times the number.

b) Find two consecutive integers such that the sum of their squares is 25.

Solutions:

a) Let n = number.

$$n^2 = 3n - 2$$
$$n^2 - 3n + 2 = 0$$
$$(n - 2)(n - 1) = 0$$
$$n = 2 \text{ or } 1$$

Ans.   2 or 1

b) Let n and $n + 1$ = consecutive integers.

$$n^2 + (n + 1)^2 = 25$$
$$n^2 + n^2 + 2n + 1 = 25$$
$$2n^2 + 2n - 24 = 0$$
$$n^2 + n - 12 = 0$$
$$(n + 4)(n - 3) = 0$$
$$n = -4 \qquad n = 3$$
$$n + 1 = -3 \quad n + 1 = 4$$

Ans.   Either -4 and -3 or 3 and 4

2.5.   AREA PROBLEMS INVOLVING QUADRATIC EQUATIONS

The length of a rectangle lot is 3 yd more than its width. Find its dimensions if the area

a) Equals 40 yds

b) Increases 70 yd^2 when the width is doubled

Solutions:

In *(a)* and *(b)*, let w = width in yards and $w + 3$ = length in yards.

a)   Area of lot = 40 yd^2

$$w(w + 3) = 40$$
$$w^2 + 3w - 40 = 0$$
$$(w - 5)(w + 8) = 0$$
$$w = 5 \quad w = -8 \quad \text{Reject since width}$$
$$l = 8 \qquad\qquad\qquad \text{cannot be negative}$$

Ans.   8 yd and 5 yd

b)   New area = old area + 70 yd^2

$$2w(w + 3) = w(w + 3) + 70$$
$$2w^2 + 6w = w^2 + 3w + 70$$
$$w^2 + 3w - 70 = 0$$
$$(w - 7)(w + 10) = 0$$
$$w = 7 \quad w = -10 \quad \text{Reject.}$$
$$l = 10 \qquad\qquad \text{Ans.}\quad 10 \text{ yd and 7 yd}$$

3.   SOLVING INCOMPLETE QUADRATIC EQUATIONS

An **incomplete quadratic equation** in one unknown lacks either **(1)** the term containing the first power of the unknown as $x^2 - 4 - 0$ *or* **(2)** the constant term as $x^2 - 4x = 0$.

RULE: If an incomplete quadratic equation lacks the constant term, then one of the roots is zero.
Thus, if $x^2 - 4x = 0$, $x = 0$ or $x = 4$.

To Solve an Incomplete Quadratic Equation Lacking the First Power of Unknown

Solve: $2(x^2 - 8) = 11 - x^2$

Procedure:

1. Express in form $ax^2 = k$, where k is a constant:

2. Divide both sides by a, obtaining $x^2 = \dfrac{k}{a}$:

3. Take the square root of both sides, obtaining $x = \pm \sqrt{k/a}$:

4. Check each root in the original equation:

Solution:

1. Tr $2x^2 - 16 = 11 - x^2$
$$3x^2 = 27$$

2. Dividing by 3,
$$x^2 = 9 \text{ (See note below.)}$$

3. Taking a square root of both sides,
$$x = \pm 3$$

4. (Check is left to the student.)

NOTE: $x^2 = 9$ may be solved by factoring as follows: $x^2 - 9 = 0$, $(x + 3)(x - 3) = 0$, $x = \pm 3$.

To Solve an Incomplete Quadratic Equation Lacking Constant Term

$$\text{Solve: } 3x^2 = 18x$$

Procedure:

1. **Express** in form $ax^2 + bx = 0$:
2. **Factor** $ax^2 + bx$:
3. **Let** each factor equal zero:
4. **Solve** each resulting equation:
5. **Check** each root in the original equation:

Solution:

1. $3x^2 - 18x = 0$
2. $3x(x - 6) = 0$
3. $3x = 0 \quad x - 6 = 0$
4. $x = 0 \qquad\quad x = 6$
5. Check in $3x^2 = 18x$:

$$\text{If } x = 0, \qquad\qquad \text{If } x = 6,$$
$$3(0^2) \overset{?}{=} 18(0) \quad 3(6^2) \overset{?}{=} 18(6)$$
$$0 = 0 \qquad\qquad 108 = 108$$

3.1. **SOLVING INCOMPLETE QUADRATIC EQUATIONS LACKING FIRST POWER OF UNKNOWN**

Solve. (Leave irrational answers in radical form.)

a) $4x^2 - 49 = 0$ b) $2y^2 = 125 - 3y^2$ c) $9x^2 - 2 = 5$ d) $\dfrac{8x}{25} = \dfrac{16}{x}$

Solutions:

a) $\mathbf{D_4} \quad 4x^2 = 49$ b) $\mathbf{D_5} \quad 5y^2 = 125$ c) $\mathbf{D_9} \quad 9x^2 = 7$ d) $\mathbf{D_8} \quad 8x^2 = 400$

Sq rt $x^2 = \dfrac{49}{4}$ **Sq rt** $y^2 = 25$ **Sq rt** $x^2 = \dfrac{7}{9}$ **Sq rt** $x^2 = 50$

Ans. $x = \pm\dfrac{7}{2}$ $y = \pm 5$ $x = \pm\dfrac{\sqrt{7}}{3}$ $x = \pm 5\sqrt{2}$

3.2. **SOLVING INCOMPLETE QUADRATIC EQUATIONS LACKING CONSTANT TERM**

Solve:

a) $3x^2 = 21x$

b) $5y(y - 6) - 8y = 2y$

c) $\dfrac{x^2}{5} = \dfrac{x}{15}$

d) $3(y^2 + 8) = 24 - 15y$

Solutions:

a) $\quad 3x^2 - 21x = 0$
$\quad\quad 3x(x - 7) = 0$
$\quad\quad 3x = 0 \quad x - 7 = 0$
$\quad\quad\quad x = 0 \qquad x = 7$

Ans. $x = 0$ or 7

b) $\quad 5y^2 - 30y - 8y = 2y$
$\quad\quad\quad 5y^2 - 40y = 0$
$\quad\quad\quad\quad 5y(y - 8) = 0$
$\quad 5y = 0 \quad y - 8 = 0$
$\quad\quad y = 0 \qquad\quad y = 8$

Ans. $y = 0$ or 8

c) $\qquad 15x^2 = 5x$
$\quad\quad 15x^2 - 5x = 0$
$\quad\quad 5x(3x - 1) = 0$
$\quad\quad 5x = 0 \quad 3x - 1 = 0$
$\quad\quad\quad x = 0 \qquad\quad x = \frac{1}{3}$

Ans. $x = 0$ or $\frac{1}{3}$

d) $\quad 3y^2 + 24 = 24 - 15y$
$\quad\quad\quad 3y^2 + 15y = 0$
$\quad\quad\quad 3y(y + 5) = 0$
$\quad\quad 3y = 0 \quad y + 5 = 0$
$\quad\quad\quad y = 0 \qquad\quad y = -5$

Ans. $y = 0$ or -5

3.3. **SOLVING INCOMPLETE QUADRATICS FOR AN INDICATED LETTER**
Solve for letter indicated:

a) Solve for x:
$$9x^2 - 64y^2 = 0$$

b) Solve for y:
$$4y^2 - 95x^2 = 5x^2$$

c) Solve for a:
$$3a^2 = b^2 - a^2$$

d) Solve for a:
$$a(a - 4) = b^2 - 4a$$

Solutions:

a) $\mathbf{D_9}$ $\quad 9x^2 = 64y^2$

Sq rt $\quad x^2 = \dfrac{64y^2}{9}$

Ans. $\qquad x = \pm \dfrac{8y}{3}$

b) $\mathbf{D_4}$ $\quad 4y^2 = 100x^2$

Sq rt $\quad y^2 = 25x^2$

Ans. $\qquad y = \pm 5x$

c) $\mathbf{D_4}$ $\qquad 4a^2 = b^2$

Sq rt $\quad a^2 = \dfrac{b^2}{4}$

Ans. $\qquad a = \pm \dfrac{b}{2}$

d) $\qquad a^2 - 4a = b^2 - 4a$

Sq rt $\quad a^2 = b^2$

Ans. $\qquad a = \pm b$

3.4. **SOLVING FORMULAS FOR AN INDICATED LETTER**
Solve for *positive* value of letter indicated:

a) Solve for r:
$$\pi r^2 = A$$

b) Solve for r:
$$\pi r^2 h = V$$

c) Solve for a:
$$a^2 + b^2 = c^2$$

d) Solve for s:
$$\dfrac{s^2}{4} = A$$

Solutions:

a) $\quad \mathbf{D_\pi}$ $\quad \dfrac{\pi r^2}{\pi} = \dfrac{A}{\pi}$

Sq rt $\quad r^2 = \dfrac{A}{\pi}$

Ans. $\qquad r = \sqrt{\dfrac{A}{\pi}}$

b) $\quad \mathbf{D_{\pi h}}$ $\quad \dfrac{\pi r^2 h}{\pi h} = \dfrac{V}{\pi h}$

Sq rt $\quad r^2 = \dfrac{V}{\pi h}$

Ans. $\qquad r = \sqrt{\dfrac{V}{\pi h}}$

c) $\quad \mathbf{S_{b^2}}$ $\quad a^2 + b^2 = c^2$

Sq rt $\quad a^2 = c^2 - b^2$

Ans. $\qquad a = \sqrt{c^2 - b^2}$

d) $\quad \mathbf{M_4}$ $\quad 4\left(\dfrac{s^2}{4}\right) = 4A$

Sq rt $\quad s^2 = 4A$

Ans. $\qquad s = 2\sqrt{A}$

4. SOLVING A QUADRATIC EQUATION BY COMPLETING THE SQUARE

The square of a binomial is a perfect trinomial square.
Thus, $x^2 + 6x + 9$ is the perfect trinomial square of $x + 3$.

RULE. If x^2 is the first term of a perfect trinomial square and the term in x is also given, the last term may be found by squaring one-half the coefficient of x.

Thus, if $x^2 + 6x$ is given, 9 is needed to complete the perfect trinomial square $x^2 + 6x + 9$. This last term, 9, is found by squaring $\frac{1}{2}$ of 6, or 3.

To Solve a Quadratic Equation by Completing the Square

Solve $x^2 + 6x - 7 = 0$ by completing the square.

Procedure:

1. Express the equation in the form of $x^2 + px = q$:
2. Square one-half the coefficient of x and add this to both sides:
3. Replace the perfect trinomial square by its binomial squared:
4. Take a square root of both sides. Set the binomial equal to plus or minus the square root of the number on the other side:
5. Solve the two resulting equations:
6. Check both roots in the original equation:

Solution:

1. Change $x^2 + 6x - 7 = 0$
    $$x^2 + 6x = 7$$
2. The square of $\frac{1}{2}(6) = 3^2 = 9$. Add 9 to get
    $$x^2 + 6x + 9 = 7 + 9$$
3. $$(x + 3)^2 = 16$$
4. **Sq rt:**
    $$x + 3 = \pm 4$$
5. $x + 3 = 4 \quad x + 3 = -4$
    $\quad\quad x = 1 \quad\quad\quad x = -7$
6. (Check is left to the student.)

Ans. $x = 1$ or -7

4.1. **COMPLETING THE PERFECT TRINOMIAL SQUARE**

Complete each perfect trinomial square and state its binomial squared:

a) $x^2 + 14x + ?$ *Ans.* The square of $\frac{1}{2}(14) = 7^2 = 49$. Add 49 to get
$$x^2 + 14x + 49 = (x + 7)^2.$$

b) $x^2 - 20x + ?$ The square of $\frac{1}{2}(-20) = (-10)^2 = 100$. Add 100 to get
$$x^2 - 20x + 100 = (x - 10)^2.$$

c) $x^2 + 5x + ?$ The square of $\frac{1}{2}(5) = (\frac{5}{2})^2 = \frac{25}{4}$. Add $\frac{25}{4}$ to get $x^2 + 5x + \frac{25}{4} = (x + \frac{5}{2})^2$.

d) $y^2 + \frac{4}{3}y + ?$ The square of $\frac{1}{2}(\frac{4}{3}) = (\frac{2}{3})^2 = \frac{4}{9}$. Add $\frac{4}{9}$ to get $y^2 + \frac{4}{3}y + \frac{4}{9} = (y + \frac{2}{3})^2$.

4.2. **SOLVING QUADRATIC EQUATIONS BY COMPLETING THE SQUARE**

Solve by completing the square: *a)* $x^2 + 14x - 32 = 0$ *b)* $x(x + 18) = 19$

Procedure:

1. Express as $x^2 + px = q$:
2. Add the square of half the coefficient of x to both sides:
3. Replace the perfect trinomial square by its binomial squared:
4. Take a square root of both sides:
5. Solve each resulting equation:
6. Check both roots:

Solutions:

1. $x^2 + 14x - 32 = 0$
    $\quad\quad x^2 + 14x = 32$
2. Square of $\frac{1}{2}(14) = 7^2 = 49$
    A_{49} $x^2 + 14x + 49 = 32 + 49$
3. $x^2 + 14x + 49 = 81$
    $\quad\quad (x + 7)^2 = 81$
4. **Sq rt** $x + 7 = \pm 9$
5. $x + 7 = 9 \quad x + 7 = -9$
    $\quad x = 2 \quad\quad x = -16$
6. (Check is left to the student.)

Ans. $x = 2$ or -16

1. $x(x + 18) = 19$
    $\quad\quad x^2 + 18x = 19$
2. Square of $\frac{1}{2}(18) = 9^2 = 81$
    A_{81} $x^2 + 18x + 81 = 19 + 81$
3. $x^2 + 18x + 81 = 100$
    $\quad\quad (x + 9)^2 = 100$
4. **Sq rt** $x + 9 = \pm 10$
5. $x + 9 = 10 \quad x + 9 = -10$
    $\quad x = 1 \quad\quad x = -19$

Ans. $x = 1$ or -19

4.3. **EQUATIONS REQUIRING FRACTIONS TO COMPLETE THE SQUARE**
Solve by completing the square:

a) $x(x - 5) = -4$

b) $5t^2 - 4t = 33$

Solutions:

a) $$x^2 - 5x = -4$$

$$\text{Square of } \frac{1}{2}(-5) = \left(-\frac{5}{2}\right)^2 = \frac{25}{4}$$

$\mathbf{A}_{25/4}$ $$x^2 - 5x + \frac{25}{4} = \frac{25}{4} - 4$$

$$\left(x - \frac{5}{2}\right)^2 = \frac{9}{4}$$

Sq rt $$x - \frac{5}{2} = \pm\frac{3}{2}$$

$$x - \frac{5}{2} = \frac{3}{2} \qquad x - \frac{5}{2} = -\frac{3}{2}$$

$$x = 4 \qquad x = 1$$

Ans. $x = 4$ or 1

b) $$t^2 - \frac{4t}{5} = \frac{33}{5}$$

$$\text{Square of } \frac{1}{2}\left(-\frac{4}{5}\right) = \left(-\frac{2}{5}\right)^2 = \frac{4}{25}$$

$\mathbf{A}_{4/25}$ $$t^2 - \frac{4t}{5} + \frac{4}{25} = \frac{33}{5} + \frac{4}{25}$$

$$\left(t - \frac{2}{5}\right)^2 = \frac{169}{25}$$

Sq rt $$t - \frac{2}{5} = \pm\frac{13}{5}$$

$$t - \frac{2}{5} = \frac{13}{5} \qquad t - \frac{2}{5} = -\frac{13}{5}$$

$$t = 3 \qquad t = -\frac{11}{5}$$

Ans. $t = 3$ or $-\frac{11}{5}$

4.4. **SOLVING THE QUADRATIC EQUATION IN STANDARD FORM**

If $ax^2 + bx + c = 0$, then $x = \dfrac{-b \pm \sqrt{b^2 - 4ac}}{2a}$. Prove this by completing the square.

Solution:

$$ax^2 + bx + c = 0$$

Tr $$ax^2 + bx = -c$$

$\mathbf{D}_a$ $$x^2 + \frac{bx}{a} = -\frac{c}{a}$$

$$\text{Square of } \frac{1}{2}\left(\frac{b}{a}\right) = \left(\frac{b}{2a}\right)^2 = \frac{b^2}{4a^2}$$

$\mathbf{A}_{b^2/4a^2}$ $$x^2 + \frac{bx}{a} + \frac{b^2}{4a^2} = \frac{b^2}{4a^2} - \frac{c}{a}$$

$$\left(x + \frac{b}{2a}\right)^2 = \frac{b^2 - 4ac}{4a^2}$$

▶ **Sq rt** $$x + \frac{b}{2a} = \pm\frac{\sqrt{b^2 - 4ac}}{2a}$$

$$x = -\frac{b}{2a} \pm \frac{\sqrt{b^2 - 4ac}}{2a}$$

By combining fractions,

$$x = \frac{b \pm \sqrt{b^2 - 4ac}}{2a} \qquad \textit{Ans.}$$

5. SOLVING A QUADRATIC EQUATION BY QUADRATIC FORMULA

Quadratic formula: If $ax^2 + bx + c = 0$, then $x = \dfrac{-b \pm \sqrt{b^2 - 4ac}}{2a}$.

(See proof of this in Example 4.4 of the previous unit.)

To Solve a Quadratic Equation by the Quadratic Formula

Solve $x^2 - 4x = -3$ by quadratic formula.

Procedure:	**Solution:**
1. Express in form of $ax^2 + bx + c = 0$:	**1.** $\quad x^2 - 4x = -3$ $\qquad x^2 - 4x + 3 = 0$
2. State the values of a, b, and c:	**2.** $\quad a = 1, b = -4, c = 3$
3. Substitute the values of a, b, and c in the formula:	**3.** $\quad x = \dfrac{-b \pm \sqrt{b^2 - 4ac}}{2a}$ $\qquad x = \dfrac{-(-4) \pm \sqrt{(-4)^2 - 4(1)(3)}}{2(1)}$
4. Solve for x:	**4.** $\quad x = \dfrac{+4 \pm \sqrt{16 - 12}}{2}$ $\qquad x = \dfrac{+4 \pm \sqrt{4}}{2}$ $\qquad x = \dfrac{4 \pm 2}{2}$ $\qquad x = \dfrac{4 + 2}{2} \quad x = \dfrac{4 - 2}{2}$ $\qquad x = 3 \qquad x = 1$
5. Check each root in the original equation:	**5.** (Check left to student.) **Ans.** $x = 3$ or 1

5.1. **FINDING $b^2 - 4ac$ IN THE QUADRATIC FORMULA**

For each equation, state the values of a, b, and c. Then find the value of $b^2 - 4ac$.

$a)$ $3x^2 + 4x - 5 = 0$ $\qquad\qquad$ $b)$ $x^2 - 2x - 10 = 0$ $\qquad\qquad$ $c)$ $3x^2 - 5x = 10$

Solutions:

$a)$ $\left.\begin{array}{l} a = 3 \\ b = 4 \\ c = -5 \end{array}\right\}$ $\begin{array}{l} b^2 - 4ac \\ 4^2 - 4(3)(-5) \\ 16 + 60 \end{array}$ $\qquad$ $b)$ $\left.\begin{array}{l} a = 1 \\ b = -2 \\ c = -10 \end{array}\right\}$ $\begin{array}{l} b^2 - 4ac \\ (-2)^2 - 4(1)(-10) \\ 4 + 40 \end{array}$

Ans.   76 $\qquad\qquad\qquad\qquad\qquad\qquad\qquad$ *Ans.*   44

$c)$ $\quad$ **Tr** $\qquad$ $3x^2 - 5x - 10 = 0$

$\left.\begin{array}{l} a = 3 \\ b = -5 \\ c = -10 \end{array}\right\}$ $\begin{array}{l} b^2 - 4ac \\ (-5)^2 - 4(3)(-10) \\ 25 + 120 \end{array}$

Ans.   145

5.2. **EXPRESSING ROOTS IN SIMPLIFIED RADICAL FORM**

Express the roots of each equation in simplified radical form, using the value of $b^2 - 4ac$ obtained in 5.1:

$a)$ $3x^2 + 4x - 5 = 0$ $\qquad\qquad$ $b)$ $x^2 - 2x - 10 = 0$ $\qquad\qquad$ $c)$ $3x^2 - 5x = 10$

Solutions:

$a)$ $\qquad\qquad\qquad x = \dfrac{-b \pm \sqrt{b^2 - 4ac}}{2a}$

$$\left.\begin{array}{l} a = 3 \\ b = 4 \\ c = -5 \end{array}\right\} \quad x = \frac{-4 \pm \sqrt{76}}{6}$$

b) $\qquad x = \dfrac{-b \pm \sqrt{b^2 - 4ac}}{2a}$

$$\left.\begin{array}{l} a = 1 \\ b = -2 \\ c = -10 \end{array}\right\} \quad x = \frac{2 \pm \sqrt{44}}{2}$$

c) $\qquad x = \dfrac{-b \pm \sqrt{b^2 - 4ac}}{2a}$

$$\left.\begin{array}{l} a = 3 \\ b = -5 \\ c = -10 \end{array}\right\} \quad x = \frac{5 \pm \sqrt{145}}{6}$$

5.3.   EVALUATING ROOTS IN RADICAL FORM

Solve for x, to the nearest tenth, using the simplified radical form obtained in 5.2. (Find the square root to two decimal places using a calculator.

a) $3x^2 + 4x - 5 = 0$ b) $x^2 - 2x - 10 = 0$ c) $3x^2 - 5x = 10$

Solutions:

a) $x = \dfrac{-4 \pm \sqrt{76}}{6}$ b) $x = \dfrac{2 \pm \sqrt{44}}{2}$ c) $x = \dfrac{5 \pm \sqrt{145}}{6}$

$x = \dfrac{-4 + 8.71}{6}$ $x = \dfrac{4 - 8.71}{6}$ $x = \dfrac{2 + 6.63}{2}$ $x = \dfrac{2 - 6.63}{2}$ $x = \dfrac{5 + 12.04}{6}$ $x = \dfrac{5 - 12.04}{6}$

$= \dfrac{4.71}{6}$ $= \dfrac{-12.71}{6}$ $= \dfrac{8.63}{2}$ $= \dfrac{-4.63}{2}$ $= \dfrac{17.04}{6}$ $= \dfrac{-7.04}{6}$

Ans. $x = 0.8$ or $x = -2.1$ *Ans.* $x = 4.3$ or $x = -2.3$ *Ans.* $x = 2.8$ or $x = -1.2$

6. SOLVING QUADRATIC EQUATIONS GRAPHICALLY

To Solve a Quadratic Equation Graphically

Solve graphically: $x^2 - 5x + 4 = 0$.

Procedure:

1. Express in form of $ax^2 + bx + c = 0$:
2. Graph the curve $y = ax^2 + bx + c$.
   (Curve is called a parabola.)

(**NOTE:**   You may use a graphing calculator to graph the curve.)

3. Find where $y = 0$ intersects $y = ax^2 + bx + c$:
   The values of x at the points of intersection are the roots of $ax^2 + bx + c = 0$.

(**NOTE:**   Think of $ax^2 + bx + c = 0$ as the result of combining $y = ax^2 + bx + c$ with $y = 0$.)

Solution:

1. $x^2 - 5x + 4 = 0$
2. See Fig. 13-1.

GRAPH of
$y = x^2 - 5x + 4$
(parabola)

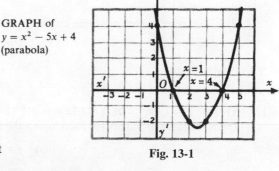

Fig. 13-1

3. $x = 1$ and $x = 4$ *Ans.*

The following is the graphic solution of a quadratic equation in greater detail.

To Solve a Quadratic Equation Graphically

Solve graphically: $x^2 - 2x = 3$

Procedure:

1. **Express in form** $ax^2 + bx + c = 0$.
2. **Graph parabola** $y = ax^2 + bx + c$.

   a) Obtain a table of values, using a

   suitable sequence of values for x:

   (This may be done by finding the

   value of $-\dfrac{b}{2a}$ and choosing values

   of x greater or smaller than $-\dfrac{b}{2a}$.)

   b) Join plotted points: (Note that

   $x = -\dfrac{b}{2a} = 1$ is the folding line

   or axis of symmetry of the parabola.)

   See Fig. 13-2.

Solution:

1. **Tr** $x^2 - 2x - 3 = 0$
2. Graph $x^2 - 2x - 3 = y$

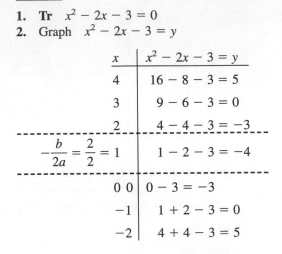

x	$x^2 - 2x - 3 = y$
4	$16 - 8 - 3 = 5$
3	$9 - 6 - 3 = 0$
2	$4 - 4 - 3 = -3$
$-\dfrac{b}{2a} = \dfrac{2}{2} = 1$	$1 - 2 - 3 = -4$
0 0	$0 - 3 = -3$
-1	$1 + 2 - 3 = 0$
-2	$4 + 4 - 3 = 5$

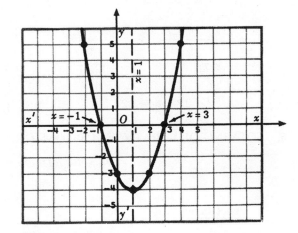

Fig. 13-2

3. Find roots where parabola crosses x axis.

3. $x = -1$ or 3 *Ans.*

6.1. **OBTAINING TABLES OF VALUES FOR A PARABOLA**

Obtain a table of values for each, using the indicated sequence of values for x:

a) $y = x^2 - 4$

   for $x = -3$ to $+3$

b) $y = x^2 - 3x$

   for $x = -1$ to 4

c) $y = x^2 - 3x - 4$

   for $x = -1$ to 5

Solutions:

a) $x^2 - 4 = y$

x	$x^2 - 4 = y$
3	$9 - 4 = 5$
2	$4 - 4 = 0$
1	$1 - 4 = -3$
→0	$0 - 4 = -4$
−1	$1 - 4 = -3$
−2	$4 - 4 = 0$
−3	$9 - 4 = 5$

b) $x^2 - 3x = y$

x	$x^2 - 3x = y$
4	$16 - 12 = 4$
3	$9 - 9 = 0$
2	$4 - 6 = -2$
$x = 1\frac{1}{2} \rightarrow$	
1	$1 - 3 = -2$
0	$0 + 0 = 0$
−1	$1 + 3 = 4$

c) $x^2 - 3x - 4 = y$

x	$x^2 - 3x - 4 = y$
5	$25 - 15 - 4 = 6$
4	$16 - 12 - 4 = 0$
3	$9 - 9 - 4 = -4$
2	$4 - 6 - 4 = -6$
$x = 1\frac{1}{2} \rightarrow$	
1	$1 - 3 - 4 = -6$
0	$0 + 0 - 4 = -4$
−1	$1 + 3 - 4 = 0$

(The arrow in each table indicates $x = -\dfrac{b}{2a}$, the axis of symmetry of the parabola.)

6.2. GRAPHING PARABOLAS $y = ax^2 + bx + c$

Graph each, using the table of values obtained in 6.1:

a) $y = x^2 - 4$ b) $y = x^2 - 3x$ c) $y = x^2 - 3x - 4$

Solutions:

See Fig. 13-3a   See Fig. 13-3b   See Fig. 13-3c

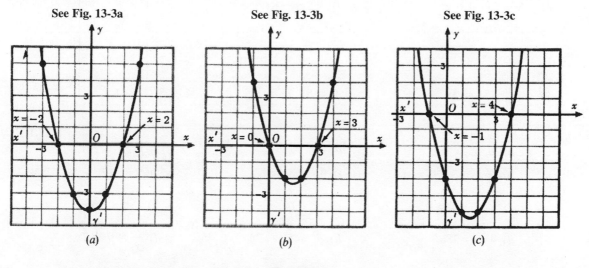

(a)   (b)   (c)

Fig. 13-3

6.3. FINDING ROOTS GRAPHICALLY

Find the roots of each, using the parabolas obtained in example 6.2:

a) $0 = x^2 - 4$ b) $0 = x^2 - 3x$ c) $0 = x^2 - 3x - 4$

Solutions:

(In each case, find the values of x at the points where the parabola crosses the x axis.)

Ans.   a) $x = -2$ or 2   b) $x = 0$ or 3   c) $x = -1$ or 4

Supplementary Problems

13.1. Express in standard quadratic equation form $ax^2 + bx + c = 0$ so that a has a positive value: **(1.1)**

a) $3(x^2 - 5) = 4x$

b) $2x(x + 4) = 42$

c) $\dfrac{7}{x} - 5 = 2x$

d) $\dfrac{6}{x} + x = \dfrac{11}{2}$

e) $3x - 9 = x^2 - 7x$

f) $x(2x - 7) = 3x^2 - 8$

g) $\sqrt{2x^2 + 3x} = x$

h) $x + 1 = \sqrt{3x + 7}$

Ans. a) $3x^2 - 4x - 15 = 0$

b) $2x^2 + 8x - 42 = 0$

c) $0 = 2x^2 + 5x - 7$

d) $2x^2 - 11x + 12 = 0$

e) $0 = x^2 - 10x + 9$

f) $0 = x^2 + 7x - 8$

g) $x^2 + 3x = 0$

h) $x^2 - x - 6 = 0$

13.2. Express in standard quadratic equation form, so that a has a positive value. Then state the values of a, b, and c.   **(1.2)**

		Standard Form	a	b	c
a) $x^2 = 5x - 4$	*Ans.* a)	$x^2 - 5x + 4 = 0$	1	-5	$+4$
b) $20 + 6x = 2x^2$	b)	$0 = 2x^2 - 6x - 20$	2	-6	-20
c) $3x^2 = -5x$	c)	$3x^2 + 5x = 0$	3	$+5$	0
d) $18 = 2x^2$	d)	$0 = 2x^2 - 18$	2	0	-18
e) $x(8 - 2x) = 6$	e)	$0 = 2x^2 - 8x + 6$	2	-8	$+6$
f) $7(x^2 - 9) = x(x - 5)$	f)	$6x^2 + 5x - 63 = 0$	6	$+5$	-63
g) $\dfrac{10}{x} + 1 = 4x$	g)	$0 = 4x^2 - x - 10$	4	-1	-10
h) $\sqrt{x^2 - 5x} = 3x$	h)	$0 = 8x^2 + 5x$	8	-1	0

13.3. Solve by factoring: **(2.1, 2.2)**

a) $x^2 - 5x + 6 = 0$

b) $y^2 + y - 20 = 0$

c) $w^2 - 64 = 0$

d) $4a^2 = 28a$

e) $4x^2 = 1$

f) $9b^2 = 3b$

g) $c^2 + 6c = -8$

h) $d^2 = 5d + 24$

i) $p(3p + 20) = 7$

j) $5 = r(2r + 3)$

k) $t + 8 = \dfrac{20}{t}$

l) $\dfrac{7}{x} = 9 - 2x$

Ans. a) $(x - 2)(x - 3) = 0$

$x = 2$ or 3

b) $(y + 5)(y - 4) = 0$

$y = -5$ or 4

c) $(w - 8)(w + 8) = 0$

$w = 8$ or -8

d) $4a(a - 7) = 0$

$a = 0$ or 7

e) $(2x + 1)(2x - 1) = 0$

$x = -\tfrac{1}{2}$ or $\tfrac{1}{2}$

f) $3b(3b - 1) = 0$

$b = 0$ or $\tfrac{1}{3}$

g) $(c + 4)(c + 2) = 0$

$c = -4$ or -2

h) $(d - 8)(d + 3) = 0$

$d = 8$ or -3

i) $(3p - 1)(p + 7) = 0$

$p = \tfrac{1}{3}$ or -7

j) $0 = (r - 1)(2r + 5)$

$r = 1$ or $-\tfrac{5}{2}$

k) $(t - 2)(t + 10) = 0$

$t = 2$ or -10

l) $(2x - 7)(x - 1) = 0$

$x = \tfrac{7}{2}$ or 1

13.4. Solve by factoring: **(2.3)**

a) $\dfrac{3x}{5} \dfrac{2}{} = \dfrac{8}{x}$

c) $(y - 2)(3y - 1) = 100$

e) $x + 1 = \sqrt{5x + 1}$

b) $\dfrac{21}{x - 3} = x - 7$

d) $(y + 3)(y - 3) = 2y - 1$

f) $\sqrt{x - 2} = x - 8$

Ans. a) $(x - 4)(3x + 10) = 0$

$x = 4$ or $-\frac{10}{3}$

c) $(y - 7)(3y + 14) = 0$

$y = 7$ or $-\frac{14}{3}$

e) $x(x - 3) = 0$

$x = 0$ or 3

b) $0 = x(x - 10)$

$x = 0$ or 10

d) $(y - 4)(y + 2) = 0$

$y = 4$ or -2

f) $0 = (x - 9)(x - 4)$

$x = 9$ (4 is an extraneous root.)

13.5. Find a number such that: **(2.4)**

a) Its square is 12 more than the number

b) Its square decreased by 3 times the number is 18

c) The product of the number and 4 less than the number is 32

d) The square of 1 more than the number is 4 more than 4 times the number

Ans.   a) $n^2 = n + 12$
4 or -3

b) $n^2 - 3n = 18$
6 or -3

c) $n(n - 4) = 32$
8 or -4

d) $(n + 1)^2 = 4n + 4$
3 or -1

13.6. Find two consecutive integers such that: **(2.4)**

a) The sum of their squares is 13

b) Their product is 30

c) The square of the first added to twice the second is 5

d) The product of the first and twice the second is 40

Ans.   a) $n^2 + (n + 1)^2 = 13$
Either 2, 3 or $-3, -2$

c) $n^2 + 2(n + 1) = 5$
Either 1, 2 or $-3, -2$

b) $n(n + 1) = 30$
Either 5, 6 or $-6, -5$

d) $n(2n + 2) = 40$
Either 4, 5 or $-5, -4$

13.7. The length of a rectangle lot is 2 km more than its width. Find the length and the width: **(2.5)**

a) If the area is 80 km^2

b) If the area is 48 km^2 when each dimension is made 2 km longer

c) If the area is 30 km^3 when the width is doubled

d) If the area is increased 70 km^2 when the width is tripled

e) If the area is decreased 7 km^2 when the length is made 5 km shorter and the width doubled

Reject all negative roots.

Ans. a) $w(w + 2) = 80$
10 and 8 km

c) $2w(w + 2) = 30$
5 and 3 cm

e) $w(w + 2) - 7 = 2w(w - 3)$
9 and 7 km

b) $(w + 2)(w + 4) = 48$
6 and 4 km

d) $w(w + 2) + 70 = 3w(w + 2)$
7 and 5 km

13.8. Solve: (Leave irrational answers in radical form.)   **(3.1)**

a) $2x^2 = 8$

c) $3a^2 - 4 = 5$

e) $\dfrac{2x}{3} = \dfrac{6}{x}$

g) $p^2 = 4(p^2 - 10) - 8$

b) $4 = 100y^2$

d) $b(b + 2) = 2b + 10$

f) $\dfrac{x + 2}{5} = \dfrac{9}{x - 2}$

h) $r(r - 6) = 3(7 - 2r)$

Ans. a) $x = \pm 2$

c) $a = \pm \sqrt{3}$

e) $x = \pm 3$

g) $p = \pm 4$

b) $y = \pm \frac{1}{5}$

d) $b = \pm \sqrt{10}$

f) $x = \pm 7$

h) $r = \pm \sqrt{21}$

13.9. Solve: (3.2)

a) $2x^2 = 8x$ c) $3a^2 - 4a = 5a$ e) $\dfrac{x^2}{40} = \dfrac{x}{5}$ g) $(r + 3)(r - 3) = 3(5r - 3)$

b) $8x^2 = 2x$ d) $2b^2 + 12b = b^2$ f) $\dfrac{y^2 + 5}{y + 1} = 5$ h) $(s + 4)(s + 7) = 2(14 - 3s)$

Ans. a) $x = 0$ or 4 c) $a = 0$ or 3 e) $x = 0$ or 8 g) $r = 0$ or 15
   b) $x = 0$ or $\frac{1}{4}$ d) $b = 0$ or -12 f) $y = 0$ or 5 h) $s = 0$ or -17

13.10. Solve for the letter indicated: (3.3)

a) $2x^2 = 50a^2$ for x c) $b^2y^2 = 49$ for y e) $aw^2 = a^2w$ for w
b) $3x^2 = 6bx$ for x d) $4y^2 = 25c^2$ for y f) $6v^2 = h^2 - 3v^2$ for v

Ans. a) $x = \pm 5a$ b) $x = 0$ or $2b$ c) $y = \pm \dfrac{7}{b}$ d) $y = \pm \dfrac{5c}{2}$

e) $w = 0$ or a f) $v = \pm \dfrac{h}{3}$

13.11. Solve for the *positive* value of the letter indicated: (3.4)

a) $S = 16t^2$ for t c) $S = 4\pi r^2$ for r e) $a^2 + b^2 = c^2$ for b

b) $K = \frac{1}{2}mv^2$ for v d) $F = \dfrac{mv^2}{r}$ for v f) $A = \pi(R^2 - r^2)$ for R

Ans. a) $t = \dfrac{\sqrt{S}}{4}$,   b) $v = \sqrt{\dfrac{2K}{m}}$ c) $r = \dfrac{1}{2}\sqrt{\dfrac{S}{\pi}}$ d) $v = \sqrt{\dfrac{Fr}{m}}$ e) $b = \sqrt{c^2 - a^2}$

f) $R = \sqrt{\dfrac{A + \pi r^2}{\pi}}$

13.12. Complete each perfect binomial square and state its binomial squared: (4.1)

a) $a^2 + 12a + ?$ c) $x^2 + 7x + ?$ e) $w^2 + 20w + ?$
b) $c^2 - 18c + ?$ d) $y^2 - 11y + ?$ f) $r^2 - 30r + ?$

Ans. a) $a^2 + 12a + 36$ c) $x^2 + 7x + \frac{49}{4}$ e) $w^2 + 20w + 100$
   $(a + 6)^2$ $(x + \frac{7}{2})^2$ $(w + 10)^2$

b) $c^2 - 18c + 81$ d) $y^2 - 11y + \frac{121}{4}$ f) $r^2 - 30r + 225$
   $(c - 9)^2$ $(y - \frac{11}{2})^2$ $(r - 15)^2$

13.13. Solve by completing the square: (Use the answers in 13.12 to help you.) (4.2, 4.3)

a) $a^2 + 12a = 45$ c) $x^2 + 7x = 8$ e) $w^2 + 20w = -19$
b) $c^2 - 18c = -65$ d) $y^2 - 11y = -28$ f) $r^2 - 30r = 99$

Ans. a) $a = 3$ or -15 c) $x = 1$ or -8 e) $w = -1$ or -19
   [from $(a + 6)^2 = 81$]   [from $(x + \frac{7}{2})^2 = \frac{81}{4}$]   [from $(w + 10)^2 = \pm 9$]

b) $c = 5$ or 13 d) $y = 4$ or 7 f) $r = -3$ or 33
   [from $(c - 9)^2 = 16$]   [from $(y - \frac{11}{2})^2 = \frac{9}{4}$]   [from $(r - 15)^2 = 324$]

13.14. For each equation, state the values of a, b, and c. Then find the value of $b^2 - 4ac$. (5.1)

a) $2x^2 + 5x + 1 = 0$ d) $2x^2 - 5x = 4$ g) $x^2 = 10 - 6x$
b) $x^2 + 7x = -5$ e) $5x^2 = 10x - 4$ h) $9x^2 = 2 - x$
c) $3x^2 - 2 = 4x$ f) $6x^2 = 10x - 3$

Ans. a) $a = 2, b = 5, c = 1$ d) $a = 2, b = -5, c = -4$ g) $a = 1, b = 6, c = -10$
   $b^2 - 4ac = 17$ $b^2 - 4ac = 57$ $b^2 - 4ac = 76$
b) $a = 1, b = 7, c = 5$ e) $a = 5, b = -10, c = 4$ h) $a = 9, b = 1, c = -2$
   $b^2 - 4ac = 29$ $b^2 - 4ac = 20$ $b^2 - 4ac = 73$
c) $a = 3, b = -4, c = -2$ f) $a = 6, b = -10, c = 3$
   $b^2 - 4ac = 40$ $b^2 - 4ac = 28$

13.15. Express the roots of each equation in simplified radical form, using the value of $b^2 - 4ac$ found in Problem 13.14:   **(5.2)**

a) $2x^2 + 5x + 1 = 0$ c) $3x^2 - 2 = 4x$ e) $5x^2 = 10x - 4$ g) $x^2 = 10 - 6x$

b) $x^2 + 7x = -5$ d) $2x^2 - 5x = 4$ f) $6x^2 = 10x - 3$ h) $9x^2 = 2 - x$

Ans. a) $x = \dfrac{-5 \pm \sqrt{17}}{4}$ c) $x = \dfrac{4 \pm \sqrt{40}}{6}$ e) $x = \dfrac{10 \pm \sqrt{20}}{10}$ g) $x = \dfrac{-6 \pm \sqrt{76}}{2}$

b) $x = \dfrac{-7 \pm \sqrt{29}}{2}$ d) $x = \dfrac{5 \pm \sqrt{57}}{4}$ f) $x = \dfrac{10 \pm \sqrt{28}}{12}$ h) $x = \dfrac{-1 \pm \sqrt{73}}{18}$

13.16. Solve for x, correct to the nearest tenth, using the simplified radical form found in Problem 13.15 and a calculator.   **(5.3)**

a) $2x^2 + 5x + 1 = 0$ c) $3x^2 - 2 = 4x$ e) $5x^2 = 10x - 4$ g) $x^2 = 10 - 6x$

b) $x^2 + 7x = -5$ d) $2x^2 - 5x = 4$ f) $6x^2 = 10x - 3$ h) $9x^2 = 2 - x$

Ans. a) $x = -2.3$ or -0.2 c) $x = 1.7$ or -0.4 e) $x = 1.4$ or 0.6 g) $x = 1.4$ or -7.4

b) $x = -6.2$ or -0.8 d) $x = 3.1$ or -0.6 f) $x = 1.3$ or 0.4 h) $x = 0.4$ or -0.5

13.17. Solve, correct to the nearest tenth: **(5.3)**

a) $2y^2 = -3y + 1$ c) $2w^2 = 3w + 5$ e) $x + 5 = \dfrac{5}{x}$

b) $x(x - 4) = -2$ d) $6v^2 + 1 = 5v$ f) $\dfrac{x^2}{2} = 4x - 1$

Ans. a) $y = 0.3$ or -1.8 c) $w = 2.5$ or -1 e) $x = 0.9$ or -5.9

b) $x = 3.4$ or 0.6 d) $v = 0.5$ or 0.3 f) $x = 7.7$ or 0.3

13.18. Solve graphically: (See sketches in Fig. 13.4(a) to (c).)   **(6.1 to 6.3.)**

a) Graph $y = x^2 - 9$ from $x = -4$ to $x = 4$ and solve $x^2 - 9 = 0$ graphically.

b) Graph $y = x^2 + 3x - 4$ from $x = -5$ to $x = 2$ and solve $x^2 + 3x = 4$ graphically.

c) Graph $y = x^2 - 6x + 9$ from $x = 1$ to $x = 5$ and solve $x^2 = 6x - 9$ graphically.

d) Graph $y = x^2 + 8x + 16$ from $x = -6$ to $x = -2$ and solve $x^2 + 8x = -16$ graphically.

e) Graph $y = x^2 - 2x + 4$ from $x = -1$ to $x = 3$ and show that $x^2 - 2x + 4 = 0$ has no real roots.

Ans. a) $x = 3$ or -3 b) $x = -4$ or 1

c) $x = 3$, equal roots since the parabola touches x axis at one point.

d) $x = -4$, equal roots since the parabola touches x axis at one point.

e) Roots are not real (imaginary) since the parabola does not meet x axis.

13.19. Solve graphically: (See sketches in Fig. 13-4(a) to (c).)   **(6.1 to 6.3)**

a) $x^2 - 1 = 0$ c) $x^2 - 4x = 0$ e) $x^2 - 2x = 8$

b) $4x^2 - 9 = 0$ d) $2x^2 + 7x = 0$ f) $4x^2 = 12x - 9$

Ans. a) $x = -1$ or 1 c) $x = 0$ or 4 e) $x = 4$ or -2

b) $x = 1\frac{1}{2}$ or $-1\frac{1}{2}$ d) $x = 0$ or $-3\frac{1}{2}$ f) $x = 1\frac{1}{2}, 1\frac{1}{2}$

Sketches of graphs needed in Problems 13.18 and 13.19, showing relationship of parabola and x axis:

13.18a, b; 13.19a to e 13.18c, d; 13.19f 13.18e

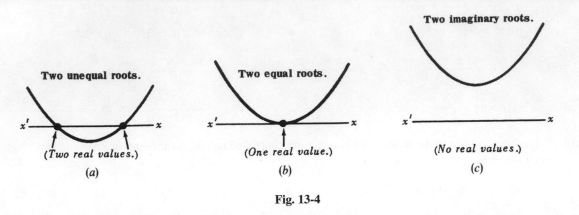

Fig. 13-4

13.20. Graph the following curves   **(6.1 to 6.3)**

a) $y = x^2$

b) $y = \sqrt{x}$

For x greater than or equal to zero, using a graphing calculator, where do these curves intersect?

Ans.   At (0, 0) and (1, 1).

CHAPTER 14

The Pythagorean Theorem and Similar Triangles

1. LAW OF PYTHAGORAS

The square of the hypotenuse:
In right triangle ABC, if C is a right angle,

$$c^2 = a^2 + b^2$$

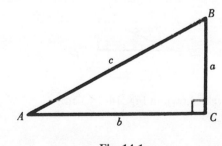

Fig. 14-1

NOTE: In a triangle, the small letter for a side should agree with the capital letter for the vertex of the angle opposite that side. Thus, side a is opposite angle A, etc.

> **Pythagorean Theorem:** In a right triangle, the **square** of the hypotenuse equals the **sum of the squares** of the two legs.

Pythagoras, a famous Greek mathematician and philosopher, lived about 500 B.C.
The square of either arm:
By transposition, $a^2 = c^2 - b^2$ and $b^2 = c^2 - a^2$

> **Transformed Pythagorean Theorem:** In a right triangle, the **square** of either leg equals the **difference of the squares** of the hypotenuse and the other leg.

To test for a right triangle, use the following rule:

> **Test Rule for a Right Triangle:** If $c^2 = a^2 + b^2$ applies to the three sides of a triangle, then the triangle is a right triangle; but if $c^2 \neq a^2 + b^2$, then the triangle is not a right triangle.

Distance between Two Points on a Graph

If d is the distance between $P_1(x_1, y_1)$, and $P_2(x_2, y_2)$,

$$d^2 = (x_2 - x_1)^2 + (y_2 - y_1)^2$$

Thus, the distance equals 5 from point $(2, 5)$ to point $(6, 8)$, as follows:

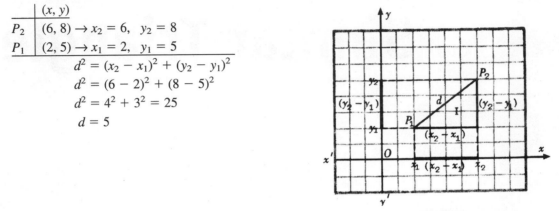

	(x, y)
P_2	$(6, 8) \rightarrow x_2 = 6,\ \ y_2 = 8$
P_1	$(2, 5) \rightarrow x_1 = 2,\ \ y_1 = 5$

$$d^2 = (x_2 - x_1)^2 + (y_2 - y_1)^2$$
$$d^2 = (6 - 2)^2 + (8 - 5)^2$$
$$d^2 = 4^2 + 3^2 = 25$$
$$d = 5$$

Fig. 14-2

In the following exercises, express each irrational answer in simplest radical form, unless otherwise indicated.

1.1. **FINDING THE HYPOTENUSE OF A RIGHT TRIANGLE:** $c^2 = a^2 + b^2$

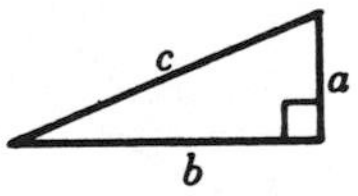

Fig. 14-3

Find hypotenuse c in the right triangle shown in Fig. 14-3 when:

a) $a = 12, b = 9$	b) $a = 3, b = 7$	c) $a = 3, b = 6$	d) $a = 3, b = 3\sqrt{3}$

<u>Solutions:</u>

a) $c^2 = a^2 + b^2$	b) $c^2 = a^2 + b^2$	c) $c^2 = a^2 + b^2$	d) $c^2 = a^2 + b^2$
$c^2 = 12^2 + 9^2$	$c^2 = 3^2 + 7^2$	$c^2 = 3^2 + 6^2$	$c^2 = 3^2 + (3\sqrt{3})^2$
$c^2 = 144 + 81$	$c^2 = 9 + 49$	$c^2 = 9 + 36$	$c^2 = 9 + 27$
$c^2 = 225$	$c^2 = 58$	$c^2 = 45$	$c^2 = 36$
$c = \sqrt{225}$	$c = \sqrt{58}$ *Ans.*	$c = \sqrt{45}$	$c = \sqrt{36}$
$c = 15$ *Ans.*		$c = 3\sqrt{5}$ *Ans.*	$c = 6$ *Ans.*

NOTE: Since the hypotenuse is to be considered as positive only, reject the negative answer obtainable in each case. Thus, if $c^2 = 225$, c may equal 15 or -15. Reject (since it represents a length), the negative (-15).

1.2.   FINDING A LEG OF A RIGHT TRIANGLE: $a^2 = c^2 - b^2$ **OR** $b^2 = c^2 - a^2$

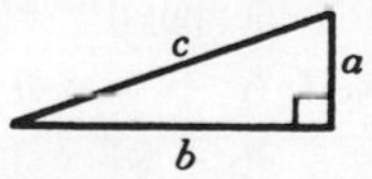

Fig. 14-4

In the right triangle shown in Fig. 14-4, find each missing leg when:

a) $b = 5, c = 13$ b) $a = 24, c = 25$ c) $b = 6, c = 8$ d) $a = 4\sqrt{3}, c = 8$

Solutions:

a) $a^2 = c^2 - b^2$ b) $b^2 = c^2 - a^2$ c) $a^2 = c^2 - b^2$ d) $b^2 = c^2 - a^2$

$\quad a^2 = 13^2 - 5^2 \qquad\quad b^2 = 25^2 - 24^2 \qquad\quad a^2 = 8^2 - 6^2 \qquad\quad b^2 = 8^2 - (1\sqrt{3})^2$

$\quad a^2 = 169 - 25 \qquad\quad b^2 = 625 - 576 \qquad\quad a^2 = 64 - 36 \qquad\quad b^2 = 64 - 48$

$\quad a^2 = 144 \qquad\qquad\quad b^2 = 49 \qquad\qquad\qquad a^2 = 28 \qquad\qquad\quad b^2 = 16$

$\quad a = \sqrt{144} \qquad\qquad b = \sqrt{49} \qquad\qquad\quad a = \sqrt{28} \qquad\qquad b = \sqrt{16}$

$\quad a = 12 \;\; Ans. \qquad\quad b = 7 \;\; Ans. \qquad\quad a = 2\sqrt{7} \;\; Ans. \qquad b = 4 \;\; Ans.$

1.3.   RATIOS IN A RIGHT TRIANGLE

In a right triangle whose hypotenuse is 20, the ratio of the two legs is 3:4. Find each leg.

Solution:

Let $3x$ and $4x$ represent the two legs of the right triangle. See Fig. 14-5.

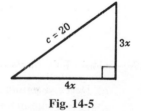

$$(3x)^2 + (4x)^2 = 20^2 \qquad\text{If } x = 4,$$
$$9x^2 + 16x^2 = 400 \qquad\quad 3x = 12$$
$$25x^2 = 400 \qquad\qquad 4x - 16$$
$$x^2 = 16$$
$$x = 4 \qquad\qquad Ans. \quad\text{Legs are 12 and 16.}$$

Fig. 14-5

1.4.   APPLYING LAW OF PYTHAGORAS TO A RECTANGLE

In a rectangle, find

a)   The diagonal if its sides are 9 and 40

b)   One side if the diagonal is 30 and the other side is 24

Solution:

The diagonal of the rectangle is the hypotenuse of a right triangle. See Fig. 14-6.

a) $d^2 = 9^2 + 40^2$, $d^2 - 1681$, $d - 41$ *Ans.*   41

b) $h^2 = 30^2 - 24^2$, $h^2 = 324$, $h = 18$ 18

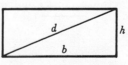

Fig. 14-6

1.5.   TESTING FOR RIGHT TRIANGLES BY USING THE TEST RULE

Using the three sides given, which triangles are right triangles?

$\triangle$ **I**: 8, 15, 17; $\triangle$ **II**: 6, 9, 11; $\triangle$ **III**: $1\frac{1}{2}, 2, 2\frac{1}{2}$.

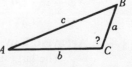

Fig. 14-7

Solutions: See Fig. 14-7.

Rule. If $c^2 = a^2 + b^2$, $\triangle ABC$ is a right triangle; but if $c^2 \neq a^2 + b^2$, $\triangle ABC$ is not.

$\triangle$ **I**: $8^2 + 15^2 \overset{?}{=} 17^2$ $\triangle$ **II**: $6^2 + 9^2 \overset{?}{=} 11^2$ $\triangle$ **III**: $(1\frac{1}{2})^2 + 2^2 \overset{?}{=} (2\frac{1}{2})^2$

$\qquad 64 + 225 \overset{?}{=} 289$ $36 + 81 \overset{?}{=} 121$ $2\frac{1}{4} + 4 \overset{?}{=} 6\frac{1}{4}$

$\qquad\qquad 289 = 289$ $117 \neq 121$ $6\frac{1}{4} = 6\frac{1}{4}$

$\triangle$ **I** is a right triangle. $\triangle$ **II** is not a right triangle. $\triangle$ **III** is a right triangle.

1.6.   FINDING DISTANCE BETWEEN TWO POINTS ON A GRAPH
Find the distance between each of the following pairs of points:

a) From $(3, 4)$ to $(6, 8)$ *b)* From $(3, 4)$ to $(6, 10)$ *c)* From $(-3, 2)$ to $(9, -3)$

Solutions:

a)

	(x, y)
P_2	$(6, 8) \rightarrow x_2 = 6, y_2 = 8$
P_1	$(3, 4) \rightarrow x_1 = 3, y_1 = 4$

$d^2 = (x_2 - x_1)^2 + (y_2 - y_1)^2$

$d^2 = (6 - 3)^2 + (8 - 4)^2$

$d^2 = 3^2 + 4^2 = 25$

$d = 5 \quad Ans.$

b)

	(x, y)
P_2	$(6, 10) \rightarrow x_2 = 6, y_2 = 10$
P_1	$(3, 4) \rightarrow x_1 = 3, y_1 = 4$

$d^2 = (x_2 - x_1)^2 + (y_2 - y_1)^2$

$d^2 = (6 - 3)^2 + (10 - 4)^2$

$d^2 = 3^2 + 6^2 = 45$

$d = 3\sqrt{5} \quad Ans.$

c)

	(x, y)
P_2	$(9, -3) \rightarrow x_2 = 9, y_2 = -3$
P_1	$(-3, 2) \rightarrow x_1 = -3, y_1 = 2$

$d^2 = (x_2 - x_1)^2 + (y_2 - y_1)^2$

$d^2 = [9 - (-3)]^2 + (-3 - 2)^2$

$d^2 = 12^2 + (-5)^2 = 169$

$d = 13 \quad Ans.$

NOTE: Since the distance is considered to be positive only, the negative answer obtainable in each case is rejected.

1.7.   USING LAW OF PYTHAGORAS TO DERIVE FORMULAS

a) Derive a formula for the diagonal d of a square in terms of any side s.

b) Derive a formula for the altitude h of any equilateral triangle in terms of any side s.

Fig. 14-8

Fig. 14-9

Solution: See Fig. 14-8.

$d^2 = s^2 + s^2$

$d^2 = 2s^2$

$d = s\sqrt{2} \quad Ans.$

Solution: See Fig. 14-9. The altitude h of the equilateral triangle bisects the base s.

$h^2 = s^2 - \left(\dfrac{s}{2}\right)^2$

$h^2 = s^2 - \dfrac{s^2}{4} = \dfrac{3s^2}{4}$

$h = \dfrac{s}{2}\sqrt{3} \quad Ans.$

1.8.   APPLYING LAW OF PYTHAGORAS TO AN INSCRIBED SQUARE
The largest possible square is to be cut from a circular piece of cardboard having a diameter of 10 cm. Find the side of the square to the nearest centimeter.

Solution: See Fig. 14-10. The diameter of the circle will be the diagonal of the square.
Hence, $s^2 + s^2 = 100$, $2s^2 = 100$.

$$s^2 = 50,\ s = 5\sqrt{2} = 7.07 \quad Ans. \quad 7\,\text{cm}$$

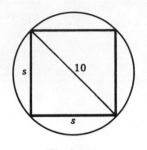

Fig. 14-10

1.9.   APPLYING PYTHAGOREAN THEOREM TO AN ISOSCELES TRIANGLE

Find the altitude of the isosceles triangle shown if : *a*) $a = 25$ and $b = 30$,   *b*) $a = 12$ and $b = 8$.

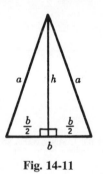

Fig. 14-11

Solution: See Fig. 14-11. The altitude h of the isosceles triangle bisects the base b.

Hence, $h^2 = a^2 - \left(\dfrac{b}{2}\right)^2$.

a) $h^2 = 25^2 - 15^2$, $h^2 = 400$, $h = 20$ *Ans.*   20

b) $h^2 = 12^2 - 4^2$, $h^2 = 128$, $h = 8\sqrt{2}$ $8\sqrt{2}$

2.   PROPORTIONS: EQUAL RATIOS

Understanding Proportions

A **proportion** is an equality of two ratios.
Thus, $2{:}5 = 4{:}10$ or $\dfrac{2}{5} = \dfrac{4}{10}$ is a proportion.
The fourth term of a proportion is the **fourth proportional** to the other three taken in order.
Thus, in $2{:}3 = 4{:}x$, x is the fourth proportional to 2, 3, and 4.
The **means** of a proportion are its middle terms, that is, its second and third terms.
The **extremes** of a proportion are its outside terms, that is, its first and fourth terms.
Thus, in $a{:}b = c{:}d$, the means are b and c, and the extremes are a and d.

> **Proportion rule:** If $a{:}b = c{:}d$, $ad = bc$.

Proof: If $\dfrac{a}{b} = \dfrac{c}{d}$, then $\not{b}d\left(\dfrac{a}{\not{b}}\right) = b\not{d}\left(\dfrac{c}{\not{d}}\right)$. Hence, $ad = bc$.

Stating the Proportion Rule in Two Forms

Fraction Form	**Colon Form**
In any proportion, the cross products are equal.	**In any proportion**, the product of the means equals the product of the extremes.

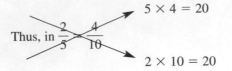

$$5 \times 4 = 20$$

Thus, in $\dfrac{2}{5}$ $\dfrac{4}{10}$

$$2 \times 10 = 20$$

$$5 \times 4 = 20$$

Thus, in $2{:}5 = 4{:}10$

$$2 \times 10 = 20$$

2.1. FINDING UNKNOWNS IN PROPORTIONS BY USING EQUAL CROSS-PRODUCTS
Solve for x:

a) $\dfrac{x}{20} = \dfrac{3}{5}$ *b)* $\dfrac{3}{x} = \dfrac{2}{5}$ *c)* $\dfrac{x}{2x-3} = \dfrac{3}{5}$ *d)* $\dfrac{32}{x} = \dfrac{x}{2}$ *e)* $\dfrac{b}{a} = \dfrac{c}{x}$

<u>Solutions:</u>

a) $5x = 60$ *b)* $2x = 15$ *c)* $5x = 6x - 9$ *d)* $x^2 = 64$ *e)* $bx = ac$

Ans. $x = 12$ $x = 7\frac{1}{2}$ $9 = x$ $x = \pm 8$ $x = \dfrac{ac}{b}$

2.2. FINDING UNKNOWNS IN PROPORTIONS BY USING MEANS AND EXTREMES
Solve for x:

Proportions	Product of Means	Product of Extremes	Product of Means = Product of Extremes	Ans.
a) $x{:}4 = 6{:}8$	$4(6) = 24$	$8x$	$8x = 24$	$x = 3$
b) $3{:}5 = x{:}12$	$5x$	$3(12) = 36$	$5x = 36$	$x = 7\frac{1}{5}$
c) $3{:}x = x{:}27$	$x \cdot x = x^2$	$3(27) = 81$	$x^2 = 81$	$x = \pm 9$
d) $x{:}5 = 2x{:}x + 3$	$5(2x) = 10x$	$x(x + 3) = x^2 + 3x$	$x^2 + 3x = 10x$	$x = 0, 7$
e) $x - 2{:}4 = 7{:}x + 2$	$4(7) = 28$	$(x - 2)(x + 2) = x^2 - 4$	$x^2 - 4 = 28$	$x = \pm 4\sqrt{2}$

2.3. SOLVING FRACTION PROBLEMS INVOLVING PROPORTIONS
The numerator of a fraction is 5 less than the denominator. If the numerator is doubled and the denominator is increased by 7, the value of the resulting fraction is $\frac{2}{3}$. Find the original fraction.

<u>Solution:</u>

Let x = denominator of the original fraction and $x - 5$ = numerator of the original fraction.

Then $\dfrac{2(x - 5)}{x + 7} = \dfrac{2}{3}$. Cross-multiply: $6x - 30 = 2x + 14$ Original fraction $= \dfrac{6}{11}$ *Ans.*

$$4x = 44, x = 11$$

3.   SIMILAR TRIANGLES

Similar Polygons Have the Same Shape

Thus, if $\triangle I$ and $\triangle I'$ are similar, then they have the same shape although they need not have the same size.

NOTE: "$\triangle I \sim \triangle I'$" is read as "triangle I is similar to triangle I prime." In the diagram in Fig. 14-12, note how the sides and angles having the same relative position are designated by using the same letters and primes. **Corresponding sides** or angles are those having the same relative position.

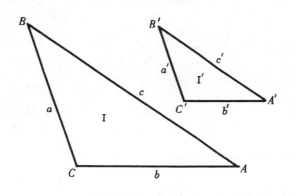

Fig. 14-12

Two Basic Properties of Similar Triangles

Rule 1:   If two triangles are similar

a) **Their corresponding angles are congruent.**

Thus, if $\triangle I \sim \triangle I'$ in Fig. 14-13,

then $\angle C' \cong \angle C$
        $\angle A' \cong \angle A$
        $\angle B' \cong \angle B$
and $m \angle C = 90°$
        $m \angle A - 19°$
        $m \angle B = 50°$

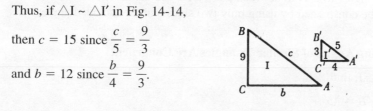

Fig. 14-13

b) **The ratios of their corresponding sides are equal.**

Thus, if $\triangle I \sim \triangle I'$ in Fig. 14-14,

then $c = 15$ since $\dfrac{c}{5} = \dfrac{9}{3}$

and $b = 12$ since $\dfrac{b}{4} = \dfrac{9}{3}$.

Fig. 14-14

Three Basic Methods of Determining Similar Triangles

Rule 2:   Two triangles are similar

a) **If two angles of one are congruent to two angles of the other.**

b) **If the three ratios of the corresponding sides are equal.**

c) **If two ratios of corresponding sides are equal and the angles between the sides are congruent.**

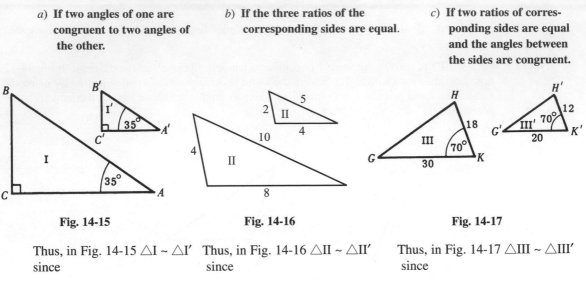

Fig. 14-15   Fig. 14-16   Fig. 14-17

Thus, in Fig. 14-15 △I ~ △I′ since

Thus, in Fig. 14-16 △II ~ △II′ since

Thus, in Fig. 14-17 △III ~ △III′ since

$\angle C \cong \angle C'$ and $\angle A \cong \angle A'$

$\frac{10}{5} = \frac{8}{4} = \frac{4}{2}$

$\frac{30}{20} = \frac{18}{12}$ and $\angle K \cong \angle K'$

Rule 3:   **A triangle is similar to any one of its scale triangles.**

Thus, if △I and △I′ are drawn to scale to represent △ ABC, then they are similar to △ ABC and also to each other; that is, in Figs. 14-18 and 14-19,

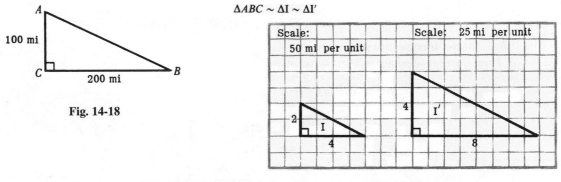

$$\triangle ABC \sim \triangle I \sim \triangle I'$$

Fig. 14-18

Fig. 14-19

NOTE.   Use **rule 2c** to show that the triangles are similar. For example, △I ~ △ ABC since

$$\frac{100}{2} = \frac{200}{4}$$

and the right angles which are between these sides are congruent. For this reason, a scale triangle may be constructed by using only two sides and the included angle.

3.1.   Rule 1a.   Corresponding Angles of Similar Triangles Are Congruent

In Fig. 14-20, if △I′ ~ △I, find $\angle C'$

a) If $m \angle A = 60°$, $m \angle B = 45°$
b) If $m \angle A + m\angle B = 110°$

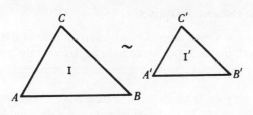

Fig. 14-20

Solutions:

Using rule 1, if $\triangle I' \sim \triangle I$, then $\angle C' = \angle C$.

a) Since the sum of the angles of a triangle equals 180°,
$$m\angle C = 180° - 60° - 45° = 75°. \text{ Hence, } m\angle C' = 75°.$$

b) Since the sum of the angles of a triangle equals 180°,
$$m\angle C = 180° - 110° = 70°. \text{ Hence, } m\angle C' = 70°.$$

3.2.   RULE 1*b*.   RATIOS OF CORRESPONDING SIDES OF SIMILAR TRIANGLES ARE EQUAL
If $\triangle II' \sim \triangle II$ in Fig. 14-21, find x and y by using the data indicated.

Solutions:

Since $\triangle II' \sim \triangle II$,

$$\frac{x}{32} = \frac{15}{20} \qquad \frac{y}{26} = \frac{15}{20}$$

$$x = \frac{15}{20}(32) \qquad y = \frac{15}{20}(26)$$

$$x = 24 \qquad y = 19\tfrac{1}{2}$$

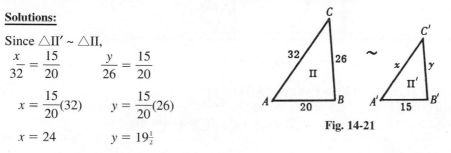

Fig. 14-21

Ans. 24, $19\tfrac{1}{2}$

3.3.   RULE 2.   DETERMINING SIMILAR TRIANGLES

    a) See Fig. 14-22.   *b)* See Fig. 14-23.   *c)* See Fig. 14-24.

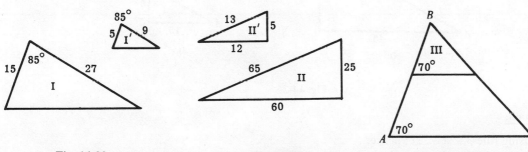

    Fig. 14-22   **Fig. 14-23**   **Fig. 14-24**

Which rule is needed in each case to show that both triangles are similar?

Solutions:

a) By rule 2*c*, $\triangle I' \sim \triangle I$ since each has an 85° angle and there are two equal ratios for the sides of these congruent angles; that is, $\frac{27}{9} = \frac{15}{5}$.

b) By rule 2*b*, $\triangle II' \sim \triangle II$ since $\frac{65}{13} = \frac{25}{5} = \frac{60}{12}$. Thus, there are three equal ratios of the corresponding sides.

c) By rule 2*a*, $\triangle III \sim \triangle ABC$ since there are two pairs of congruent angles. Each triangle has $\angle B$ and a 70° angle.

3.4.   FINDING HEIGHTS BY USING GROUND SHADOWS

A tree casts a 15-ft shadow at a time when a nearby upright pole of 6 ft casts a shadow of 2 ft. Find the height h of the tree if both tree and pole make right angles with the ground.

<u>Solution:</u>

See Fig. 14-25. At the same time in localities near each other, the rays of the sun strike the ground at equal angles; hence $\angle B \cong \angle B'$. Since the tree and pole make right angles with the ground, $\angle C \cong \angle C'$.

Since there are two pairs of congruent angles, $\triangle I' \sim \triangle I$. Hence, $\dfrac{h}{6} = \dfrac{15}{2}$, $h = \dfrac{15}{2}(6) = 45$.

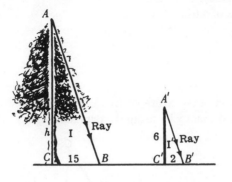

Ans.   45 ft

Fig. 14-25

3.5.   USING A SCALE TRIANGLE TO FIND PARTS OF A TRIANGLE

If $\triangle I'$ is a scale triangle of $\triangle I$ in Fig. 14-26*a* and *b*,

a) Find a and c when $b = 45$,   *b*) Show that $\triangle ABC$ is a right triangle.

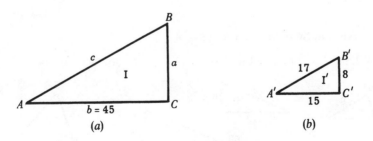

Fig. 14-26

<u>Solution:</u>

a) Using rule 3, $\triangle I \sim \triangle I'$; hence:

$$\frac{a}{8} = \frac{45}{15} \qquad \text{and} \qquad \frac{c}{17} = \frac{45}{15}$$
$$a = 3(8) = 24 \qquad\qquad c = 3(17) = 51 \quad \textit{Ans.} \quad 24, 51$$

b) Since $8^2 + 15^2 = 17^2$, $\triangle I'$ is a right triangle.

  Since $\triangle I \sim \triangle I'$, $\triangle I$ is a right triangle.

3.6.   USING A SCALE TRIANGLE TO FIND A DISTANCE

An aviator traveled east a distance of 150 mi from A to C. She then traveled north for 50 mi to B. Using a scale of 50 mi per unit, find her distance from A, to the nearest mile.

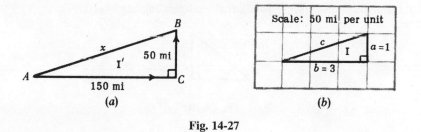

Fig. 14-27

Solution:

See Fig. 14-27a and b. By rule 3, a triangle is similar to any one of its scale triangles.

Let x = length of $\overline{AB}$ in miles.

In $\triangle$I, the scale triangle of $\triangle$I′, $c^2 = 3^2 + 1^2 = 10$. Hence, $c = \sqrt{10}$.

Since $\triangle$I′ ~ $\triangle$I, $\dfrac{x}{c} = \dfrac{50}{1}$

$$x = 50c$$

$$x = 50\sqrt{10} = 50(3.162) = 158.1 \quad Ans. \quad 158 \text{ mi}$$

3.7.   APPLYING A SCALE TO A SQUARE

A baseball diamond is a square 90 ft on each side. Using a scale of 90 ft per unit, find the distance from home plate to second base, to the nearest foot.

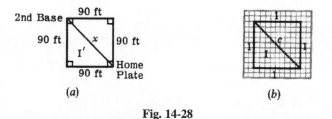

Fig. 14-28

Solution:

See Fig. 14-28a and b. Using a scale of 90 ft per unit, each side of the new square will be 1 unit.

Let x = distance from home plate to second base in feet.

In $\triangle$I, the scale triangle of $\triangle$I′, $c^2 = 1^2 + 1^2 = 2$. Hence, $c = \sqrt{2}$.

Since $\triangle$I′ ~ $\triangle$I, $\dfrac{x}{c} = \dfrac{90}{1}$

$$x = 90c$$

$$x - 90\sqrt{2} = 90(1.414) = 127.26 \quad Ans. \quad 127 \text{ ft}$$

Supplementary Problems

14.1.   In a right triangle whose legs are a and b, find the hypotenuse c when: **(1.1)**

a) $a = 15, b = 20$	*Ans.* 25	c) $a = 5, b = 4$	*Ans.* $\sqrt{41}$	e) $a = 7, b = 7$	*Ans.* $7\sqrt{2}$
b) $a = 15, b = 36$	39	d) $a = 5, b = 5\sqrt{3}$	10	f) $a = 8, b = 4$	$4\sqrt{5}$

14.2. In the right triangle shown in Fig. 14-29, find each missing leg when: **(1.2)**

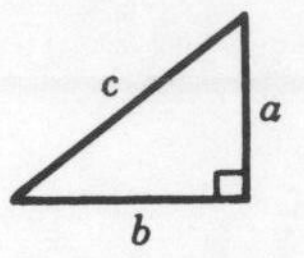

Fig. 14-29

a) $a = 12, c = 20$ c) $b = 15, c = 17$ e) $a = 5\sqrt{2}, c = 10$

b) $b = 6, c = 8$ d) $a = 2, c = 4$ f) $a = \sqrt{5}, c = 2\sqrt{2}$

Ans. a) $b = 16$ b) $a = 2\sqrt{7}$ c) $a = 8$ d) $b = 2\sqrt{3}$ e) $b = 5\sqrt{2}$ f) $b = \sqrt{3}$

14.3. Find the legs of a right triangle whose hypotenuse is c if these legs have a ratio of: **(1.3)**

a) 3:4 and $c = 15$,   b) 5:12 and $c = 26$,   c) 8:15 and $c = 170$,   d) 1:2 and $c = 10$.

Ans. a) 9,12   b) 10, 24   c) 80, 150   d) $2\sqrt{5}, 4\sqrt{5}$

14.4. In a rectangle, find the diagonal if its sides are: **(1.4)**

a) 30 and 40   b) 9 and 40   c) 5 and 10   d) 2 and 6.

Ans. a) 50   b) 41   c) $5\sqrt{5}$ d) $2\sqrt{10}$

14.5. In a rectangle, find one side if the diagonal is 15 and the other side is: **(1.4)**

a) 9   b) 5   c) 10   d) 12.

Ans. a) 12   b) $10\sqrt{2}$ c) $5\sqrt{5}$ d) 9

14.6. Using the three sides given, which triangles are right triangles?   **(1.5)**

a) 33, 55, 44   c) 4, $7\frac{1}{2}$, $8\frac{1}{2}$ e) 5 in, 1 ft, 1 ft 1 in   g) 11 mi, 60 mi, 61 mi

b) 120, 130, 50   d) 25, 7, 24   f) 1 yd, 1 yd 1 ft, 1 yd 2 ft   h) 5 m, 5 m, 7 m

Ans.   Only (h) is not a right triangle since $5^2 + 5^2 \neq 7^2$. In all the other cases, the square of the largest side equals the sum of the squares of the other two.

14.7. Find the distance between each of the following pairs of points: **(1.6)**

a) (4, 1) and (7, 5)   c) (1, 7) and (10, 7)   e) (2, 3) and $(-10, -2)$

b) (3, 3) and (3, 5)   d) $(-3, -6)$ and (3, 2)   f) (2, 2) and (5, 5)

Ans. a) 5   b) 2   c) 9   d) 10   e) 13   f) $3\sqrt{2}$

14.8. Find the lengths of the sides of triangle *DEF* if its vertices are $D(2, 5)$, $E(6, 5)$, and $F(2, 8)$. Show that $\triangle DEF$ is a right triangle.   **(1.5, 1.6)**

Ans. $DE = 4$, $DF = 3$, and $EF = 5$. Since $3^2 + 4^2 = 5^2$, $\triangle DEF$ is a right triangle.

14.9. Derive a formula for the side s of a square in terms of its diagonal d.   **(1.7)**

Ans. $s = \dfrac{d}{\sqrt{2}}$ or $s = \dfrac{d}{2}\sqrt{2}$

14.10. Using formula $d = s\sqrt{2}$, express in radical form the diagonal of a square whose side is: **(1.7)**

a) 5   b) 7.2   c) $\sqrt{3}$ d) 90 ft   e) 3.47 yd

Ans. a) $5\sqrt{2}$ b) $7.2\sqrt{2}$ c) $\sqrt{6}$ d) $90\sqrt{2}$ ft   e) $3.47\sqrt{2}$ yd

14.11. Using $h = \frac{s}{2}\sqrt{3}$, express in radical form the altitude of an equilateral triangle whose side is: **(1.7)**

  a) 6   b) 20   c) 11   d) 90 in   e) 4.6 yd.

  Ans. a) $3\sqrt{3}$ b) $10\sqrt{3}$ c) $\frac{11}{2}\sqrt{3}$ d) $45\sqrt{3}$ in   e) $2.3\sqrt{3}$ yd

14.12. The largest possible square is to be cut from a circular piece of wood. Find the side of the square, to the nearest inch, if the diameter of the circle is   **(1.8)**

    a) 30 in   b) 14 in   c) 17 in.
  Ans. a) 21 in   b) 10 in   c) 12 in

14.13. Find the altitude of an isosceles triangle if one of its two congruent sides is 10 and its base is   **(1.9)**

    a) 12   b) 16   c) 18   d) 10

  Ans. a) 8   b) 6   c) $\sqrt{19}$ d) $5\sqrt{3}$

14.14. Solve for x: **(2.1)**

  a) $\dfrac{5}{7} = \dfrac{15}{x}$ c) $\dfrac{3}{x} = \dfrac{x}{12}$ e) $x + \dfrac{2}{5} = \dfrac{6}{3}$ g) $\dfrac{2x}{x+7} = \dfrac{3}{5}$

  b) $\dfrac{7}{x} = \dfrac{3}{2}$ d) $\dfrac{x}{5} = \dfrac{15}{x}$ f) $\dfrac{x-1}{3} = \dfrac{5}{x+1}$ h) $\dfrac{a}{x} = \dfrac{x}{b}$

  Ans. a) 21   b) $4\frac{2}{3}$ c) ± 6 d) $\pm 5\sqrt{3}$ e) 8   f) ± 4 g) 3   h) $\pm\sqrt{ab}$

14.15. Solve for x: **(2.2)**

  a) $x{:}6 = 8{:}3$ d) $x{:}2 = 10{:}x$ g) $a{:}b = c{:}x$
  b) $5{:}4 = 20{:}x$ e) $(x+4){:}(3 = 3){:}(x-4)$ h) $x{:}2y = 18y{:}x$
  c) $9{:}x = x{:}4$ f) $(2x+8){:}(x+2 = 2x+5){:}(x+1)$

  Ans. a) 16   b) 16   c) ± 6 d) $\pm 2\sqrt{5}$ e) ± 5 f) 2   g) $\dfrac{bc}{a}$ h) $\pm 6y$

14.16. a) The denominator of a fraction is 1 more than twice the numerator. If 2 is added to both the numerator and denominator, the value of the fraction is $\frac{3}{5}$. Find the original fraction.   **(2.3)**
   b) A certain fraction is equivalent to $\frac{2}{5}$. If the numerator of this fraction is decreased by 2 and its denominator is increased by 1, the resulting fraction is equivalent to $\frac{1}{3}$. Find the original fraction.   **(2.3)**

  Ans.   a) $\dfrac{1}{3}\left[\text{Solve: } \dfrac{x+2}{(2x+1)+2} = \dfrac{3}{5}\right]$,   b) $\dfrac{14}{35}\left(\text{Solve: } \dfrac{2x-2}{5x+1} = \dfrac{1}{3}\right)$

14.17. In Fig. 14-30, if $\triangle\mathrm{I} \sim \triangle\mathrm{I'}$, find $m\angle B$.   **(3.1)**

  a) If $m\angle A' = 120°$ and $m\angle C' = 25°$,
  b) If $m\angle A' + m\angle C' = 127°$.

  Ans. a) 35°   b) 53°

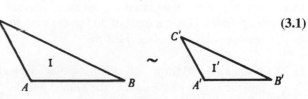

Fig. 14-30

14.18. In Fig. 14-31, if $\triangle\mathrm{I} \sim \triangle\mathrm{I'}$, using the data shown,   **(3.2)**

  a) Find a if $c = 24$.   *Ans.* $\dfrac{a}{4} = \dfrac{24}{6}$, $a = 16$

  b) Find b if $a = 20$. $\dfrac{b}{3} = \dfrac{20}{4}$, $b = 15$

  c) Find c if $b = 63$. $\dfrac{c}{6} = \dfrac{63}{3}$, $c = 126$

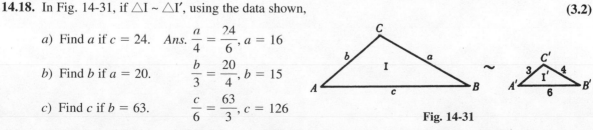

Fig. 14-31

14.19. Which rule is needed in each case to show both triangles are similar?   **(3.3)**

 a) Show △I ~ △*ABC*.   *b*) Show △II ~ △*PQR*.   *c*) Show △III ~ △*FGH*.
  (See Fig. 14-32*a*.)   (See Fig. 14-32*b*.)   (See Fig. 14-32*c*.)

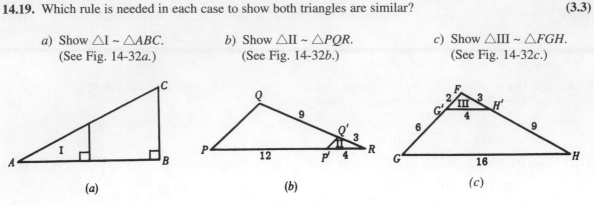

   (*a*)   (*b*)   (*c*)

Fig. 14-32

Ans.   *a*) Rule 2*a*.   *b*) Rule 2*c*, since $\frac{3}{12} = \frac{1}{16}$ and each triangle contains ∠ *R*.   *c*) Rule 2*b*, since $\frac{2}{8} = \frac{3}{12} = \frac{4}{16}$.

14.20. A 7-ft upright pole near a vertical tree casts a 6-ft shadow. At that time,   **(3.4)**

 a) Find the height of the tree if its shadow is 36 ft.   *Ans.*   42 ft
 b) Find the shadow of the tree if its height is 77 ft.   66 ft

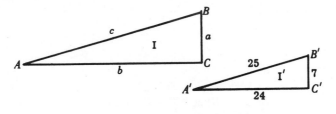

Fig. 14-33

14.21. If △I′ is a scale triangle of △I in Fig. 14-33,   **(3.5)**

 a) Find *a* and *b* when *c* = 125,
 b) Show that △I is a right triangle.

 Ans.   *a*) 35, 120

   b) △I is a right triangle since $7^2 + 24^2 = 25^2$ and △I ~ △I′.

14.22. Two planes leave an airport at the same time, one going due east at 250 mi/h and the other due north at 150 mi/h. Using a scale of 200 mi per unit, find the distance between them at the end of 4 h, to the nearest mile.   *Ans.*   1166 mi   **(3.6)**

14.23. A square lot 50 ft on each side has a diagonal path. Using a square drawn to a scale of 50 ft per unit, find the length of the path to the nearest foot.   *Ans.*   71 ft   **(3.7)**

Introduction to Trigonometry

1. UNDERSTANDING TRIGONOMETRIC RATIOS

Trigonometry means "measurement of triangles." Consider its parts: *tri* means 3, *gon* means "angle," and *metry* means "measure." Thus, in trigonometry we study the measurement of triangles.

2. TRIGONOMETRIC FUNCTIONS OF AN ACUTE ANGLE

In dealing with any right triangle, it will be convenient (see Fig. 15-1) to denote the vertices as A, B, C such that C is the vertex of the right triangle; to denote the angles of the triangle as A, B, C such that $m \angle C = 90°$; and the sides opposite the angles as a, b, c, respectively. With respect to angle A, a will be called the *opposite side* and b will be called the *adjacent side*; with respect to angle B, a will be called the *adjacent side* and b the *opposite side*. Side c will always be called the *hypotenuse*.

Fig. 15-1

Fig. 15-2

If now the right triangle is placed in a coordinate system (Fig. 15-2) so that angle A is in standard position, point B on the terminal side of angle A has coordinates (b, a) and distance $c = a^2 + b^2$. Then the trigonometric functions of angle A may be defined in terms of the sides of the right triangle, as

$$\sin A = \frac{a}{c} = \frac{\text{length of opposite side}}{\text{length of hypotenuse}}$$

$$\cos A = \frac{b}{c} = \frac{\text{length of adjacent side}}{\text{length of hypotenuse}}$$

$$\tan A = \frac{a}{b} = \frac{\text{length of opposite side}}{\text{length of adjacent side}}$$

$$\cot A = \frac{b}{a} = \frac{\text{length of adjacent side}}{\text{length of opposite side}}$$

$$\sec A = \frac{c}{b} = \frac{\text{length of hypotenuse}}{\text{length of adjacent side}}$$

$$\csc A = \frac{c}{a} = \frac{\text{length of hypotenuse}}{\text{length of opposite side}}$$

315

3.   TRIGONOMETRIC FUNCTIONS OF COMPLEMENTARY ANGLES

The acute angles A and B of the right triangle ABC are complementary, that is, m $\angle A$ + m $\angle B$ = 90°. From Fig. 15-1 we have:

$$\sin B = \frac{b}{c} = \cos A \qquad\qquad \cot B = \frac{a}{b} = \tan A$$

$$\cos B = \frac{a}{c} = \sin A \qquad\qquad \sec B = \frac{c}{a} = \csc A$$

$$\tan B = \frac{b}{a} = \cot A \qquad\qquad \csc B = \frac{c}{b} = \sec A$$

These relations associate the functions in pairs—sine and cosine, tangent and cotangent, secant and cosecant—each function of a pair being called the *cofunction* of the other. Thus, any function of an acute angle is equal to the corresponding cofunction of the complementary angle.

4.   TRIGONOMETRIC FUNCTIONS OF 30°, 45°, 60°

The results in Table 15.1 are obtained in Problems 15.8 to 15.9.

Table 15.1

Angle θ	$\sin \theta$	$\cos \theta$	$\tan \theta$	$\cot \theta$	$\sec \theta$	$\csc \theta$
30°	$\frac{1}{2}$	$\frac{1}{2}\sqrt{3}$	$\frac{1}{3}\sqrt{3}$	$\sqrt{3}$	$\frac{2}{3}\sqrt{3}$	2
40°	$\frac{1}{2}\sqrt{2}$	$\frac{1}{2}\sqrt{2}$	1	1	$\sqrt{2}$	$\sqrt{2}$
60°	$\frac{1}{2}\sqrt{3}$	$\frac{1}{2}$	$\frac{1}{\sqrt{3}}$	$\frac{1}{3}\sqrt{3}$	2	$\frac{2}{3}\sqrt{3}$

Problems 15.10 to 15.16 illustrate a number of simple applications of the trigonometric functions. For this purpose, Table 15.2 will be used. You may also use a calculator to find these values.

Table 15.2

Angle θ	$\sin \theta$	$\cos \theta$	$\tan \theta$	$\cot \theta$	$\sec \theta$	$\csc \theta$
15°	0.26	0.97	0.27	3.7	1.0	3.9
20°	0.34	0.94	0.36	2.7	1.1	2.9
30°	0.50	0.87	0.58	1.7	1.2	2.0
40°	0.64	0.77	0.84	1.2	1.3	1.6
45°	0.71	0.71	1.0	1.0	1.4	1.4
50°	0.77	0.64	1.0	0.84	1.6	1.3
60°	0.87	0.50	1.7	0.58	2.0	1.2
70°	0.94	0.34	2.7	0.36	2.9	1.1
75°	0.97	0.26	3.7	0.27	3.9	1.0

Supplementary Problems

NOTE. We will write AB (or c) to denote the length of AB, and $\overline{AB}$ to denote "the segment AB" and $\overrightarrow{AB}$ denote "the line AB."

15.1. Find the values of the trigonometric functions of the acute angles of the right triangle ABC given $b = 24$ and $c = 25$.

Since $a^2 = c^2 - b^2 = (25)^2 - (24)^2 = 49$, $a = 7$. See Fig. 15-3. Then

$$\sin A = \frac{\text{opposite side}}{\text{hypotenuse}} = \frac{7}{25} \qquad \cos A = \frac{\text{adjacent side}}{\text{opposite side}} = \frac{24}{7}$$

$$\cos A = \frac{\text{adjacent side}}{\text{hypotenuse}} = \frac{24}{25} \qquad \sec A = \frac{\text{hypotenuse}}{\text{adjacent side}} = \frac{25}{24}$$

$$\tan A = \frac{\text{opposite side}}{\text{adjacent side}} = \frac{7}{24} \qquad \csc A = \frac{\text{hypotenuse}}{\text{opposite side}} = \frac{25}{7}$$

and
$$\sin B = \tfrac{24}{25} \qquad \cot B = \tfrac{7}{24}$$

$$\cos B = \tfrac{7}{25} \qquad \sec B = \tfrac{25}{7}$$

$$\tan B = \tfrac{24}{7} \qquad \csc B = \tfrac{25}{24}$$

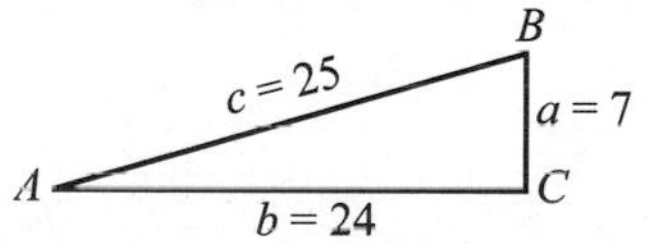

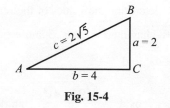

Fig. 15-3

15.2. Find the values of the trigonometric functions of the acute angles of the right triangle ABC, given $a = 2$, $c = 2\sqrt{5}$.

Since $b^2 = c^2 - a^2 = (2\sqrt{5})^2 - 2^2 = 20 - 4 = 16$, $b = 4$. See Fig. 15-4. Then

$$\sin A = \frac{2}{2\sqrt{5}} = \frac{\sqrt{5}}{5} = \cos B \qquad \cot A = \tfrac{4}{2} = 2 = \tan B$$

$$\cos A = \frac{4}{2\sqrt{5}} = \frac{2\sqrt{5}}{5} = \sin B \qquad \sec A = \frac{2\sqrt{5}}{4} = \frac{\sqrt{5}}{2} = \csc B$$

$$\tan A = \tfrac{2}{4} = \tfrac{1}{2} = \cot B \qquad \csc A = \frac{2\sqrt{5}}{2} = \sqrt{5} = \sec B$$

Fig. 15-4

15.3. Find the values of the trigonometric functions of the acute angle A, given $\sin A, = \tfrac{3}{7}$.

Construct the right triangle ABC having $a = 3$, $c = 7$, and $b = \sqrt{7^2 - 3^2} = 2\sqrt{10}$ units. See Fig. 15-5. Then

$$\sin A = \tfrac{3}{7} \qquad\qquad\qquad \cot A = \frac{2\sqrt{10}}{3}$$

$$\cos A = \frac{2\sqrt{10}}{7} \qquad\qquad \sec A = \frac{7}{2\sqrt{10}} = \frac{7\sqrt{10}}{20}$$

$$\tan A = \frac{3}{2\sqrt{10}} = \frac{3\sqrt{10}}{20} \qquad \csc A = \frac{7}{3}$$

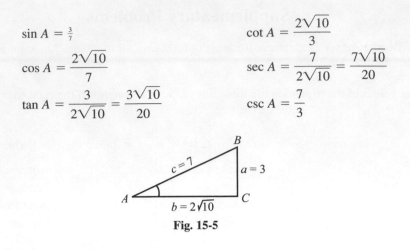

Fig. 15-5

15.4.   Find the values of the trigonometric functions of the acute angle B, given $\tan B = 1.5$.

Refer to Fig. 15-6. Construct the right triangle ABC having $b = 15$ and $a = 10$ units. (Note that $1.5 = \tfrac{3}{2}$ and a right triangle with $b = 3$, $a = 2$ will serve equally well.)

Then $c = \sqrt{a^2 + b^2} = \sqrt{10^2 + 15^2} = 5\sqrt{13}$ and

$$\sin B = \frac{15}{5\sqrt{13}} = \frac{3\sqrt{13}}{13} \qquad\qquad \cot B = \tfrac{2}{3}$$

$$\cos B = \frac{10}{5\sqrt{13}} = \frac{2\sqrt{13}}{13} \qquad\qquad \sec B = \frac{5\sqrt{13}}{10} = \frac{\sqrt{13}}{2}$$

$$\tan B = \tfrac{15}{10} = \tfrac{3}{2} \qquad\qquad\qquad \csc B = \frac{5\sqrt{13}}{15} = \frac{\sqrt{13}}{3}$$

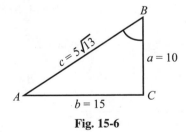

Fig. 15-6

15.5.   If A is acute and $\sin A = 2x/3$, determine the values of the remaining functions.

Construct the right triangle ABC having $a = 2x < 3$ and $c = 3$, as in Fig. 15-7.

Then $b = \sqrt{c^2 - a^2} = \sqrt{9 - 4x^2}$ and

$$\sin A = \frac{2x}{3}, \quad \cos A = \frac{\sqrt{9 - 4x^2}}{3}, \quad \tan A = \frac{2x}{\sqrt{9 - 4x^2}}, \quad \cot A = \frac{\sqrt{9 - 4x^2}}{2x},$$

$$\sec A = \frac{3}{\sqrt{9 - 4x^2}}, \quad \csc A = \frac{3}{2x}.$$

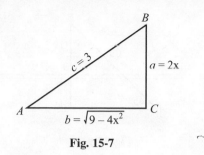

Fig. 15-7

15.6. If A is acute and $\tan A = x = x/1$, determine the values of the remaining functions.

Construct the right triangle ABC having $a = x$ and $b = 1$, as in Fig. 15-8. Then $c = \sqrt{x^2 + 1}$ and

$$\sin A = \frac{x}{\sqrt{x^2 + 1}}, \cos A = \frac{1}{\sqrt{x^2 + 1}}, \tan A = x, \cot A = \frac{1}{x}, \sec A = \sqrt{x^2 + 1}, \csc A = \frac{\sqrt{x^2 + 1}}{x}.$$

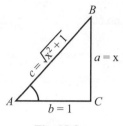

Fig. 15-8

15.7. If A is an acute angle:

(a) Why is $\sin A < 1$?   (d) Why is $\sin A < \tan A$?

(b) When is $\sin A = \cos A$?   (e) When is $\sin A < \cos A$?

(c) Why is $\sin A < \csc A$?   (f) When is $\tan A > 1$?

In any right triangle ABC:

(a) Side $a <$ side c; therefore, $\sin A = a/c < 1$.

(b) $\sin A = \cos A$ when $a/c = b/c$; then $a = b$, $A = B$, and $A = 45°$.

(c) $\sin A < 1$ (above) and $\csc A = 1/\sin A > 1$.

(d) $\sin A = a/c$, $\tan A = a/b$, and $b < c$; therefore $a/c < a/b$ or $\sin A < \tan A$.

(e) $\sin A < \cos A$ when $a < b$; then $A < B$ or $A < 90° - A$, and $A < 45°$.

(f) $\tan A = a/b > 1$ when $a > b$; then $A > B$ and $A > 45°$.

15.8. Find the value of the trigonometric functions of 45°.

In any isosceles right triangle ABC, $A = B = 45°$ and $a = b$. See Fig. 15-9. Let $a = b = 1$; then
$c = \sqrt{1 + 1} = \sqrt{2}$ and

$$\sin 45° = \frac{1}{\sqrt{2}} = \frac{1}{2}\sqrt{2} \qquad\qquad \cot 45° = 1$$

$$\cos 45° = \frac{1}{\sqrt{2}} = \frac{1}{2}\sqrt{2} \qquad\qquad \sec 45° = \sqrt{2}$$

$$\tan 45° = \frac{1}{1} = 1 \qquad\qquad\qquad \csc 45° = \sqrt{2}$$

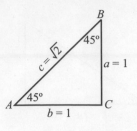

Fig. 15-9

15.9. Find the values of the trigonometric functions of 30° and 60°.

In any equilateral triangle *ABD* (see Fig. 15-10), each angle is 60°. The bisector of any angle, as B, is the perpendicular bisector of the opposite side. Let the sides of the equilateral triangle be of length 2 units. Then in the right triangle *ABC*, $AB = 2$, $AC = 1$, and $BC = \sqrt{2^2 - 1^2} = \sqrt{3}$.

$$\sin 30° = \tfrac{1}{2} = \cos 60° \qquad\qquad \cot 30° = \sqrt{3} = \tan 60°$$

$$\cos 30° = \frac{\sqrt{3}}{2} = \sin 60° \qquad\qquad \sec 30° = \frac{2}{\sqrt{3}} = \frac{2\sqrt{3}}{3} = \csc 60°$$

$$\tan 30° = 1\backslash\sqrt{3} = \frac{\sqrt{3}}{3} = \cot 60° \qquad \csc 30° = 2 = \sec 60°$$

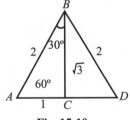

Fig. 15-10

15.10. When the sun is 20° above the horizon, how long is the shadow cast by a building 150 ft high?

In Fig. 15-11 $A = 20°$ and $CB = 150$. Then $\cot A = AC/CB$ and $AC = CB \cot A = 150 \cot 20° = 150(2.7) = 405$ ft.

15.11. A tree 100 ft tall casts a shadow 120 ft long. Find the measure of the angle of elevation of the sun.

In Fig. 15-12, $CB = 100$ and $AC = 120$. Then $\tan A = CB/AC = \frac{100}{120} = 0.83$ and m $\angle = 40°$.

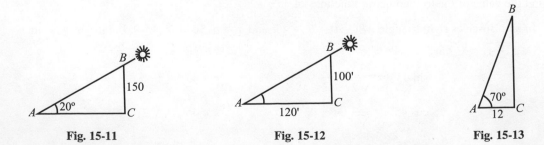

Fig. 15-11   **Fig. 15-12**   **Fig. 15-13**

15.12. A ladder leans against the side of a building with its foot 12 m from the building. How far from the ground is the top of the ladder and how long is the ladder if it makes an angle of 70° with the ground?

From Fig. 15-13, tan A = CB/AC; then CB = AC tan A = 12 tan 70° = 12(2, 7) = 32.4. The top of the ladder is 32 m above the ground.

Sec A = AB/AC; then AB = AC sec A = 12 sec 70° = 12(2,9) = 34.8. The ladder is 35 m long.

15.13. From the top of a lighthouse, 120 ft above the sea, the angle of depression of a boat is 15°. How far is the boat from the lighthouse?

In Fig. 15-14, the right triangle ABC has A measuring 15° and CB = 120°; then cot A = AC/CB and AC = CB cot A = 120 cot 15° = 120(3.7) = 444 ft.

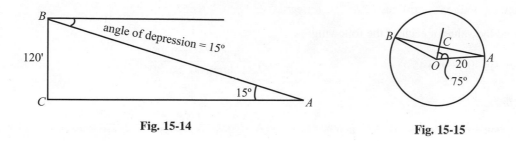

Fig. 15-14   Fig. 15-15

15.14. Find the length of the chord of a circle of radius 20 cm subtended by a central angle of 150°.

In Fig. 15-15, OC bisects $<AOB$. Then BC = AC and OAC is a right triangle. In $\triangle OAC$,
$$\sin \angle COA = \frac{AC}{OA} \text{ and } AC = OA \sin \angle COA = 20 \sin 75° = 20(0.97) = 19.4$$

Then BA = 38.8 and the length of the chord is 39 cm.

15.15. Find the height of a tree if the angle of elevation of its top changes from 20° to 40° as the observer advances 75 ft toward its base. See Fig. 15-16.

In the right triangle ABC, cot A = AC/CB; then AC = CB cot A or DC + 75 = CB cot 20°.

In the right triangle DBC, cot D = DC/CB; then DC = CB cot 40°.

Then: DC = CB cot 20° − 75 = CB cot 40°, CB(cot 20° − cot 40°) = 75.

$$CB(2.7 - 1.2) = 75 \quad \text{and} \quad CB = \frac{75}{1.5} = 50 \text{ ft.}$$

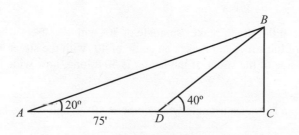

Fig. 15-16

15.16. Find the values of the trigonometric functions of the acute angles of the right triangle ABC, given:

(a) $a = 3, b = 1$ (b) $a = 2, c = 5$ (c) $b = \sqrt{7}, c = 4$

Ans.   (a) A: $3/\sqrt{10}$, $1/\sqrt{10}$, 3, $\frac{1}{3}$, $\sqrt{10}$, $\sqrt{10}/3$; B: $1/\sqrt{10}$, $3/\sqrt{10}$, $\frac{1}{3}$, 3, $\sqrt{10}/3$, $\sqrt{10}$

(b) A: $\frac{2}{3}$, $\sqrt{21}/5$, $2/\sqrt{21}$, $\sqrt{21}/2$, $5/\sqrt{21}$, $\frac{5}{2}$; B: $\sqrt{21}/5$, $\frac{2}{3}$, $\sqrt{21}/2$, $2/\sqrt{21}$, $\frac{5}{2}$, $5/\sqrt{21}$

(c) A: $\frac{3}{4}$, $\sqrt{7}/4$, $3/\sqrt{7}$, $\sqrt{7}/3$, $4/\sqrt{7}$, $\frac{4}{3}$; B: $\sqrt{7}/4$, $\frac{3}{4}$, $\sqrt{7}/3$, $3/\sqrt{7}$, $\frac{4}{3}$, $4/\sqrt{7}$

15.17. Which is the greater and why?

(a) $\sin 55°$ or $\cos 55°$?   (c) $\tan 15°$ or $\cot 15°$?
(b) $\sin 40°$ or $\cos 40°$?   (d) $\sec 55°$ or $\csc 55°$?

Hint: Consider a right triangle having as acute angle the given angle.

Ans.   (a) $\sin 55°$ (b) $\cos 40°$ (c) $\cot 15°$ (d) $\sec 55°$

15.18. Find the value of each of the following:

(a) $\sin 30° + \tan 45°$

(e) $\dfrac{\tan 60° - \tan 30°}{1 + \tan 60° \tan 30°}$

(b) $\cot 45° + \cos 60°$

(c) $\sin 30° \cos 60° + \cos 30° \sin 60°$

(f) $\dfrac{\csc 30° + \csc 60° + \csc 90°}{\sec 0° + \sec 30° + \sec 60°}$

(d) $\cos 30° \cos 60° - \sin 30° \sin 60°$

Ans.   (a) $\frac{3}{2}$ (b) $\frac{3}{2}$ (c) 1 (d) 0 (e) $1/\sqrt{3}$ (f) 1

15.19. A man drives 500 ft along a road which is inclined 20° to the horizontal. How high above his starting point is he?

Ans.   170 ft

15.20. A tree broken over by the wind forms a right triangle with the ground. If the broken part makes an angle of 50° with the ground and if the top of the tree is now 20 ft from its base, how tall was the tree?

Ans.   56 ft

15.21. Two straight roads intersect to form an angle of 75°. Find the shortest distance from one road to a gas station on the other road 1000 m from the junction.

Ans.   970 m

15.22. Two buildings with flat roofs are 60 ft apart. From the roof of the shorter building, 40 ft in height, the angle of elevation to the edge of the roof of the taller building is 40°. How high is the taller building?

Ans.   90 ft

15.23. A ladder, with its foot in the street, makes an angle of 30° with the street when its top rests on a building on one side of the street and makes an angle of 40° with the street when its top rests on a building on the other side of the street. If the ladder is 50 ft long, how wide is the street?

Ans.   82 ft

15.24. Find the perimeter of an isosceles triangle whose base is 40 cm and whose base angle is 70°.

Ans.   156 cm

CHAPTER 16

Introduction to Geometry

1. SOLVING PROBLEMS GRAPHICALLY

Problems may be solved graphically. For this purpose, obtain a table of values directly from the problem relationships rather than from equations.

1.1. SOLVING A WORK PROBLEM GRAPHICALLY

Solve graphically: Abe can dig a certain ditch in 6 h, while Naomi requires 3 h. If both start together from opposite ends, how long would they take to complete the digging of the ditch?

Solution:

Table of Values for Abe		Table of Values for Naomi	
Hours Worked	**Position Along Ditch**	**Hours Worked**	**Position Along Ditch**
0	O	0	1
3	$\frac{1}{2}$	$1\frac{1}{2}$	$\frac{1}{2}$
6	1	3	0

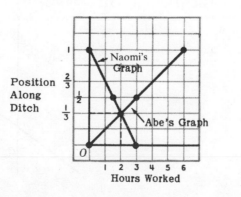

Fig. 16-1

323

In the table and on the graph in Fig. 16-1, O represents the beginning of the ditch where Abe started digging, while 1 represents the end of the ditch where Naomi began.

The common solution shows that they required 2 h. At that time Abe had completed $\frac{1}{3}$ of the job, while Naomi finished the remaining $\frac{2}{3}$ of the work.   *Ans.*   2 h

1.2.   SOLVING A MOTION PROBLEM GRAPHICALLY

Solve graphically: At 12 noon, Mr. Pabst left New York. After traveling at 30 mi/h for 2 h, he rested for 1 h and then continued at 35 mi/h. Mrs. Mayer wishes to overtake him. If Mrs. Mayer starts at 3 P.M. from the same place and travels along the same road at 50 mi/h, when will they meet?

__Solution:__   See Fig. 16-2.

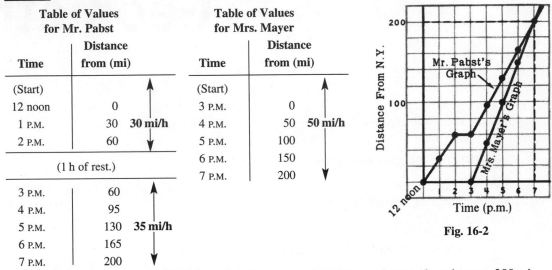

Table of Values for Mr. Pabst

Time	Distance from (mi)	
(Start)		
12 noon	0	
1 P.M.	30	30 mi/h
2 P.M.	60	
(1 h of rest.)		
3 P.M.	60	
4 P.M.	95	
5 P.M.	130	35 mi/h
6 P.M.	165	
7 P.M.	200	

Table of Values for Mrs. Mayer

Time	Distance from (mi)	
(Start)		
3 P.M.	0	
4 P.M.	50	50 mi/h
5 P.M.	100	
6 P.M.	150	
7 P.M.	200	

Fig. 16-2

The common solution shows that Mrs. Mayer will overtake Mr. Pabst at 7 P.M. when they are 200 mi from New York.   Ans.   7 P.M.

2.   UNDERSTANDING THE SLOPE OF A LINE

THINK! $$\text{Slope of Line} = \frac{y_2 - y_1}{x_2 - x_1} = \frac{\Delta y}{\Delta x} = m = \tan i$$

The following are three slope rules for the **slope of a line through two points:**

(1) By definition:

If line passes through $P_1(x_1, y_1)$ and $P_2(x_2, y_2)$ (as in Fig. 16-3):

Rule 1. $$\text{Slope of } P_1P_2 = \frac{y_2 - y_1}{x_2 - x_1}$$

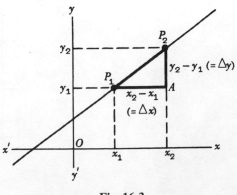

Fig. 16-3

NOTE: Now $y_2 - y_1$ is the difference of the y values. In the diagram, the length of $\overline{P_2A}$ equals $y_2 - y_1$. Similarly, $x_2 - x_1$ is the corresponding difference of the x values. The length of $\overline{P_1A}$ equals $x_2 - x_1$.

Delta Form of Slope Definition

$$\boxed{\textbf{Slope of } P_1P_2 = \frac{\Delta y}{\Delta x}}$$

Δy means $y_2 - y_1$
Δx means $x_2 - x_1$.

Delta (Δ), the fourth letter of the Greek alphabet, corresponds to d, the fourth letter of the English alphabet. Read Δy as "delta y" and Δx as "delta x." If you think of delta y as "y difference," you will see why Δy is used to replace $y_2 - y_1$.

(2) Slope of a line is m if its equation is in form $y = mx + b$.

Rule 2. $\boxed{m = \textbf{Slope of a line whose equation is } y = mx + b}$

Thus, the slope of $y = 3x + 2$ equals 3.

(3) Slope of a line = tangent of its inclination.

The inclination of a line is the angle it makes with the positive direction of the x axis. If i is the inclination of P_1P_2:

Rule 3. $\boxed{\textbf{Slope of } P_1P_2 = \tan i}$

Note in Fig. 16-4 that $\tan i = \dfrac{\Delta y}{\Delta x}$.

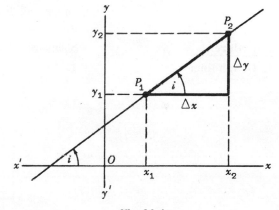

Fig. 16-4

2.1.   APPLYING RULE 1: SLOPE OF LINE $= \dfrac{y_2 - y_1}{x_2 - x_1} = \dfrac{\Delta y}{\Delta x}$

Graph and find the slope of the line through
  a) (0, 0) and (2, 4)   *b)* $(-2, -1)$ and (4, 3)

Procedure: **Solutions:**   See Fig. 16-5*a* and *b*.

1. Plot the points   **1.**   **1.**
and draw the
line which pas-
ses through
them.

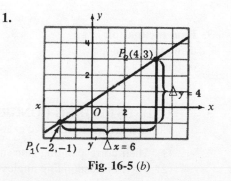

Fig. 16-5 (*a*)   **Fig. 16-5** (*b*)

2. Find Δy **and** Δx, the corresponding differences of y and x:

2. $P_2(2, 4) \rightarrow x_2 = 2, y_2 = 4$

$\underline{P_1(0, 0) \rightarrow x_1 = 0, \; y_1 = 0}$

$\Delta x = 2, \Delta y = 4$

2. $\qquad P_2(4, 3) \rightarrow x_2 = 4, y_2 = 3$

$\underline{P_1(-2, -1) \rightarrow x_1 = -2, \; y_1 = -1}$

$\Delta x = 6, \quad \Delta y = 4$

3. Find slope of line $= \dfrac{\Delta y}{\Delta x}$:

3. Slope of $P_1P_2 = \dfrac{\Delta y}{\Delta x} = \dfrac{4}{2} = 2$

Ans.   2

3. Slope of $P_1P_2 = \dfrac{\Delta y}{\Delta x} = \dfrac{4}{6} = \dfrac{2}{3}$

Ans. $\dfrac{2}{3}$

2.2.   APPLYING RULE 2: $m = $ **SLOPE OF LINE** $y = mx + b$

Find the slope of the line whose equation is

a) $2y = 6x - 8$ b) $3y - 4x = 15$

Procedure:

1. Transform equation into the form, $y = mx + b$:

2. Find slope of line $= m$, the coefficient of x:

Solutions:

1. D$_2$ $\begin{aligned} 2y &= 6x - 8 \\ y &= 3x - 4 \end{aligned}$

2. Slope $= 3$.

Ans.   3

1. Tr $3y - 4x = 15$

D$_3$ $\qquad \begin{aligned} 3y &= 4x + 15 \\ y &= \tfrac{4}{3}x + 5 \end{aligned}$

2. Slope $= \tfrac{4}{3}$.

Ans. $\tfrac{4}{3}$

2.3.   APPLYING RULE 3:   **SLOPE OF LINE OF ITS INCLINATION**

Graph each line and find its inclination, to the nearest degree:

a) $y = 2x + 3$ b) $y = \tfrac{1}{2}x - 1$

Procedure:

1. Graph line:

Solutions: See Fig. 16-6(a) and (b).

1.

y	3	5
x	0	1

1.

y	-1	0
x	0	2

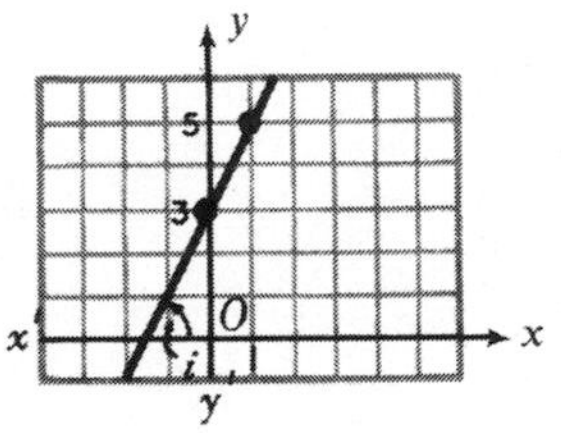

Fig. 16-6 (a) 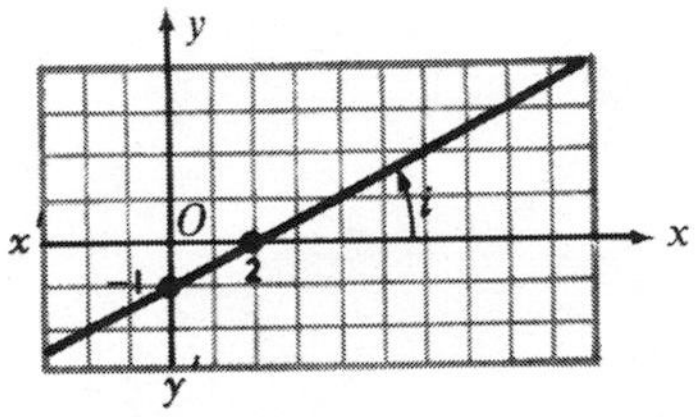   **Fig. 16-6 (b)**

2. Find slope of line, using $m = $ slope of $y = mx + b$:

3. Find inclination i, using $\tan i = $ slope of line:

2.   Slope $= 2$

3. $\tan i = 2.0000$

To nearest degree, $i = 63°$.

Ans.   63°

2.   Slope $= \tfrac{1}{2}$

3. $\tan i = \tfrac{1}{2} = 0.5000$

To nearest degree, $i = 27°$.

Ans.   27°

---

3.   UNDERSTANDING CONGRUENT TRIANGLES

Congruent triangles have exactly the same size and shape. If triangles are congruent,

(1) Their corresponding sides are congruent.

(2) Their corresponding angles are congruent.

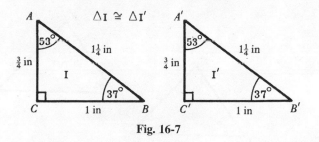

Fig. 16-7

Thus, in Fig. 16-7 congruent triangles I and I′ have congruent corresponding sides and angles—that is, three pairs of congruent sides and three pairs of congruent angles.

Read △I ≅ △I′ as "triangle I is congruent to triangle I prime."

Congruent triangles are exact duplicates of each other and may be made to fit together or coincide with one another. The aim of mass production is to produce exact duplicates or congruent objects.

Thus, if △I′ above is cut out and placed on △I, each side or angle of △I′ may be made to fit the corresponding part of △I.

The following three tests are used to determine when two triangles are congruent to each other.

Rule 1: [SSS = SSS]

Two triangles are **congruent** if three sides of one triangle are congruent, respectively, to three sides of the other.

Thus, △II ≅ △II′ (Fig. 16-8) since

(*1*) $DE = D'E'$ (= 4 in)
(*2*) $DF = D'F'$ (= 5 in)
(*3*) $EF = E'F'$ (= 6 in)

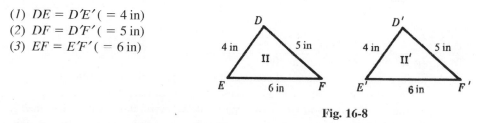

Fig. 16-8

Rule 2: [ASA = ASA]

Two triangles are **congruent** if two angles and the included side of one triangle are congruent, respectively, to two angles and the included side of the other.

Thus, △III ≅ △III′ (Fig. 16-9) since

(*1*) $\angle H \cong \angle H'$ (= 85°)
(*2*) $\angle J \cong \angle J'$ (= 25°)
(*3*) $EF = E'F'$ (= 6 in)

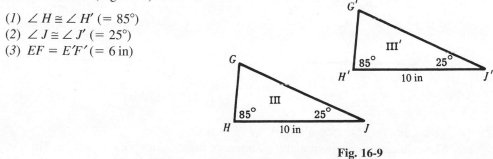

Fig. 16-9

Rule 3: [SAS = SAS]

Two triangles are **congruent** if two sides and the included angle of one triangle are congruent, respectively, to two sides and the included angle of the other.

Thus, △IV ≅ △IV′ (Fig. 16-10) since

(*1*) $KL = K'L'$ (= 9 in)
(*2*) $\angle L \cong \angle L'$ (= 55°)
(*3*) $LM = L'M'$ (= 12 in)

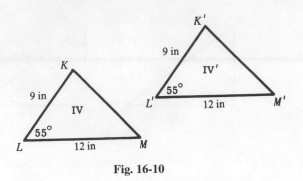

Fig. 16-10

Duplicating Lines, Angles, and Triangles

1.   **To duplicate or copy a given line:**

A line segment $\overline{A'B'}$ may be constructed congruent to a given line segment $\overline{AB}$, as follows: Using A' on a working line as a center, draw an arc of a circle whose radius equals AB.

Thus, in Fig. 16-11, $\overline{A'B'} \cong \overline{AB}$, by construction.

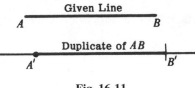

Fig. 16-11

2.   **To duplicate or copy a given angle:**

An angle A' may be constructed congruent to a given angle A, as follows:

(*1*) Using A as center and with a convenient radius, draw arc a, intersecting AP in B and AQ in C.

(*2*) Using A' as center and with same radius, draw arc a', intersecting $A'P'$ in B'.

(*3*) Using B' as center and with BC as a radius, draw arc b', intersecting arc a' in C'.

(*4*) Draw $A'C'$.

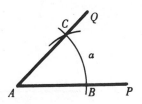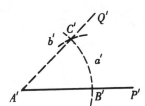

Fig. 16-12

Thus, in Fig. 16-12, $\angle A' \cong \angle A$, by construction.

To duplicate an angle of a given number of degrees, use your protractor. Thus, to duplicate an angle of 30° at A on BC, set the protractor as shown in Fig 16-13.

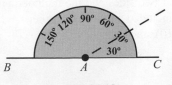

Fig. 16-13

3.   To duplicate or copy a given triangle:

   Using each of the three rules for showing that triangles are congruent, a given $\triangle ABC$ may be duplicated. See Figs. 16-14*a* and *b*, 16-15*a* and *b*, and 16-16*a* and *b*.

   Using Rule 1, $\triangle I \cong \triangle ABC$, by SSS $\cong$ SSS

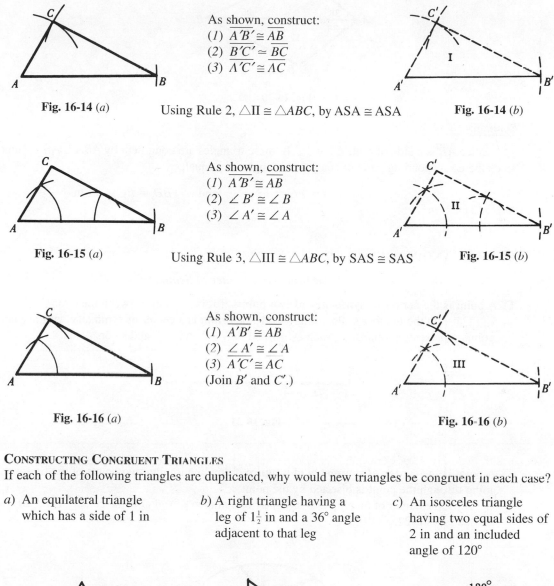

As shown, construct:
(*1*) $\overline{A'B'} \cong \overline{AB}$
(*2*) $\overline{B'C'} \cong \overline{BC}$
(*3*) $\overline{A'C'} \cong \overline{AC}$

Fig. 16-14 (*a*)   Fig. 16-14 (*b*)

   Using Rule 2, $\triangle II \cong \triangle ABC$, by ASA $\cong$ ASA

As shown, construct:
(*1*) $\overline{A'B'} \cong \overline{AB}$
(*2*) $\angle B' \cong \angle B$
(*3*) $\angle A' \cong \angle A$

Fig. 16-15 (*a*)   Fig. 16-15 (*b*)

   Using Rule 3, $\triangle III \cong \triangle ABC$, by SAS $\cong$ SAS

As shown, construct:
(*1*) $\overline{A'B'} \cong \overline{AB}$
(*2*) $\angle A' \cong \angle A$
(*3*) $\overline{A'C'} \cong AC$
(Join B' and C'.)

Fig. 16-16 (*a*)   Fig. 16-16 (*b*)

3.1.   CONSTRUCTING CONGRUENT TRIANGLES

   If each of the following triangles are duplicated, why would new triangles be congruent in each case?

a) An equilateral triangle which has a side of 1 in

b) A right triangle having a leg of $1\frac{1}{2}$ in and a 36° angle adjacent to that leg

c) An isosceles triangle having two equal sides of 2 in and an included angle of 120°

Fig. 16-17

Fig. 16-18

Fig. 16-19

Solutions:

a) A duplicate equilateral triangle would be congruent to $\triangle I$ by SSS $\cong$ SSS. See Fig. 16-17.
b) A duplicate right triangle would be congruent to $\triangle II$ by ASA $\cong$ ASA. See Fig. 16-18.
c) A duplicate isosceles triangle would be congruent to $\triangle III$ by SAS $\cong$ SAS. See Fig. 16-19.

3.2.   DETERMINING LENGTHS AND ANGLES INDIRECTLY

Using the data indicated in Fig. 16-20, find AD and DB.

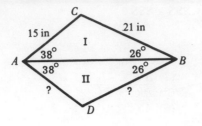

Fig. 16-20

<u>Solution:</u>

Since $\overline{AB}$ is a side of both $\triangle$I and $\triangle$II, these triangles are congruent by ASA $\cong$ ASA. Hence, since the corresponding parts of congruent triangles are equal,

$$AD = AC = 15 \text{ in} \qquad \text{and} \qquad BD = BC = 21 \text{ in}$$

Ans.   15 in, 21 in

4.   UNDERSTANDING SYMMETRY

Point Symmetry: Center of Symmetry

1) A point is the **center of symmetry of two points** if it is midway between the two.

   Thus, on the number scale in Fig. 16-21, the origin is a center of symmetry with respect to the points representing opposites, such as $+1$ and -1, $+2$ and -2, and so on.

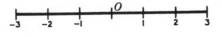

Fig. 16-21

2) A point is the **center of symmetry of a figure** if any line joining two points of the figure and going through the point is bisected by the point.

   Thus, the center of an ellipse is its center of symmetry. And O is the center of AC, BD, and so on. See Fig. 16-22.

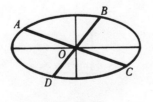

Fig. 16-22

Line Symmetry: Axis of Symmetry

1) A line is the **axis of symmetry of two points** if it is the perpendicular bisector of the line joining the two points. See Fig. 16-23.

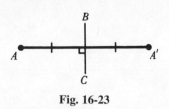

Fig. 16-23

Thus, *BC* is the axis of symmetry of points *A* and *A'*.

   2) A line is an **axis of symmetry of a figure** if any line perpendicular to it is bisected by the figure. Thus, *FE* is the axis of symmetry of the curve (parabola) shown in Fig. 16-24, $\overline{FE}$ is perpendicular to and bisects $\overline{AA'}$, $\overline{BB'}$, etc. *FE* is the "folding line" of the parabola.

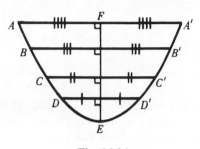

Fig. 16-24

5.   TRANSFORMATIONS

INTRODUCTION TO TRANSFORMATIONS

   If you look back at the previous chapters on geometry, you will notice that while we have concentrated on different topics from chapter to chapter, all the material had one very important thing in common: The *positions* of all the geometric figures were *fixed*. In other words, when we considered a triangle such as △*ABC* in Fig. 16-25, we did not move it. In this section, we consider objects in geometry as they change position. These objects (such as triangles, lines, points, and circles) will move as a result of transformations of the plane.

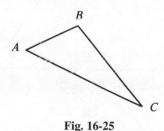

Fig. 16-25

DEFINITION: By a transformation of the plane, we mean a rule that assigns to each point in the plane a different point or the point itself.

   Note that *each* point in the plane is assigned to *exactly one point*. Points that are assigned to themselves are called *fixed* points. If point *P* is assigned to point *Q*, then we say the *image* of *P* is *Q*, and the image of *Q* is *P*.

5.1.   REFLECTIONS

   Imagine that a mirror is placed along line *m* in Fig. 16-26. What would be the image of point *S* in the mirror? How would you describe *S'*, the image of *S*? If we actually placed a mirror along *m*, we would see that the image of *S* lies on *l*, on the other side of *m*, and that the distance from *S* to *O* is equal to the distance from *O* to *S'* (see Fig. 16-27). We say that *S'* is the image of *S* under a reflection in line *m*. Notice that, under this reflection, *O* is the image of *O*.

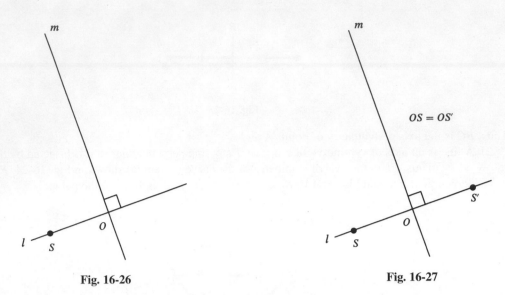

Fig. 16-26   Fig. 16-27

DEFINITION: A **reflection in line** m is a transformation of the plane having the property that the image of any point S not on m is S', where m is the perpendicular bisector of $\overline{SS'}$; the image of any point O on m is O itself.

We write $R_m(S) = S'$ to mean S' is the image of S under the reflection in line m.

EXAMPLE 1:   Image of a Point   Find the image of (a), A, (b) B, (c) C, (d) $\overline{AC}$, and (e) $\triangle DAC$ under the reflection in line t indicated in Fig. 16-28.

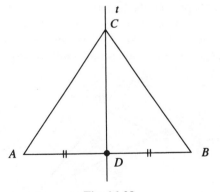

Fig. 16-28

SOLUTIONS

(a) B, because t is the perpendicular bisector of $\overline{AB}$

(b) A

(c) C, because C is on t

(d) $\overline{BC}$ (Why?)

(e) $\triangle DBC$, because D and C are fixed and $R_t(B) = A$

EXAMPLE 2:   Image of a Triangle   What is the image of $\triangle ABC$ in Fig. 16-28 under a reflection in line t?

SOLUTION

We saw that $R_t(A) = B$, $R_t(B) = A$, and $R_t(C) = C$; thus, $\triangle ABC$ is its own image.

5.2. LINE SYMMETRY

Notice that the images of angles are angles and the images of segments are segments under a reflection in a line. When a figure is its own image under a reflection in a line (like $\triangle ABC$ in Fig. 16-28), we say the figure has *line symmetry*.

DEFINITION: A figure F exhibits **line symmetry** if there exists a line l such that the image of F under a reflection in line l is F itself. In this case, l is called a *line of symmetry* or an *axis of symmetry*.

Notice that when a figure exhibits line symmetry, not all its points are necessarily fixed. In Fig. 16-28, only points C and D are fixed in triangle ABC.

EXAMPLE 3: Finding the Axis of Symmetry In Fig. 16-29, find all axes of symmetry for regular hexagon *ABCDEF*.

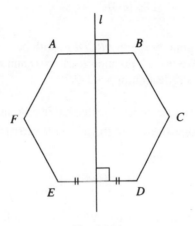

Fig. 16-29

SOLUTION

Here $\overline{AD}$, $\overline{FC}$, $\overline{BE}$, and the indicated line l are all axes of symmetry. Find two others.

EXAMPLE 4: Discovering Line Symmetry Which of the objects in Fig. 16-30 exhibit line symmetry?

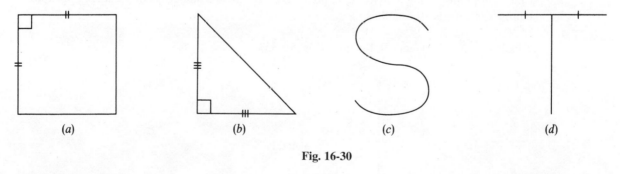

Fig. 16-30

SOLUTION

All except Fig. 16-30*c*.

5.3. POINT SYMMETRY

Not only can we transform the plane by reflections in a line, but also we can reflect in a point P. In Fig. 16-31, for example, we can reflect Q in the point P by finding the point Q' such that $QP = PQ'$.

DEFINITION: A **reflection in the point** P is a transformation of the plane such that the image of any point Q except P is Q', where $QP = PQ'$, and Q, P, and Q' are collinear, and the image of P is P (that is, P is fixed). If figure F is its own image under such a transformation, then we say F exhibits **point symmetry**.

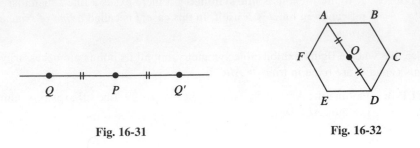

Fig. 16-31   Fig. 16-32

Figure 16-32 shows a regular hexagon $ABCDEF$, with $AO = OD$. Notice that A is the image of D under the reflection in O. We use the notation $R_o(A) = D$ and $R_o(D) = A$ to indicate that A and D are each other's images under a reflection in point O.

EXAMPLE 5: Finding Images under a Reflection in a Point   Referring to Fig. 16-32, find (a) $R_o(B)$, (b) $R_o(C)$, (c) $R_o(\overline{AD})$, (d) $R_o(\triangle AOB)$, and (e) $R_o(ABCDEF)$.

SOLUTIONS

(a) E (b) F (c) $\overline{AD}$ (d) $\triangle DOE$ (e) Hexagon $ABCDEF$ (Thus, $ABCDEF$ exhibits point symmetry.)

EXAMPLE 6: Finding Point Symmetry   Which of the following exhibit point symmetry?

(a) Squares (b) Rhombuses (c) Scalene triangles (d) A regular octagon

SOLUTIONS

All except (c).

Since points can change position in transformational geometry, analytic geometry is a particularly useful tool for these transformations. Recall that in analytic geometry, we deal extensively with the positions of points; being able to locate points and determine distances is of great help in exploring the properties of transformations.

EXAMPLE 7: Images Under Reflections (Fig. 16-33)

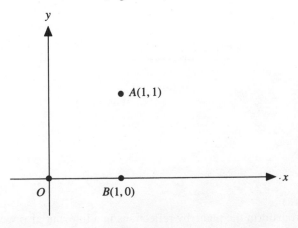

Fig. 16-33

(a) What is the image of point A under a reflection in the x axis? The y axis?

(b) What is the image of B under a reflection in the y axis?

(c) What is the image of O under a reflection in point O?

(d) What is the image of B under a reflection in the line $y = x$?

(e) What is the image of A under a reflection in the line $x = -1$?

(f) What is the image of $\triangle AOB$ under a reflection in the y axis? Under a reflection in O?

SOLUTIONS

(a) Point A' in Fig. 16-34 is the image of A under a reflection in the x axis; the coordinates of A' are $(1, -1)$. Point A'' is the image of A under a reflection in the y axis: $A'' = (-1, 1)$.

(b) Point B' in Fig. 16-35 is the image of B under a reflection in the y axis. Its coordinates are $(-1, 0)$.

(c) Point O is a fixed point. The point in which we reflect is always fixed.

(d) $R_l(B) = B'(0, 1)$ in Fig. 16-36. Notice that line l is the perpendicular bisector of $\overline{BB'}$.

(e) $R_m(A) = A'(-3, 1)$ in Fig. 16-37. Note that m is the perpendicular bisector of AA'.

(f) The image of $\triangle AOB$ under a reflection in the y axis is $\triangle A'B'O$ in Fig 16-38a, where $A' = (-1, 1)$, $B' = (-1, 0)$, and $O = (0, 0)$. The image under a reflection in the origin is $-A''B''O$ in Fig. 16-38b, where $A'' = (-1, 1)$, $B'' = (-1, 0)$, and $O = (0, 0)$.

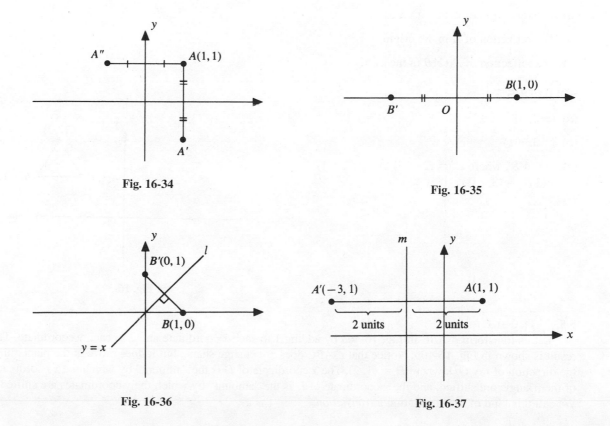

Fig. 16-34

Fig. 16-35

Fig. 16-36

Fig. 16-37

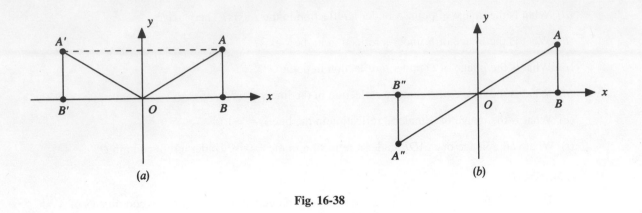

Fig. 16-38

5.4.   PATTERNS IN REFLECTIONS

We can observe several patterns in the results of the last solved problem:

1. The distance from A' to B' in Fig. 16-38a equals the distance from A to B. In other words, distance is *preserved* under a reflection. Observe that measures of angles are also preserved. In other words, $m[\angle 357]$ $BAO = m[\angle 357]B'A'O$ in Fig. 16-38a, and that property appears to hold for other reflections. As you will see, other properties are preserved as well.

2. Under a reflection in the x axis, the point (a, b) moves to $(a, -b)$; under a reflection in the y axis, (a, b) moves to $(-a, b)$; and under a reflection in the origin, (a, b) moves to $(-a, -b)$. These patterns hold only for these reflections.

EXAMPLE 8:   More Images Under Reflections   In Fig. 16-39, find

(a) The reflection of C in the y axis

(b) The reflection of B in the origin

(c) The reflection of $\triangle CAB$ in the x axis

SOLUTIONS

(a) $(-2, 3)$

(b) $(-3, -1)$

(c) $\triangle C'A'B'$, where $C' = (2, -3)$, $A' = (1, -1)$, and $B' = (3, -1)$

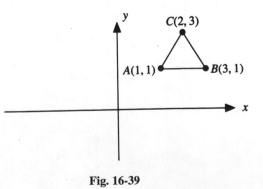

Fig. 16-39

5.5.   TRANSLATIONS

Let us transform $\triangle ABC$ in Fig. 16-40a by adding 1 to each x coordinate and 2 to each y coordinate. The result is shown in Fig. 16-40b. Notice that $\triangle ABC$ does not change shape, but it does move in the plane, in the direction of ray $\overrightarrow{OD}$, where $D = (1, 2)$. The x coordinate of D is the "amount" by which the x coordinates of the triangle are shifted, and the y coordinate of D is the "amount" by which the y coordinates are shifted. We call this kind of transformation a *translation*.

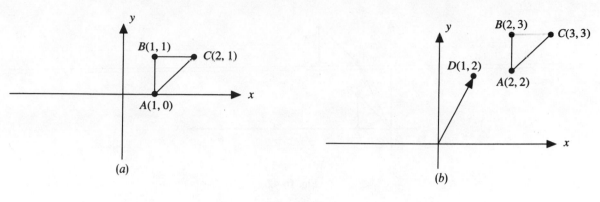

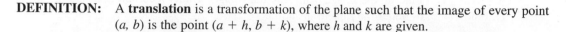

Fig. 16-40

DEFINITION: A **translation** is a transformation of the plane such that the image of every point (a, b) is the point $(a + h, b + k)$, where h and k are given.

A translation has the effect of moving every point the *same distance* in the *same direction*. We use the notation $T_{(h, k)}(a, b)$ to mean the image of (a, b) under a translation of h units in the x direction and k units in the y direction.

As in a reflection, distance and angle measure are preserved in a translation.

EXAMPLE 9:   Finding the Image of a Point   Find $T_{(-1, 1)}(1, 4)$ and $T_{(-1, 1)}(-1, 2)$.

SOLUTION

$$T_{(-1, 1)}(1, 4) = (1 + (-1), 4 + 1) = (0, 5)$$

$$T_{(-1, 1)}(-1, 2) = (-1 + (-1), 2 + 1) = (-2, 3)$$

Notice in Fig. 16-41 that $(1, 4)$ and $(-1, 2)$ are translated the same number of units in the same direction by the same translation T.

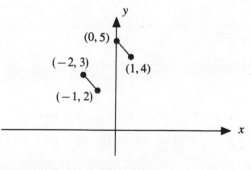

Fig. 16-41

EXAMPLE 10:   Finding the Image of a Triangle   Find the image of $\triangle ABC$ under the translation $T_{(1, 2)}$, where $A = (0, 0)$, $B = (1, 1)$, and $C = (1, 0)$.

SOLUTION

$$T_{(1, 2)}(0, 0) = (1, 2) \qquad T_{(1, 2)}(1, 1) = (2, 3) \qquad \text{and} \qquad T_{(1, 2)}(1, 0) = (2, 2)$$

Hence, the image of $\triangle ABC$ is $\triangle A'B'C'$ in Fig. 16-42. All points are translated along ray $\overrightarrow{AA'} = \overrightarrow{OA'}$, where $A'(1, 2)$ has the coordinates of the translation.

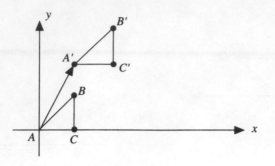

Fig. 16-42

EXAMPLE 11:   Finding the Image from Another Image   Under a certain translation, $T(5, 2) = (7, 1)$. Find $T(-3, 6)$ under the same translation.

SOLUTION

We have $T_{(h, k)}(5, 2) = (7, 1)$. Thus $5 + h = 7$, or $h = 2$; and $2 + k = 1$, or $k = -1$.

Then $T_{(2, -1)}(-3, 6) = (2 + (-3), -1 + 6) = (-1, 5)$.

EXAMPLE 12:   Finding Various Images under Translation

(a) Find $T_{(-1, 0)}(6, 2)$.

(b) Find h and k if $T_{(h, k)}(1, 7) = (0, 0)$.

(c) Find the image of square $ABCD$ under the translation $T_{(1, 1)}$, where $A = (0, 0)$, $B = (1, 0)$, $C = (0, 1)$, and $D = (1, 1)$.

(d) Find $T_{(h, k)}(1, 6)$ if $T_{(h, k)}(4, 1) = (0, -7)$.

(e) Find all fixed points under $T_{(-1, 4)}$.

SOLUTIONS

(a) $T(6, 2) = (6 + (-1), 2 + 0) = (5, 2)$

(b) $h = 0 - 1 = -1$; $k = 0 - 7 = -7$

(c) $A'B'C'D'$, where $A' = (1, 1)$, $B' = (2, 1)$, $C' = (1, 2)$, and $D' = (2, 2)$

(d) $h = 0 - 4 = -4$ and $k = -7 - 1 = -8$, so $T(1, 6) = (-3, -2)$

(e) Only $T_{(0, 0)}$ has fixed points. Any other translation, including $T_{(-1, 4)}$, has none.

EXAMPLE 13:   Finding Images of Figures   Let $A = (1, 1)$, $B = (2, 2)$, and $C = (3, 1)$. Find the image under $T_{(2, -1)}$ of (a) $\overline{AB}$, (b) $\triangle ABC$, and (c) CBA.

SOLUTIONS

(a) $\overline{A'B'}$, where $A' = (3, 0)$ and $B' = (4, 1)$

(b) $\triangle A'B'C'$, with $C' = (5, 0)$

(c) $C'B'A'$

Supplementary Problems

16.1. Solve graphically. How long will it take Tom and George to mow a lawn together if Tom takes 12 h to do the job alone while George requires *a)* 6 h, *b)* 4 h, *c)* 3 h? **(1.1)**

    Ans.   *a)* 4 h,   *b)* 3 h,   *c)* $2\frac{2}{5}$ h

16.2. Solve graphically. A motorist traveling at 60 mi/h seeks to overtake another who started 3 h earlier. Both are traveling along the same road. In how many hours will the slower motorist be overtaken if he is traveling at *a)* 40 mi/h, *b)* 30 mi/h, *c)* 20 mi/h? **(1.2)**

    Ans.   *a)* 6 h,   *b)* 3 h,   *c)* $1\frac{1}{2}$ h

16.3. Solve graphically: Towns A and B are 170 mi apart. Mr. Cahill left town A and is traveling to town B at an average rate of 40 mi/h. Mr. Fanning left town B one hour later and is traveling to town A along the same road at an average rate of 25 mi/h. When and where will they meet? **(1.2)**

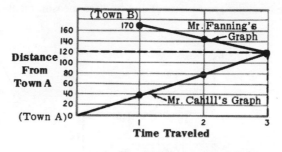

Fig. 16-43

    Ans.   They will meet, as shown graphically in Fig. 16-43, 120 mi from A, three hours after Mr. Cahill started.

16.4. Find the slope of the line through: **(2.1)**

    a) (0, 0) and (6, 15)   *c)* (3, −4) and (5, 6)   *e)* (8, 10) and (0, −2)

    b) (2, 3) and (6, 15)   *d)* (−2, −3) and (2, 1)   *f)* (−1, 2) and (5, 14)

Ans. *a)* $\dfrac{15-0}{6-0}=\dfrac{15}{6}=\dfrac{5}{2}$ *c)* $\dfrac{6-(-4)}{5-3}=\dfrac{10}{2}=5$ *e)* $\dfrac{-2-10}{0-8}=\dfrac{-12}{-8}=\dfrac{3}{2}$

      b) $\dfrac{15-3}{6-2}=\dfrac{12}{4}=3$ *d)* $\dfrac{1-(-3)}{2-(-2)}=\dfrac{4}{4}=1$ *f)* $\dfrac{14-2}{5-(-1)}=\dfrac{12}{6}=2$

16.5. Find the slope of the line whose equation is: **(2.2)**

    a) $y = 3x - 4$ *d)* $2y = 6x - 10$ *g)* $\dfrac{y}{2} = x - 3$

    b) $y = 5x$ *e)* $3y = 21 - 6x$ *h)* $\dfrac{y}{3} + 2x = 8$

    c) $y = 5$ *f)* $4y - 3x = 16$ *i)* $x - \frac{2}{3}y = 12$

    a) $y = 3x - 4$ *Ans.* 3   *d)* $y = 3x - 5$ *Ans.* 3   *g)* $y = 2x - 6$ *Ans.* 2

    b) $y = 5x + 0$ 5   *e)* $y = -2x + 7$ −2   *h)* $y = -6x + 24$ −6

    c) $y = 0x + 5$ 0   *f)* $y = \frac{3}{4}x + 4$ $\frac{1}{4}$ *i)* $y = \frac{3}{2}x - 18$ $\frac{3}{2}$

16.6. Find the inclination, to the nearest degree, of each line: **(2.3)**

    a) $y = 3x - 1$ *c)* $y = \frac{5}{2}x + 5$ *e)* $5y = 5x - 3$

    b) $y = \frac{1}{3}x - 1$ *d)* $y = \frac{2}{5}x + 5$ *f)* $y = -3$

a) tan *i* = 3   *Ans.* 72°   *c*) tan *i* = 2.5   *Ans.* 68°   *e*) tan *i* = 1   *Ans.* 45°

b) tan *i* = 0.3333   18°   *d*) tan *i* = 0.4   22°   *f*) tan *i* = 0   0°

16.7. If each of the following triangles are duplicated, why will the new triangles be congruent in each case?   **(3.1)**

   a) A right triangle whose legs are 2 and $1\frac{3}{4}$ in

   b) An isosceles triangle having a base of 3 in and congruent base angles of 35°

   c) An isosceles triangle having congruent sides of 4 in and a vertical angle of 50°

   d) A right triangle having a leg of $3\frac{1}{2}$ in and a 55° angle opposite that leg

   e) A triangle having its sides in the ratio of 2:3:4 and 1 in for its smallest side

Ans. *a*) SAS ≅ SAS   *b*) ASA ≅ ASA *c*) SAS ≅ SAS
   d) ASA ≅ ASA, using 35° angle   *e*) SSS ≅ SSS, using sides of 1, $1\frac{1}{2}$, and 2 in

16.8. In each, determine why the triangles are congruent and find the measurements indicated with a question mark.   **(3.2)**

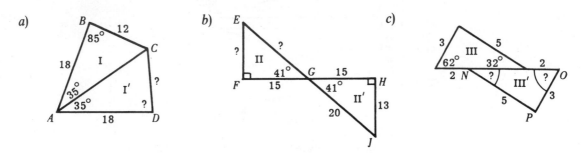

Fig. 16-44

Ans.
 a) △I ≅ △I′ by SAS ≅ SAS *b*) △II ≅ △II′ by ASA ≅ ASA *c*) △III ≅ △III′ by SSS ≅ SSS
   $m\angle D = 85°$, $CD = 12$ $EF = 13$, $EG = 20$ $m\angle O = 62°$, $m\angle ONP = 32°$

16.9. State the point and line symmetry of each: **(4.5)**
  a) Circle *b*) Parallelogram *c*) Rhombus *d*) Isosceles triangle

Diagram	 ⬭	 ▱	 ◇	 △
Figure	*a*) Circle	*b*) Parallelogram	*c*) Rhombus	*d*) Isosceles triangle
Center of Symmetry	Center of circle	Point of intersection of the diagonals	Point of intersection of the diagonals	None
Axis Symmetry	Any diameter	None	Each diagonal	Altitude from vertex

Fig. 16-45

16.10. Find the image of $\overline{AB}$ in Fig. 16-46 under a reflection in the *x* axis; then in the *y* axis.   **(5.1)**

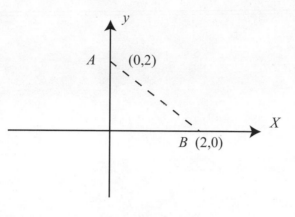

Fig. 16-46

Ans.   The segment connecting $(0, -2)$ and $(2, 0)$; the segment connecting $(0, 2)$ and $(-2, 0)$.

16.11. Given the square that follows, find: **(5.3)**

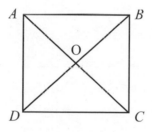

Fig. 16-47

a) $R_o\,(\overline{AC})$ b) $R_o\,(\triangle AOB)$

Ans.   a) $\overline{AC}$ b) $\triangle DOC$

16.12. Find: a) $T_{(1,\,3)}$ $(-3, 6)$

b) $T_{(0,\,4)}$ (h, k) **(5.5)**

Ans.   a) $(-2, 9)$

b) $(h, 4 + k)$

APPENDIX A

Review of Arithmetic

1. THE WHOLE NUMBERS

A. *Understanding Whole Numbers*

(1) **Whole numbers are the numbers used in counting.**
Thus, 3, 35, and 357 are whole numbers, but −3 and 4/3 are not.

(2) **Place value of each digit of a whole number.**
Depending on its **place in the whole number**, each digit of the number **represents** a number of units, tens, hundreds, thousands, and so on.
Thus, in 35, 3 represents 3 tens, while in 357, 3 represents 3 hundreds.

(3) **Reading whole numbers by using place value.**
In reading a whole number, give place value to its digits. Thus, read 4444 as "four **thousand**, four **hundred**, forty (four **tens**) four."

(4) **Rounding a whole number: approximate value of a number.**
In rounding a whole number to the nearest ten, replace the number by the multiple of 10 to which it is closest.
Thus, 37 to the nearest ten becomes 40. Here, 37 is between 30 and 40. However, 37 is closer to 40. Similarly, round a number to the nearest hundred, nearest thousand, and so on.

NOTE: When a whole number is exactly halfway between two possible answers, round it off to the larger. Thus, round 235 to the nearest ten as 240, not 230. Round 1500 to the nearest thousand as 2000.

(5) **One way in which to provide an approximate value of a whole number** is to round the whole number. Thus, 40 is an approximate value of 37.

(6) **Estimating answers by using approximate values of whole numbers.**

Estimate: *a*) 38 × 72 *b*) 4485/503 *c*) 7495-2043

Procedure:

1. Round each number to a suitable multiple of 10, 100, etc.:	**1.** Round 38 to 40, 72 to 70.	**1.** Round 4485 to 4500, 503 to 500.	**1.** Round 7495 to 7500, 2403 to 2000.

2. Perform the indicated operations on the resulting **approximations**	**2.** $40 \times 70 = 2800$	**2.** $4500/500 = 9$	**2.**	$\begin{array}{r} 7500 \\ -\,2000 \\ \hline 5500 \end{array}$
	Ans. 2800	*Ans.* 9	*Ans.* 5500	

B.   Terms Used in the Fundamental Operations

Addition, subtraction, multiplication, and **division** are the four fundamental operations performed on numbers. The names used in each operation must be memorized. Note them in each of the following:

(1) **Addition:** Add 77 and 20.

Terms Used in Addition		
Addends are the numbers that are added.	**Addend**	77
Sum is the answer obtained in addition.	**Addend**	$+20$
	Sum	$\overline{97}$

(2) **Subtraction:** Subtract 20 from 77.

Terms Used in Subtraction		
Minuend is the number from which we subtract.	**Minuend**	77
Subtrahend is the number being subtracted.	**Subtrahead**	-20
Difference is the answer obtained in subtraction.	**Difference**	$\overline{57}$

(3) **Multiplication:** Multiply 77 by 20.

Terms Used in Multiplication		
Multiplicand is the number being multiplied.	**Multiplicand**	77
Multiplier is the number by which we multiply.	**Multiplier**	$\times 20$
Product is the answer obtained in multiplication.	**Product**	$\overline{1540}$

(Numbers being multiplied are also **factors** of their product.)

(4) **Division:** Divide 77 by 20.

Terms Used in Division		
Dividend is the number being divided.	**Dividend**	$\frac{77}{20}$
Divisor is the number by which we divide.	**Divisor**	
Quotient is the answer obtained in division.	**Quotient** $=$	$3\frac{17}{20}$

(Call the quotient a total quotient to distinguish it from 3, the partial quotient.)

RULE: Final Quotient = Partial Quotient + $\dfrac{\text{Remainder}}{\text{Divisor}}$

Thus, in $3\frac{17}{20}$ above, 3 is the partial quotient, and 17 is the remainder.

Keep in mind, $3\frac{17}{20} = 3 + \frac{17}{20}$.

C.   Checking the Fundamental Operations

(1) **Add** 25, 32, and 81. **Check** the addition.
Add either down the column or up the column.
To check addition, add the column in the reverse direction.

	Add down	$\begin{array}{r} 25 \\ 32 \\ 81 \\ \hline 138 \end{array}$	Check up

(2) **Subtract** 32 from 85. **Check** the subtraction.
To check subtraction, add the difference and subtrahend. The answer thus obtained should be the minuend.

	$\begin{array}{r} 85 \\ -\,32 \\ \hline 53 \end{array}$	Check:	$\begin{array}{r} 53 \\ +\,32 \\ \hline 85 \end{array}$

(3) Multiply 85 by 32. **Check** the multiplication.
    To check multiplication, multiply the numbers after interchanging them.

$$\begin{array}{r} 85 \\ \times 32 \\ \hline 170 \\ 255 \\ \hline 2720 \end{array}$$

Check:

$$\begin{array}{r} 32 \\ \times 85 \\ \hline 160 \\ 256 \\ \hline 2720 \end{array}$$

---

(4) **Divide** 85 by 32. **Check** the division.
    To check division, multiply the divisor by the partial quotient. To the result obtained, **add** the remainder. The final answer should be the dividend.

$$\begin{array}{r} 2 \\ 32\overline{)85} \\ 64 \\ \hline 21 \end{array}$$ Remainder

Check:

$$\begin{array}{r} 32 \\ \text{Partial quotient} \rightarrow \times\, 2 \\ \hline 64 \\ +21 \\ \hline 85 \end{array}$$

Ans. $2\frac{21}{32}$

2. FRACTIONS

A. *Understanding Fractions*

Proper and Improper Fractions

(1) A **proper fraction** is a fraction whose value is less than 1. In a proper fraction, the numerator is less than the denominator.

   Thus, $\frac{5}{8}$ and $\frac{69}{70}$ are proper fractions. Each has a value less than 1.

(2) An **improper fraction** is a fraction whose value is equal to or greater than 1. In an improper fraction, the numerator is equal to or greater than the denominator.

   Thus, $\frac{8}{5}$ and $\frac{15}{15}$ are improper fractions; $\frac{8}{5}$ is greater than 1, and $\frac{15}{15}$ equals 1.

(3) The **terms of a fraction** are its numerator and denominator.

   Thus, the terms of $\frac{25}{30}$ are 25 and 30.

(4) **Equivalent fractions** are fractions having the same value.

   Thus, $\frac{1}{5}$, $\frac{2}{10}$, $\frac{5}{25}$, and $\frac{20}{100}$ are equivalent fractions: that is, $\frac{1}{5} = \frac{2}{10} = \frac{5}{25} = \frac{20}{100}$.

B. *Changing Forms of Mixed Numbers and Improper Fractions*

(1) A **mixed number** equals a whole number plus a fraction. Thus, $17\frac{2}{5} = 17 + \frac{2}{5}$.

(2) **Changing a mixed number to an improper fraction.**

   Change to an improper fraction: *a)* $17\frac{2}{5}$ *b)* $101\frac{2}{9}$

Procedure:	Solutions:	
1. Multiply whole number by denominator:	**1.** $17 \times 5 = 85$	**1.** $101 \times 9 = 909$
2. Add numerator to product:	**2.** $\begin{array}{r}+2\\\hline 87\end{array}$	**2.** $\begin{array}{r}+2\\\hline 911\end{array}$
3. Form fraction by placing result over denominator:	**3.** *Ans.* $\dfrac{87}{5}$	**3.** *Ans.* $\dfrac{911}{9}$

(3) **To change an improper fraction to a mixed number**, divide the numerator by the denominator.

   Thus, $\frac{87}{5} = 87 \div 5 = 17\frac{2}{5}$

C. *Changing a Fraction to an Equivalent Fraction*

(1) **Fundamental law of fractions:**

   To change a fraction to an equivalent fraction, multiply or divide both the numerator and denominator by the same number, other than zero, of course.

   Thus, $\dfrac{2}{5} = \dfrac{2 \times 10}{5 \times 10} = \dfrac{20}{50}$ In turn, $\dfrac{20}{50} = \dfrac{20 \div 10}{50 \div 10} = \dfrac{2}{5}$

(2) To raise **a fraction to higher terms**, multiply its terms by the same whole number. Thus, when $\frac{2}{5}$ is changed to $\frac{20}{50}$, it has been changed to higher terms.

(3) **To reduce a fraction to lower terms**, divide its terms by the same whole number. Thus, when $\frac{20}{50}$ is changed to $\frac{2}{5}$, it has been changed to lower terms.

(4) **To reduce a fraction to lowest terms**, divide both numerator and denominator by the highest common factor.

Thus, $\frac{20}{50}$ can be reduced to $\frac{10}{25}$, $\frac{4}{10}$, or $\frac{2}{5}$: but to reduce it to lowest terms, use $\frac{2}{5}$. To obtain $\frac{2}{5}$, divide terms by 10, the highest common factor of 20 and 50.

(5) **To change a fraction to a new specified fraction:**

Change to new fraction: *a*) $\frac{3}{5} = \frac{?}{45}$ *b*) $\frac{3}{5} = \frac{45}{?}$

Procedure:

1. **Divide** new denominator by old denominator, or divide new numerator by old numerator:
2. **Multiply** the result by old numerator or old denominator:
3. **Form fraction** needed:

Solutions:

1. $45 \div 5 = 9$ **1.** $45 \div 3 = 15$

2. $\frac{\times 3}{27}$ **2.** $\frac{\times 5}{75}$

Ans. $\frac{27}{45}$ *Ans.* $\frac{45}{75}$

(6) **To compare the size of fractions:**

Compare $\frac{5}{9}$ and $\frac{7}{12}$.

Procedure:

1. **Change to equivalent fractions** having the same denominator. Use the lowest common denominator (**LCD**).

2. **Compare** new fractions:

Solutions:

1. Change: $\frac{5}{9} = \frac{?}{36}$ and $\frac{7}{12} = \frac{?}{36}$.

    Here, $\frac{5}{9} = \frac{20}{36}$ and $\frac{7}{12} = \frac{21}{36}$.

2. Since $\frac{20}{36}$ is less than $\frac{21}{36}$,

    then $\frac{5}{9}$ is less than $\frac{7}{12}$.

D.   *Multiplying and Dividing Mixed Numbers and Fractions*

Multiply and divide as indicated: *a*) $1\frac{1}{6} \div 1\frac{5}{9}$ *b*) $1\frac{1}{7} \times 2\frac{11}{12} \div \frac{8}{9}$

Procedure:

1. **Change** mixed numbers to fractions:

2. **Multiply** by each inverted divisor:

3. **Divide** a numerator and a denominator by a common factor of both:

Solutions:

1. $\frac{7}{6} \div \frac{14}{9}$ 1. $\frac{8}{7} \times \frac{35}{12} \div \frac{8}{9}$

2. $\frac{7}{6} \times \frac{9}{14}$ 2. $\frac{8}{7} \times \frac{35}{12} \times \frac{9}{8}$

3. $\dfrac{\overset{1}{\cancel{7}}}{\underset{2}{\cancel{6}}} \times \dfrac{\overset{3}{\cancel{9}}}{\underset{2}{\cancel{14}}}$ 3. $\dfrac{\overset{1}{\cancel{8}}}{\underset{1}{\cancel{7}}} \times \dfrac{\overset{5}{\cancel{35}}}{\underset{4}{\cancel{12}}} \times \dfrac{\overset{3}{\cancel{9}}}{\underset{1}{\cancel{8}}}$

4. Multiply the remaining factors in the numerator and denominator separately:

$$4.\ \frac{3}{2 \times 2} = \frac{3}{4} \quad Ans. \qquad 4.\ \frac{5 \times 3}{4} = \frac{15}{4} \text{ or } 3\tfrac{3}{4} \quad Ans.$$

E. Adding and Subtracting Fractions

(1) Combining fractions having the same denominator:

Combine:　　　$a)\ \dfrac{5}{12} + \dfrac{11}{12} - \dfrac{7}{12}$　　　$b)\ \dfrac{15}{7} - \dfrac{3}{7} + \dfrac{2}{7}$

Procedure:　　　　　　　　　　　　　　　　　　**Solutions:**

1. Keep the common denominator and add or subtract the numerator as indicated:

$$1.\ \frac{5 + 11 - 7}{12} \qquad\qquad 1.\ \frac{15 - 3 + 2}{7}$$

2. Reduce result to the lowest terms:

$$2.\ \frac{9}{12} = \frac{3}{4} \quad Ans. \qquad 2.\ \frac{14}{7} = 2 \quad Ans.$$

(2) Combining fractions having different denominators:

Combine:　　　$\dfrac{5}{6} - \dfrac{7}{12} + \dfrac{23}{36}$

Procedure:　　　　　　　　　　　　　　　　　　**Solutions:**

1. Change the fractions to equivalent fractions having the same denominator, using the LCD:

$$1.\quad \text{LCD} = 36$$

$$\frac{30}{36} - \frac{21}{36} + \frac{23}{36}$$

2. Combine the resulting fractions:

$$2.\quad \frac{30 - 21}{36} + 23$$

3. Reduce to lowest terms:

$$3.\quad \frac{32}{36} = \frac{8}{9} \quad Ans.$$

F. Adding and Subtracting Mixed Numbers

Add or subtract as indicated:　　　$a)\ 15\tfrac{7}{8} + 9\tfrac{1}{4}$　　　$b)\ 15\tfrac{7}{8} - 9\tfrac{1}{4}$

Procedure:　　　　　　　　　　　　　　　　　　**Solutions:**

1. Add the fractions or **subtract** the fractions as indicated:

$$
\begin{array}{ll}
1.\ \ 15\tfrac{7}{8} \rightarrow \tfrac{7}{8} & 1.\ \ 15\tfrac{7}{8} \rightarrow \tfrac{7}{8} \\
\ \ +9\tfrac{1}{4} \rightarrow \tfrac{2}{8} & \ \ -9\tfrac{1}{4} \rightarrow \tfrac{2}{8} \\
\hline
\qquad\quad 1\tfrac{1}{8} & \qquad\quad \tfrac{5}{8}
\end{array}
$$

2. Add the whole numbers or **subtract** the whole numbers as indicated:

3. Add both results:

$$
\begin{array}{ll}
2.\ \ 15 & 2.\ \ 15 \\
3.\ \ \dfrac{+9}{24} + 1\tfrac{1}{8} & 3.\ \dfrac{-9}{6} + \tfrac{5}{8} \\
\\
\qquad 25\tfrac{1}{8} \ \ Ans. & \qquad 6\tfrac{5}{8} \ \ Ans.
\end{array}
$$

Here are the complete solutions of (*a*) and (*b*) above!

$$
\begin{array}{ll}
a)\ 15\tfrac{7}{8} \rightarrow \tfrac{7}{8} & b)\ 15\tfrac{7}{8} \rightarrow \tfrac{7}{8} \\
\ \ +9\tfrac{1}{4} \rightarrow \tfrac{2}{8} & \ \ -9\tfrac{1}{4} \rightarrow \tfrac{2}{8} \\
\hline
\ \ 24 + 1\tfrac{1}{8} & \ \ 6 + \tfrac{5}{8} \\
\\
\quad 25\tfrac{1}{8} \ \ Ans. & \quad 6\tfrac{5}{8} \ \ Ans.
\end{array}
$$

NOTE: When the fraction of the minuend is smaller than the fraction of the subtrahend, borrow one unit from the minuend to increase the smaller fraction.

Thus, $5\frac{2}{7} - 1\frac{5}{7} = 4\frac{9}{7} - 1\frac{5}{7} = 3\frac{4}{7}$ $\longrightarrow$

$$\begin{array}{r} 5\frac{2}{7} = 4\frac{9}{7} \\ -1\frac{5}{7} = -1\frac{5}{7} \\ \hline 3\frac{4}{7} \end{array}$$

3. DECIMALS

A. *Understanding Decimals*

(1) **A decimal or decimal fraction** is a fraction whose denominator is 10, 100, 1000, or some other power of 10.

   Thus, $\frac{3}{100}$ may be written in decimal form as 0.03

(2) **Rounding off a decimal: approximate value of a decimal:**

   In rounding a decimal to the nearest tenth, replace the decimal by the tenth to which it is closest. Thus, 0.37 to the nearest tenth becomes 0.4. Here, 0.37 is between 0.30 and 0.40. However, 0.37 is closer to 0.40 or 0.4.

   Similarly, round a decimal to the nearest hundredth, nearest thousandth, and so on.

NOTE.   When a decimal is exactly halfway between two possible answers, round it off to the larger. Thus, round 0.25 to the nearest tenth as 0.3, not 0.2.

(3) **An approximate value of a decimal** is a value obtained by rounding off the decimal.

   Thus, 0.4 is an approximate value of 0.37.

(4) **To estimate answers** of exercises or problems involving decimals, use rounded values of these decimals.

   Thus, to estimate 0.38×0.72, round to 0.4×0.7 and use 0.28 as the estimated answer.

B. *Changing Forms of Fractions and Decimals*

(1) **To change a common fraction to a decimal fraction**, divide the numerator by the denominator. Carry the answer to the desired number of places if a remainder exists.

   Thus, $\frac{3}{8} = \frac{3.000}{8} = 0.375,$ $\frac{5}{16} = \frac{5.0000}{16} = 0.3125.$

Equivalent Fractions and Decimals

$\frac{1}{2} = 0.5$	$\frac{1}{3} = 0.33\frac{1}{3}$	$\frac{1}{8} = 0.125$	$\frac{1}{5} = 0.2$	$\frac{1}{7} = 0.14\frac{2}{7}$
$\frac{1}{4} = 0.25$	$\frac{2}{3} = 0.66\frac{2}{3}$	$\frac{3}{8} = 0.375$	$\frac{2}{5} = 0.4$	$\frac{1}{9} = 0.11\frac{1}{9}$
$\frac{3}{4} = 0.75$	$\frac{1}{6} = 0.16\frac{2}{3}$	$\frac{5}{8} = 0.625$	$\frac{3}{5} = 0.6$	$\frac{1}{12} = 0.08\frac{1}{3}$
$\frac{1}{10} = 0.1$	$\frac{5}{6} = 0.83\frac{1}{3}$	$\frac{7}{8} = 0.875$	$\frac{4}{5} = 0.8$	$\frac{1}{16} = 0.06\frac{1}{4}$

(2) **To change a decimal fraction to a common fraction**, change the decimal to a common fraction and reduce to lowest terms.

   Thus: $0.65 = \frac{65}{100} = \frac{13}{20}.$

C. Adding and Subtracting Decimals

| | a) Add 1.35 and 0.952 | b) Subtract 0.952 from 1.35 |

Procedure:

Solutions:

1. **Arrange the decimals** vertically with decimal points directly under each other:

    1.   1.35   **1.**   1.350
        +0.952   −0.952

2. **Add or subtract** as in whole numbers, placing the decimal point in the result directly under the other decimal points:

    2.   2.302   **2.**   0.398

    Ans. 2.302   *Ans.* 0.398

D. Multiplying and Dividing Decimals

(1) **Multiplying and dividing numbers by** 10, 100, 1000, or some power of 10.

To multiply a decimal by a power of 10, move the decimal point as many places to the right as there are zeros in the power.

Thus, to multiply 5.75 by **1000**, move decimal point **three** places to the right.

    Ans.   5750

To divide a decimal by a power of 10, move the decimal point as many places to the left as there are zeros in the power.

Thus, to divide 5.75 by **1000**, move decimal point **three** places to the left.

    Ans.   0.00575

(2) **Multiplying decimals:**

| | Multiply: | a) 1.1 by 0.05 | b) 3.71 × 0.014 |

Procedure:

Solutions:

1. **Multiply** as in whole numbers:

    1.   1.1   (one place)   **1.**   3.71   (two places)
        ×0.05 (two places)   ×0.014 (three places)

2. **Mark off** decimal places in the product equal to the sum of the decimal places in the numbers multiplied:

    2. 0.055 (three places)   1484
                         371
                        2. 0.05194 (five places)

    Ans. 0.055   *Ans.* 0.05194

(3) **Dividing decimals:**

| | Divide: | a) 0.824 by 0.04 | b) 5.194 by 1.4 |

Procedure:

Solutions:

Vertically arranged

1. **Move** the decimal point of the divisor to the right to make the divisor a whole number:

    1. $\dfrac{0.824}{0.04}$

2. **Move** the decimal point of the dividend the same number of places to the right:

    2. $\dfrac{0.824}{0.04}$

3. **Divide** as in whole numbers and mark off decimal places in the quotient equal to new number of places in the dividend:

    3. $\dfrac{82.4}{4}$
         = 20.6

```
          3.71
    1.4 )5.1 94
          42
          99
          98
          14
          14
```

NOTE.   If a remainder exists, the quotient may be carried to additional decimal places by adding zeros to the dividend.

$$\text{Thus, } \frac{1}{7} = \frac{1.0000000000000}{7} = 0.142857142857\tfrac{1}{7}$$

4. REVIEWING PERCENTS AND PERCENTAGE

A. *Understanding Percents and Percentage*

Percent means hundredths. The percent symbol is %.
Thus, 7% of a number means 0.07 or $\frac{7}{100}$ of the number.

RULE: Percentage = Rate $\times$ Base

The **percentage** is the answer obtained when a percent is taken of a number.

The **rate** is the percent taken of a number.

The **base** is the number of which a percent is being taken.

Thus, since 2 percent of 400 = 8, 8 is the percentage, 2 percent is the rate, and 400 is the base.

B. *Interchanging Forms of Percents, Decimals, and Fractions*

(1) **Interchanging forms of percents and fractions:**

To change a percent to a fraction:
1. **Omit** the percent sign
2. **Divide** the number by 100.
3. **Reduce** to lowest terms.

Thus, $150\% = \frac{150}{100} = \frac{3}{2}$

Also, $2\tfrac{1}{2}\% = \frac{\overset{1}{\cancel{5}}}{2} \times \frac{1}{\underset{20}{\cancel{100}}} = \frac{1}{40}$

To change a fraction to a percent:
1. **Add** the percent sign
2. **Multiply** the number by 100.

Thus, $\frac{3}{2} = (\frac{3}{2} \times 100)\% = 150\%$

Also, $\frac{1}{40} = (\frac{1}{40} \times 100)\% = 2\tfrac{1}{2}\%$

Equivalent Percents and Fractions

$\frac{1}{2} = 50\%$	$\frac{1}{3} = 33\tfrac{1}{3}\%$	$\frac{1}{8} = 12\tfrac{1}{2}\%$	$\frac{1}{5} = 20\%$	$\frac{1}{7} = 14\tfrac{2}{7}\%$
$\frac{1}{4} = 25\%$	$\frac{2}{3} = 66\tfrac{2}{3}\%$	$\frac{3}{8} = 37\tfrac{1}{2}\%$	$\frac{2}{5} = 40\%$	$\frac{1}{9} = 11\tfrac{1}{9}\%$
$\frac{3}{4} = 75\%$	$\frac{1}{6} = 16\tfrac{2}{3}\%$	$\frac{5}{8} = 62\tfrac{1}{2}\%$	$\frac{3}{5} = 60\%$	$\frac{1}{12} = 8\tfrac{1}{3}\%$
$\frac{1}{10} = 10\%$	$\frac{5}{6} = 83\tfrac{1}{3}\%$	$\frac{7}{8} = 87\tfrac{1}{2}\%$	$\frac{4}{5} = 80\%$	$\frac{1}{16} = 6\tfrac{1}{4}\%$

(2) **Interchanging forms of percents and decimals:**

To change a percent to a decimal:
1. **Omit** the percent sign
2. **Move** the decimal point two places to the left.

Thus, 175% = 1.75
Also, 12.5% = 0.125

To change a decimal to a percent:
1. **Add** the percent sign
2. **Move** the decimal point two places to the right.

Thus, 1.75 = 175%
Also, 0.125 = 12.5%

C. *Percentage Problems*

Three Types of Percentage Problems

Types and Their Rules	Problems and Their Solutions	Problems and Their Solutions
(1) Finding percentage **Rule: Rate $\times$ Base =** **Percentage**	*a)* Find 5 percent of 400. $0.05 \times 400 = 20$	*d)* Find 30 percent of 80. $0.30 \times 80 = 24$
(2) Finding base **Rule: Base $= \dfrac{\text{Percentage}}{\text{Rate}}$**	*b)* 5 percent of what number is 20? $\dfrac{20}{0.05} = \dfrac{2000}{5} = 400$	*e)* 30 percent of what number is 24? $\dfrac{24}{0.30} = \dfrac{240}{3} = 80$
(3) Finding rate **Rule: Rate $= \dfrac{\text{Percentage}}{\text{Base}}$**	*c)* 20 is what percent of 400? $\dfrac{20}{400} = \dfrac{1}{20} = 0.05 = 5\%$	*f)* 24 is what percent of 80? $\dfrac{24}{80} = \dfrac{3}{10} = 0.3 = 30\%$

APPENDIX B

The Use of Calculators

1. USING A CALCULATOR TO PERFORM THE FUNDAMENTAL OPERATIONS OF ARITHMETIC

The calculator can be used to perform arithmetic calculations with extreme ease. We assume, for this text, that the student will use a basic calculator which accommodates the four basic operations, as well as signed numbers, squares, and square roots. Figure B-1 illustrates such a calculator.

OFF	√	%	÷
Mrc	M−	M+	*
7	8	9	−
4	3	2	+
1	2	3	=
On/C	0	.	

Fig. B-1

B.1 SIMPLE CALCULATIONS

Evaluate each using a calculator:

(a) $430 + 920$ (b) 671×284 (c) $5063 \div 29$

Solutions:

(*a*) Press: [4] [3] [0] [+] [9] [2] [0] [=]

Answer on screen is 1350.

(*b*) Press: [6] [7] [1] [*] [2] [8] [4] [=]

Answer on screen is 190,564.

(*c*) Press: [5] [0] [6] [3] [÷] [2] [9] [=]

Answer on screen is 174.58621.

B.2 DIVISIBILITY TESTS

Use a calculator to determine whether: (*a*) 8754 is divisible by 8   (*b*) 1234789 is divisible by 7.

Solutions

(*a*) Press: [8] [7] [5] [4] [÷] [8] [=]

Answer (*a*) on screen shows the division yields a remainder.

(*b*) Press: [1] [2] [3] [4] [7] [8] [9] [÷] [7] [=]

Answer (*b*) on screen shows the division yields a remainder.

(*c*) Press: [3] [0] [1] [8] [8] [4] [÷] [6] [=]

Answer (*c*) on screen proves the division does not leave a remainder.

2. INTRODUCTION TO THE GRAPHING CALCULATOR

In this text, a number of opportunities to utilize calculators have been provided to the reader. In many of these cases, a traditional hand-held calculator would be sufficient. However, hand-held algebraic-entry graphing calculators have become quite prevalent, and although some of the applications of such calculators are well above the level of the mathematics in this book, it would be worthwhile to introduce these devices at this point. The goal here is to make you familiar with such calculators, and to make you at ease with them. In that way, when you engage in more advanced mathematics such as calculus, it will be easier for you to use graphing calculators.

First, what does such a calculator look like? An example is indicated in Fig. B-2. This diagram illustrates the keyboard of the Hewlett-Packard HP 38G calculator.

While there are other such calculators available on the market, I have found that this particular one offers the easiest retrieval of answers and meshes particularly well with the needs of the mathematics student at both the precalculus level and more advanced levels. In this appendix, you are presented with a brief overview of the calculator and of its use in solving equations and graphing functions.

You turn the calculator on by pressing the [ON] key. Note that the [ON] key is also the [CANCEL] key. Also, if you wish to lighten or darken your calculator's screen, simply hold down the [ON] key while you press the [−] or [+] keys. (*Note*: Please refer to the *Reference Manual* provided with the HP 38G for a more detailed description of the many uses of the calculator).

The HOME screen is the main area in which you will work. Press [HOME] to find this area. Note that the [HOME] screen is divided into two main parts: the large rectangular area is the space in which entries and answers are indicated; the smaller rectangular area is the editline. For example, if you press the keystrokes:

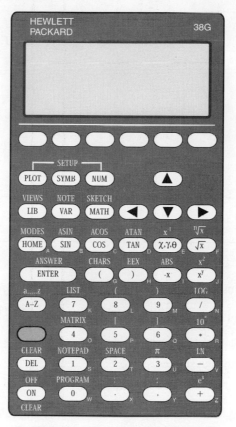

**Fig. B-2 (Reprinted with permission of
the Hewlett-Packard Company.)**

| 7 | | 5 | | ENTER |

you will find the following on the HOME screen:

HOME
75   75
75 will appear here before your press **ENTER**

If you press the keystrokes:

| 7 | | 5 | | / | | 5 | | ENTER |

you will find the following on the HOME screen:

```
┌──────────────────────────────────────────────────────┐
│                                                        │
│                        HOME                            │
│                                                        │
│                                                        │
│       75/5                                             │
│                                                        │
│                                                        │
│       15                                               │
│                                                        │
│                                                        │
│    ──────────────────────────────────────────────     │
│                                                        │
│    ──────────────────────────                          │
│                                                        │
└──────────────────────────────────────────────────────┘
```

Notice that the HP 36G uses algebraic-entry notation. Thus, to perform a calculation, you first press the operations required, and then press $\boxed{\text{ENTER}}$. Again, if you wish to find $689-231$, press the keystrokes:

$\boxed{6}$ $\boxed{8}$ $\boxed{9}$ $\boxed{-}$ $\boxed{2}$ $\boxed{3}$ $\boxed{1}$ $\boxed{\text{ENTER}}$

To find 5 squared (5^2), press the keystrokes:

$\boxed{5}$ $\boxed{x^y}$ $\boxed{2}$ $\boxed{\text{ENTER}}$

Evaluate each of the following using the HP 36G or similar graphing calculator:

1. $483 + 286$
2. $47 - 81$
3. $843 * 35$
4. $75/21$
5. $45x^y2$, or $(45)^2$
6. $\sqrt{15}$
7. The reciprocal of 21
8. The absolute value of -45

The HP 38G can be used to solve simple equations and to graph elementary functions. Let's look at some examples. Please refer to the HP 38G *Reference Manual* for more detailed instructions.

Let us solve the equation $X - 2 = 9$. At any one of the equation lines (marked E1, E2, etc.), enter the equation $X - 2 + 9$. Note that "X" is entered by pressing the $\boxed{\text{"A...Z"}}$ key, and then pressing the $\boxed{\bullet}$ key. Also note that the "$=$" sign is entered by pressing the key under the "$=$" sign on the lower, darkened area of the screen. After the equation appears correctly, press $\boxed{\text{ENTER}}$. A check mark should appear next to the equation you have entered. This check mark indicates that when you attempt to solve, the equation checked is the one you will be solving. Next, press the $\boxed{\text{NUM}}$ key, and then the key under the word "SOLVE" in the lower, darkened area of the screen. The number 11 should appear next to the symbol X.

Now try to solve the equation $2X - 11 = 13$ using the HP 38G and the following series of steps. Press $\boxed{\text{LIB}}$, use the arrow keys to scroll to Solve, press $\boxed{\text{ENTER}}$, go to any of the "E" lines, and enter the equation $2X - 11 = 13$. Press $\boxed{\text{ENTER}}$. Press the key under the $\boxed{\sqrt{\text{CHK}}}$ box on the lower, darkened area of the screen. Press the $\boxed{\text{NUM}}$ key, and then press the key under the word SOLVE in the lower, darkened area of the screen. You should see the answer 12 appear next to the symbol X.

Next, solve each of the following using the HP 38G or similar graphing calculator:

1. $2X - 15 = 26$
2. $3X = 4X + 4$
3. $5Y = 6Y + 12.5$

Let's now investigate the graph of the equation $Y = X - 4$. Press the $\boxed{\text{LIB}}$ key, and use the arrow keys to locate "FUNCTION." Press $\boxed{\text{ENTER}}$, and at any one of the "F" lines, enter the equation $X - 4$.

Make certain that you press $\boxed{\text{ENTER}}$. Note that you may enter the "X" by pressing the key under the symbol X in the lower, darkened area of the screen. Make certain that the equation you have entered is "checked," and then press the $\boxed{\text{PLOT}}$ key. The graph of the line will appear on a set of coordinate axes. Consult the *Reference Manual* for more details concerning the use of the HP 38G for graphing purposes.

Now try to graph the equation $Y = 2X - 11$ using the HP 38G and the following series of steps. Press $\boxed{\text{LIB}}$, use the arrow keys to scroll to Function, and press $\boxed{\text{ENTER}}$. Go to any of the "F" lines, and enter $2X - 11$. Press $\boxed{\text{ENTER}}$. Make certain that the expression is "checked" and that no other expressions are "checked." All expressions checked will be plotted when you press $\boxed{\text{PLOT}}$. Now press $\boxed{\text{PLOT}}$. The graph will appear on the coordinate axes on the calculator's screen.

Now graph each of the following using the HP 38G or similar calculator:

1. $Y = 2X - 5$
2. $Y = -5X = 13$
3. $2Y = 3X - 7$ (*Hint*: Divide both sides of the equation by 2.)
4. $X = \sqrt{y}$

INDEX

366INDEX